Electronic Packaging Interconnect Technology Symposium (EPITS 2019)

IOP Conference Series: Materials Science and Engineering
Volume 701

Penang, Malaysia
24 - 25 November 2019

ISBN: 978-1-7138-1031-5
ISSN: 1757-8981

Printed from e-media with permission by:

Curran Associates, Inc.
57 Morehouse Lane
Red Hook, NY 12571

CURRAN ASSOCIATES INC. proceedings.com

Some format issues inherent in the e-media version may also appear in this print version.

Institute of Physics
Dirac House, Temple Back
Bristol BS1 6BE UK

Phone: 44 1 17 929 7481
Fax: 44 1 17 920 0979

techtracking@iop.org

Additional copies of this publication are available from:

Curran Associates, Inc.
57 Morehouse Lane
Red Hook, NY 12571 USA
Phone: 845-758-0400
Fax: 845-758-2633
Email: curran@proceedings.com
Web: www.proceedings.com

TABLE OF CONTENTS

IOP Conf. Series: Materials Science and Engineering **701** (2019) 011001 doi:10.1088/1757-899X/701/1/011001

The Electronic Packaging Interconnect Technology Symposium (EPITS 2019) 24-25 November 2019

Preface

EPITS 2019 provides an excellent opportunity for intellectual discourse on demonstration of advances and research results in the fields of electronic packaging interconnect technology and green materials. This conference brings together all academics, researchers, engineers and scientists around the world contributing to the global debate on these fields. Topics of interest or submission include, but are not limited to;

- Challenges in improving the reliability of Tin (Sn) base Pb-free solder alloys.
- Challenges in implementing the Restriction of Hazardous Substances (RoHS).
- Emerging interconnect and green materials technologies.
- Non-solder interconnect materials at chip and package levels.
- Fundamental materials behavior for electronic packaging materials and green materials.
- Electromigration, thermomigration, stress-migration and mechanical effects on materials developments.
- Advanced characterization methods as applied to electronic packaging technology and materials processing and technology
- Whisker growth in tin, tin-based alloys and other metallic systems related to electronic packaging materials.
- Developments in high temperature Pb-free solders and associated interconnects for automotive and power electronics

Peer review statement

All papers published in this volume of *IOP Conference Series: Materials Science and Engineering* have been peer reviewed through processes administered by the proceedings Editors. Reviews were conducted by expert referees to the professional and scientific standards expected of a proceedings journal published by IOP Publishing.

Effects of Gallium addition on the thermal properties and whiskers growth under electrical current stressing

Noor Zaimah Mohd Mokhtar[1,*], *Mohd Arif Anuar* Mohd Salleh[1], *Aimi Noorliyana* Hashim[1] and *Sufian Firdaus* Nazri[1]

[1]School of Materials Engineering, University Malaysia Perlis, 02600, Jejawi, Arau Perlis, Malaysia

Abstract. This study investigates the effects of Gallium addition to the thermal behaviour and whiskers growth under electrical current stressing to the microstructure and intermetallic (IMCs) between solder and copper substrate. The samples were dipped using Sn-0.7Cu and Sn-0.7-0.05Ga solder and further continued with electromigration process for whiskers growth observation. The characterization were then analysed using Scanning Electron Microscopy (SEM) on the surface and intermetallic layer. From the observation, the addition of 0.05% Ga has lowered the melting point of Sn-0.7Cu-0.05Ga Copper and refined the microstructure. There is no indication of new phase form after 0.05%Ga addition. Furthermore, whiskers were observed to grow at both Sn-0.7Cu and Sn-0.7Cu-0.05Ga. However, the whiskers growth on both cathode and anode sides have been reduced with the Ga addition. It is also observed that the hardness increased with the addition of Ga element to the solder.

1 Introduction

Due to increasing environmental concerns, the microelectronics industries have been replacing Pb-containing solders with lead-free solder elements [1-5]. Among the most promising Pb-free candidates to replace the conventional SnPb includes Sn-0.7Cu [6-9]. The use of Sn-0.7Cu will enhance the growth of Sn whiskers [10-11]. Sn whiskers are an electrically conductive single or polycrystalline structure that spontaneously formed on the Sn coating. The issues of whiskers are they may grow long enough and cause short circuiting and interfere with other devices in application [12].

Several reports have studied the growth of Sn whiskers with the addition of this rare earth element. Nd has been added to the Sn-0.7Cu with the formation of Sn_3Nd and cause the growth of tin whiskers [13]. Furthermore, Sn-Zn has been used to change the microstructure to the equiaxed grain that will incubate the whiskers growth using pulsed plating technique [14]. Moreover, Sn-2Bi has shown improvement in mitigating Sn whiskers growth by refining the grain structure and altering the microstructure from

* Corresponding author: noorzaimah3287@gmail.com

columnar to equiaxed grain structure [15]. This mitigating process is due to the releasing internal stress with the Bi addition.

Other than that, several method have been proposed to mitigate the whiskers formation such as annealing at 150°C for 1 hour, mitigating the whiskers formation by inserting the Ni nanocones, conformal coating as it can stop the whiskers from penetrating barrier and forming short, reduce the thickness of plating, and by controlling the environmental effects [16-19]. But the mechanism of whiskers growth and all these mitigation methods are still have not yet been fully understood.

The previous research have found that the electrical current stressing had effects on the whiskers growth and IMCs formation [12]. But there are limited study on the effect of Ga addition to the whiskers growth under electromigration in term of whiskers growth and IMCs refinement. In this study, we focus on the thermal, mechanical and microstructural properties of Ga addition which allow the reduction on the formation of whiskers.

2 Methodology

2.1 Raw Materials Preparation

The Sn-0.7Cu and Sn-0.7Cu-0.05Ga solders were obtained in the form of ingots from Nihon Superior (M) Sdn. Bhd. The raw materials were melted in the solder pot at 350°C for 1 hour by casting method. Stirring was then performed every 1 minute during the whole duration of 15 minutes to obtain homogenized solder alloys. A flat square-shaped copper plate with dimensions of 15 mm x 15 mm size and 1 mm thickness was used as a substrate. The substrates were cleaned using acid cleaning liquid that contain of 5g (35%) of hydrochloric acid with 95g of deionized water (1.75%) to remove surface oxides and contaminations.

2.2 Dipping Procedure

After the cleaning process, the dipping procedure was performed using a hot soldering machine. The process is chosen as it can save cost and reduce time consuming compared to the electroplating technique. Furthermore, the preparation of the solder will also be easier. Each substrate was hanged using a crocodile clip at the dipping machine. The dipping machine is equipped with a direct current (DC) motor with 12V power supply and 531 rpm speed. The substrate was first covered with flux before dipped in the solder pot containing molten bath for 20 seconds and blown at a pressure of 300 Mpa to get an even thickness. The thickness for each samples were measured at 50-70μm each. Then, the samples were cleaned again with acetone for 3 minutes using ultrasonic cleaner to remove the excessive flux.

2.3 Electromigration

After the dipping process, the samples were continued with the electromigration testing. The EM test for the dipping sample was performed by applying a direct current to both ends of sample with crocodile clips at the anode and cathode areas. Current density of 8.0×10^4 A/cm^2 had been applied to the sample for 14 days at a room temperature by supplying a constant electric current to the solder joint. The testing parameters for dipping and electromigration testing are presented in Table 1.

Table 1. Testing parameter for dipping and electromigration.

Solder	Sn-0.7Cu, Sn-0.7Cu-0.05Ga
Substrate	Copper
Dipping time	20s
Dipping temperature	350°C-400°C
Coating thickness	50-60 μm
Current density	8.0x10^4 A/cm^2
Time for current stressing	14days
Current temperature	30°C

3 Result and Discussion

3.1 Differential Scanning Calorimeter (DSC)

DSC analysis was performed to measure the melting characteristic and undercooling measurement of the solder alloy on the copper substrate. Fig. 1 shows the DSC curves of Sn-0.7Cu alloys, Sn-0.7Cu solder on Cu substrate, Sn-0.7Cu-0.05Ga and Sn-0.7Cu-0.05Ga solder on Cu substrate. From the below figure, the melting temperature of Sn-0.7Cu seems to be on the par with Ga addition. This is because the number of Ga addition is quite low compared to the previous research using 0.1% Ga [20]. Moreover, the melting temperature of both alloy on Cu substrate show the value has been decreased. This is because, the atom from the substrate had diffused from Cu to solder alloy. This resulted in a microstructure refining and reducing intermetallic compound thickness.

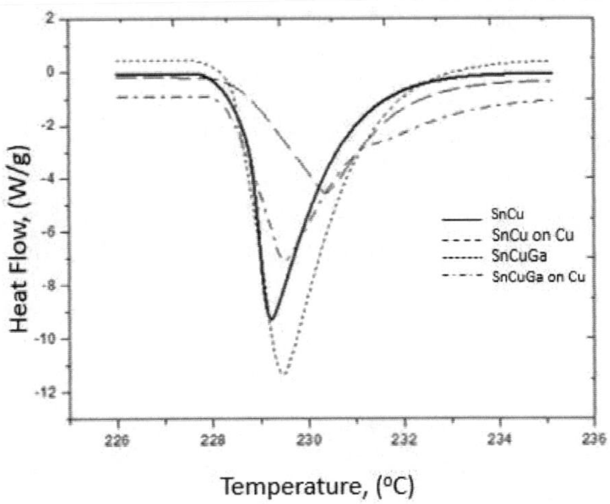

Fig. 1. DSC curve of the alloys.

3.2 X-ray Diffraction (XRD)

XRD were performed on the bulk samples to identify all the existing phase using High Score software. From XRD curve, as the addition of Ga is only 0.05%, the secondary phases would not form. Ga element acts as a solid solution strengthening the elements that already diffused into the Sn atom.

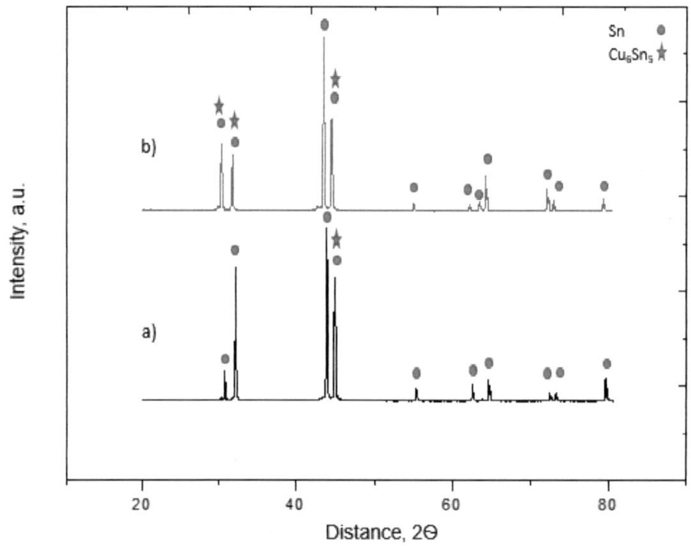

Fig. 2. XRD pattern of a) Sn-0.7Cu b) Sn-0.7Cu-0.05Ga.

3.3 Hardness Test

The hardness test of solder on copper substrate was tested with a Vickers hardness instrument. The load was tested at 1kg.f and dwelled in 20s. The test has revealed that the hardness value increases with the addition of Ga element compared to Sn-0.7Cu solder. The increase of the hardness value is likely due to the refinement of microstructure and uniform distribution of IMC phase in the solder alloy [21].

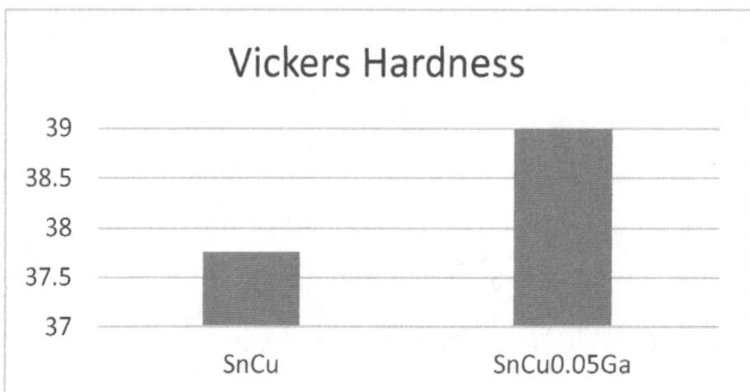

Fig. 3. Hardness value of the a) Sn-0.7Cu b) Sn-0.7Cu-0.05Ga.

IOP Conf. Series: Materials Science and Engineering **701** (2019) 012001 doi:10.1088/1757-899X/701/1/012001

3.4 Whiskers growth

Electrical current tests were performed to investigate the whisker growth. Fig. 4 shows the surface of Sn-0.7Cu and Sn-0.7Cu-0.05Ga after $8.0x10^4$ A/cm^2 current stressing. The hillocks were observed on both solder alloy but the growth of hillock shows lower in length after Ga addition. This aligns with the previous study, where the growth of hillocks can release the compressive stress during electrical current stressing [5] [22]. Reduction in the hillock size is due to the small addition of Ga by altering the IMC growth during current stressing.

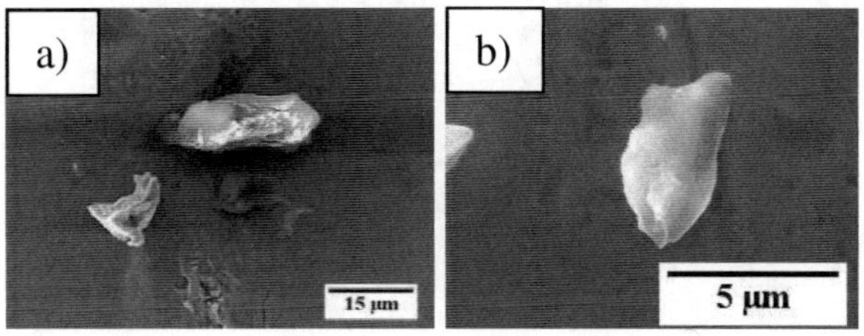

Fig. 4. The whiskers/hillock form on the solder a) Sn-0.7Cu and b) Sn-0.7Cu-0.05Ga after current stressing at $8.0x10^4$ A/cm^2.

3.5 Microstructure of primary IMC and interfacial IMC

Fig. 5 shows the optical microscope image of Sn-0.7Cu and Sn-0.7Cu-0.05Ga solder. From the figure, the addition of Ga refines the grains structure and IMCs formation. Ga reduces the surface energy and allows for the smaller grains [20]. Also, Ga element will diffuse slowly and restricts the growth of the IMC particles and formation of secondary phases.

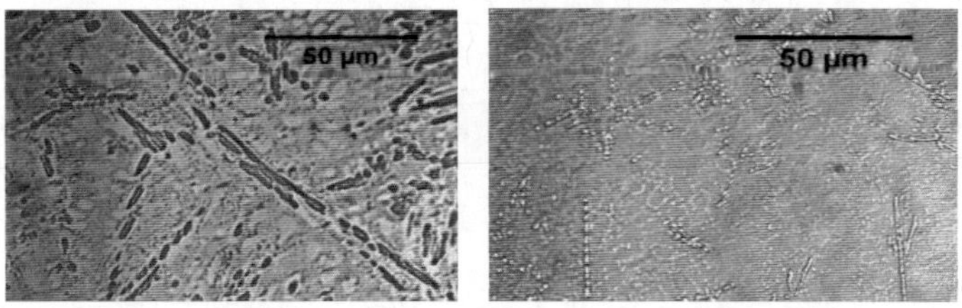

a) b)

Fig. 5. Optical Microscope of primary IMC and ß-Sn of solder alloys a) Sn-0.7Cu and b) Sn-0.7Cu-0.05Ga solder.

Fig. 6 shows the IMCs formation before and after current stressing. IMCs layer is important for good bonding. It is essential to have good metallurgical contact between solder and substrate. In this study, the IMC formation is focused on the anode, middle and cathode sides in order to understand the effect of current stressing to the IMC and whiskers growth. From the figure, it can be seen that the IMC layer shows a refinement at the cathode side. This is mainly because the Ga element acts as a solid strengthening [20] that diffuses to the Sn atom during the electric current that flows from cathode to anode. The higher IMC thickness on the anode side is due to the fact that electron moves away from the cathode towards the anode [22]. Cu substrate was consumed along with the IMC growth. This resulted in the increment of Cu supply for the IMC growth at the anode. By electromigration, the Cu atoms has transported from the cathode to anode side which improved the interfacial IMC growth at the anode. However, under electrical current stressing, the addition of Ga to the solder inhibits the diffusion of Cu and Sn to the anode which furthermore reduces whiskers formation.

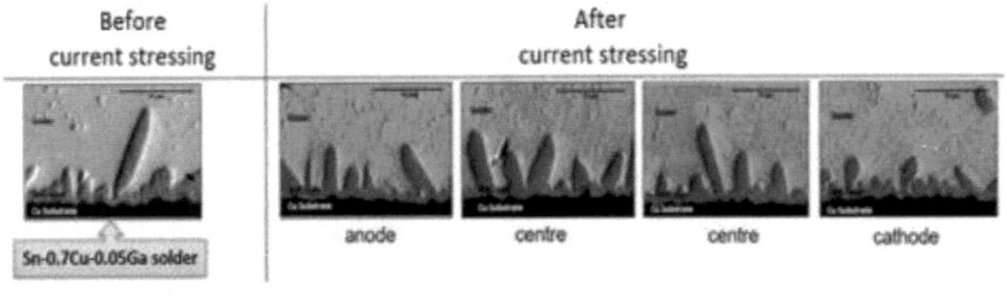

Fig. 6. Interfacial IMCs of Sn-0.7Cu-0.05Ga before and after electromigration.

4 Conclusion

As a summary, this paper investigates the relation of electrical current testing on the tin whiskers formation. Moreover, the results of addition of Ga as the microalloying to the Sn-0.7Cu were systematically presented in this paper. Other results are summarised as below:

- Thermal properties of Sn-0.7Cu has been improved with the addition Ga elements.
- XRD phase analysis shows the present of Cu_6Sn_5 element in the Sn-0.7Cu and Sn-0.7Cu-0.05Ga solder alloy
- The initial IMC has scallop-type morphology and Cu_6Sn_5 forms and grows with or without current.
- Whiskers length have increased with the Ga addition.
- The hardness value increased with the Ga addition which refined the grains and IMC formation.

The authors would like to acknowledge the support from University Malaysia Perlis and Nihon Superior Co. Ltd. The work was financially supported by Collaborative Research in Engineering, Science and Technology (CREST) under grant No. P14C1-17/001) with Nihon Superior Co.Ltd. (Japan).

References

1. M.I.I. Ramli, M.A.A.M. Salleh, F.A.M. Sobri, P. Narayanan, K. Sweatman, and K. Nogita, Journal of Materials Science: Materials in Electronics. **30**(4) (2019
2. N.M. Nasir, N. Saud, M.N. Deraman, A.A.M. Salleh, M.I.I. Ramli, and R.M. Said, Materials Science Forum. **803** (2015
3. R.M. Said, M.A.A.M. Salleh, M.I.I. Ramli, and N. Saud, Solid State Phenomena. **280** (2018)
4. M.A.A.M. Salleh, S.D. McDonald, and K. Nogita, Journal of Materials Processing Technology. **242** (2017)
5. N. M. Mokhtar and M. A. A. M. Salleh, in Solid State Phenomena, *Sustainable Materials*, (2018), **280**, pp. 175–180.

6. F. Somidin, H. Maeno, M.M. Salleh, X.Q. Tran, S.D. McDonald, S. Matsumura, and K. Nogita, Materials Characterization. **138** (2018)
7. M.A.A.M. Salleh, M.H. Hazizi, M.A.B.M. Abdullah, N. Noriman, R. Mayapan, and Z.A. Ahmad, Advanced Materials Research. **626** (2012)
8. K. Nogita, M.A.A.M. Salleh, S. Smith, Y. Q Wu, S.D. McDonald, A.G.A. Razak, T. Akaiwa, and T. Nishimura, 2017 International Conference on Electronics Packaging (ICEP)
9. K. Nogita, M.A.A.M. Salleh, E. Tanaka, G. Zeng, S.D. McDonald, and S. Matsumura, JOM. **68**(11) (2016
10. M.A.A.M. Salleh, A.M.M.A. Bakri, F. Somidin, and H. Kamarudin, International Review of Mechanical Engineering
11. M.A.A.M. Salleh, M.M.A.B. Abdullah, F. Somidin, A.V. Sandu, N. Saud, K. Hussin, S.D. McDonald, and K. Nogita, Revista de Chimie **64**(7) (2013)
12. N. Zaimah, M. Mokhtar, M. Arif, A. Mohd, and N. Hashim, (2019) IOP Conf. Ser.: Mater. Sci. Eng. 551 012096
13. M. Liu and A. P. Xian, J. Electron. Mater., **38,** no. 11, pp. 2353–2361, (2009).
14. F. Wang, L. Liu, L. Zhou, J. Wang, M. Wu, and X. Wang, *Mater. Trans.* **58**, no. 11, pp. 1593–1600, (2017).
15. Z. J. Zhang, J. Mater. Sci., (2019). **54,** no.10,pp. 7975-7986
16. C. C. Wei, P. C. Liu, and C. Chen, J. Mater. Res. **23**, no.7, pp. 2017–2022, (2008)..
17. M. Sun *et al.*, *Mater. Charact.*, **134**, 2017.
18. L. W. Condra, S. J. Meschter, D. A. Pinsky, and A. J. Rafanelli, *Proceedings of the Custom Integrated Circuits Conference*, (2009).
19. F. He *et al.*, pp. 102–106, (2017). *12th International Microsystems, Packaging, Assembly and Circuits Technology Conference (IMPACT)* (pp. 102-106). IEEE.
20. S. F. Nazri and M. A. A. M. Salleh, *Solid State Phenomena,* **280**, pp. 187–193, (2018).
21. M. Hasnine and M. J. Bozack, (2019). J. Electron. Mater, **48,** no.6, 3970-3978.
22. Z. X. Yao, L. M. Yin, Y. H. Lu, W. Gang, Z. G. Chen, and Z. X. Yao, *Proc. Electron. Packag. Technol. Conf. EPTC*, pp. 1096–1099, (2014).

The effect of chloride ion diffusion on corrosion activity of kaolin geopolymer paste in artificial seawater

Farah Farhana Zainal[1,*], *Sahrul* Abd Hakim[1], *Yusrina* Mat Daud[1], *Mohammad Firdaus* Abu Hashim[1], *Mohd Mustafa Al Bakri* Abdullah[1], *Hasri*[2] and *Hartati*[2]

[1]Center of Excellence Geopolymer and Green Technology (CEGeoGTech), School of Materials Engineering, Universiti Malaysia Perlis (UniMAP), P.O. Box 77, D/A Pejabat Pos Besar, 01000 Kangar, Perlis, Malaysia

[2]Fakultas Matematika & Ilmu Pengetahuan Alam, Universitas Negeri Makassar, Kampus FMIPA UNM, Parang Tambung, JL. Mallengkeri, Makassar, 90224, Indonesia

Abstract. This paper presents the effect of chloride ion diffusion on the corrosion activity of kaolin geopolymer paste in artificial seawater. An aluminosilicate material that rich with aluminium and silicon such as kaolin is selected as a raw material in the geopolymerization of geopolymer. Kaolin then mixed with an alkaline activator that consist mixture of sodium hydroxide (NaOH) solution with sodium silicate (Na_2SiO_3) solution. Two types of samples are used in this study, which 7 days and 14 days sample. A few testing is conducted in order to test the chloride ion diffusion on the samples such as Rapid Chloride Permeability Test, Rapid Migration Test and Salinity Test. The result shows that sample for 7 days consist the higher penetration of chloride ion compare sample of 14 days. This is proven with the result of salinity. A longer curing time of the sample cause a lesser penetration of Cl^- ion.

1 Introduction

Seawater that contains high amount of salt has greatly pretentious the strength, durability and life span of the building and cement for construction especially for offshore building [1, 2]. Corrosion in concrete, mainly occurs due to the chloride ion and carbon dioxide, besides that, the diffusion characteristic of chloride and other corrosion-inducing substances is depending on the porosity and pore of the concrete, indirectly it may influence the initiation and the rate of corrosion [1, 3]. Penetration of chloride ion depends on a single parameter that is the coefficient of diffusion. According to Wang et al. [4], the value of chloride ion diffusion was similar for geopolymer and OPC concrete. Diffusion of chloride occurs when there are differences of mobility of ions or in a simple way there is a different concentration gradient of ion in the pore solution. According to [1] the source chloride ion came from 2 ways that is from internal sources, that chloride exist as contaminant in raw material and another is from external sources such as seawater or de-icing salt.

[*] Corresponding author: farahfarhana@unimap.edu.my

Generally, geopolymer has been used as an alternative material to conventional Portland cement and concrete composite [5]. Currently, many studies involve geopolymer grows rapidly and some of it has recorded a good performance of geopolymer that can be obtained by mixing of binder material such as kaolin and fly ash with alkaline activator [6].

2 Materials and Methodology

2.1 Materials and Sample Preparation

Geopolymer paste in this study was made from a mixture of kaolin, sodium hydroxide (NaOH) solution and sodium silicate (Na_2SiO_3) solution. Kaolin is a type of clay that can be found in the nature [7]. It is chemically a hydrous aluminium silicate characterized as very fine powder [8, 9]. The NaOH pellet was used in this research under the brand name of Merck Chemicals by the molar mass of 40.00 g/mol. Distilled water was used to dilute NaOH pellet to produce the NaOH solution. Na_2SiO_3 solution used in this study was obtained from South Pacific Chemicals Industries Sendirian Berhad (SPCI), Malaysia.

An alkaline activator was prepared 24 hrs prior to use. Then, the prepared alkaline activator was mixed with the kaolin. After that, the mixture was placed in molds. The sample in the mold was compacted 25 times by using a rod in three layers of equal weight. The sample remained in the mold until the sample was hardened after 24 hrs, the sample was taken out from the mold and cured at 80°C in the oven for 24 hrs. The sample was then tested after 7 and 14 days. Artificial seawater in this study was prepared by mixing the sodium chloride (NaCl) and distilled water according to ASTM D1141-98 (2013) [10].

2.2 Testing

A few testing is conducted in order to test the chloride ion diffusion on the samples such as Rapid Chloride Permeability Test, Rapid Migration Test and Salinity Test.

3 Results and Discussion

This section will contain all the results and discussion approximately from all the testing that has been conducted to determine the effect of chloride ion diffusion on the geopolymer paste in artificial seawater.

3.1 Rapid Chloride Permeability Test

After 6 hours of testing, the amount of charges passed through the sample was recorded by using a digital multimeter. Based on the results obtained, it shows that most of the samples at 7 days consist range of potential between -170 mV to -190 mV. While for samples at 14 days it consists range of potential between -150 mV to -170 mV. Then the result was compared with the value of ASTM C1202 and AASHTO T277-83 standard in order to determine the rate of chloride permeability of the concrete. It was found out that one joule of energy per coulombs of charge is equal to one volt of potential, in that case for -185.2 mV of potential is equal to -185.2 coulombs of charges. After conversion of the unit, it shows that both samples were categorized for the charge below 100 coulombs. Based on Table 1, if the chloride permeability for the charge below than 100 coulombs the rate of chloride permeability is negligible.

Table 1. Rate chloride of concrete.

Chloride Permeability	Charge (Coulombs)	Type of Concrete
High	>4000	High water-to-cement ratio (>0.6) conventional Portland cement
Moderate	2000-4000	Moderate water-to-cement ratio (0.4-0.5) conventional Portland cement concrete
Low	1000-2000	Moderate water-to-cement ratio (<0.4) conventional Portland cement concrete
Very Low	100-1000	Latex modified concrete, internally sealed, concrete
Negligible	<100	Polymer impregnated concrete, polymer concrete

3.2 Rapid Migration Test

After 8 hours immersion, the sample was split in two, the sample than freshly sprayed using calorimetric indicator ($AgNO_3$ solution) in order to determine the chloride penetration depth. The calorimetric boundary between the region that contain chloride ion and without chloride ion is shown in Figure 1. The area that contains chloride ion will form a white precipitate that indicates the presences of chloride molecule in the sample while the area that does not contain chloride ion it will form a dark precipitate. This is also being supported by the researcher of Elfmarkova et al. [11], it states that the formation white precipitate in the presence of chloride ion is due to the chemical reaction of of Ag^+ with Cl^- or OH^-.

(a) **(b)**

Fig. 1. (a) Before and (b) after geopolymer paste surface being sprayed with $AgNO_3$ solution.

It also states that the value of non-steady migration state coefficient increase with the increase of penetration depth or when the original colour of the concrete is darker. Based on

Table 2 states that the reading of the non-steady state migration coefficient for sample in 14 days is much lower compared to the sample in 7 days.

Table 2. Reading of non-steady state migration coefficient after 8 hours testing.

Sample	No. of sample	Applied voltage, (V)	Time (hours)	Average temperature (^{0}C)	Depth penetration, X_d, (mm)	Non-steady state migration coefficient (m^2/s)
7	1	30	8	20.50	31	53.68×10^{-12}
	2	30	8	20.05	22	37.27×10^{-12}
	3	30	8	20.00	23	39.07×10^{-12}
14	1	30	8	19.96	13	21.15×10^{-12}
	2	30	8	20.05	17	28.28×10^{-12}
	3	30	8	20.45	15	24.74×10^{-12}

3.3 Salinity Test

Through 6 hours of testing, the salinity of the distilled water was measured using a salinity meter every 1 hour and the resulted was shown in Figure 2 and Figure 3 for sample 7 and 14 days. Based on Figure 2, it shows how the salinity permeability of chloride ion from seawater into deionized water was slowly increased from 0.0 ppt to 1.0 ppt. For sample 1 and 2, it only required one hour to reach a transition state while for sample 3 it required two hours to reach a transition state. This indicates that for a kaolin geopolymer paste that samples 7 days required a minimum of two hours for the chloride ion to fully penetrate the sample and in touch or diffuse into the distilled water. It has been reported that the transition state only begins when there is a change in concentration of salinity [12]. In order for transition state alternate into steady state it requires a steady increase of salinity proportional with time [12]. Based on Figure 2, steady state for all the sample starts after 2 hours where the salinity starts to increase proportional to time.

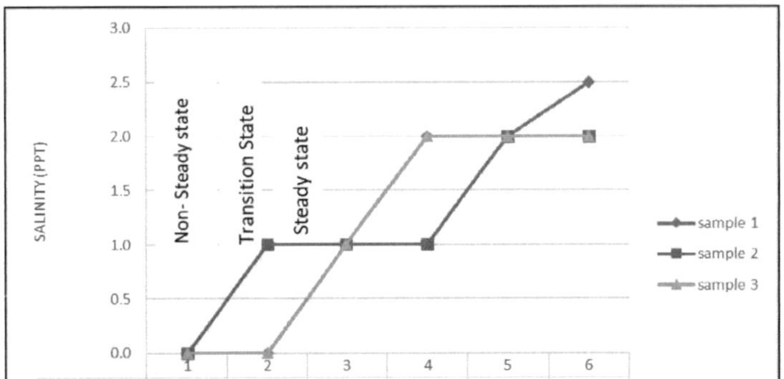

	1	2	3	4	5	6
sample 1	0.0	0.0	1.0	2.0	2.0	2.5
sample 2	0.0	1.0	1.0	1.0	2.0	2.0
sample 3	0.0	0.0	1.0	2.0	2.0	2.0

Fig. 2. Salinity test result for samples 7 days.

For samples at 14 days, the salinity result was shown in Figure 3. During the first three hours, the chloride just starts to diffuse into the sample but not yet reach the deionized water this is known as non-steady state where there is no change in concentration of the distilled water [6]. All the sample cured in 14 days start to experience transition state after 3 hours exposed at artificial seawater and change to steady state in the next hour.

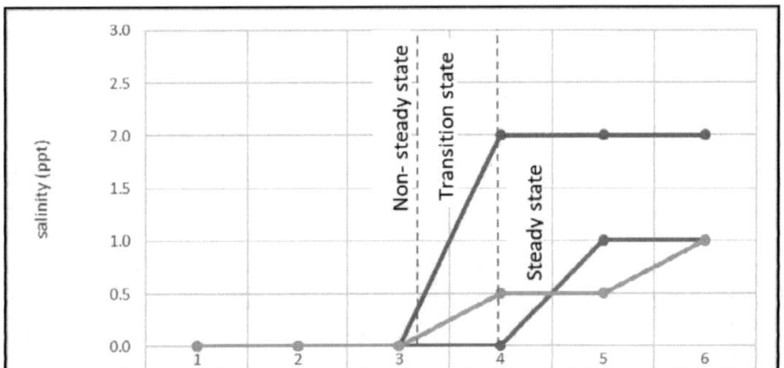

	1	2	3	4	5	6
sample 1	0.0	0.0	0.0	0.0	1.0	1.0
sample 2	0.0	0.0	0.0	2.0	2.0	2.0
sample 3	0.0	0.0	0.0	0.5	0.5	1.0

Fig. 3. Salinity test result for samples 14 days.

Conclusion

Detailed work has been done to achieve the objectives of this study.

i. Chloride permeability of geopolymer paste is negligible because the charges less than 100 Coulombs.
ii. Sample for 14 days shows a low migration of chloride due to the lower reading of non-steady state migration coefficient after 8 hours testing compare to sample 7 days.
iii. All the sample cured in 14 days start to experience transition state after 3 hours exposed at artificial seawater, while for 7 days sample required a minimum of 2 hours for the chloride ion to fully penetrate.

The authors would like to thank the staffs of Center of Excellence Geopolymer & Green Technology (CEGeoGTech), School of Materials Engineering, Universiti Malaysia Perlis (UniMAP) for their involvement in the research. This work was supported and funded by the Universitas Negeri Makassar Kampus FMIPA UNM, Makassar, Indonesia. The authors also would like to acknowledge the support from the Fundamental Research Grant Scheme (FRGS) under a grant number of **FRGS/1/2018/TK06/UNIMAP/02/2** from the Ministry of Education Malaysia.

References

1. Chindaprasirt, P., W. Chalee. Effect of Sodium Hydroxide Concentration on Chloride Penetration and Steel Corrosion of Fly Ash-Based Geopolymer Concrete under Marine Site. *Construction and Building Materials*, vol. 63, pp. 303–310 (2014).
2. Rozyanty, A.R., Mohammed, M.M., Musa, L., Shahnaz, S.B.S., Zuliahani, A. The weathering effect in natural environment on hybrid kenaf/glass fiber unsaturated polyester composite (2017) AIP Conference Proceedings, 1835, art. no. 020015.
3. Zainal, F.F., Amli, S.F.M., Hussin, K., Rahmat, A., Abdullah, M.M.A.B. Corrosion Studies of Fly Ash and Fly Ash-Slag Based Geopolymer (2017) IOP Conference Series: Materials Science and Engineering, 209 (1), art. no. 012026.
4. Wang, H, Dai, J., Sun, X. Chloride transport in concrete under sustainable flexural loading. Magazine of Concrete Research, 69(5), 245-254 (2017).
5. Salem, I.A.S., Rozyanty, A.R., Betar, B.O., Adam, T., Mohammed, M., Mohammed, A.M. Study of the effect of surface treatment of kenaf fiber on chemical structure and water absorption of kenaf filled unsaturated polyester composite (2017) Journal of Physics: Conference Series, 908 (1), art. no. 012001.
6. Hashim, M.F.A., Abdullah, M.M.A.B., Sandu, A.V., Puskas, A., Daud, Y.M., Zainal, F.F., Faris, M.A., Hasri, Hartati, Advanced glass reinforced epoxy filled fly ash based geopolymer filler: Preparation and characterization on piping materials (2019) IOP Conference Series: Materials Science and Engineering , 572 (1), art. no. 012037.
7. Zainal, F.F., Hussin, K., Rahmat, A., Abdullah, M.M.A.B., Shamsudin, S.R. A study on hardness behavior of geopolymer paste in different condition (2016), AIP Conference Proceedings, 1756, art. no. 040001.
8. Heah, C., Kamarudin, H., Bakri, A. M., Binhussain, M., Luqman, M., Nizar, I. K. Liew, Y. Effect of Curing Profile on Kaolin-based Geopolymers. Physics Procedia, 22, 305-311 (2011).
9. Mat Daud, Y., Fazlina Osman, A., Farhana Zainal, F., Firdaus Abu Hashim, M., Si, H.M., Si, H.S., Soo Li, N., Arzuria Adnan, S. Effect of Mixing Parameter on Epoxy-Organo Montmorillonite Composites: Flexural and Microscopy Analysis (2019) IOP Conference Series: Materials Science and Engineering, 551 (1), art. no. 012013.

10. ASTM D1141-98 (2013), Standard Practice for the Preparation of Substitute Ocean Water, American Society for Testing and Materials (2013).
11. Elfmarkova, V., Spiesz, P., Brouwers, H. Determination of the chloride diffusion coefficient in blended cement mortars. Cement and Concrete Research, 78, 190-199 (2015).
12. Muntingh, Y. Durability and diffusive behaviour evaluation of geopolymeric material Unpublished Master's Thesis, (2006).

Synthesis and charactherization of TiO_2 doped SnO_2 thin film prepared by sol-gel method

Dewi Suriyani Che Halin[1*], *Kamrosni* Abdul Razak[1,2], *Azliza* Azani[1], *Mohd Mustafa Al Bakri* Abdullah[1], *Mohd Arif Anuar* Mohd Salleh[1], *Norsuria* Mahmed[1], *Nur Syahirah* Shafee[1], *Muhammad Mahyiddin* Ramli[2], *Ayu Wazira* Azhari[3],and *Varistha* Chobpattana[4]

[1]Center of Excellence Geopolymer & Green Technology (CEGeoGTech), School of Materials Engineering, Universiti Malaysia Perlis, (UniMAP), 02600 Jalan Kangar-Arau, Perlis.
[2]School of Microelectronic Engineering, Universiti Malaysia Perlis, Pauh Putra Campus, 02600, Arau, Perlis,Malaysia.
[3]Water Research Group (WAREG), School of Environmental Engineering, Universiti Malaysia Perlis 02600 Arau, Perlis, Malaysia.
[4]Department of Materials and Metallurgical Engineering, Faculty of Engineering, Rajamangala University of Technology Thanyaburi (RMUTT), Thailand.

Abstract. In this work, preparation of titanium dioxide doped with tin oxide, SnO_2/TiO_2 thin films deposited onto silicon wafer via sol-gel method. Different amount of SnO_2 was added (5 ml, 10ml and 15 ml) into parent solution. The obtained films were annealed at different temperature which is 400°C, 500°C and 600°C for 1 hour. Morphological and surface topography of the SnO_2 doped TiO_2 thin films were studied using Scanning Electron Microscope (SEM) and Atomic Force Microscope (AFM). The annealed films shows non-uniform crack due to the mismatch of coefficient of thermal expansion (CTE) between SnO_2/TiO_2 thin films and silicon wafer.

1 Introduction

Nowadays, there are numerous application of titanium dioxide (TiO_2) that benefits the society such as cosmetics, foods, restraining and sterilizing virus, electrical devices and photocatalytic oxidation (PCO) under ultraviolet (UV) and visible light also known as a pollution control technology [1, 2]. Self-cleaning acitvity will occur when TiO_2 is being exposed under light source which will display a photocatalytically induced superhydrophilicity properites. The hydrophobic surface of the substrate will transform to hydrophilic surface after coated with TiO_2 where it will cause a uniform water film. This will prevent the adhesion of organic or inorganic components on the surface. Therefore, the cleanliness can be retained. TiO_2 thin film can be coated on the surfaces of various substrates such as metals, textiles, ceramics, fibres and glass that will exhibits self-sterilisation and self-cleaning properties when it is exposed to the light source.

* Corresponding author : dewisuriyani@unimap.edu.my

When TiO_2 is doped with metal oxide such as SnO_2, it will improve the efficiency of TiO_2 photocatalyst. By doping SnO_2 into TiO_2 thin films, hydrophilicity and photocatalytic activity of composite thin films would be improved due to the reduction of TiO_2 particle growth rate [3-6].

In this work, TiO_2 doped with SnO_2 thin films coated on silicon substrate were fabricated using a sol-gel dip coating technique. The thin films were annealed at 400°C, 500°C and 600 °C for 1 hour with the heating rate of 5 °C/min. X-ray Diffractometer (XRD) was used to study the phase composition of SnO_2 doped TiO_2 thin films while Scanning Electron Microscope (SEM) and Atomic force microscope (AFM) was used to observe the morphology and the surface topography of SnO_2 doped TiO_2 thin films, respectively.

2 Methodology

2.1 Materials

The main raw materials used to produce TiO_2 sol in these studies are titanium (IV) isoproxide (TTIP), $Ti(OC_3H_7)_4$, isopropanol, C_3H_8O and acetic acid, CH_3COOH. Tin-tetrachloride penta-hydrate ($SnC_4.5H_2O$) powder was used as the source of SnO_2 solution. The silicon wafer was used as a substrate.

2.2 Methods

Firstly, 0.5 ml titanium (IV) isoproxide (TTIP) and 10 ml isopropanol, C_3H_8O were stirred at room temperature for 20 minutes using magnetic stirrer. Then, 2.5ml of acetic acid was added into the solution while stirring to speed up the chemical reaction. SnO_2 solution was produced by adding 50ml of distilled water and 50 ml of ethylene glycol into 3.51g tin-tetrachloride penta-hydrate ($SnCl_4.5H_2O$) powder. The solution was dissolved for 10 minutes to make a 100 ml precursor mixture. The TiO_2 solution was mixed together with 5ml, 10ml and 15ml of SnO_2 respectively. The mixture was vigorously stirred for another 30 minutes at room temperature. The substrate used is silicon wafer (1 cm x 1 cm). The silicon wafers were washed in acetone using ultrasonic cleaner for 30 minutes. The TiO_2/SnO_2 sol were deposited on silicon wafer by using spin coating at 1500rpm for 30s. The coated samples were dried at 40°C for 10 minutes. After that, the coated samples were annealed in muffle furnace for 1 hour with heating rate of 5°C/min at 400°C, 500°C and 600°C. The TiO_2 doped with SnO_2 thin films were prepared by sol-gel method according to the amount of SnO_2 doping. In this study, the characterization of TiO_2 thin films was investigated by using Scanning Electron Microscope (SEM) and Atomic Force Microscopes (AFM).

3 Results and Discussions

Fig. 1 shows the XRD pattern of SnO_2 doped TiO_2 thin films. It can be seen that the diffraction pattern shows dual-phase of anatase and rutile with no existence of SnO_2. The samples were annealed at three different temperatures which are, 400°C, 500°C and 600°C, respectively. The angle 2θ shifts towards lower angles with the increase of annealing temperature from 400 °C to 600 °C.

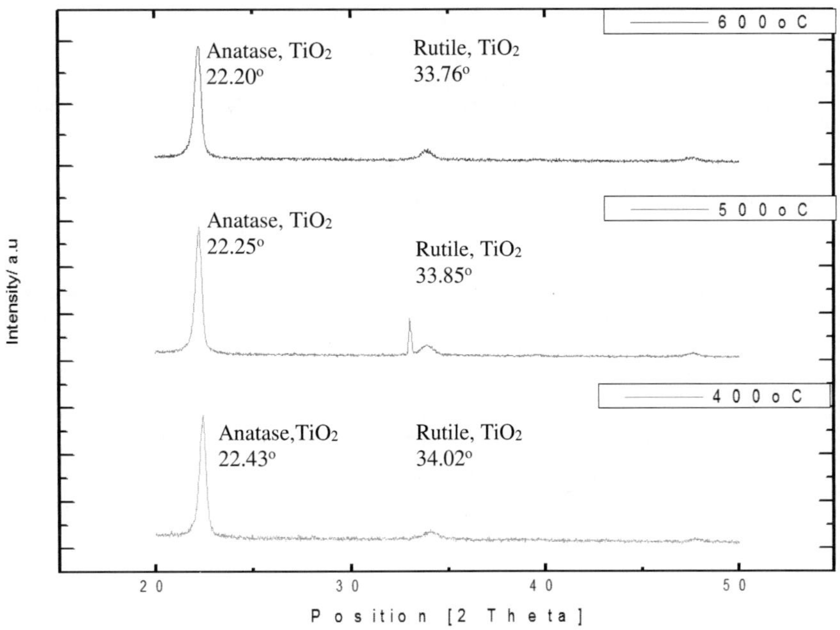

Fig. 1. XRD pattern of SnO_2 doped TiO_2 thin films which was annealed at 400°C, 500°C and 600°C.

Fig. 2 and Fig. 3 show the micrographs of SnO_2/TiO_2 thin films that were coated on silicon wafer with different amount of SnO_2 added and annealed in muffle furnace at 400°C, 500°C and 600°C for 1 hour. The micrographs show a non-uniform films with large flaky and large cracks film. It might due to the surface tension between the film and the air during the drying process [7]. It can be seen that as the annealing temperature increases, the distance between cracks become smaller. The higher the annealing temperature, the least cracks formed. To prevent crack from forming when annealed at high temperature, more layers of coating were deposited to ensure the coverage of substrate become better and formed a thicker film. It is well agree with the previous work by Abdul Razak et al [8]. From Fig. 2, it was observed that 5 ml SnO_2 doped with TiO_2 annealed at 400°C has the most agglomerated region as compared to the samples annealed at 500°C and 600°C. Meanwhile, in Fig. 3 shows the micrograph of 10 ml SnO_2 doped with TiO_2 annealed at 500°C has the most agglomerated region as compared to the samples annealed at 400°C and 600°C. The micrographs show a Volmer-Weber growth. Volmer–Weber growth of SnO_2 doped TiO_2 thin films includes nucleation of 3-D islands, growth, impingement and coalescence of islands [4].

The surface topography of SnO_2/TiO_2 thin films was investigated by Atomic Force Microscopes (AFM). Typical topographical images of the surface of SnO_2/TiO_2 coatings are represented in 3D images. Fig. 4 shows the surface topography of 5ml SnO_2 doped TiO_2 thin films at different annealing temperature of 400°C, 500°C and 600°C respectively. Fig. 5 shows the surface topography of 10ml SnO_2 doped TiO_2 thin films at different annealing temperature 400°C, 500°C and 600°C respectively.

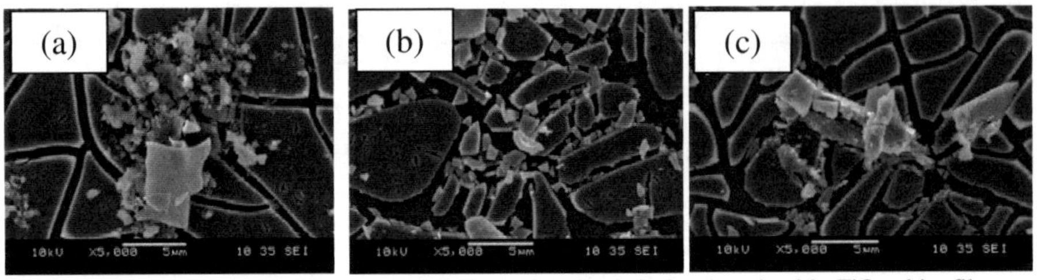

Fig. 2. SEM image of the surface morphology for 5 ml SnO_2 doped with TiO_2 thin films at temperature of (a) 400°C, (b) 500°C and (c) 600°C.

Fig. 3. SEM image of the surface morphology for 10 ml SnO_2 doped with TiO_2 thin film at temperature of (a) 400°C, (b) 500°C and (c) 600°C.

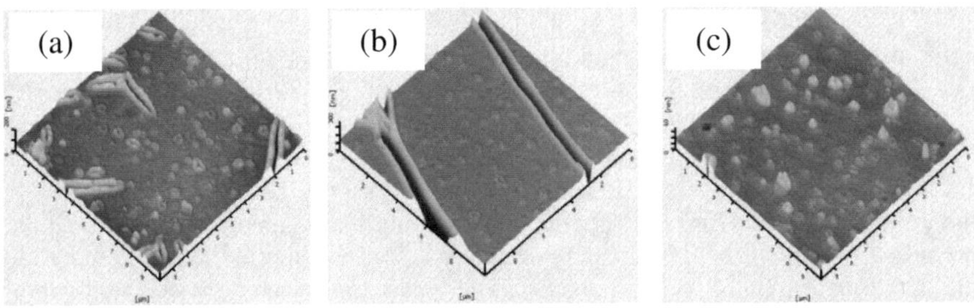

RMS=1.155nm RMS=3.677nm RMS=3.796nm

Fig. 4. The surface topography and 3D images of 5ml SnO_2 dope TiO_2 thin films at different annealing temperature of (a) 400°C, (b) 500°C and (c) 600°C.

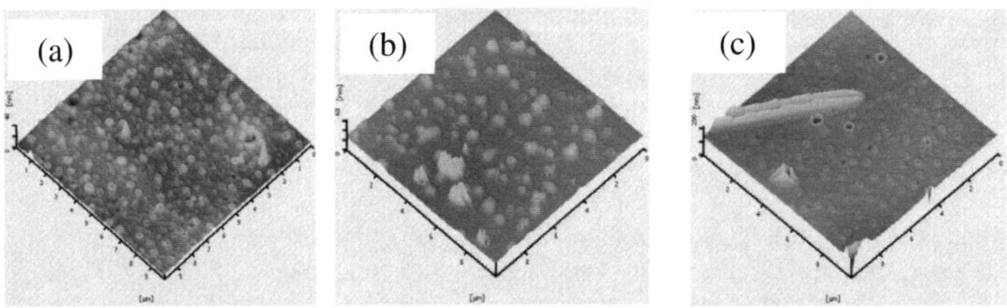

RMS=1.570nm RMS=2.995nm RMS=9.937nm

Fig. 5. The surface topography and 3D images of 10 ml SnO_2 dope TiO_2 thin films at different annealing temperature of (a) 400°C, (b) 500°C and (c) 600°C.

It can be observed that the sample which was annealed at 600°C produce the highest value of roughness (RMS) while the sample annealed at 400°C produced the lowest value of roughness (RMS). This showed that the surface roughness (RMS) of the samples increased with the annealing temperature. It is well agreed that the presences of SnO_2 influenced the RMS value. As the amount of SnO_2 added increased, the RMS value also increased. This revealed that the RMS value was affected by annealing temperature and the amount of SnO_2 added into the parent solution.

4 Conclusion

In this study, SnO_2 doped TiO_2 thin films were prepared by sol-gel method. The XRD pattern shows the phase composition of SnO_2 doped TiO_2 thin films were dual-phase of anatase and rutile with no existence of SnO_2. The morphological characteristics of SnO_2 doped TiO_2 thin films have non-uniform cracks, agglomerate and the formation of Volmer-Weber growth. When the annealing temperature and the amount of SnO_2 increased, the surface roughness (RMS) value increased.

The author would like to acknowledge the support from the Fundamental Research Grant Scheme (FRGS) under a grant number of FRGS/1/2017/TK07/UNIMAP/02/6 from the Ministry of Education Malaysia and Tin Solder Technology Research Grant (TSTRG) under grant number of 9002-00082 from Tin Industry (Research and Development) Board. The authors wish to thank the Center of Excellence Geopolymer & Green Technology (CEGeoGTech), School of Materials Engineering, Universiti Malaysia Perlis, UniMAP for their partial support.

References

1. W. Sangchay, Industrial Technology Lampang Rajabhat University Journal **2**, (2016)
2. W. Sangchay, Energy Procedia **89**, (2016).
3. S. Karuppuchamy, K. Nonomura, T. Yoshida, T. Sugiura, H. Minoura, Solid State Ionics **151**, 19 (2002).
4. H. Z. Yu and C. V Thompson, Acta Mater. **67**, (2014).
5. B. K. Kaleji and R. Sarraf-mamoory, Materials Research Bulletin **47**(2), (2012).
6. M. Shipochka A.Eliyas, I.Stambolova, V.Blaskov, S.Vassilev, S.Simeonova, K.Balashev, Materials Chemistry and Physics **220**, 249–259 (2018).
7. D S C Halin, M M A B Abdullah, N Mahmed, S N A Abdul Malek, P Vizureanu and A W Azhari, IOP Conf. Series: Materials Science and Engineering **209**, (2017).
8. K. Abdul Razak, D.S. Che Halin, M.M.A. B. Abdullah, M.A.A. Mohd Salleh, N. Mahmed, N.S. Danial, Solid State Phenomena **280**, (2018).

IOP Conf. Series: Materials Science and Engineering **701** (2019) 012004 doi:10.1088/1757-899X/701/1/012004

Microstructural studies of doped PEG Ag/TiO_2 thin film

Kamrosni Abdul Razak[1,2*], *Dewi Suriyani* Che Halin[1], *Azliza* Azani[1], *Mohd Mustafa Al Bakri* Abdullah[1], *Mohd Arif Anuar* Mohd Salleh[1], *Norsuria* Mahmed[1], *Varistha* Chobpattana[3], and *Ayu Wazira* Azhari [1,4]

[1]Center of Excellence Geopolymer & Green Technology (CEGeoGTech), School of Materials Engineering, Universiti Malaysia Perlis, (UniMAP), 02600 Jalan Kangar-Arau, Perlis,
[2]Center for Diploma Studies, Universiti Malaysia Perlis, Unicity Alam Campus, Perlis,Malaysia
[3]Department of Materials and Metallurgical Engineering, Faculty of Engineering, Rajamangala University of Technology Thanyaburi (RMUTT), Thailand
[4]Water Research Group (WAREG), School of Environmental Engineering, Universiti Malaysia Perlis 02600 Arau, Perlis, Malaysia.

Abstract. Ag/TiO_2 thin film was prepared by the sol-gel method through the hydrolysis of titanium tetraisopropoxide and silver nitrate solution. Various amount of PEG was doped into the solution preparation to study the effect on crystalline state and microstructural of the prepared thin films. Spin coating method was used to get uniform film on ITO glass substrate followed by annealing process for 1 hour. The obtained thin films were analysed using XRD test, SEM and AFM. Results showed that all the prepared thin films are in anatase TiO_2 Increasing the PEG amount into the solution could increase the thickness and surface roughness of the obtained thin films.

1 Introduction

Titanium dioxide known as titanium (IV) dioxide or titania has a molecular weight of 79.87 g/mol which is a natural titanium oxide with TiO_2 chemical formula. Titanium dioxide is a substance commonly used as white pigment due to its remarkable properties such as high purity, brightness, and non-toxicity. It is also known for its excellent chemical stability, biologically non-toxic and low cost [1]. It is an important inorganic functional material with good physical properties, making it ideal for thin film applications. TiO_2 thin films are one of the most important oxides due to their attractive chemical, electrical and optical properties [2-4]. TiO_2 has been studied by many researchers because of its many applications in various industries [5-7]. Due to its high corrosion resistance and chemical

Corresponding author:dewisuriyani@unimap.edu.my

stability, excellent optical transparency near infrared and visible areas, it provides a high refractive index which makes it useful for anti-reflection coating in optical devices [8].

Additives are used to enhance the performance and simultaneously improve the coating of the thin film. Additives can be in many types of materials or chemicals. The additive was found to improve the surface area whereby increase the surface area resulting the great efficiency of prepared thin films. Polyethylene glycol (PEG) is made up of various sizes and functional groups that are synthetic polymers [9]. These attractive additives are also known as polymer additives which have several advantages over the resulting coating such as adding active surface area with porous surface structure due to pore formation and reducing energy gap in TiO_2 [10]. Porous films have been obtained by using PEG as a chelating agent [11]. In this research, we focused on the effect of the addition of PEG into Ag/TiO_2 thin films.

2 Materials and Preparation of Ag/Tio_2 Thin Films

Titanium (IV) isopropoxide 97% (TTIP) and silver nitrate ($AgNO_3$) powder were obtained from Sigma Aldrich. . Propan-2-ol was purchased from QReC Chemicals and acetic acid (99.5%) and methylene blue were obtained from Daejung Reagent Chemicals and respectively. All the chemicals were analytical reagents and were used as received without further purification.

The Ag/TiO_2 thin film without PEG was prepared as in previous study [12]. Then, for the Ag/TiO_2 thin film with PEG, a different amount of PEG (0.05g, 0.10g, 0.15g and 0.20g) was added into the solution after the 25 min sol stirring. The resultant alkoxide solution was kept stirring at room temperature until a clear sol produced, without any precipitation. ITO coated glass (15mm×15mm×0.5mm) were used as the substrates for the deposition of films.

The prepared Ag/TiO_2 thin films were characterised structurally through XRD and morphologically by SEM. The phase composition of the synthesized thin film were determined by using an X-ray diffraction (XRD) over a 2θ range 10°- 65° using Cu Kα (λ=1.5046). Analysis of the XRD patterns was carried out using Diffract Eva Software. Surface morphology and surface roughness of prepared the thin films were observed via SEM and AFM.

3 Results and Discussions

3.1 Phase and Crystallite Size Analysis

X-ray diffraction analysis of Ag/TiO_2 thin films with different amount of PEG deposited on ITO substrate is shown in **Fig. 1**. All the films present five strong diffraction peaks located at 21.51°, 30.57°, 35.40°, 51.04°and 60.70°. These peaks are attributed to the ITO substrates, SiO_2 and In_2O_3 peaks. As can be seen, a small peaks at 25.30° which are assigned to TiO_2 anatase (101) plane, referring to JCPDS JCPDS PDF-021-1272. No diffraction peak of silver is detected which can be explained by the dominance of the ITO peaks in the spectrum with their high intensities, making it difficult to detect silver existing in low amounts [13]. A small peak located at 33.66° is assigned to silver oxides, Ag_2O (100) plane, according to JCPDS No 01-072-2108. However, the X-ray diffraction pattern for the doped PEG Ag/TiO_2 thin film showed the intensity of the diffraction lines is gradually decrease with the increasing amount of PEG. The addition of the PEG 2000 in the Ag/TiO_2 influenced on the unit cell parameters of anatase [14]. The intensity of the diffraction peak decreases and the amplitude of the diffraction peak decreases, indicating that TiO_2 crystalline size is decreasing, consistent with studies conducted by previous researchers [15-17].

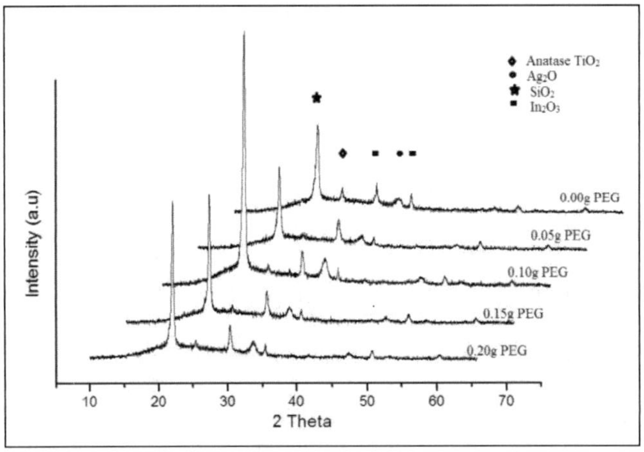

Fig. 1: XRD pattern of Ag/TiO_2 thin films with different amounts of polyethylene glycol

As can be seen, the intensity of the diffraction peak decreases and the apex aperture decreases, indicating that TiO_2 crystalline size decreases, consistent with a study conducted by previous researchers [15-17]. The graph in **Fig. 2** shows the crystalline size distribution of Ag/TiO_2 thin films produced with different PEG amounts. The crystalline size is calculated from the Debye-Scherer formula using the XRD extension. Overall, the TiO_2 crystalline size decreased with the addition of PEG additives. According to [16], the presence of PEG in the Ag/TiO_2 thin film layer reduced the size of the crystals, but did not reduce the amount of TiO_2 anatase crystals that were an important factor in the photocatalytic process. The Ag/TiO_2 thin film crystals were reduced from 40.39 nm to 29.78 nm, 23.88 nm, 19.62 nm and 23.63 nm with the addition of PEG weights of 0.05g, 0.10 g, 0.15 g and 0.20 g, respectively. The smallest crystals recorded were 0.15 g of PEG, 19.62 nm, which is approximately 105% of the Ag/TiO_2 thin film structure that was not added to PEG, 40.39 nm.

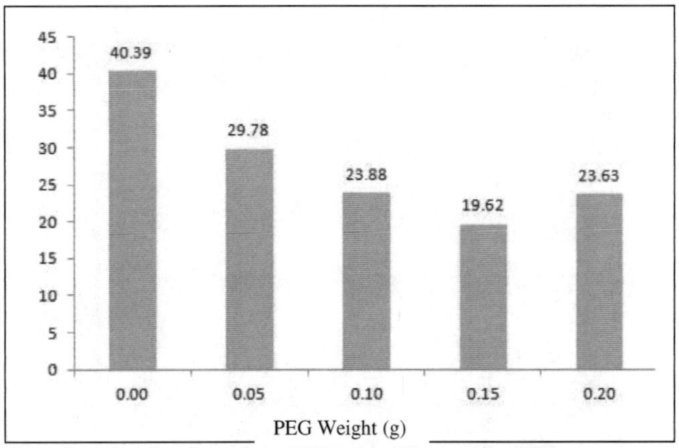

Fig. 2: Crystallite size distribution of Ag/TiO_2 thin films with different amounts of polyethylene glycol

According to Petrik et al. [18] during the Ag/TiO_2 thin film synthesis process, the PEG molecules are absorbed externally into the -Ti-O-Ti- oligomers and form a hydrogen bond. The subsequent formation and polymerization process consists of the formation of an oligomer/PEG composite, in which PEG determines the size of the crystals formed.

3.2 Morphological Observations

The surface morphology of Ag/TiO_2 thin films with different amounts of PEG was observed under SEM as shown in **Fig. 3**. All the prepared Ag/TiO_2 thin films are uniform and evenly distributed in uneven shape and porous structure. From **Fig. 3**(b), Ag/TiO_2 thin film with the addition of 0.05g of PEG shows a thin film with a series of a small, coarse and loose crystalline structures compared to the non-PEG thin film structure (**Fig. 3**(a)), with a larger and thicker structure, in line with the crystalline size obtained from XRD analysis. The PEG 2000 additives in the Ag/TiO_2 solvents act as a dispersion agents, where the low TiO_2 particle aggregation can be explained by the presence of a steric barrier. PEG additives ensure that the soluble particles are separated in solution due to the dispersion of polymers as long-chain molecules [17]. However, no significant changes were observed in the surface morphology of the Ag/TiO_2 thin films when the amount of PEG was increased from 0.10g to 0.15 and 0.20g as shown in **Fig. 3** (c) to **Fig. 3** (e). Structural changes on the surface of the resulting thin films were due to the thermal decomposition of PEG during the annealing process which led to the formation of coarse and porous structures. A similar situation has been reported by previous researchers [13].

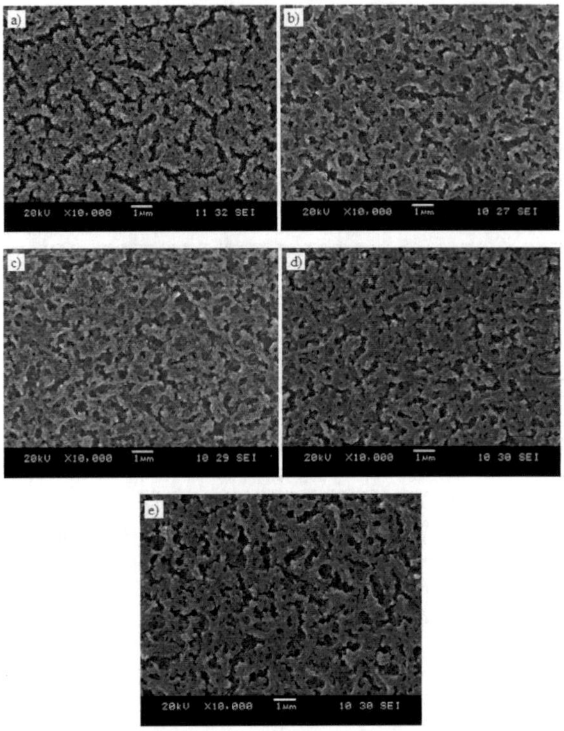

Fig. 3 : SEM image of Ag/TiO_2 thin films with different amounts of polyethylene glycol

In order to investigate the effect of PEG addition on the morphology and surface roughness of the resulting Ag/TiO_2 thin films, AFM analysis was performed. **Fig. 4** shows the topography and 3D views of the surface structure of the thin films produced, while **Table 1** shows the surface roughness of Ag/TO_2 thin films with different PEG amounts. It can be clearly seen that the structure of the Ag/TiO_2 thin film without the addition of PEG is smooth, uniform and covers the entire surface of the substrate, **Fig. 4** (a). The maximum height of this film was 14.98nm with an average roughness of 1.81nm. When 0.05g of PEG was added, the resulting film structure changed to a coarse, non-uniform form, as shown in **Fig. 4** (b). The maximum height of this film was 36.29nm with the average roughness slightly increase to 5.88nm.

Furthermore, when the added PEG weight is increased to 0.10g, 0.15g and 0.20g, the resulting Ag/TiO_2 thin film structures becomes rough and porous. The Ag/TiO_2 films thickness were also increased with the maximum layer thickness recorded at 229.67 nm, 279.49 nm and 300.80 nm. The root mean square roughness of the film also increased with increasing PEG weight to 39.5nm, 44.71nm and 49.37nm. The differences in the surface morphology of the resulting films can be explained by the formation of pores within the Ag/TiO_2 film layer due to PEG thermal decomposition during the annealing process [13]. When more amount of PEG added to the sol solution, the viscosity of the sol increased. The solution attached to the substrate increase and make the resulting thin film layer thicker. Tiwari et al., [19] in their study also found that the TiO_2 thin film coated with PEG was thicker than the non-PEG coated film.

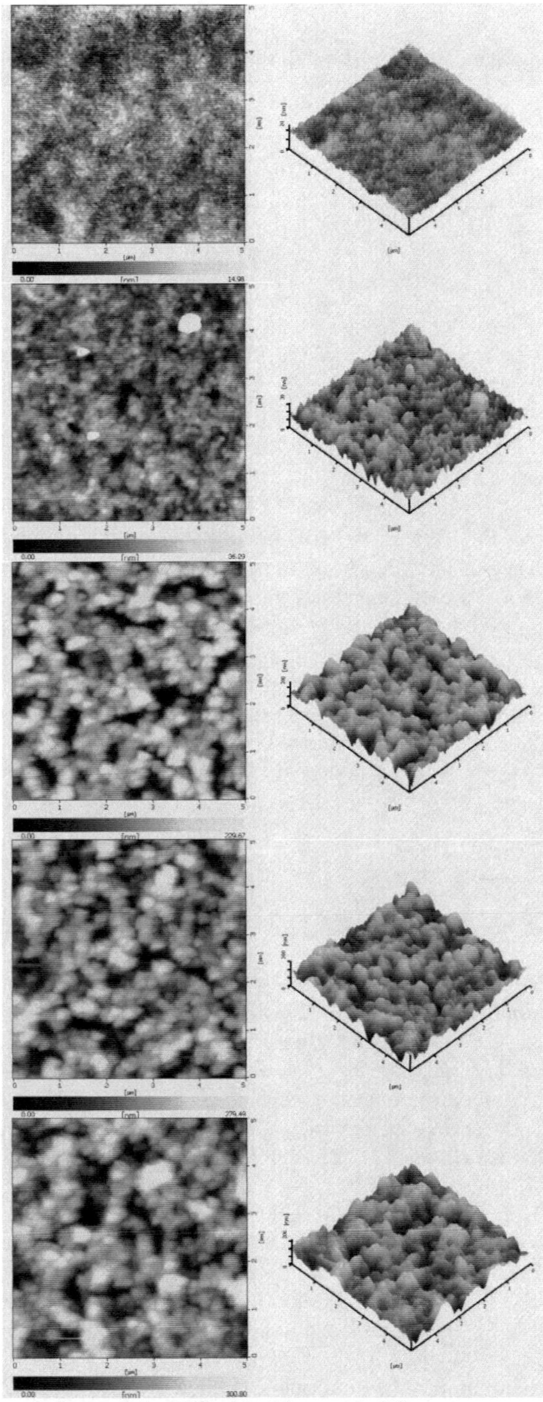

Fig. 4 : Surface morphology of Ag/TiO_2 thin films with different amounts of polyethylene glycol

Table 1 : Surface roughness of Ag/TO_2 thin film with different amount of polyethylene glycol

Ag/TiO_2 thin film	PEG weight (g)	Surface roughness, Ra (nm)	Surface roughness, RMS (nm)
(a)	0.00	1.38	1.81
(b)	0.05	3.60	5.88
(c)	0.10	30.71	39.5
(d)	0.15	35.55	44.71
(e)	0.20	38.78	49.37

4 Conclusion

This work observes the effect of various PEG amounts on morphological and optical properties of Ag/TiO_2 thin films. The XRD pattern showed the presence of anatase TiO_2 and Ag_2O phases. The Ag/TiO_2 thin films shows a rough and porous surface morphology. The surface roughness of the film increased with the increasing of PEG amounts.

The author would like to acknowledge the support from the Fundamental Research Grant Scheme (FRGS) under a grant number of FRGS/1/2017/TK07/UNIMAP/02/6 from the Ministry of Education Malaysia and Tin Solder Technology Research Grant (TSTRG) under grant number of 9002-00082 from Tin Industry (Research and Development) Board. The authors wish to thank the Center of Excellence Geopolymer & Green Technology (CEGeoGTech), School of Materials Engineering, Universiti Malaysia Perlis, UniMAP for their partial support.

References

1. Y. Liang, S. Sun, T. Deng, H. Ding, W. Chen, & Y. Chen, Materials, **11**(3), 1–12 (2018)
2. A. Marzec, M. Radecka, W. Maziarz, A. Kusior and Z. Pedzich, Journal of the European Ceramic Society **36** 2981- 2989 (2016)
3. A. Lewkowicz, A. Synak, B. Grobelna, P. Bojarski, R. Bogdanowicz, J. Karczewski, K Szczodrowski and M. Behrendt , Optical Materials **36** 1739 – 1744 (2014)
4. N. Arconada, Y. Castro, A. Duran, S. Suarez, R. Portela, J. M. Coronado and B. Sanchez Applied Catalysis B: Environmental **86** 1-7 (2009)
5. M. Kumar, M. Kumar and D. Kumar, *Microelectronic Engineering* **87** 447–450 (2010)
6. A. Azani, D. S. C. Halin, K. A. Razak, M. M. A. B. Abdullah, M.A.A. Mohd Salleh, N. Mahmed , M. M. Ramli, S. Sepeai, IOP Conf. Series: Materials Science and Engineering **551** (2019)
7. A. Azani, D. S. C. Halin, K. A. Razak, M. M. A. B. Abdullah, M.A.A. Mohd Salleh, N. Mahmed , M. M. Ramli, S. Sepeai, V. Choptabbana, AIP Conference Proceedings **2129**, 020062 (2019)
8. A. Elfanaoui, E. Elhamri, L. Boulkaddat, A. Ihlal, K. Bouabid, L. Laanab, A. Taleb and X. Portier, International Journal of Hydrogen Energy **36** 4130-4133 (2010)

9. Z. Karimi, L. Karimi and H. Shokrollahi, *Materials Science & Engineering C*, *33*(5), 2465–2475 (2013)
10. A. R. Nurhamizah, M. R. Zulkifli and J. M. Juoi, *www.scientific.net/KEM* **694** 160–164 (2016)
11. K. A. Razak, D. S. C. Halin, M. M. A. B. Abdullah, M.A.A. Mohd Salleh, N. Mahmed and N.S. Danial. Solid State Phenomena, ISSN: 1662-9779, Vol. **280**, pp 26-30 (2018)
12. K. A. Razak, D. S. C. Halin, A. Azani, M. M. A. B. Abdullah, M.A.A. Mohd Salleh, N. Mahmed , M. M. Ramli, S. Sepeai, IOP Conf. Series: Materials Science and Engineering **551** (2019) 012098
13. D. Guitoume, S. Achour, N. Sobti, M. Boudissa, N. Souami, Y. Messaoudi, Optik 154 182–191 (2018)
14. V. Vetrivel, *UGC Sponsored National Seminar on Emerging Trends in Plasma Technology and Its Applications* (2014)
15. Calderon Moreno, J.M., Popa, M., Ivanescu, S. et al. Met. Mater. Int. **20**,I1,pp 177–187 (2014)
16. J. C. Morales Mejia, L. Angeles, R. Almanza, Computational Water, Energy, and Environmental Engineering, 2014, **3**, 36-40 (2014)
17. H. S. Bhojya Naik, P. N. Prashanth Kumar, K. N. Harish and R. Vishwanath, European Journal of Applied Engineering and Scientific Research, **2** (2):1-7 (2013)
18. I. Petrik., E. Frolova, A. Turchin, N. Smirnova, and A. Eremenko, Research on Chemical Intermediates, (2019)
19. D. Tiwari, C. Lalhriatpuia, Lalhmunsiama, S. M. Lee and S. H. Kong, Applied Surface Science **353** 275–283 (2015)

IOP Conf. Series: Materials Science and Engineering **701** (2019) 012005 doi:10.1088/1757-899X/701/1/012005

Tin (Sn) whisker growth from electroplated Sn finished

Aimi Noorliyana Hashim[*], *Mohd Arif Anuar Mohd Salleh, Noor Zaimah Mohd Mokhtar* and *Shahida Idris*

Centre of Excellence Geopolymer and Green Technology, School of Materials Engineering, Universiti Malaysia Perlis (UniMAP), Taman Muhibah 02600, Jejawi, Arau, Perlis, Malaysia

Abstract. Electroplated tin (Sn)Limeited finishes are widely used in the electronics industry due to their excellent solderability, electrical conductivity and corrosion resistance. However, since prohibiting the use of lead (Pb) in electronics, the issue of tin (Sn) whisker has become one of the most imperative issues that need to be resolved. Moreover, with the increasing demand for electronics miniaturization, Sn whisker growth is a severe threat to the reliability of microelectronic device. This paper investigates the effect of electroplating current density (10, 30 and 50 mA/cm^2) on whisker growth at room temperature. A simple micro-indentation test was used to accelerate continuously the growth of Sn whisker at room temperature storage for 24 h. Morphology analysis of Sn whisker growth was done using scanning electron microscope (SEM) and image-J Software. Axial length of the Sn whisker was calculated by using JEDEC Standards (JESD22-A121A).

1 Introduction

Electroplated Sn finishes are widely used in the electronics industry due to their excellent solderability, electrical conductivity and corrosion resistance [1-4]. However, since the implementation of the Restriction of Hazardous Substances Directive (RoHS) in 2006 prohibiting the use of lead in electronics, the issue of tin (Sn) whisker has become one of the most imperative issues that need to be resolved. Moreover, with the increasing demand for electronics miniaturization, Sn whisker growth is a severe threat to the reliability of microelectronic device [1, 5, 6]. Sn whisker is a conductive tin structure that grows spontaneously from Sn finished surfaces, which can lead to well-documented system failures in electronics industries [7-11].

It has been believed that compressive stress in Sn solder coating is a necessary factor for tin whisker growth. Many factors may contribute to the compressive stress level in Sn solder finished, including residual stresses generated during plating process, grain size of coating, formation of intermetallic compound (IMC) due to interfacial reaction between Sn solder and copper substrate, mechanical loading, electroplating currents, and mismatch of coefficient of thermal expansion (CTE) between solder and substrate [5, 12-14].

[*] Corresponding author: aimiliyana@unimap.edu.my

Limited study was discovered on the effect of electrical current on the growth of tin whiskers [15]. M.A. Ashworth et al. [1] studied the effect of electroplating parameters on tin whisker formation, they observed the electroplating current density strongly influenced the Sn grain structure of the electrodeposited coating and whisker morphology. As the deposition current density is increased, the deposit becomes increasingly columnar and the width of the columnar grains reduces, reflecting the reduction in grain size observed on the surface of the deposit. K.S. Kim et al. [16] have shown that the whisker has grain boundaries connected with the grain/whisker root to the intermetallic region and provides a path for highly-compressed intermetallic region tin atoms to diffuse toward the whisker grain boundaries.

This paper investigates the effect of electroplating current density on whisker growth at room temperature. A simple micro-indentation test was used to accelerate continuously the growth of Sn whisker at room temperature storage for 24 h. Morphology analysis of Sn whisker growth was done using scanning electron microscope (SEM) and image-J Software. Axial length of the Sn whisker was calculated by using Joint Electron Device Engineering Council (JEDEC) Whisker Standards (JESD22-A121A).

2 Methodology

In this experiment, pure Sn was electrodeposited onto Cu substrate (>99.9% Cu) using sulphate-based electroplating solution. Sheared Cu substrate 1.5 cm x 1.5 cm squares of 1.0 cm in thickness was immersed into acid liquid bath that contain of 5 g (35%) of hydrochloric acid with 95 g of deionized water (1.75%) for 5 s to remove surface oxides, rinsed with acetone followed by distilled water and dried. Electroplating was performed at different current density of 10, 30 and 50 mA/cm^2 calculated using Eq. 1. The time constant has been set up to 20 minutes.

$$\textit{Current density, mA/cm}^2 = \textit{Current, mA / cross-sectional area, cm}^2 \quad (1)$$

Micro-indentation test, shown in Fig.1 is consisted of a 200 g weight was used to accelerate the growth of Sn whisker at room temperature storage. After the samples were indented for 24 hours, morphology analysis of Sn whisker growth was done using scanning electron microscope (SEM) and image-J Software. Length and density of the Sn whisker was calculated by using Joint Electron Device Engineering Council (JEDEC) Whisker Standards (JESD22-A121A). Based on JESD22-A121A, the axial length of whisker was measured from the electroplated surface to the whisker tip, shown in Fig.2.

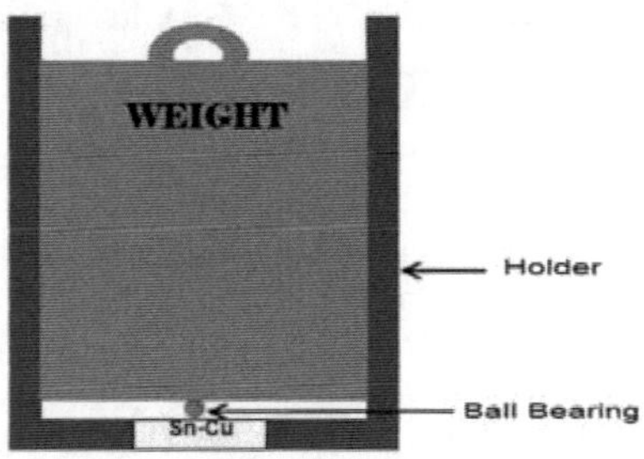

Fig. 1. Schematic of micro-indentation test apparatus for accelerated Sn whisker growth.

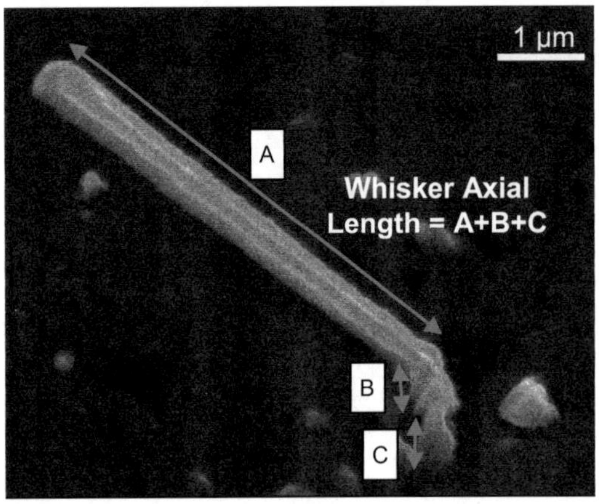

Fig. 2. Axial methods for calculating whisker length based on JESD22-A121A.

3 Results and Discussion

Tin whiskers growth from Sn electroplated finished with different current density 10, 30, 50 mA/cm^2 has been observed after 24 hours of mechanical indentation test. Most of the observed whiskers had striations along the outer surface. Fig.3 show the SEM images of Sn grain structure coating and type of Sn whiskers. It is evident that the grain structure of the electrodeposited tin coating is strongly influenced by deposition current density [1]. As the deposition current density is increased from 10 to 50 mA/cm^2 the grain size of the electrodeposited tin decreases. The increased grain size might be as a result of hydrogen evolution which was observed at the deposit surface during electroplating. Increasing levels of hydrogen evolution may hinder the adsorption of the plating additives onto the sample surface of the tin during deposition and result in an increased tin grain size. Hydrogen evolution may also increase mass transport and encourage grain growth rather than fresh nucleation. At a deposition current density of 100 mA/cm^2, the tin grains frequently have a cubic morphology. As the deposition current density increases the tin grain boundaries become progressively less facetted [1].

The length of whisker becomes slightly lessen as the deposition current density is increased. The longest filament needle shape whisker found is 29.6μm with the average diameter is 1.0μm for 10 mA/cm^2 as shown in Fig.3(b). In Fig.3 (d), the length of filament whiskers growth on substrate for 30 mA/cm^2 current density is 9.93μm and its diameter is 0.64μm. For 50 mA/cm^2 current density (Fig.3(f)), the ribbon shape whisker was initiate with average length 2.2μm-3.9μm and the diameter is 1.1μm-1.2μm. Studied by B. Jiang et al. [17], indicate that the ribbon like whiskers morphology often observed on finish plated with sulfate-based electrolyte and has a smooth surface with size is shorter and larger [17].

W.J. Boettinger et al. [18] believed the growth of hillocks is also a factor of compressive residual stress similar to the growth of whiskers. Growth of the hillock is often said to be localized diffusional creep or grain boundary sliding phenomenon which relieves the compressive stress. The differences between whiskers and hillock is surely about the shape. The whiskers shape emerges outward from the surfaces with very thin such as single filament or hair-like protrusion. The hillocks shape is growth of precursor to whiskers, but due to its size is not harmful, it is not considered to be whiskers [18].

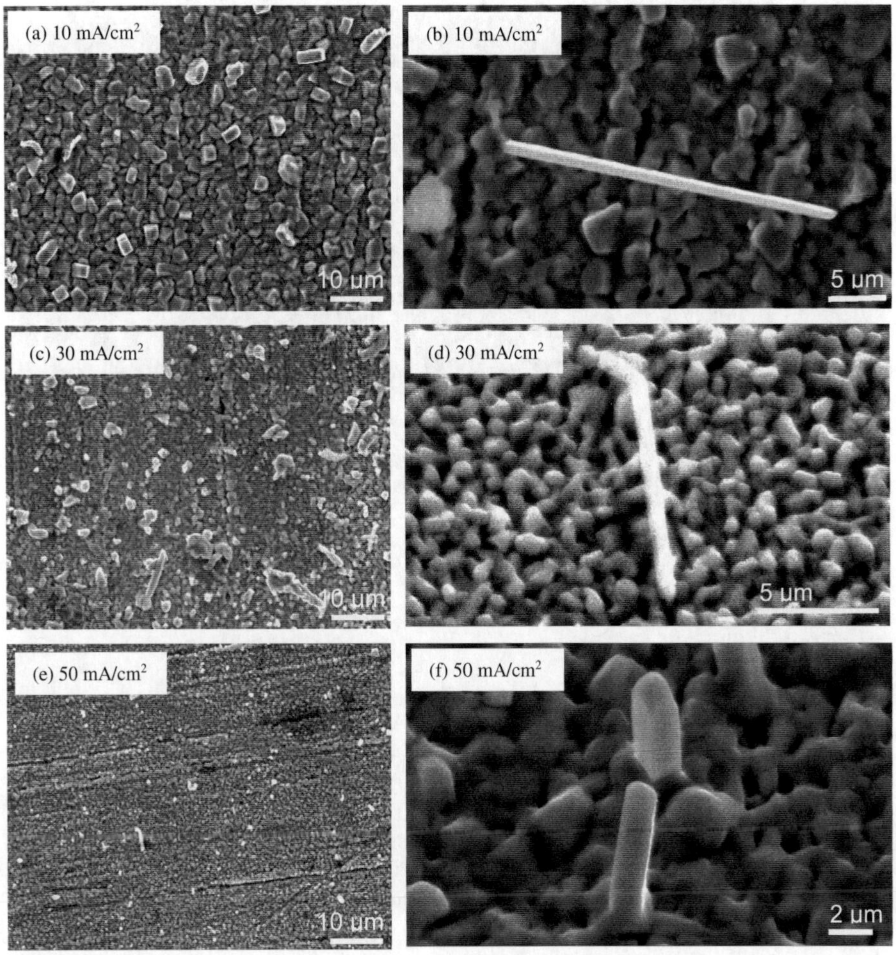

Fig. 3. SEM images of Sn whisker growth from Sn electrodeposition coating with different current density (a) Sn deposition at 10 mA/cm^2 (b) Sn whisker at 10 mA/cm^2 (c) Sn deposition at 30 mA/cm^2 (d) Sn whisker at 30 mA/cm^2 (e) Sn deposition at 50 mA/cm^2 (f) Sn whisker at 50 mA/cm^2

The driving force for whisker formation is generally accepted to be stress generated by the formation of Cu-Sn intermetallic (IMC) between the Sn and Cu. High compressive average residual stress was present in thinner films and the stress decreased with increase the film thickness. Fig.4 show the formation of Sn–Cu intermetallic compounds (IMC) at the interface between the electroplated Sn solder and the Cu substrate for 50 mA/cm^2 current density. The average thickness of Sn solder deposition is 19.4μm. and the average thickness of IMC is 0.42μm. The presence of IMC occurs when Cu seeps into the film Sn. The density changes between Cu_6Sn_5 and Cu causes the volume expansion on the IMC layer. Thus, the compressive stress is generated by the expansion of the volume in the vertical direction of the Cu and Sn interface. Therefore, compressive stresses can enhance to accelerate the growth of whiskers. Basically, IMC is able to promote whisker growth, but whisker growth is not directly proportional to IMC thickness [16, 19, 20].

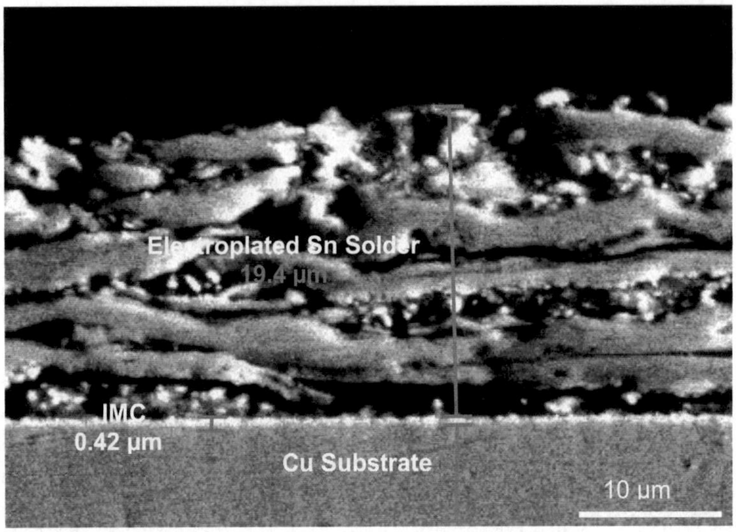

Fig. 4. SEM images of the formation of Sn–Cu intermetallic compounds (IMC) at the interface between the Sn electrodeposit and the Cu substrate for 50 mA/cm^2 current density.

4 Conclusions

From the research, the following conclusions have been obtained;

i. After 24 h continuously indented by mechanical compressive stress with different current density 10, 30, 50 mA/cm^2, the tin whiskers growth from Sn electroplated finished. The grain structure of the electrodeposited tin coating is strongly influenced by deposition current density. As the deposition current density is increased from 10 to 50 mA/cm^2 the grain size of the electrodeposited tin decreases.

ii. The length of whisker becomes slightly lessen as the deposition current density is increased. The longest filament needle shape whisker found is 29.6µm with the average diameter is 1.0µm for 10 mA/cm^2. Meanwhile for 50 mA/cm^2 current density, the ribbon shape whisker was initiate with average length 2.2µm-3.9µm and the diameter is 1.1µm-1.2µm.

iii. Sn whiskers significantly is the most dangerous to circuit reliability as the conductive Sn whiskers can grow to several hundred microns, which are long enough to cause electrical shorts circuit by bridging neighbouring conductors or may break off and cause damage elsewhere in the device.

This work was financially supported by Unimap-Nihon Superior Research Colaboration (Grant No: 2016/10/0001) and Tin Solder Technology Researchs Group (TSTRG) Research Grant 2019.

IOP Conf. Series: Materials Science and Engineering **701** (2019) 012005 doi:10.1088/1757-899X/701/1/012005

References

1. M.A. Ashworth, G.D. Wilcox, R.L. Higginson, R.J. Heath, C. Liu, and R.J. Mortimer, Microelectronics Reliability. **55**(1) (2015)
2. M.I.I. Ramli, M.A.A.M. Salleh, F.A.M. Sobri, P. Narayanan, K. Sweatman, and K. Nogita, Journal of Materials Science: Materials in Electronics. **30**(4) (2019)
3. N.M. Nasir, N. Saud, M.N. Deraman, A.A.M. Salleh, M.I.I. Ramli, and R.M. Said, Materials Science Forum. **803** (2015)
4. M.I.I. Ramli, M.S.S. Yusof, M.A.A.M. Salleh, R.M. Said, and K. Nogita, Solid State Phenomena. **273** (2018)
5. A. Noorliyana Hashim, M. Arif Anuar Mohd Salleh, and N. Zaimah Mohd Mokhtar, IOP Conference Series: Materials Science and Engineering. **551** (2019)
6. F. Somidin, H. Maeno, M.A.A.M. Salleh, X.Q. Tran, S.D. McDonald, S. Matsumura, and K. Nogita, Materials Characterization. **138** (2018)
7. N. Jadhav, M. Williams, F. Pei, G. Stafford, and E. Chason, Journal of Electronic Materials. **42**(2) (2013)
8. A. Skwarek, M. Pluska, J. Ratajczak, A. Czerwinski, K. Witek, and D. Szwagierczak, Materials Science and Engineering: B. **176**(4) (2011)
9. E. Chason, N. Jadhav, F. Pei, E. Buchovecky, and A. Bower, Progress in Surface Science. **88**(2) (2013)
10. N. Zaimah Mohd Mokhtar, M. Arif Anuar Mohd Salleh, and A. Noorliyana Hashim, IOP Conference Series: Materials Science and Engineering. **551** (2019)
11. M.A.A.M. Salleh, M.H. Hazizi, M.A.B.M. Abdullah, N. Noriman, R. Mayapan, and Z.A. Ahmad, Advanced Materials Research. **626** (2012)
12. C.-K. Lin and T.-H. Lin, Microelectronics Reliability. **48**(10) (2008)
13. K. Nogita, M.A.A.M. Salleh, S. Smith, Y.Q. Wu, S. D.McDonald, A.G.A. Razak, T. Akaiwa, and T. Nishimura. *Effects of Bi in Sn-Cu based lead-free solder alloys and interconnects*. in *Electronics Packaging (ICEP), 2017 International Conference on*. 2017. IEEE.
14. K. Nogita, M.A.A.M. Salleh, E. Tanaka, G. Zeng, S.D. McDonald, and S. Matsumura, JOM. **68**(11) (2016)
15. Y. Fukuda, M. Osterman, and M. Pecht, Microelectronics Reliability. **47**(1) (2007)
16. K.S. Kim, C.H. Yu, S.W. Han, K.C. Yang, and J.H. Kim, Microelectronics Reliability. **48**(1) (2008)
17. B. Jiang and A.-P. Xian, *Observations of ribbon-like whiskers on tin finish surface*. Vol. 18. 2007. 513-518.
18. W.J. Boettinger, C.E. Johnson, L.A. Bendersky, K.W. Moon, M.E. Williams, and G.R. Stafford, Acta Materialia. **53**(19) (2005)
19. M.A.A.M. Salleh, A.M.M.A. Bakri, F. Somidin, and H. Kamarudin, International Review of Mechanical Engineering. (2013)
20. M.A.A.M. Salleh, M.M.A.B. Abdullah, F. Somidin, A.V. Sandu, N. Saud, K. Hussin, S.D. McDonald, and K. Nogita, Revista de Chimie **64**(7) (2013)

IOP Conf. Series: Materials Science and Engineering **701** (2019) 012006 doi:10.1088/1757-899X/701/1/012006

Influence of cooling conditions on the interfacial Cu_6Sn_5 intermetallic compound in Sn-37Pb/Cu solder joints during reflow

Flora Somidin[1, 2, *], *Stuart* McDonald[1], *Tetsuya* Akaiwa[3], *Tetsuro* Nishimura[3], and *Kazuhiro* Nogita[1]

[1]Nihon Superior Centre for the Manufacture of Electronic Materials (NS CMEM), School of Mechanical and Mining Engineering, The University of Queensland (UQ), Brisbane, QLD 4072, Australia
[2]Centre of Excellence Geopolymer and Green Technology, School of Materials Engineering, Universiti Malaysia Perlis, Taman Muhibbah, 02600 Jejawi, Arau, Perlis, Malaysia
[3]Nihon Superior Co. Ltd., Suita City, Osaka 564-0063, Japan

Abstract. Despite a shift to Pb-free solders in the majority of electronic assemblies, Pb-containing solders are still used in a variety of applications. Recent research has revealed the complexities of the intermetallic Cu_6Sn_5 in terms of its crystal structure and stability when present as the reaction product between Pb-free solders and common substrates. In light of these findings, the effect of multiple reflow cycles with different cooling conditions on the microstructure of the interfacial Cu_6Sn_5 intermetallic compound (IMC) between Sn-Pb solders and Cu substrates was re-examined. For this purpose, the Sn-37Pb (in wt.%) solder was reflowed on a Cu substrate at a temperature of around 240 °C and cooled at a rate of around 30 °C/min under each of two conditions; (i) direct-cooling to room temperature, and (ii) interrupted-cooling with extended isothermal holding periods of 30 and 180 seconds at 140 °C. This second condition was chosen to encourage the polymorphic transformation that occurs in the Cu_6Sn_5 (at an equilibrium temperature of 186 °C) at a temperature which minimises the stresses involved. Smaller cracks and a thinner interfacial IMC layer were observed in the solder joints that were reflowed repeatedly with the interrupted-cooling conditions. However, despite the higher number of cracks in the samples undergoing conventional reflow, the shear strength was not significantly affected.

1 Introduction

Previously, eutectic Sn-37Pb (in wt.%) solder with a melting point of 183 °C found widespread use in the electronic assembly process due to its unique combination of properties and lower cost [1]. However, there are now major restriction on the use of Pb-based solder alloys due to environmental and toxicity concerns [2]. Legislation surrounding the use of Pb-free alloys has granted exemptions [3] for some products whose applications require high long-term reliability, e.g. defense, medical, instrumentation, etc. However, since most electronics companies have already moved towards Pb-free assembly, it is expected that some componentry may be assembled with a combination of both Sn-Pb and Pb-free solders.

* Corresponding author: f.somidin@uq.net.au

This means the lower melting point Sn-Pb solder joint may be exposed to higher temperatures with multiple thermal cycles in an assembly process designed for Pb-free alloys.

One of the major concern with the thermal history during processing is it's impact on the stability of the phases of the solder joint, especially the intermetallic Cu_6Sn_5. Above 186 °C the stable form of Cu_6Sn_5 is the hexagonal η phase [4]. At temperatures below 186 °C the stable form of Cu_6Sn_5 is the monoclinic η' phase [5]. It has been hypothesised that when the Cu_6Sn_5 is constrained within a solder matrix the approximately 2.15% volume change that is associated with the η↔η' transformation may result in the buildup of stresses that could lead to cracking of Cu_6Sn_5, particularly when the joint is subjected to multiple reflow cycles [6], [7]. Another concern is that coarsening of the Pb phase as a function of time, temperature and mechanical loading, may affect the reliability of the solder joints [8]. Thus, the present study, examines the effect of cooling conditions during the reflow of Sn-37Pb solders on copper (Cu) substrates on the shear strength and initiation of damage in the interfacial intermetallic compounds (IMCs).

2 Materials and Methodology

Solder balls of Sn-37Pb (in wt.%) alloy of approximately 500 µm diameter were reflowed onto 500 µm diameter Cu pads that had an organic solderability preservative (OSP) surface finish. Reflow was carried out in a benchtop reflow simulator with the aid of a small amount of no-clean flux. The time above 220 °C was fixed at ~110 s for all reflow profiles. In the reference reflow profile (see Fig. 1a) the reflowed solder ball cooled from 240 °C to room temperature at a rate of about 30 °C/min.

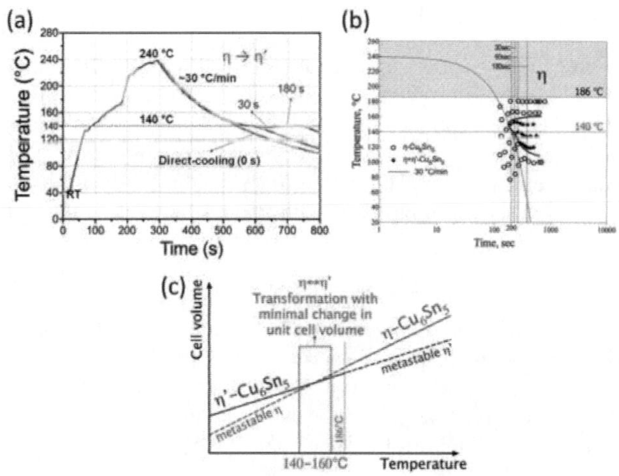

Fig. 1. (a) Reflow temperature profiles used in this study; conventional direct-cooling (0 s) and interrupted-cooling with an extended hold at 140 °C for around 30 s and 180 s; (b) TTT-curve of η↔η+η' [9]; (c) Schematic illustration of a possible technique for minimizing the stress generated by the Cu_6Sn_5 η→η' polymorphic transformation based on previous dilatometry measurements [10].

The cooling rate of 30 °C/min is based on the time-temperature transformation (TTT) curve η↔η+η' shown in Fig. 2(b) which ensures the solder joint reaches 140 °C prior to the η→η+η' polymorphic transformation commencing, as determined from the TTT diagram [9]. There will be minimal change in the unit cell volume of Cu_6Sn_5 as it transforms from the η to the η' form in the temperature range between 140-160 °C as schematically illustrated in Fig. 1(c) [11]. The objective is to confirm that the transformation occurs and to determine the time required for the completion of the transformation at that temperature. To determine the effect of multiple reflows using the cooling-condition stated in Fig. 1(a), the samples were reflowed 2 and 8 times.

Reflowed samples were cross-sectioned and fine polished perpendicular to the solder/Cu interface. The solder joints were observed using a desktop scanning electron microscope (SEM) (TM3030, Hitachi, Japan). The average thickness of the total IMC layer, d_{ave}, was determined using the following equation:

$$d_{ave} = \frac{A}{L} \quad (1)$$

where A is the total hatched IMC area, and L is the IMC length in the horizontal length of the measurement window. The total crack area was normalized by the total measured area of IMCs in the horizontal length the solder joint, which was typically close to 300 µm. The quantitative image analysis was done using ImageJ software (NIH, USA) [12].

Ball shear tests were performed at a shear speed of 2000 mm/s using a DAGE-4000s bond microtester (Nordson, USA). At least 20 reflowed solder balls were tested for each condition. The hammer height above the substrate was fixed at 50 µm.

The TEM lamella samples were extracted from solder joints using a focused ion beam (FIB) system (FEI Scios Dual-beam FIB/Scanning Electron Microscope, Thermo Fisher Scientific Waltham, MA, USA) and welded with Pt deposition on Cu TEM grids. Energy dispersive x-ray spectroscopy (EDS) mapping was also conducted. TEM characterisation was performed with a JEM-1300NEF (JEOL, Akishima, Japan) at an accelerating voltage of 1250 kV.

3 Results and Discussion

Fig. 2(a, b) show SEM images of the interfacial region of Sn-37Pb/Cu in the single reflowed (1st cycle) and after eight reflow cycles with both direct-cooling (0 s holding at 140 °C) and interrupted-cooling (30 s and 180 s holding at 140 °C). It can be seen that there no obvious differences in the microstructure between the two reflow cooling conditions in the 1st reflow cycle and subsequent reflow cycles. The average interfacial IMC layer thickness growth over multiple reflow cycles is shown in Fig. 2(b). Multiple reflow cycles, in general, increase the interfacial IMC thickness for all cooling conditions. Interestingly, the samples reflowed with interrupted-cooling show less IMC growth after 8 reflow cycles when compared to the direct-cooled samples. As shown in Fig. 2(d), minor cracking was also observed in the interfacial IMC when the solder joints were reflowed repeatedly with longer interrupted-cooling cycles. The results of the shear impact testing at a speed of 2000 mm/s indicate that the variation in reflow cooling conditions does not have any significant effect on the maximum force in Sn-37Pb/Cu solder joints (see Fig. 2e). This can be due to the soft Pb distributed along the Cu_6Sn_5 grain boundaries in the reaction layer [12] that could have accommodated the stress near the IMC area during shear test.

Fig. 2(a) shows the TEM lamella sample that was extracted from the Sn-37Pb/Cu joint after 8 reflow cycles with interrupted-cooling for 180 s at 140 °C. On-zone bright-field TEM images of the sample are shown in Fig. 2(b). As shown in Fig. 2(C), a combination of strong and weak reflections were observed in the selected electron diffraction pattern (SADP) of the chosen Cu_6Sn_5 grain. The stronger reflections can be indexed as hexagonal η-Cu_6Sn_5 ($P6_3/mmc$) along the [120] zone-axis. The weaker reflections can be indexed as monoclinic η′-Cu_6Sn_5 (C2/1) in the [010]//[0-10] zone-axis. The EDS mapping analysis on the solder joint sample is shown in Fig. 2(d). The formation of the monoclinic phase in the grain shows that the interrupted-cooling for 180 s at 140 °C allows for transformation of the metastable hexagonal structure to a stable monoclinic structure during cooling. Because very minor cracking was observed in samples held for 180 s, it is assumed the polymorphic transformation has occurred in a controlled fashion in this condition during which the volume change in the unit cell of Cu_6Sn_5 is at a minimum.

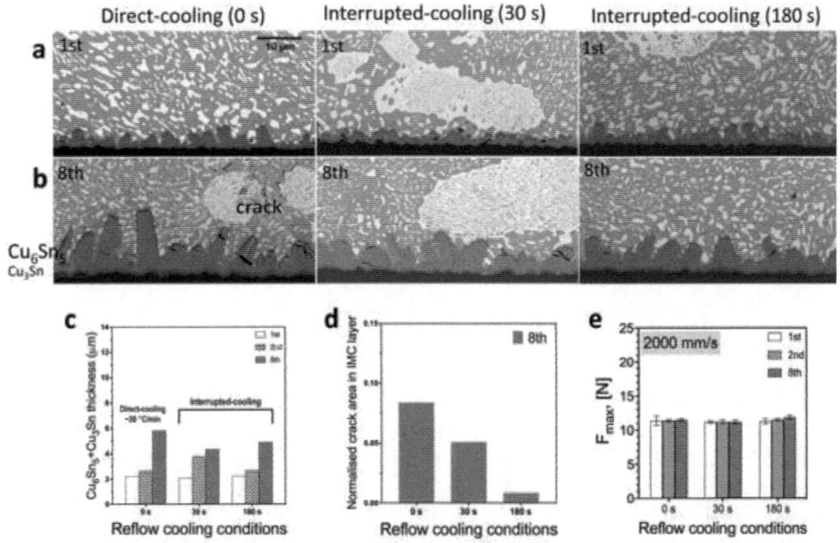

Fig. 2. Representative backscattered SEM micrographs of cross-sections of Sn-37Pb/Cu after the (a) 1st reflow and (b) 8th reflow cycle; (c) average interfacial IMC thickness; (d) normalized crack area; (e) maximum force in shear impact at a shear speed of 2000 mm/s.

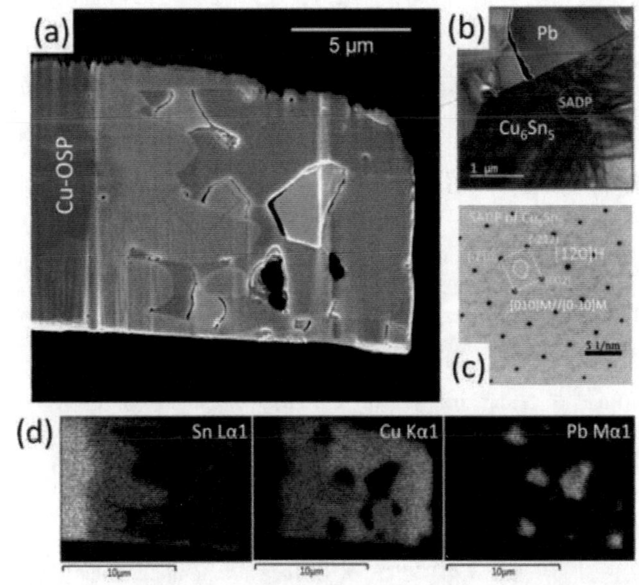

Fig. 3. (a) SEM image of the FIB sample obtained from Sn-37Pb/Cu joint after 8 reflow cycles with interrupted-cooling for 180 s at 140 °C; (b) Bright-field TEM image; (c) corresponding selected white circle area diffraction pattern (SADP) in (b); (d) elemental mapping analysis of the FIB sample.

4 Conclusion

The cooling condition with an isothermal holding step of 180 s at ~140 °C during cooling from 240 °C at a cooling rate of 30 °C/min is proposed for Sn-37Pb/Cu solder joints to minimise cracking in solder joints that occurs after exposure to multiple reflow cycles. It is found that multiple reflow cycles with interrupted-cooling cycles are associated with less cracking and thinner IMC layers than what occurs in direct-cooled (uninterrupted) reflow conditions. However, despite the high number of cracks in samples undergoing conventional reflow, the shear strength was not significantly affected in these Pb containing alloys.

The authors gratefully acknowledge financial support from the University of Queensland-Nihon Superior (2016001895) collaborative research program. The authors acknowledge the facilities, and the scientific and technical assistance at the Centre for Microscopy and Microanalysis, UQ and the Ultramicroscopy Research Center, Kyushu University. F. Somidin is financially supported by the Ministry of Education Malaysia and Universiti Malaysia Perlis (UniMAP).

References

1. J. Bath, Ed., *Lead-Free Soldering*. Boston, MA: Springer US, 2007.
2. World Health Organization, "Lead poisoning and health," 23-Aug-2018. [Online]. Available: https://www.who.int/news-room/fact-sheets/detail/lead-poisoning-and-health. [Accessed: 01-Aug-2019].
3. "RoHS Dispensation update - exemption validity until July 2021," *Retronix*, 30-Apr-2019.
4. A.-K. Larsson, L. Stenberg, and S. Lidin, "The superstructure of domain-twinned η'-Cu_6Sn_5," *Acta Crystallographica Section B Structural Science*, vol. 50, no. 6, pp. 636–643, Dec. 1994.
5. O. B. Karlsen *et al.*, "The Ternary System Au-Cu-Sn.," *Acta Chemica Scandinavica*, vol. 46, pp. 147–156, 1992.
6. D. K. Mu, S. D. McDonald, J. Read, H. Huang, and K. Nogita, "Critical properties of Cu_6Sn_5 in electronic devices: Recent progress and a review," *Current Opinion in Solid State and Materials Science*, vol. 20, no. 2, pp. 55–76, Apr. 2016.
7. K. Nogita, C. M. Gourlay, and T. Nishimura, "Cracking and phase stability in reaction layers between Sn-Cu-Ni solders and Cu substrates," *JOM*, vol. 61, no. 6, pp. 45–51, Jun. 2009.
8. H. Ye, C. Basaran, and D. C. Hopkins, "Pb phase coarsening in eutectic Pb/Sn flip chip solder joints under electric current stressing," *International Journal of Solids and Structures*, vol. 41, no. 9–10, pp. 2743–2755, May 2004.
9. F. Somidin *et al.*, "Characterising the polymorphic phase transformation at a localised point on a Cu_6Sn_5 grain," *Materials Characterization*, vol. 138, pp. 113–119, Apr. 2018.
10. D. Mu, J. Read, Y. Yang, and K. Nogita, "Thermal expansion of Cu_6Sn_5 and $(Cu,Ni)_6Sn_5$," *Journal of Materials Research*, vol. 26, no. 20, pp. 2660–2664, Oct. 2011.
11. K. Nogita *et al.*, "Effect of reflow conditions on the intermetallic layer in solder joints," in *2018 International Conference on Electronics Packaging and iMAPS All Asia Conference (ICEP-IAAC)*, Mie, Japan, 2018, pp. 94–96.
12. K. Nogita *et al.*, "Synchrotron Micro-XRF Measurements of Trace Element Distributions in BGA Type Solders and Solder Joints," *Transactions of The Japan Institute of Electronics Packaging*, vol. 3, no. 1, pp. 40–46, 2010.

IOP Conf. Series: Materials Science and Engineering **701** (2019) 012007 doi:10.1088/1757-899X/701/1/012007

Temperature dependency of the growth rate of $(Cu,Ni)_6Sn_5$ on Cu-xNi substrates

Xin Fu Tan[1,*], *Nurul Razliana* Abdul Razak[1], *Muhammad Hafiz* Husain[1], *Stuart* McDonald[1], and *Kazuhiro* Nogita[1]

[1]Nihon Superior Centre for the Manufacture of Electronic Materials (NS CMEM), School of Mechanical and Mining Engineering, The University of Queensland, Brisbane, Qld 4072, Australia

Abstract. Cu_6Sn_5 is the most common intermetallic compound (IMC) in lead-free solder joints, and is also a promising anode material for advanced lithium-ion batteries. It has been reported that the growth rates of $(Cu,Ni)_6Sn_5$ in Sn-based solder alloy/Cu-xNi substrate couples are greatly accelerated compared to the intermetallic layers that form on a pure Cu substrate. Due to the faster growth rates, solidification can progress through a transient-liquid-phase reaction in commercially relevant timeframes. The complete reaction of the liquid Sn in the couple will result in a joint that is stable at high temperatures. Furthermore, the presence of Ni in the intermetallic suppresses the formation of a Cu_3Sn layer and stabilises the $(Cu,Ni)_6Sn_5$ preventing a polymorphic transformation on cooling along with the associated induced strains. This study investigates the growth rates of $(Cu,Ni)_6Sn_5$ in Sn/Cu-xNi couples, where $0 <= x <= 14$ wt%, and reports on the effect of temperature on the growth rates at temperatures above the melting point of Sn (231.9°C) between 232°C and 300°C. The results show the highest growth rates occur in the Sn/Cu-6Ni couples, and there is an increase of the $(Cu,Ni)_6Sn_5$ growth rate with increasing temperature due to faster diffusion and reaction rates.

1 Introduction

Cu_6Sn_5 is a near ubiquitous intermetallic compound (IMC) in Sn-based lead-free solder joints made with copper substrates and plays an important role in the mechanical integrity of the joint [1]. Cu_6Sn_5 also has potential functional applications as an anode material in lithium-ion batteries due to its higher energy capacity relative to conventional carbon-based anodes [2, 3].

With the miniaturisation and the rapid increase in the density of electronic packaging, local service temperatures in the integrated circuits increase accordingly, creating a demand for high temperature bonding materials. Transient-liquid-phase (TLP) bonding is an attractive method for high temperatures applications as the fabrication can be carried out at relatively low temperatures, while the bonded joints can operate at much higher temperatures. Cu_6Sn_5 IMCs forms naturally at the interface of Cu substrates and Sn-based solder joints during typical reflow temperatures below 250°C. The liquidus temperature of Cu_6Sn_5 is

* Corresponding author: xin.tan@uq.edu.au

408˚C, and it has better electromigration resistance [4] and a higher strength compared to bulk solder alloys and is therefore an ideal candidate material for the end product of a TLP bond.

However, the growth rate of Cu_6Sn_5 in a Cu/Sn-based solder alloy couple is often sluggish, and different methods have been therefore been explored to shorten the bonding time. These include enhancing the Cu/Sn interfacial reaction by applying an electric current [4], ultrasonic-assisted bonding [5] and the addition of Ni in the Cu substrates [6-8].

In this study, Cu-xNi substrates with varying concentrations of Ni, where x = 0, 2, 6, 10, and 14 wt%, were examined with respect to the growth rate of $(Cu,Ni)_6Sn_5$ that forms on this substrate during isothermal contact with liquid Sn. Furthermore, the effect of temperature on the growth rates was investigated at commercially relevant temperatures of 232˚C, 250˚C, 270˚C and 300˚C. It was hypothesised that the combined effect of optimising the temperature and Ni concentration in the substrate would greatly increase the rate of formation of $(Cu,Ni)_6Sn_5$.

2 Methods

Soldering flux was applied to the surface of Cu-xNi sheets, where x = 0, 2, 6, 10, and 14 wt%, to remove surface oxides. Sn (99.9%) sheets rolled to approximately 100 μm in thickness were placed on the Cu-xNi sheets. The Cu-xNi/Sn couples were sandwiched between two glass slides and clamped together with a metal clip. The samples were placed in an annealing furnace preheated to a temperature of either 250˚C, 270˚C or 300˚C for 3, 5, 7 and 10 minutes. For comparison, a set of control samples were held at 232˚C for 5 and 10 minutes. The Sn sheets liquefied and reacted with the Cu-xNi sheets to form (Cu,Ni)6Sn5 at the Cu-xNi/Sn interface. The samples were then removed from the furnace and cooled in air, and subsequently rinsed in ethanol.

To observe the cross-sections, the samples were cold mounted in epoxy resin. Standard metallography procedures were used to polish the surface of the samples for scanning electron microscopy (SEM) observations. SEM was conducted under backscatter electron (BSE) mode at an accelerating voltage of 15 kV.

3 Results and discussions

3.1 Growth rates at 232˚C

At a reaction time of 5 minutes at 232˚C, just above the melting point of Sn of 231.9˚C, very little bonding occurs between the Cu-xNi/Sn couples, and most of the couples detached as they were removed from the glass slides.

Fig. 1 shows the cross-section SEM images of the Cu-xNi/Sn couples placed at 232˚C for 10 minutes. Under BSE imaging mode, the top layers with the brightest shade of grey are the Sn sheets (highest Z-contrast), the bottom layers with the darkest shade are the Cu-xNi substrates, and a layer of $(Cu,Ni)_6Sn_5$ can be seen at the Cu-xNi/Sn interface. The thickness of the $(Cu,Ni)_6Sn_5$ layer varies dramatically with the Ni content in the substrate. In a pure Cu/Sn couple, approximately 4 μm of Cu_6Sn_5 with a scallop-like morphology is grown. As the concentration of Ni in the substrate is increased to 2 wt% Ni, the morphology of the $(Cu,Ni)_6Sn_5$ changes to a rod-like structure, although the growth rate remains relatively constant at around 4 μm in 10 minutes. As the amount of Ni in the substrate is increased to 6 wt%, the growth rate of $(Cu,Ni)_6Sn_5$ increases dramatically. Very little Sn remained in the Cu-6Ni/Sn couple and 48 μm of $(Cu,Ni)_6Sn_5$ was formed. However, further increasing the content of Ni results in a decrease of the growth kinetics and the same conditions see

$(Cu,Ni)_6Sn_5$ layers of about 28 µm and 10 µm form in the Cu-10Ni/Sn and the Cu-14Ni/Sn couple respectively.

The results show that the presence of 6-14 wt% Ni in the Cu substrate can significantly increase the growth rates of the interfacial $(Cu,Ni)_6Sn_5$ relative to Ni-free substrates, with a maximum growth rate observed in the Cu-6Ni/Sn couple. This result matches that found by Choquette [9], who reported a maximum growth rate at 5.5 wt% Ni in a Cu-xNi/Sn-0.7Cu-0.05Ni system, while they deviates slightly from the results of Vuorinen et. al., where the fastest rate was observed on Cu-10Ni substrates [7], both conducted at 240°C. Furthermore, the minimal amount of bonding in the samples held for 5 minutes, including the Cu-6Ni/Sn couple, suggests that an incubation period between 5-10 minutes is required before the $(Cu,Ni)_6Sn_5$ formation reaction proceeds at this temperature.

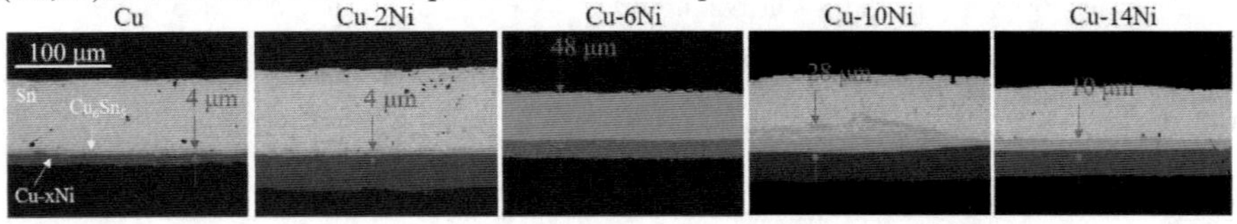

Fig. 1. Cross-section SEM images of (Cu, $Ni)_6Sn_5$ grown between Sn and Cu-xNi, where x = 0, 2, 6, 10 and 14, at 232°C for 10 minutes. Images are shown at the same scale.

3.2 Growth rates at 250°C, 270°C and 300°C

SEM images of cross-section of the coupes showing the $(Cu,Ni)_6Sn_5$ grown at 250°C, 270°C and 300°C are seen in Fig. 2. The average thicknesses of the $(Cu,Ni)_6Sn_5$ layers for the different samples and conditions are plotted in Fig. 3. In comparison to Fig. 1, the thickness of the interfacial $(Cu,Ni)_6Sn_5$ after 10 minutes of growth has increased with the increasing temperature. Increasing temperature has also reduced the incubation time for bonding, as shown by the formation of a layer of $(Cu,Ni)_6Sn_5$ in all the samples held at 250°C, 270°C and 300°C.

Minimal bonding was observed in the Cu/Sn couple held for 3 minutes at 250°C and part of the sample detached when the metal clip used to clamp the sample was removed, entrapping contaminants at the Cu/Sn interface as can be seen in Fig. 2 (a). At this initial bonding stage, instead of having a maximum growth rate in the 6 wt% Ni sample, the 3 minutes samples bonded at 250°C show an increasing thickness of $(Cu,Ni)_6Sn_5$ with increasing amounts of Ni in the substrates, indicating that the addition of Ni in the substrates may reduce incubation time. At 5 minutes, the interface layer of $(Cu,Ni)_6Sn_5$ in the Cu-6Ni, Cu-10Ni and Cu-14Ni samples were significantly thicker than those in the Cu and Cu-2Ni samples. At 7 and 10 minutes, accelerated growth of $(Cu,Ni)_6Sn_5$ is also observed in the Cu-2Ni sample, while the Cu-6Ni sample again shows the maximum growth rate. In the Cu-6Ni and Cu-10Ni samples, $(Cu,Ni)_6Sn_5$ grains have grown to the top of the Sn layer within 7 minutes, and in the Cu-6Ni sample, the Sn was fully consumed within 10 minutes, forming a layer of $(Cu,Ni)_6Sn_5$ of about 34 µm.

For the samples reacted at 270°C (Fig. 2 (b)) and 300°C (Fig. 2 (c)), significantly shorter incubation times were observed with increasing temperature. Accelerated growth of $(Cu,Ni)_6Sn_5$ was observed in the Cu-6Ni sample within 3 minutes at 270°C, while at 300°C the Cu-10Ni and Cu-14Ni also showed accelerated growth. The Cu-6Ni sample formed a homogeneous layer of $(Cu,Ni)_6Sn_5$ of about 32 µm after 7 minutes at 300°C. The grains had grown to the top of the Sn layer within 3 minutes, and Sn was fully consumed within 7 minutes.

As the thickness of the Sn layers in this study limits the growth of the $(Cu,Ni)_6Sn_5$ layers, it is difficult to quantify the growth rates. However, it is expected that the relationship

IOP Conf. Series: Materials Science and Engineering **701** (2019) 012007 doi:10.1088/1757-899X/701/1/012007

between the thickness of the $(Cu,Ni)_6Sn_5$ layer and the bonding temperature and time will follow an Arrhenius-type equation:

$$h = At^n exp\left(\frac{-Q}{RT}\right) \tag{1}$$

where h is the thickness at time, t; A is a numerical constant derived empirically; n is the time constant; Q is the activation energy; R is the universal gas constant; and T is the temperature [10]. The equation can be rewritten in the form:

$$\ln h = \ln A + n \ln t - \frac{Q}{RT} \tag{2}$$

, or

$$\ln h = \frac{-Q}{R}\left(\frac{1}{T}\right) + \ln A + n \ln t \tag{3}$$

which has the form of the linear equation y = kx + c. As a result, the equation predicts that ln h changes linearly with -(1/T), while h also changes parabolically with tn. It is often

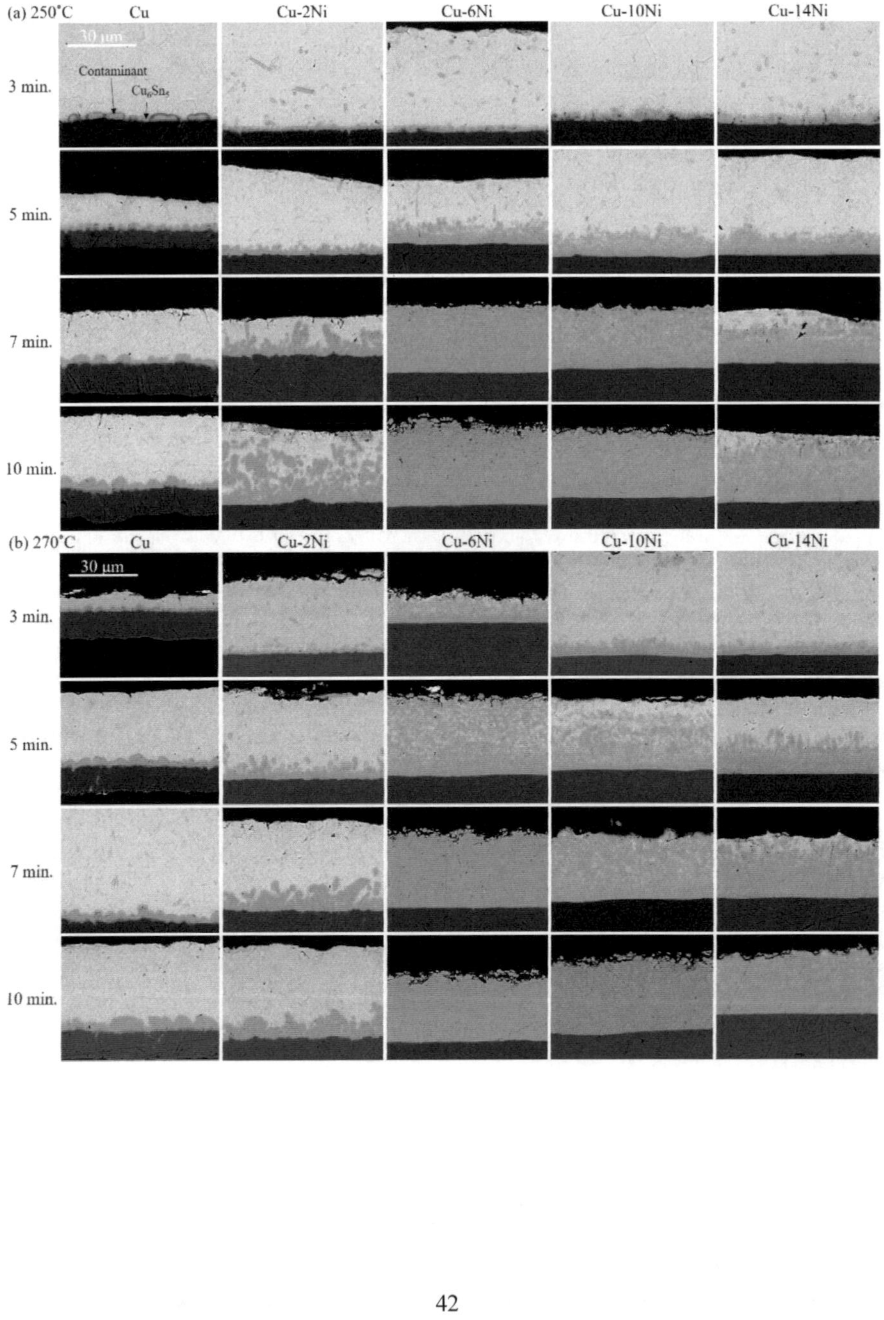

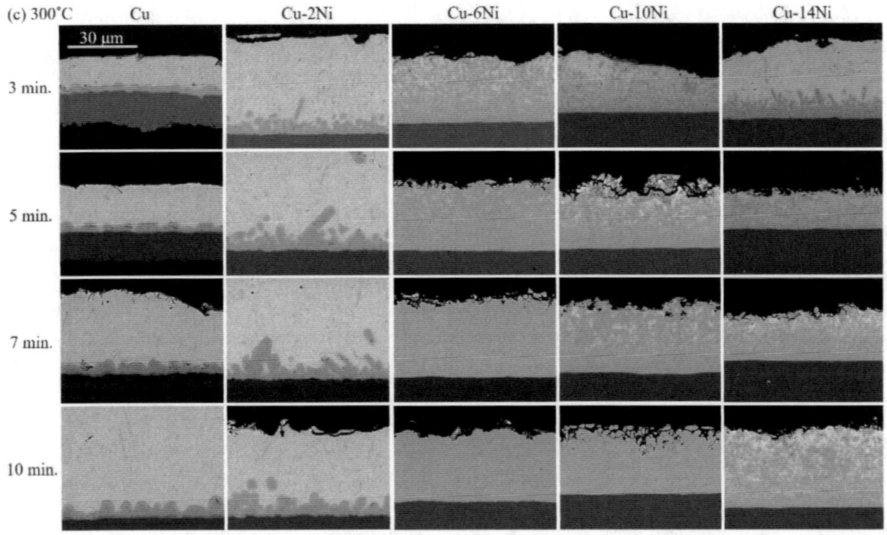

Fig. 2. Cross-section SEM images of $(Cu,Ni)_6Sn_5$ grown between Sn and Cu-xNi, where x = 0, 2, 6, 10 and 14, at (a) 250°C, (b) 270°C, and (c) 300°C for 3, 5, 7 and 10 minutes. Images are shown at the same scale.

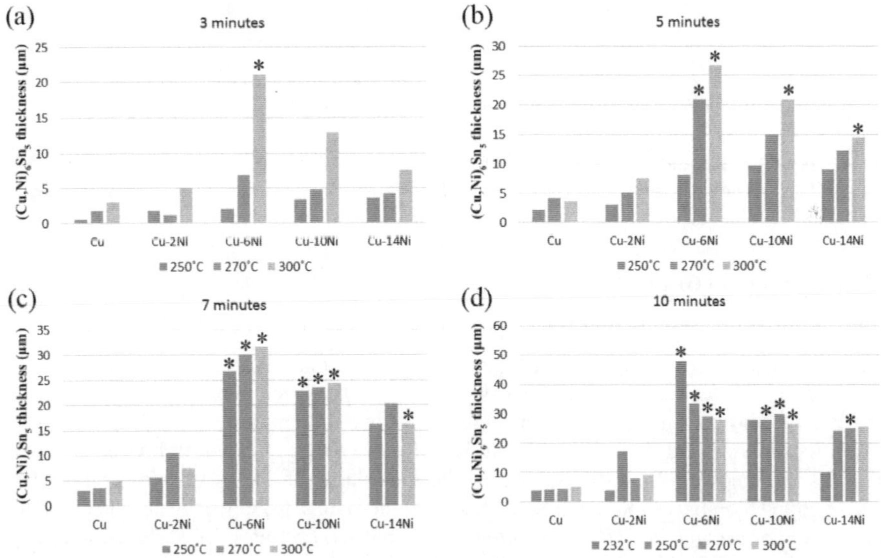

Fig. 3. The average thickness of $(Cu,Ni)_6Sn_5$ grown between Sn and Cu-xNi, where x = 0, 2, 6, 10 and 14, at different temperatures for (a) 3, (b) 5, (c) 7 and (d) 10 minutes. Samples where further $(Cu,Ni)_6Sn_5$ growth is limited due to the consumption of all available Sn in the solder by are marked "*".

described that time constant of $n = 0.33$ is related to grain boundary diffusion controlled reactions, $n = 0.5$ is related to volume diffusion controlled reactions, while at $n = 1$ the reaction is not limited by diffusion [11]. It is expected that the constants A and n are different for the substrates with different amounts of Ni, and this will affect the growth rate of the $(Cu,Ni)_6Sn_5$ layer.

Another factor which may affect the incubation time and growth rate is the effect of temperature on the shifts in the equilibrium phase boundaries in the Sn-Cu-Ni ternary phase diagram as presented by Gourlay et al [12] (Fig. 4). As Cu and Ni dissolve in liquid Sn, the liquid composition of the remaining solder coupled with each Cu-xNi substrate follows the a line of constant Cu:Ni ratio as shown in Fig. 4. The plot suggests that for Cu and Cu-2Ni substrates, lower concentrations of Cu and Ni are required for the formation of $(Cu,Ni)_6Sn_5$ at 250°C compared to that at 268°C, which is likely to reduce incubation time and increase the growth rate at the lower temperature. This factor competes with the effect of temperature, and results in the abnormality observed in the Cu-2Ni samples, where a thicker layer of $(Cu,Ni)_6Sn_5$ is observed in the 250°C sample compared to those at 270°C and 300°C in Fig. 3 (d). At x = 6, the constant Cu:Ni ratio line crosses the liquid/liquid + $(Cu,Ni)_6Sn_5$ equilibrium phase boundaries for 250°C and 268°C at similar Cu and Ni compositions, while at x = 10 and 14, the plot predicts lower concentration of Cu and Ni are required to form $(Cu,Ni)_6Sn_5$ at the higher temperature, enhancing the growth rates of the IMCs.

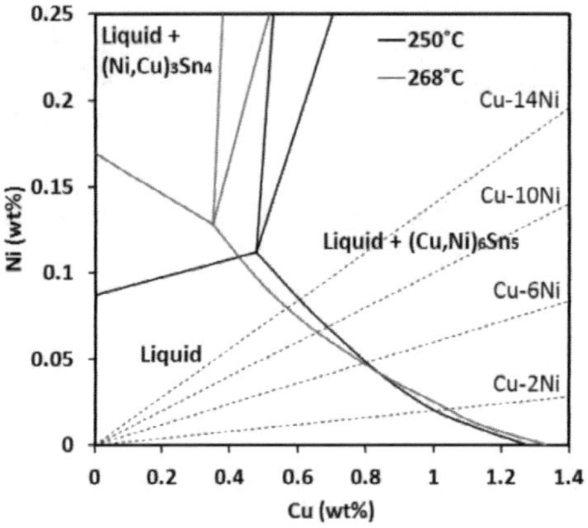

Fig. 4. Very Sn-rich corner of Sn-Cu-Ni ternary phase diagram showing isothermal sections at 250°C and 268°C reconstructed from [12]. Dashed lines are constant Cu:Ni ratio lines for Cu-2Ni, Cu-6Ni, Cu-10Ni and Cu-14Ni.

3.3 Suppression of Cu_3Sn

Fig. 5 shows the Cu/Sn, Cu-2Ni/Sn and Cu-6Ni/Sn interfaces after 10 minutes at 300°C at higher magnifications. A thin layer of intermetallic compound (IMC) is observed between $(Cu,Ni)_6Sn_5$ and the substrates in the Cu/Sn and Cu-2Ni/Sn couples. The formation of a Cu_3Sn layer between Cu_6Sn_5 and Cu substrates is commonly reported for Cu/Sn-based solder alloys. As shown in Fig. 5, the formation of the $(Cu,Ni)_3Sn$ layer is noticeably reduced with an increasing amount of Ni in the substrate, where the thickness of Cu_3Sn is lower in the Cu-2Ni/Sn couple compared to the Cu/Sn couple, and the $(Cu,Ni)_3Sn$ layer is not visible in the Cu-6Ni/Sn couple and the couples with higher Ni content substrates. The effect of Ni in the suppression of Cu_3Sn formation is similar to what is reported in reactions between pure Cu substrates and Ni containing Sn-based solder alloys [13, 14], indicating that a similar effect is achieved regardless of the source of Ni.

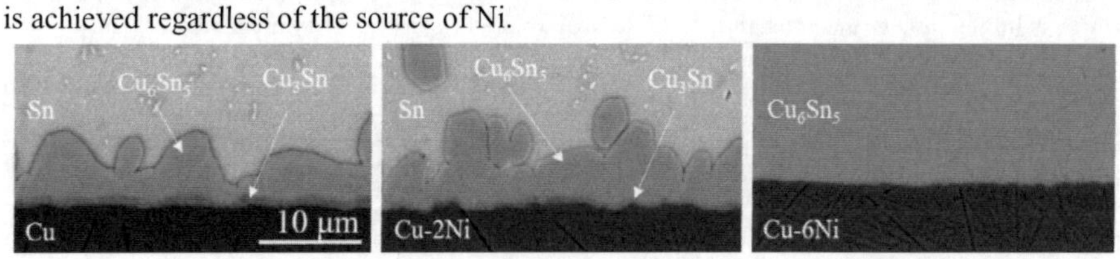

Fig. 5. Cross-section SEM images of $(Cu,Ni)_6Sn_5$ and $(Cu,Ni)_3Sn$ grown between Sn and Cu-xNi, where x = 0, 2, and 6 at 300°C for 10 minutes. Images are shown at the same scale.

4 Conclusions

This study shows the growth rate of $(Cu,Ni)_6Sn_5$ can be greatly accelerated by optimising the Ni content in the Cu-xNi substrate and increasing the bonding temperature. At temperatures above 250°C, a homogeneous, Sn and Cu_3Sn-free layer of $(Cu,Ni)_6Sn_5$ can be obtained within 10 minutes. This enables TLP bonding and anode fabrication at commercially relevant times and temperatures. The results also show there are shorter incubation times for the initial formation of $(Cu,Ni)_6Sn_5$ with increasing Ni content in the substrates.

Quantitative analysis of the growth rates of $(Cu,Ni)_6Sn_5$ is not possible in this experiment as the thickness of the $(Cu,Ni)_6Sn_5$ layer in many samples were limited by the amount of Sn available. However this study does demonstrate the formation of a Sn-free $(Cu,Ni)_6Sn_5$ layer is possible if composition and processing variables are optimised. The relationship between these variables and growth rates will be quantified in future work.

This work was supported by the UQ-Nihon Superior cooperative research grant at the University of Queensland (UQ) [grant number 2016001895], Australia; an Australian Government Research Training Program (RTP) grant from the University of Queensland, Australia.

References

1. D.K. Mu, S.D. Mcdonald, J. Read, H. Huang, K. Nogita, Curr Opin Solid St M., **20**(2), 55 (2016).
2. X.F. Tan, S.D. Mcdonald, Q. Gu, Y. Hu, L. Wang, S. Matsumura, T. Nishimura, K. Nogita, J. Power Sources, **415**, 50 (2018).
3. K.D. Kepler, Electrochem. Solid-State Lett., **2**(7), 307 (1999).
4. J. Feng, C. Hang, Y. Tian, B. Liu, C. Wang, Sci Rep, **8**(1), 1775 (2018).
5. Z. Li, M. Li, Y. Xiao, C. Wang, Ultrason Sonochem, **21**(3), 924 (2014).
6. S.M. Choquette, I.E. Anderson, Int J Powder Metall., **51**(4), 9 (2015).
7. V. Vuorinen, H. Yu, T. Laurila, J.K. Kivilahti, J. Electron. Mater., **37**(6), 792 (2008).
8. V.A. Baheti, S. Islam, P. Kumar, R. Ravi, R. Narayanan, D. Hongqun, V. Vuorinen, T. Laurila, A. Paul, Philos Mag., **96**(1), 15 (2015).
9. S.M. Choquette, Materials Science And Engineering. 2018, Iowa State University.
10. P.T. Vianco, K.L. Erickson, P.L. Hopkins, J. Electron. Mater., **23**(8), 721 (1994).
11. N. Zhao, Y. Zhong, M.L. Huang, H.T. Ma, W. Dong, Sci Rep, **5**, 13491 (2015).
12. C.M. Gourlay, K. Nogita, J. Read, A.K. Dahle, J. Electron. Mater., **39**(1), 56 (2009).
13. F. Gao, T. Takemoto, H. Nishikawa, Mat Sci Eng A, **420**(1-2), 39 (2006).
14. S.U. Mehreen, K. Nogita, S. Mcdonald, H. Yasuda, D. Stjohn, J. Alloys Compd, **766**, 1003 (2018).

Interfacial reactions between different Sn-based lead-free solder alloys and CuNi substrates

Nurul Razliana Abdul Razak [1,2,3 *], *Xin Fu* Tan [1], *Stuart D.* McDonald [1], *Kazuhiro* Nogita [1]

[1]Nihon Superior Centre for the Manufacture of Electronic Materials (NS CMEM), School of Mechanical and Mining Engineering, The University of Queensland, Brisbane, QLD 4072, Australia
[2]Faculty of Engineering Technology, Universiti Malaysia Perlis, Kampus UniCITI Alam, Sungai Chuchuh, 02100 Padang Besar, Perlis, Malaysia
[3]Centre of Excellence Geopolymer and Green Technology, School of Materials Engineering, Universiti Malaysia Perlis, Taman Muhibbah, 02600 Jejawi, Arau, Perlis, Malaysia

Abstract. The effect of the presence of Ni in Cu substrates on the interface reaction during soldering and the growth of Sn-Cu intermetallic compounds (IMCs) was studied. The study involved reflow soldering of Sn-0.7Cu, Sn-3Ag-0.5Cu and Sn-0.7Cu-0.05Ni lead-free solders on pure Cu and Cu-6Ni substrates with holding times of 5, 10 and 15 minutes at 232 °C. The presence of 6 wt% Ni in the Cu substrate strongly affects the microstructural evolution and the growth of the IMCs. The presence of Ni in the substrate significantly increased the overall thickness of the $(Cu,Ni)_6Sn_5$ layer while suppressing the formation of the Cu_3Sn layer. The thickness of the $(Cu,Ni)_6Sn_5$ layer increased with reflow time for both Cu and Cu-6Ni substrate.

1 Introduction

Recently, there has been interest in the potential to make joints containing high melting point intermetallic compounds (IMCs) by allowing reactions between low melting point materials and high melting point materials with consideration of appropriate pressures and temperatures [1]. These high melting point IMCs are expected to function as reliable interconnection materials operating at high temperature environments and could be suitable for a wide range of applications. One of the methods to form a full IMC is low temperature transient liquid phase (TLP) bonding by interdiffusion between low melting point materials such as tin (Sn) and indium (In) with high melting point materials like copper (Cu), nickel (Ni) and silver (Ag). Due to the high melting temperature of the IMCs, they are thermally stable [2]. Cu_3Sn and Cu_6Sn_5 are brittle IMCs that grow at the interface between Cu substrate and Sn-based solder alloys during the soldering process. These phases grow further by solid-state diffusion and it has been shown that the presence of Ni in the Cu system can greatly influence the growth rate of the Cu_3Sn and Cu_6Sn_5 [3]. It has been reported that addition of Ni in Cu reduces the growth of the Cu_3Sn phase and greatly accelerates that of Cu_6Sn_5 [4-6]. Based on a thermodynamic analysis, the change in Ni content alone is insufficient to explain these changes. An analysis of the IMC growth with respect to the activation energy of the diffusion coefficient also indicates a strong role of grain boundary diffusion in the growth of Cu_6Sn_5. An analysis of the IMC composition shows the Ni incorporated into the intermetallic, forming $(Cu,Ni)_6Sn_5$ during the interfacial reaction between the solder and Cu substrate. This study focuses on the interfacial reaction and thicknesses of the IMCs grown between three

* Corresponding author: nurulrazliana.abdulrazak@uq.net.au

different solder alloy: Sn-0.7Cu, Sn-3Ag-0.5Cu and Sn-0.7Cu-0.05Ni with pure Cu and Cu-6 wt% Ni substrates during soldering.

2 Materials and Methodology

2.1 Fabrication of Sn-based solder alloy/Cu-xNi

Sn-0.7Cu, Sn-3Ag-0.5Cu and Sn-0.7Cu-0.05Ni (herein after referred to as the Sn-based solder alloy) (supplied by Nihon Superior) were used as the solder alloys, while Cu-*x*Ni (x = 0 and 6 wt%) foils were used as the substrates. The Sn-based solder alloys were rolled into sheets with thickness about 100 μm, then cut into dimensions of 10 mm x 10 mm. Cu-xNi foils of 30 - 40 μm in thickness were also cut into the same dimensions. The soldering flux (Pacweld 62) was applied on each Sn-based based solder alloy/ Cu-*x*Ni interface to remove the surface oxides on the Cu-xNi and assist the soldering process. The Sn-based solder alloys were set on the top of the Cu-*x*Ni substrates and clamped between glass slides using metal clips. The heating process of the samples was performed as follow: first, the oven was heated from room temperature to 232 °C; second, the prepared samples were placed in the oven and reflowed at 232 °C for 5, 10 and 15 minutes; lastly, the reflowed samples were cooled in air. The Sn-based solder alloy/ Cu-*x*Ni couple was then removed from the glass slide and cleaned using acetone to remove the excess flux. The samples were ground with silicon carbide (SiC) paper and then polished with 3 and 1 μm diamond suspension, and final polishing by using oxide polishing suspensions (OPS). The cross-section of the samples was observed using a Hitachi TM3030 scanning electron microscope (SEM). An analysis of the elements present in selected areas was performed using energy dispersive X-ray spectroscopy (EDS). The thickness of the IMCs was measured using image analysis.

3 Result and discussion

3.1 Microstructure evolution of interfacial IMC

Fig 1 shows the interface cross-section of Sn-0.7Cu, Sn-3Ag-0.5Cu and Sn-0.7Cu-0.05Ni on Cu and Cu-6Ni substrates solder joints reflowed at 232 °C for 5 minutes. Previous literature [5, 7, 8] has shown that the $(Cu,Ni)_6Sn_5$ intermetallic phase forms and grows faster when Ni is present in the Cu substrate at concentrations between 5 and 10 %. After reflowing for 5 minutes, three IMC morphologies can be observed clearly, namely, a scalloped morphology at the Sn-0.7Cu/Cu interface, a combination of scalloped and columnar morphologies for Sn-3Ag-0.5Cu/Cu and a near-flat and stable IMC growth at the interface of Sn-0.7Cu-0.05Ni/Cu, as seen in Fig 1(a-c). The formation of the scalloped Cu_6Sn_5 intermetallics is the results of a combination of kinetics processes including ripening and interfacial reactions [5] while the stable formation of the IMC for Sn-0.7Cu-0.05Ni/Cu may be due to the role of Ni in suppressing the polymorphic transformation of Cu_6Sn_5 [9, 10]. From the cross-section of Sn-0.7Cu/Cu-6Ni, Sn-3Ag-0.5Cu/Cu-6Ni and Sn-0.7Cu-0.05Ni/Cu-6Ni shown in Fig 1(d-f), it is clearly seen that the growth of the IMCs is accelerated and becomes thicker compared to the pure Cu substrate, however the IMCs are no longer continuous and in full contact with the substrate (at least in two dimensions) with the exception of the Sn-0.7Cu-0.05Ni/Cu-6Ni joint. The formation of Cu_6Sn_5 intermetallics is reliant on the diffusion of Cu from the substrate to the Sn-based solder/Cu interface [2, 11] and this interfacial reaction is accelerated in the case where the substrate contains 6 wt% Ni. The discontinuity of the interfacial IMC (in the Ni containing substrates and the Sn-0.7Cu and Sn-3Ag-0.5Cu solder alloys) may lead to massive spalling of the Cu-Ni-Sn ternary compounds during growth [5, 8].

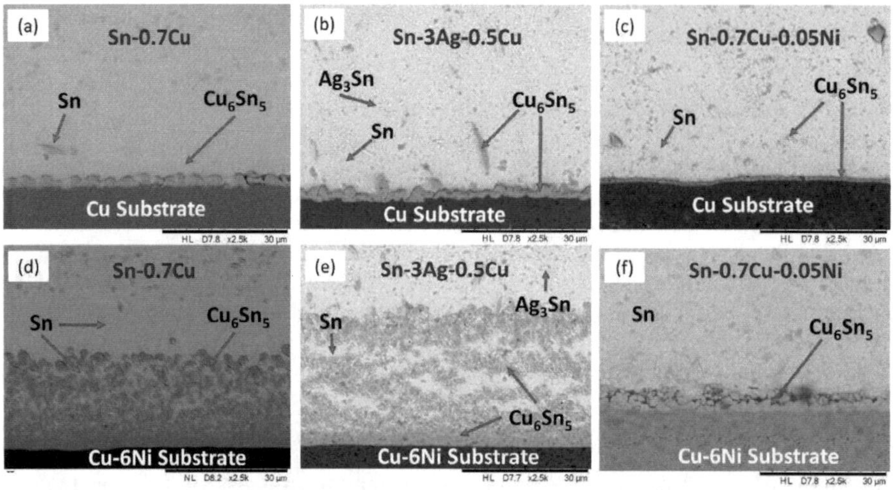

Fig. 1. SEM images showing the interfacial microstructures of Sn-0.7Cu, Sn-3Ag-0.5Cu and Sn-0.7Cu-0.05Ni after reflowed at 5 minutes at 232 °C on (a-c) Cu substrates (d-f) Cu-6Ni substrates

As the reflow time increased from 5 to 15 minutes, the Cu_6Sn_5 intermetallic layer grew thicker with a scalloped morphology and a layer of Cu_3Sn formed between the Cu_6Sn_5 and Cu substrate for Sn-0.7Cu/Cu (see Fig 2(a)), while for Sn-0.7Cu/Cu-6Ni, the formation of $(Cu,Ni)_6Sn_5$ intermetallic consumed the Sn-0.7Cu solder entirely (see Fig 2(b)). A reduction of the thickness of the Cu-6Ni substrates also results from the reaction process. For Sn-3Ag-0.5Cu on Cu and Cu-6Ni, Ag_3Sn intermetallics were found at the interface above the CuSnNi layer, however due to the thin layers of the intermetallic, it is difficult to quantitatively analyze this phase [8]. An elemental mapping analysis was done to confirm the $(Cu,Ni)_6Sn_5$ intermetallic phases formed. From previous reported literature, the Cu_6Sn_5 compound can extend to ternary compositions due to the presence of Ni forming $(Cu,Ni)_6Sn_5$ ternary intermetallic compounds [8, 12, 13]. Similar intermetallic growth was also found for Sn-3Ag-0.5Cu and Sn-0.7Cu-0.05Ni when reflowed on Cu substrates. Addition of Ni atoms are believed to suppress the formation of the Cu_3Sn [5] and no trace of Cu_3Sn layer was found in Fig 2(b).

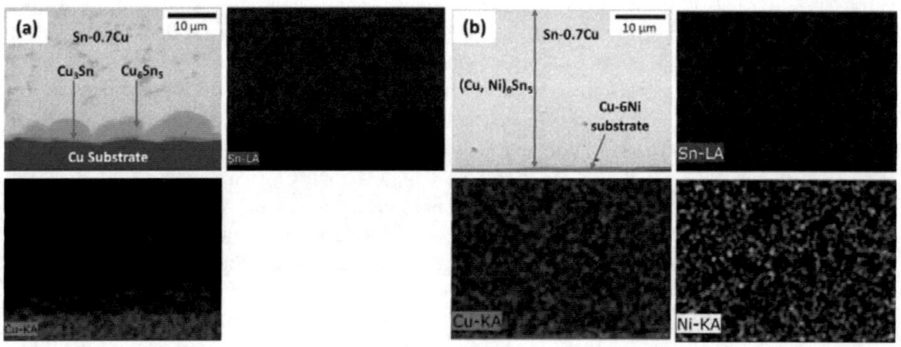

Fig. 2. SEM morphology and elemental mapping analysis of (a) Sn-0.7Cu/Cu and (b) Sn-0.7Cu/Cu-6Ni interface reflowed for 15 minutes at 232 °C

3.2 Thickness of IMCs

The relationship between the total interfacial IMC thickness and reflow time for all three Sn-based solders; Sn-0.7Cu, Sn-3Ag-0.5Cu and Sn-0.7Cu-0.05Ni on Cu and Cu-6Ni substrates are shown in Fig 3. As shown in Fig 3(a), the IMCs on the pure Cu substrate grow continuously throughout the 15 minutes of reflow. When reacting with pure Cu substrate, the thickness of the IMC grew to a maximum of 5 µm with the Sn-3Ag-0.5Cu and this was the highest compared to the other two solders. When 6 wt% of Ni was present into the Cu substrate, the IMC growth accelerated and the maximum thickness was much larger than that present in reactions with the Ni-free substrates. Similar to the example shown in Fig 2(b), the full formation of the IMC was observed at reflow time of 15 minutes for all Sn-based solders on the Cu-6Ni substrate.

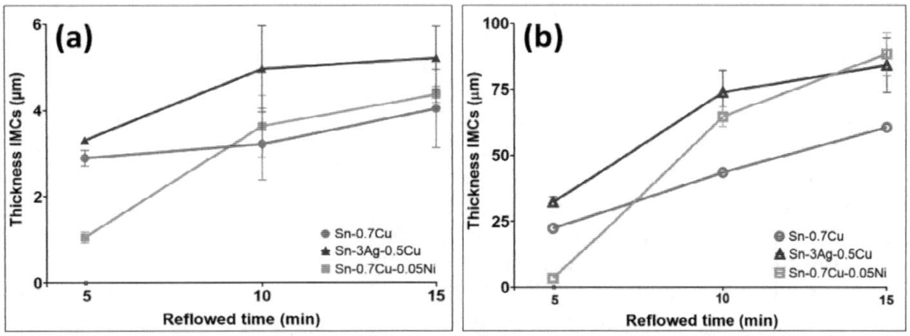

Fig. 3. Total IMC thickness versus the reflow time with Sn-0.7Cu, Sn-3Ag-0.5Cu and Sn-0.7Cu-0.05Ni solder on different substrate: (a) Cu and (b) Cu-6Ni substrates

4 Conclusion

The effects of having Ni present in the Cu substrate on the interfacial reaction and IMC growth when reacting with Sn-0.7Cu, Sn-3Ag-0.5Cu and Sn-0.7Cu-0.05Ni solders were studied. The following conclusions can be made:

1. With a reflow time of 5 minutes, different morphologies of Cu_6Sn_5intermetallic formed when three different solders reacted with pure Cu substrates; scalloped IMCs formed in the Sn-0.7Cu/Cu couple, while scalloped and columnar IMCs formed in the Sn-3Ag-0.5Cu/Cu couple and near-planar IMC growth occurred at the interface of the Sn-0.7Cu-0.05Ni/Cu couple. With the presence of 6 wt% Ni in the Cu substrate, when reflowed for 5 minutes, the Cu_6Sn_5 intermetallic growth accelerated significantly for Sn-0.7Cu and Sn-3Ag-0.5Cu while for Sn-0.7Cu-0.05Ni, the Cu_6Sn_5 was slightly thicker compared to Sn-0.7Cu-0.05Ni/Cu.
2. With an increase in the reflow time to 15 minutes, the planar-like Cu_3Sn layers grew between the Cu_6Sn_5 and Cu substrate but when Ni was present in the Cu substrate, the Cu_3Sn layer was not present at the intermetallic/substrate interface.
3. The presence of 6 wt% Ni increased the growth of the Cu_6Sn_5 intermetallic for all Sn-based solder alloys and a complete transformation of the 100 µm solder sheet occurred after 15 minutes of reflow time.

The authors gratefully acknowledge financial support from the University of Queensland-Nihon Superior (2016001895) collaborative research program. The authors acknowledge the facilities, and the scientific and technical assistance at the School of Mechanical and Mining Engineering, UQ. N. R. A. Razak is financially supported by the Ministry of Education Malaysia and Universiti Malaysia Perlis (UniMAP).

References

1. Y. Bao, A. Wu, H. Shao, Y. Zhao, L. Liu, and G. Zou, J Mater Sci, **54**, no. 1, 765-776 (2018)
2. L. Sun, M.-h. Chen, and L. Zhang, J Alloy Compd, **786**, 677-687 (2019)
3. H. Ma *et al.*, J Mater Sci-Mater El, **29**, 602-613 (2017)
4. V. A. Baheti, S. Kashyap, P. Kumar, K. Chattopadhyay, and A. Paul, Philos Mag, **97**, 1782--1802, 2017,
5. X. Zhang, X. Hu, X. Jiang, and Y. Li, Appl Phys A, **124** (2018)
6. Y. W. Wang, Y. W. Lin, C. T. Tu, and C. R. Kao, J Alloy Compd, **478,** 121-127 (2009)
7. V. Vuorinen, T. Laurila, T. Mattila, E. Heikinheimo, and J. K. Kivilahti, J Electron Mater, **36**, 1355-1362, (2007)
8. T. M. Korhonen, P. Su, S. J. Hong, M. A. Korhonen, and C. Y. Li, J Electron Mater, **29**, 1194-1199 (2000)
9. K. Nogita, Intermetallics, **18**, 145-149 (2010)
10. W. N. C. Weng, *Evolution of Pb-Free Solders in Recent Progress in Soldering Materials*, A. A. Mohamad (InTech,Croatia, 2017)
11. N. Zhao, Y. Zhong, M. L. Huang, H. T. Ma, and W. Dong, *Sci Rep-UK,* **5**, 1-12 (2015)
12. C. H. Wang, W. H. Lai, and S. W. Chen, J Electron Mater, **43**, 195-203 (2014)
13. V. Vuorinen, H. Yu, T. Laurila, and J. K. Kivilahti, J Electron Mater, **37**, 792-805 (2008)

IOP Conf. Series: Materials Science and Engineering **701** (2019) 012009 doi:10.1088/1757-899X/701/1/012009

Effect of Zn addition on Cu_3Sn formation in Sn-10Cu alloys

Syeda U. Mehreen[1,2*], *Kazuhiro* Nogita[1,2] , *Stuart* McDonald[1,2], and *David* StJohn[1]

[1]Nihon Superior Centre for the Manufacture of Electronic Materials, Australia
[2]School of Mechanical & Mining Engineering, The University of Queensland, 4072, Australia

Abstract. Sn-Cu alloys have generated interest as potential candidates for high-temperature soldering applications, as they are economically feasible and environmentally safe. Sn-Cu alloys with copper contents beyond 7.6wt.% solidify with a primary Cu_3Sn phase, followed by peritectic and eutectic reactions. Typically, phases which form as elongated needles are undesirable, as they tend to be more brittle and in turn reduce the strength and toughness of the alloy. The effect of Zn on the peritectic reaction and phase formation in a Sn-10wt.%Cu alloy is investigated through cooling curves and microstructural analysis using optical and electron microscopy. Zn was alloyed in the range of 0.1-1wt.% to the base Sn-10Cu alloy system. The results indicate that the amount of the primary Cu_3Sn phase that forms is inversely proportional to the concentration of Zn. Although complete suppression of Cu_3Sn was not observed, there was a clear effect of Zn reducing the amount of the primary Cu_3Sn phase, with the greatest reduction of Cu_3Sn occurring with a 1wt.% Zn addition. The reduction of the elongated needle-shaped primary Cu_3Sn phase may find application in the development of more reliable bulk solder microstructures.

1 Introduction

Due to the RoHS (Restriction of Hazardous Substances) and WEEE (Waste Electrical and Electronic Equipment) directives aimed at regulating the use of Pb-based solders in electronic assembly and manufacturing, there has been momentum to transition into Pb-free solders in many electronic applications [1]. Many candidate alloy systems have been investigated as potential replacements for conventional Pb-based solders [2]. One such candidate is the Sn-Cu system, in particular Sn-0.7 wt.%Cu, which is widely recognized as an alloy that has excellent wetting properties much similar to the conventional Sn-Pb eutectic alloy [3]. Suitable high temperature Sn-Cu alloy solders require increased copper contents (greater than 7.6wt.% Cu). These higher Cu contents causes the alloy to undergo a peritectic reaction, which involves the formation of a needle-shaped Cu_3Sn phase, encapsulated by Cu_6Sn_5 phase. Peritectic reactions typically do not go to completion, and as a result the primary Cu_3Sn phase remains in the microstructure. In peritectic solidification, from a nomenclature perspective, Cu_3Sn is known as the primary phase and the Cu_6Sn_5 that forms around the Cu_3Sn is called the peritetic phase (or secondary phase as it forms after

* Corresponding author: syeda.mehreen@uqconnect.edu.au

the formation of Cu_3Sn). Needle-shaped structures are not desirable from a microstructural perspective, as they may pose a threat to the homogeneity and mechanical properties of a solder. More rounded/ compact morphologies are preferred as they absorb elastic stresses much better compared to needle-shaped structures [4]. Previous research has shown that the presence of Ni at a concentration of 0.06wt.% or more in a Sn-10Cu alloy is required to completely suppress the needle-shaped Cu_3Sn, and the Cu_6Sn_5 phase can directly nucleate from the melt [5]. Direct formation of phases which would have originally been peritectic phases has also been reported in literature previously as result of varying the cooling rate [6] or applying a direct current field [7]. The aim of this study is to investigate if Zn, can also act to reduce the Cu_3Sn phase when alloyed to the Sn-10Cu alloy. Due to its low cost and desirable mechanical properties, Zn has been proposed as an alloying element for soldering applications, in particular for low-temperature soldering (Sn-Zn alloys) [8]. This research will help uncover the potential for Zn to be used as an alloying element to Sn-Cu solders for high-temperature soldering applications.

2 Experimental

The compositions of the ternary Sn-Cu alloys in this research were analysed by inductively coupled plasma mass spectrometry (ICPMS) and are given in Table 1. The samples included the base alloy Sn-10Cu with 0, 0.1 and 1 wt.% Zn. The compositions listed in Table 1 were prepared by melting Sn and Cu of 99.9wt.% purity and Sn-1Zn master alloy in boron nitride coated clay-graphite crucibles at 600°C in a Nabotherm electric resistance furnace which was allowed to equilibrate for 1.5 hours. Samples were taken from the melt for metallography and cooling curve analysis (CCA) in boron nitride coated steel cups as per the method described elsewhere [5,9]. Prior to the first solidification event, the average cooling rate for all the samples was 1°C/s. The metallography samples collected for microstructural observation were prepared and analysed as described previously [5].

Table 1. Analysed compositions of the Sn-10Cu-xZn alloys

Sample	Sn	Cu	Zn
Sn-10Cu	Balance	9.40	0.00
Sn-10Cu-0.1Zn	Balance	9.19	0.08
Sn-10Cu-1Zn	Balance	10.01	1.01

3 Results and Discussion

The cooling curve and the associated derivative curve (dT/dt) for the Sn-10Cu base alloy are shown in Figure 1a. The inflections in the associated derivative curve (three peaks) of Figure 1a suggest three reactions occur. The first peak (Peak 1) is associated with the formation of the Cu_3Sn phase, the second peak (Peak 2) is associated with the formation of Cu_6Sn_5 phase and the third (Peak 3) with the long Sn-Cu_6Sn_5 eutectic arrest. Both the Sn-10Cu base alloy and Sn-10Cu-0.1Zn showed these three reactions. However, for the Sn-10Cu-1Zn sample, the eutectic reaction was preceded by only one reaction peak (Figure 1b), i.e. one peak is absent from the cooling curve of Sn-10Cu-1Zn in comparison with the other two alloys. Upon closer investigation of the peaks for the eutectic reaction, it can be observed that the eutectic peak height appears to be directly correlated to the concentration of Zn added to the base Sn-10Cu alloy (Figure 1c). The reaction temperatures at which nucleation events have occurred are summarized in Figure 1d. It can be seen that the three nucleation temperatures (where present) increase with an increasing concentration of Zn.

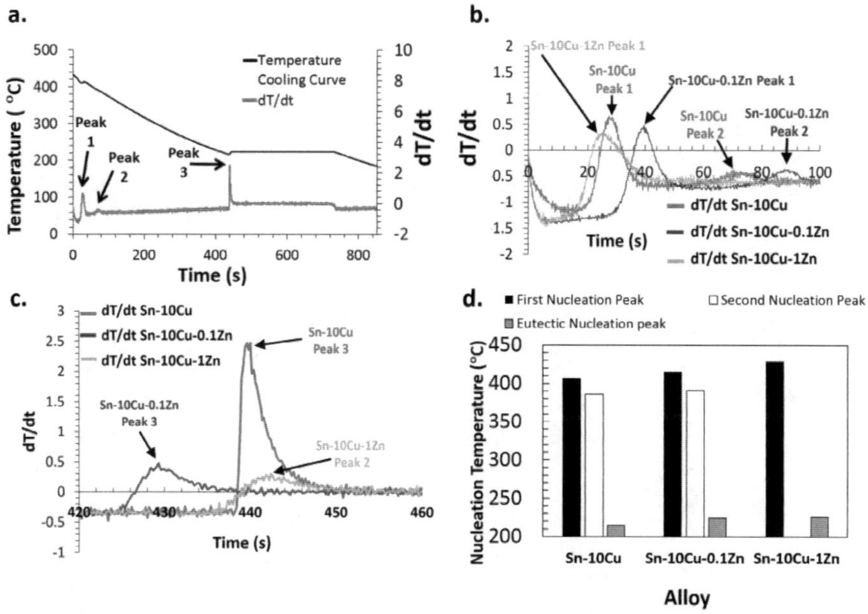

Fig. 1. a. Cooling curve of Sn-10Cu b. Cooling curves of all alloys showing the peaks in the derivative curve for reactions before the eutectic reaction c. Eutectic Nucleation Peak Temperature for Sn-10Cu-xZn (x = 0-1.0wt.%) and d. Comparison of all the nucleation peaks

Figure 2 shows the microstructure of all three experimental alloys. With increasing additions of Zn, it can be observed that there is a clear transition from a needle-shaped morphology to a more compact dendritic morphology. The darker shaded primary Cu_3Sn phase can be observed surrounded by the peritectic phase (peritectic Cu_6Sn_5) which is surrounded by a mixture of eutectic (composed of Cu_6Sn_5 + Sn). Furthermore, the thickness of Cu_3Sn appears to decrease with an increasing concentration Zn. Additionally, it was also noted that the amount of eutectic also decreased with an increasing concentration of Zn. Figure 3 shows the energy dispersive spectroscopy (EDS) analysis results from a Sn-10Cu sample where Points 1, 2 and 3 correspond to Cu_3Sn, peritectic Cu_6Sn_5 and the Sn-rich matrix, respectively. Elemental mapping was conducted on the Sn-10Cu sample to understand the distribution of the elements across the phases in Figure 3a. These results are presented in Figure 3b and c which show that the Sn is mainly concentrated in the matrix and Cu is concentrated in the Cu_3Sn/Cu_6Sn_5 IMCs.

Figure 4 shows the microstructure in backscatter secondary electron (BSE) micrographs. As the concentration of Zn in the Sn-10Cu increases, the reduction of Cu_3Sn can be clearly seen in the BSE images presented in Figure 4. The larger the concentration of Zn the lower the amount of Cu_3Sn. Figure 4c and d are BSE images taken from a Sn-10Cu-1Zn sample which show how thin the Cu_3Sn has become. Furthermore, much of the structure of the Sn-10Cu-1Zn sample is dominated by morphologies similar to Figure 4d which include only the Cu_6Sn_5 phase in a Sn-rich matrix, with Cu_3Sn being completely suppressed. This directly nucleating Cu_6Sn_5 phase is dominant and widespread in the Sn-10Cu-1Zn sample.

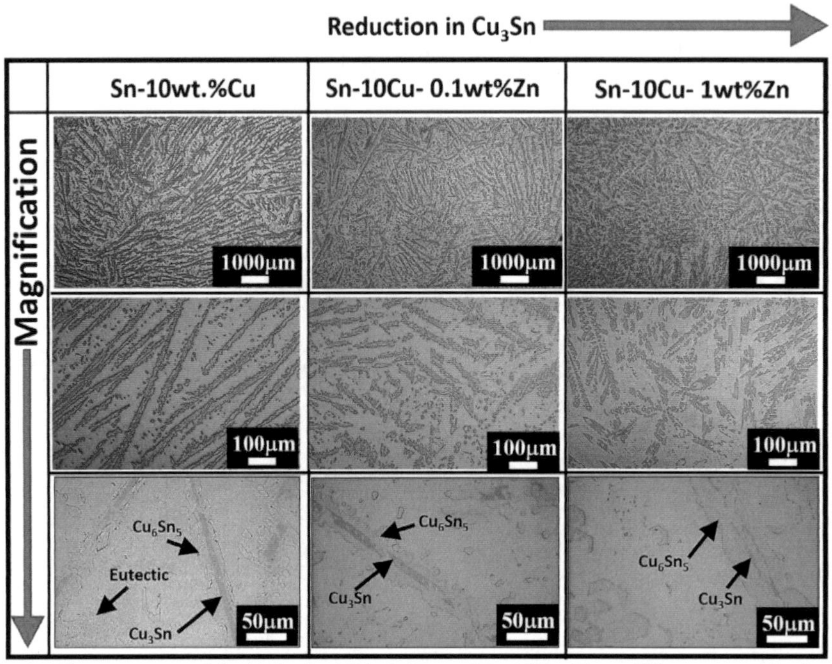

Fig. 2. Microstructures of Sn-10Cu, Sn-10Cu-0.1Zn and Sn-10Cu-1Zn at different magnifications

In order to further investigate the distribution of Zn through the Sn-10Cu-1Zn sample which was dominated by the Cu_6Sn_5 phase in a Sn-rich matrix, multiple EDS points were investigated (Figure 5). It can be observed that those regions corresponding to Cu_6Sn_5 were the regions with larger atomic percentages of Zn compared to those regions with the Sn-rich matrix (Figure 5b). Table 2 shows the atomic percentage of all the elements in the Sn-10Cu-1Zn sample shown in Figure 5a. The presence of Zn in the Cu_6Sn_5 phase suggests that it may play a vital role in Cu_3Sn suppression. Previously, Sn-10Cu-xNi (x=0-1wt.%) alloy showed complete Cu_3Sn suppression [5] at Ni contents greater than 0.06 wt.%. At concentrations below this threshold point, it was observed that the Cu_3Sn thickness reduced with increasing Ni content. The proposed mechanism suggested that a shift in the phase boundary of Cu_3Sn caused by the Ni addition changes the sequence of phase formation events during solidification, thus promoting the direct nucleation of Cu_6Sn_5 from the melt. This ease of direct Cu_6Sn_5 nucleation is progressively made easier by the increasing additions of Ni. As part of this research, it has been established that alloying Zn to Sn-10Cu alloys affects the microstructural formation of Sn-10Cu alloys and shows similar Cu_3Sn suppression phenomena as occurs in Sn-10Cu-xNi alloys [5].

IOP Conf. Series: Materials Science and Engineering **701** (2019) 012009 doi:10.1088/1757-899X/701/1/012009

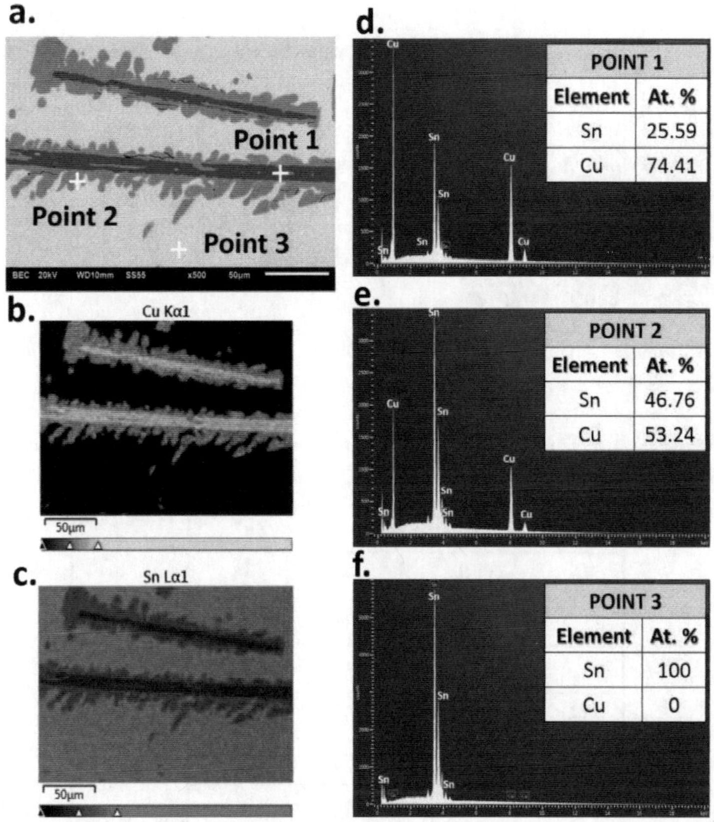

Fig. 3. a. BSE image of Sn-10Cu, b. Elemental Mapping of Sn-10Cu with respect to Cu distribution, c. Elemental Mapping of Sn-10Cu with respect to Sn distribution, d. Atomic Percent Distribution of Sn-10Cu BSE Image Point 1, e. Atomic Percent Distribution of Sn-10Cu BSE Image Point 2, and f. Atomic Percent Distribution of Sn-10Cu BSE Image Point 3

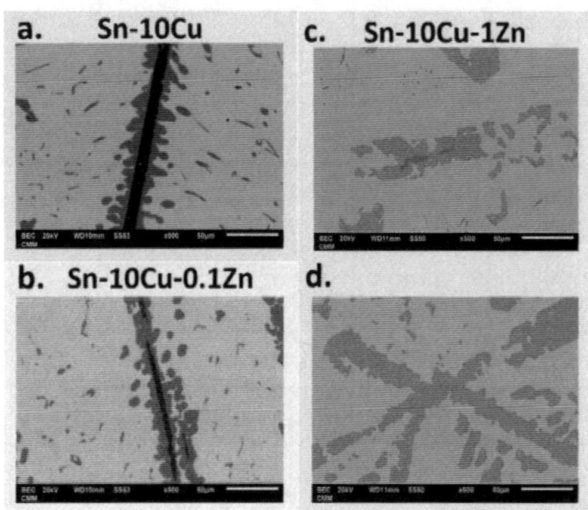

Fig. 4. BSE images of a. Sn-10Cu, b. Sn-10Cu-0.1Zn, c. Sn-10Cu-1Zn (reduced Cu_3Sn), and d. Sn-10Cu-1Zn (Cu_3Sn suppressed)

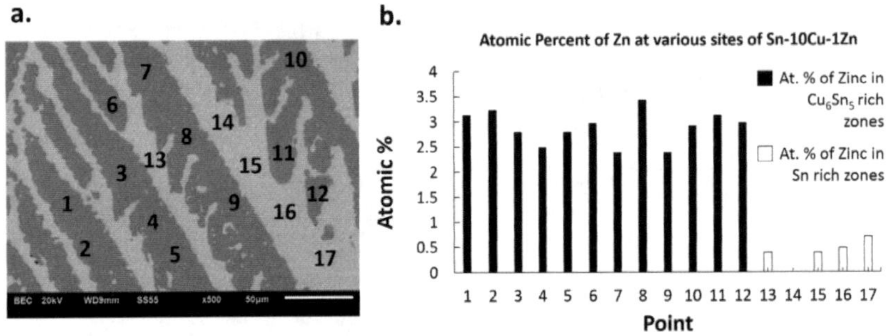

Fig. 5. a. BSE image with embedded Points for EDS and b. Graph indicating Zn at both Cu6Sn5 rich zone and Sn-rich zones

Table 2. Atomic percentage of all elements at various Points marked on Figure 5a

Point	**Atomic Percentage of Elements**			**Zone**
	Sn	**Cu**	**Zn**	
Point 1	42.96	53.90	3.14	**Cu_6Sn_5-rich zones**
Point 2	42.91	53.85	3.24	
Point 3	42.49	54.71	2.80	
Point 4	42.57	54.93	2.50	
Point 5	41.65	55.56	2.80	
Point 6	42.10	54.93	2.97	
Point 7	42.39	55.22	2.39	
Point 8	41.60	56.96	3.44	
Point 9	41.16	56.45	2.39	
Point 10	40.96	56.12	2.92	
Point 11	41.80	55.07	3.13	
Point 12	42.50	54.52	2.98	
Point 13	98.66	0.96	0.38	**Sn-rich zones**
Point 14	99.06	0.94	0.00	
Point 15	99.62	0.00	0.38	
Point 16	98.33	1.18	0.48	
Point 17	98.48	0.82	0.7	

4 Conclusion

The effect of Zn on the amount of Cu_3Sn formed in a Sn-10Cu base alloy with Zn addition of 0.1 & 1 wt.% was investigated for the absence of Cu_3Sn. It can be concluded that Zn additions have a significant effect on the reduction of the amount of Cu_3Sn in the Sn-10Cu base alloy investigated. There was a clear trend associated with not only the phases present but also the morphology of those phases i.e. the transition from needle-shaped morphology dominated by Cu_3Sn needles into a more dendritic structure of the Cu_6Sn_5 rich phases. Furthermore, EDS mapping revealed that the Zn element is concentrated in the Cu_6Sn_5 regions which suggests that perhaps there is a link between Zn and directly nucleating Cu_6Sn_5. The results from this study can be used to further investigate the solidification events leading to the suppression of Cu_3Sn due to the addition of Zn. Promoting phases with more desirable morphologies such as Cu_6Sn_5 may provide an avenue for designing more mechanically favourable microstructures.

The authors gratefully acknowledge financial support from the University of Queensland (UQ)-Nihon Superior (NS) collaboration research project (2016001895) and ARC Linkage project (LP140100485). SEM imaging and EDS mapping were performed at the Centre for Microscopy and Microanalysis (CMM), and DSC experiments were carried out at ANFF-Q facilities, The University of Queensland. Syeda U. Mehreen is financially supported by an Australian Post-graduate Award (APA).

References

1. M.K Pal, G. Gergely, D.K Horváth, Z Gácsi, Microstructural investigations and mechanical properties of pure lead-free (Sn–3.0 Ag–0.5 Cu and Sn–4.0 Ag–0.5 Cu) solder alloy. Metal. Mater. Eng., **24** (2018)
2. K. Suganuma, Advances in lead-free electronics soldering. Curr. Opin. Solid State Mat. Sci, **5,** 1 (2001)
3. T. Ventura, C.M Gourlay, K Nogita, T. Nishimura, M. Rappaz, A.K. Dahle, The influence of 0–0.1 wt.% Ni on the microstructure and fluidity length of Sn-0.7 Cu-xNi. J. Elec. Mat., **37**, 1 (2008)
4. J. Zhao, J. Zhang, W. Liu, G. Wu, L. Zhang, Effect of Y content on microstructure and mechanical properties of as-cast Mg–8Li–3Al–2Zn alloy with duplex structure. Mat. Sci. Eng. A, **650** (2016)
5. S.U. Mehreen, K. Nogita, S McDonald, H. Yasuda, D StJohn, Suppression of Cu_3Sn in the Sn-10Cu peritectic alloy by the addition of Ni. J Alloy. Compd., **766** (2018)
6. W. Zhai, B.Wei, Direct nucleation and growth of peritectic phase induced by substantial undercooling condition. Mater. Lett., **108** (2013).
7. Z. Xuan, F. Mao, Z. Cao, T. Wang, L. Zou, In situ observation on the solidification of Sn-10Cu hyperperitectic alloy under direct current field by synchrotron microradiography. J Alloy. Compd., **721** (2017).
8. T. Laurila, V. Vuorinen, M. Paulasto-Kröckel, Impurity and alloying effects on interfacial reaction layers in Pb-free soldering. Materials Science and Engineering: R: Reports, **68**(1-2) (2010)
9. S. McDonald, K. Nogita, A. Dahle, Eutectic nucleation in Al-Si alloys Acta Mater., **52**, 14 (2004).

IOP Conf. Series: Materials Science and Engineering **701** (2019) 012010 doi:10.1088/1757-899X/701/1/012010

Effect of different ratio of geopolymer paste based fly ash-metakaolin on compressive strength and water absorption

Nor Shida Dalila Mohd Azhar[1], *Farah Farhana* Zainal[1], *Mohd Mustafa Al Bakri* Abdullah[1]

[1]Center of Excellence Geopolymer and Green Technology (CEGeoGTech), School of Materials Engineering, Universiti Malaysia Perlis (UniMAP), P.O. Box 77, D/A Pejabat Pos Besar, 01000 Kangar, Perlis, Malaysia

Abstract. This paper presents the effect of different solid to solid (S/S) ratio by mixing the fly ash and metakaolin as geopolymer paste on the compressive strength and water absorption. Fly ash is a waste materials produced by combustion of coal at power plant while metakaolin was obtained by calcining the kaolin at 750 °C for 4 hours. Different weight percentage of S/S ratio of fly ash/metakaolin used were 50%:50% (Mix 1), 70%:30% (Mix 2) and 90%:10% (Mix 3). Alkaline activators that act as binder were sodium hydroxide (NaOH) and sodium silicate (Na_2SiO_3) solutions. The ratio of Na_2SiO_3/NaOH is 2.5 to 1. The mixture then placed into cylinder mould and left for 7, 14 and 28 days before it be tested. The product obtained were observed and it shows that Mix 2 has highest compressive strength for 28 days that was 41.22 MPa and has lower water absorption. The pore size decreases and permeability also decreases hence the strength potentially improve.

1 Introduction

Geopolymers as an alternative binder systems are becoming one of the main interest in research and development nowadays. Geopolymers can display outstanding technical properties, such as high strength, low water absorption, high acid resistance and also high temperature resistance [1,2]. It has been reported that geopolymer also can give a good performance by utilized the secondary raw materials (industrial wastes like fly ash). This explains the strong interest in this development from countries with growing industrialisation [3]. The use of waste for geopolymer production could not only solve a waste problem, but also reduce the consumption of primary raw materials.

Geopolymers are inorganic aluminosilicate polymers that form solid ceramic-like materials at near ambient temperature. The geopolymer cement is produced as an alternative way to replace the Ordinary Portland Cement (OPC). This is because the production of one ton of Portland cement emits approximately one ton of carbon dioxide (CO_2) an it will lead to global warming [4, 5]. Hence, the use of geopolymer technology not only will reduce the CO_2 emissions by the cement industries, but also enhancing the properties of the geopolymer paste.

One of the possible sources for making geopolymer binders is fly ash because it is available abundantly worldwide, and yet it is rarely used [6]. Therefore, consumption of fly ash in the manufacture of geopolymers is an important strategy in making concrete more environmentally friendly. Other than that, metakaolin is also one of the geopolymer binder. Metakaolin is obtain from the calcination of kaolin at a certain time and high temperature to remove its moisture. There have been some interests in the use of metakaolin as it probably the most effective pozzolanic material for use in concrete [7]. For this reason, fly ash and metakaolin have been chosen as a base material for this project in order to utilize this industrial waste and to improve the paste properties such as compressive strength value and water absorption percentage.

2.1 Materials Selection

The materials that used to prepare the geopolymer paste was made from the mixture of fly ash and metakaolin. Fly ash used in this study was obtained from Lumut Power Plant, Perak, Malaysia while metakaolin used was purchased from Associated Kaolin Industries Sdn. Bhd. Malaysia. Fly ash and metakaolin undergo X-ray Fluorescence (XRF) analysis to characterize their chemical composition. Table 1 shows the compound composition for both materials.

Table 1. The compound composition of fly ash and metakaolin.

Compound	SiO_2	Al_2O_3	Fe_2O_3	CaO	Na_2O	TiO_3	ZrO_2
Fly ash, wt (%)	65.89	24.75	3.89	1.49	2.17	1.49	0.139
Metakaolin, wt (%)	55.70	38.60	2.03	-	2.43	0.78	0.0035

2.2 Sample Preparation

Samples were prepared and undergo mixing process for solid to liquid (S/L) ratio that was 1.2:1. Na_2SiO_3 and NaOH solution are most common alkaline activator that had been used in geopolymer paste. Firstly, the NaOH solution was prepared by dissolving the NaOH pellets in 1 L of distilled water and kept it at room temperature for 24 hours [8,9]. Then, the NaOH solution was mixed with Na_2SiO_3 solution with ratio of the mixture of Na_2SiO_3/ NaOH is 2.5:1. This mixture was prepared 24 hours prior to use [9]. The alkaline activator was later mixed with the fly ash and metakaolin as geopolymer based.

Fly ash and metakaolin were weighted and mixed according to the different S/S ratio named as Mix 1, Mix 2 and Mix 3. Table 2 shows the mixing proportion ratio for geopolymer paste based fly ash and metakaolin. The solid and liquid were mixed using automatic mixer then placed it in 50mm x 50mm cylinder mould, compacted and kept in the mould until it hardened and taken out after 24 hours. Then, the samples were cured at 60 °C in the oven for 24 hours [9]. The main purpose of this process was to remove the excessive moisture and give more strength to the sample [10]. The sample were then left for 7, 14 and 28 days at ambient temperature before undergo compressive strength and water absorption testing.

Table 2. Mixing proportion ratio of fly ash-metakaolin (S/S) geopolymer paste.

Mix No.	NaOH concentration (M)	Na_2SiO_3/NaOH	S/L ratio	S/S ratio
1	12	2.5:1	1.2:1	50:50
2	12	2.5:1	1.2:1	70:30
3	12	2.5:1	1.2:1	90:10

2.3 Compressive Strength

Shimadzu machine was used for compressive strength test. Each sample was carried out by pressing the maximum load enforced to it by cross-sectional area, and outcomes were recorded in the unit N/mm^2 or MPa by using.

2.4 Water Absorption

The water absorption tests were conducted according to ASTM C642-06 [9]. The percent of water absorption for these samples were determined by Eq. (1).

$$\% \ Water\ Absorption = \frac{Wet\ weight - Dr\ \ weight}{Dry\ weight} \times 100\% \tag{1}$$

3 Results and Discussion

3.1 Compressive Strength

Fig.1 shows the effect of different S/S ratios on compressive strength of fly ash-metakaolin geopolymer paste in units of MPa at 7, 14 and 28 days. From Fig.1, it was observed that the overall results from Mix 1-3 shows increasing value in strength along with the curing time. The compressive strength initially increased up from S/S ratio of 50:50 (Mix 1) to 70 :30 (Mix 2) but decreased with the S/S of 90:10 (Mix 3). Mix 2 exhibit highest compressive strength at 7, 14 and 28 days which were 21.65 MPa, 28.57 and 41.22 MPa respectively compared to Mix 1 and Mix 3. The strength increment in the fly ash-metakaolin geopolymer paste at the Mix 1 and Mix 2 might be due to the optimum quantities of binder and alkaline materials content that leads to higher degree of dissolution of source material and the formation of reaction products with lesser pores and more uniform aluminosilicate matrix. Hence, the mixture leads to optimum particle packing [11,12].

At Mix 3, the compressive strength was very low which was 15.61 MPa after 28 days. This is because the amount of liquid content was greater than the solid content which causes the mixture to become watery and the contact becomes limited between activating solution and reactive materials [8,12]. Thus, the dissolution rate of the fly ash-metakaolin became slower and difficult to be harden. Besides, there were many bubbles released which were not able to escape from the mixture may create pores thus producing porous structure which may weaken the geopolymer structure [13]. Hence, low compressive strength was reported at 90:10 (S/S) ratios.

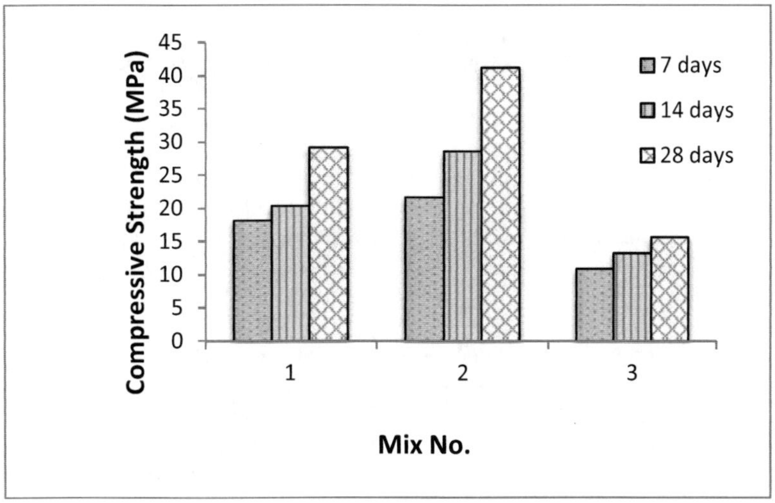

Fig.1. Compressive strength of different ratios of mixed fly ash-metakaolin geopolymer paste at 7,14 and 28 days.

3.2 Water Absorption

Fig.2 shows the results of water absorption in units of percent (%) for fly ash-metakaolin geopolymer paste after 7 days, 14 days and 28 days. Water absorption can be calculated by using Eq.1 [9].

$$\% \; Water \; Absorption = \frac{Wet\ weight - Dry\ weight}{Dry\ weight} \times 100\% \qquad (2)$$

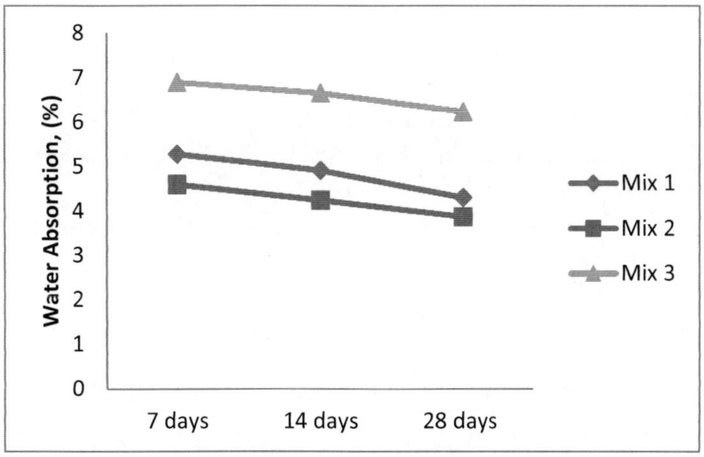

Fig.2. Water absorption of different ratios of mixed fly ash-metakaolin geopolymer paste at 7,14 and 28 days.

From Fig.2, it was observed that at Mix 2, the percentage of water absorption was the lowest after day 7, 14 and day 28 where the average absorption were 4.59%, 4.24% and 3.87% respectively. It proved that this mixture ratio of geopolymer paste was in good condition because has lower than 5% of water absorption. Low water absorption gives a good indicator of less open porosity [14]. Porosity can lead to sample crack and low compressive strength. Meanwhile, Mix 3 ratio has highest water absorption value compared to Mix 1 and Mix 2, may caused by the excessive alkaline activator content supply for fly ash-metakaolin geopolymer paste. Thus, it leads to low geopolymerization process [15,16].

Conclusions

Based on this research, it can be said that the Mix 2 which was 70:30 (S/S) ratio of fly ash-metakaolin geopolymer paste has achieved its highest compressive strength value and shows acceptable water absorption. It is due to sufficient amount of solid and liquid supplied. With this result, it shows that a balanced ratio between the fly ash, metakaolin and alkaline activator content was very important. This is because to obtain optimum dissolution and less unreacted particles left in the geopolymer matrix also can prevent cracking from occur.

The authors would like to thank the staffs of Center of Excellence Geopolymer & Green Technology (CEGeoGTech), School of Materials Engineering, Universiti Malaysia Perlis (UniMAP) for their involvement in the research. This work was supported and funded by the Fundamental Research Grant Scheme (FRGS) under a grant number of FRGS/1/2018/TK06/UNIMAP/02/2 from the Ministry of Education Malaysia.

References

1. M. Weil, K. Dombrowski, A. Buchwald, *Geopolymers*, (2009)
2. S. N. Fifinatasha *et al., Bandung Indones. Adv. Environ. Biol,* **7**, (2013)
3. D. M. J. Sumajouw, D. Hardjito, S. E. Wallah, and B. V. Rangan, J. Mater. Sci, **42**, (2011)
4. N. S. M. F. Ahmed, M. F. Nuruddin, *World Academy of Science, Engineering and Technology International Journal of Civil & Environmental Engineering*, **5**, (2011)
5. Z. F. Farhana, H. Kamarudin, A. Rahmat, M. M. A Bakri and S. R. Shamsudin , *AIP Conference Proceedings*, (2016).
6. S.E. Wallah and B.V Rangan , (2013)
7. J.M. Khatib, O. Baalbaki, A.A Elkordi, Metakaolin. *Waste and Supplementary Cementitious Materials in Concrete*, (2018)
8. S. Aparna, D. Sathyan, K. Anand, *Materials Today: Proceedings*, **5**, (2018)
9. Z. F. Farhana, H. Kamarudin, A. Rahmat, M. M. A Bakri, *Materials Science Forum*, ***803***, (2014)
10. I.H Aziz, M. M. A Bakri, H.C. Yong, L.Y. Ming, H. Kamarudin, *Matec Web of Conferences*, **78**, (2016)
11. M. Ibrahim, M. A. M. Johari, M. K. Rahman, M. Maslehuddin, Const. Buil. Mater., (2017)
12. N.A. Jaya, L. Y. Ming, H. C. Yong, M. M. A Bakri, (2018)
13. K. Zulkifly, H.C. Yong, L.Y. Ming, M. M. A. Bakri, S. F. A. Abdullah, (2018)

14. F. H. A. Zaidi, R. Ahmad, M. M. A. B. Abdullah, M. F. M. Tahir, Z. Yahya, W. M. W. Ibrahim, A. S. Sauffi, *IOP Conference Series: Materials Science and Engineering*, (2019)
15. F. Zainal, S. Amli, K. Hussin, A. Rahmat and M. Abdullah, *IOP Conference Series: Materials Science and Engineering*, **209**, (2017)
16. N. S. D. M. Azhar, F. F. Zainal, M. M. A. B. Abdullah, *AIP Conference Proceedings*, (2019)

Effect of graphene oxide on microstructure and optical properties of TiO_2 thin film

Azliza Azani[1], *Dewi Suriyani* Che Halin[1,*], *Kamrosni* Abdul Razak[1], *Mohd Mustafa Al Bakri* Abdullah[1], *Mohd Arif Anuar* Mohd Salleh[1], *Mohd Fairul Sharin* Abdul Razak[1], *Norsuria* Mahmed[1], *Muhammad Mahyiddin* Ramli[1,2], *Ayu Wazira* Azhari[3], *V Chobpattana*[4]

[1]Center of Excellence Geopolymer & Green Technology (CEGeoGTech), School of Materials Engineering, Universiti Malaysia Perlis, (UniMAP), 02600 Jalan Kangar-Arau, Perlis.
[2]School of Microelectronic Engineering, Universiti Malaysia Perlis, Pauh Putra Campus, 02600, Arau, Perlis,Malaysia.
[3]Water Research Group (WAREG), School of Environmental Engineering, Universiti Malaysia Perlis 02600 Arau, Perlis, Malaysia.
[4]Department of Materials and Metallurgical Engineering, Faculty of Engineering, Rajamangala University of Technology Thanyaburi (RMUTT), Thailand

Abstract. GO/TiO_2 thin films have been synthesized from titanium (IV) isopropoxide (TTIP) by a sol-gel method. The films were deposited onto a glass substrate using spin coating deposition technique then were subjected to annealed process at 350 °C. The different amount of graphene oxide (GO) was added into the parent solution of sol in order to investigate the microstructure, topography, optical band gap and photocatalytic activity of the thin films. The prepared thin films were characterized by atomic force microscopy (AFM), scanning electron microscopy (SEM), UV-VIS spectrophotometry and degradation of methylene blue (MB). AFM images reveal a rougher surface of GO/TiO_2 thin film than bare TiO_2 thin film due to GO particles. Moreover, the SEM images showed the formation of semispherical microstructure of bare TiO_2 changes to some larger combined molecules with GO addition. The UV–Vis spectrophotometer results show that with optical direct energy gap decreases from 3.30 to 3.18 eV after GO addition due to the effect of high surface roughness and bigger grain size. Furthermore, the optical results also indicated that GO improved the optical properties of TiO_2 in the visible range region.

1 Introduction

Titanium dioxide (TiO_2) belongs to the family of transition metal oxides and is also occurring as a mineral in the nature. TiO_2 has received a great deal of attention due to its chemical stability, non-toxicity, low cost and other advantageous properties [1]. According to, the physical properties of TiO_2 make it suitable for thin film applications.
There are three types of crystal phases of TiO_2 which are anatase, brookite and rutile [2][3]. Among the three phases of TiO_2, anatase is known to be the most active under UV irradiation. However, the photooxidation process of anatase is still restricted for wide application because

* Corresponding author : dewisuriyani@unimap.edu.my

of low activity under visible light irradiation. To overcome these issues, several experiments have been made to enhance the reactivity and improve the visible photocatalytic activity of TiO_2. Numerous efforts was done such as noble metal deposition, cationic and anionic doping, sensitization and addition of sacrificial agents [4], [5]. Though the noble metal deposition has several advantages over others, high cost and low abundance of the noble metals restrict their use in large scale applications [6]. In this study, TiO_2 thin film was modified by adding various amount of graphene oxide to improve the photocatalytic activity performance of the thin film.

2 Experimental

An amount of GO (5, 10, 15 and 20) mg powder was mixed with some amount of ethanol and sonicate in ultrasonic bath for 30 minutes. In different beaker, titanium (IV) isopropoxide (TTIP) was mixed with ethanol with a ratio 1:20 (TTIP: ethanol) by using magnetic stirrer. The solution was stirred for about 5 minutes. Then the sonicated GO solution was added to the TTIP solution and the solution was continued vigorous stirring for 1 hour. Afterward, a few drops of acetic acid (~ 0.30ml) that act as a catalyst to the hydrolysis process were slowly dropped into the solution with vigorous stirring until clear solution is formed.

When GO/TiO_2 sol-gel solution was ready, the GO/TiO_2 with and without additives sol-gel were deposited onto the clean glass substrate by spin coating technique at 800 rpm for 30 seconds by using VTC-50 desktop spin coater. The glass substrate was coated with the GO/TiO_2 solution for 3 layers to make sure that all the surface covered with the solution. Next, the coated glass substrate will be annealed in a muffle furnace at 350 °C for 1 hour soaking time with annealing rate of 10°C/minute

3 Results and Discusssion

3.1 Phase Analysis

The XRD pattern of aTiO_2, aGO_5, aGO_{10}, aGO_{15} and aGO_{20} thin films with different amount of GO, 0 mg, 5 mg, 10 mg, 15 mg and 20 mg GO respectively, were presented in Fig. 1. Based on XRD pattern of GO/TiO_2 thin films, the results indicated that all the thin film sample presented only brookite phase. According to the standard ICDD card No. 00-029-1360, characteristic diffraction peaks for brookite phase of TiO_2 are shown at $2\theta = 33.5°$ which corresponded to (121) plane. The diffraction peaks of aGO_5, aGO_{10} and aGO_{15} showed mixed phase of anatase and brookite, where 2θ values of 24.8° can be attributed to (101) plane of anatase while 2θ value of 33.5° can be attribute to (121) plane of brookite TiO_2 (ICDD Card No. 01-070-8501). It can also be seen in that there is a sharp peak at $2\theta = 21.9°$ which indicate the silicone(Si) peak resulted from the substrate XRD pattern.

This occured due to the annealing temperature factor where the brookite TiO_2 will form an anatase structure slowly with further annealing to 500-600°C [7]. As from the previous study by K. Fischer *et al* [8], anatase phase of TiO_2 will growth better at higher temperature. Meanwhile, D.S.C Halin *et al* have characterization of TiO_2/SiO_2 thin films formed brookite phase at higher annealing temperature (400-500°C) [3]. Interestingly, the diffraction pattern of aGO_5, aGO_{10} and aGO_{15} shows an anatase phase of TiO_2 and the intensity is slightly reduced when amount of GO increases and later at aGO_{20} thin films the phase showed brookite $TiO_2$was formed again.

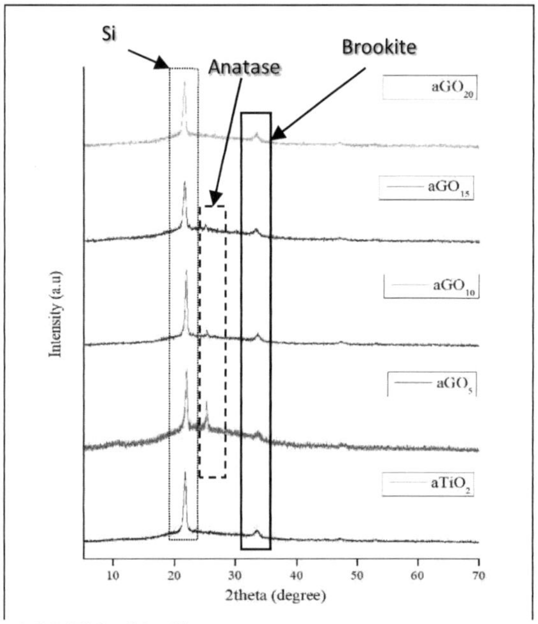

Fig. 1. XRD pattern of GO/TiO_2 thin films.

3.2 Surface Roughness and Topography Analysis

Surface roughness and topography analysis was examined by using Atomic Force Microscope (AFM) for all the produced thin films. The width the scan area was fixed at 5μm x 5μm. Ra and RMS values of thin film surfaces were calculated between microscopic peaks and valleys. Table 1 presents the roughness average (Ra) and roughness root mean square (RMS) of annealed GO/TiO_2 film samples. The $aTiO_2$ film having the RMS value of 2.34×10^2. After the addition of 5 mg GO (aGO_5) to the film, the RMS value shows a very significant increment which is 3.16×10^2. The RMS value was found to be the highest which indicates that by addition of 5mg GO, the surface become rougher. However, when 10mg, 15mg and 20 mg GO (aGO_{10}, aGO_{15} and aGO_{20}) added to the film, the RMS values decreases with RMS values of 1.87×10^2, 1.88×10^2 and 2.23×10^2 respectively.

After the addition of GO, the RMS of the surface increases which indicates the grains of regular shapes develop on the surface [9]. A significant increase in RMS value for GO/TiO_2 film with 5mg GO is observed due to the transformation process of TiO_2 to anatase phase, which involves the combination of smaller particles into bigger ones [10]. As reported by previous study by [11], high RMS indicates that GO/TiO_2 thin film exhibited high surface area, which in turn will give a greater absorption of the dye.

Table 1. Roughness average (Ra) and roughness root mean square (RMS) of thin films at various amount of GO.

Samples	Roughness Average (Ra)	Roughness Root Mean Square (RMS)
TiO_2	1.83×10^2	2.34×10^2
GO_5	2.57×10^2	3.16×10^2
GO_{10}	1.52×10^2	1.87×10^2
GO_{15}	1.36×10^2	1.88×10^2
GO_{20}	1.81×10^2	2.23×10^2

IOP Conf. Series: Materials Science and Engineering **701** (2019) 012011 doi:10.1088/1757-899X/701/1/012011

Fig. 2 shows the topography view and also three dimensional view of (a)aTiO_2, (b)aGO_5, (c)aGO_{10}, (d)aGO_{15} and (e)aGO_{20} thin films. From the Fig. 2(a), the molecular arrangement of the crystallites can be seen in TiO_2 film. The molecule is big and rough with the maximum height of 1395 nm. After the addition of 5mg GO as shown in Fig. 2**Fig.** (b), the surface morphology does not show the round shape molecule but the molecule is attached together and arrange in layers with the maximum thickness of 1740 nm. Meanwhile in aGO_{10}, aGO_{15} and aGO_{20} shows the round shape of TiO_2 molecule are slightly arrange in layers as the amount of GO increases. The maximum heights of the film also increase when the amount of GO increases which are 1083 nm, 1192 nm and 1388 nm. It can be seen that the formation of few layers of films and the stacking of GO/TiO_2 layers during formation of the thin films was also confirmed using AFM.

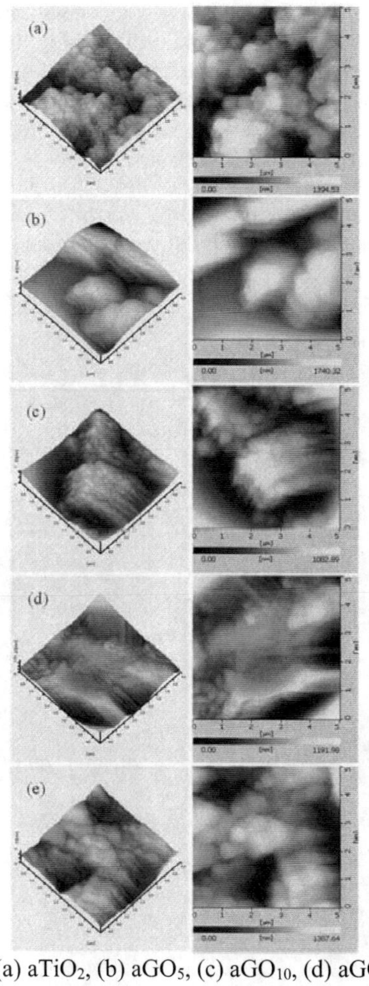

Fig. 2. AFM images of (a) aTiO_2, (b) aGO_5, (c) aGO_{10}, (d) aGO_{15} and (e) aGO_{20} thin film annealed at 350°C.

3.3 Microstructure Analysis

Microstructure of all the GO/TiO_2 thin films prepared were analysed through scanning electron microscope (SEM) at 10000 times magnification with 20kV energy. The SEM analysis performed on annealed GO/TiO_2 films at different GO content, as can be seen in Fig. 3. From SEM images, it can be seen that sol-gel process has a homogeneous TiO_2 distribution in the film samples with the spherical TiO_2 particles. Fig. 3(a) shows a basic morphology of spherical grains presenting different sizes combined with a flower-like random formation synthesized TiO_2 photocatalyst [12] while as showed in Fig. 3(b) to (e), show that the different amount of GO in the composite obviously affect the microstructure of the films. Fig. 3(b) showed the particles of TiO_2 were coated by GO. While Fig. 3(c) and (d) the particles of TiO_2 become bigger grains size and less agglomerates were formed. It was noticed that, the molecules of the TiO_2 with coated GO shows better dispersion with even surfaces at higher loading of GO as shown in Fig. 3(d) and Fig. 3(e) respectively. Overall, the SEM shows that with addition of GO the grain size of TiO_2 becomes bigger as the amount of GO increase and aGO_5 exhibits unique microstructure. The findings were consistent by research done by [13] on microstructure of TiO_2 with GO addition shown that GO covered tightly the TiO_2 surface and the covered TiO_2 surface area increases with the GO content. Proper introduction of GO will both enhance the light absorption and the separation of photogenerated electrons and holes. In contrast, too much introduced GO will disturb the light absorption of TiO_2 and reduce the mobility of photo carriers [14].

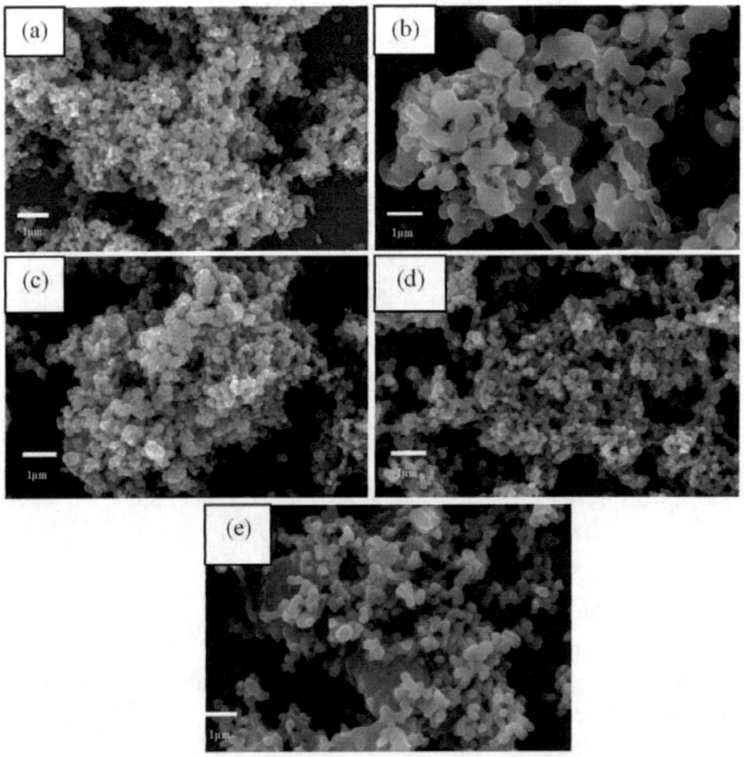

Fig. 3. SEM images for annealed thin films at 350°C (a) $aTiO_2$, (b) aGO_5, (c) aGO_{10}, (d) aGO_{15} and (e) aGO_{20}.

IOP Conf. Series: Materials Science and Engineering **701** (2019) 012011 doi:10.1088/1757-899X/701/1/012011

3.4 Optical properties

Optical analysis was conducted on to study the optical properties of the bare TiO_2 and GO/TiO_2 thin films. Optical properties of GO/TiO_2 thin films includes absorption or transmission rate, emission percentage and also energy band gap at 300 to 500 nm wavelength using visible-visible ultraviolet spectrophotometer as shown in Fig. 4. The absorption of $aTiO_2$ thin films is around 0.7 (arbitary unit) at a wavelength of 380 nm. With the addition of 5 mg GO (aGO_5), the absorption at a wavelength of 380 nm was about 1.7 (arbitrary unit). Highest absorption is produced by aGO_5 thin film which contain 5mg GO. The addition of GO particles has a significant effect on the absorption of light [15]. When the amount of GO increased, the absorption edge shifts in the absorption spectra. The cut-off wavelength has been expanded into visible regions with the addition GO. The cut-off wavelength can be determined by drawing a tangent along the edge of the absorption line. The intersection point with the wavelength axis on the absorption spectra and the tangent edge gives the cut-off wavelength value [16].

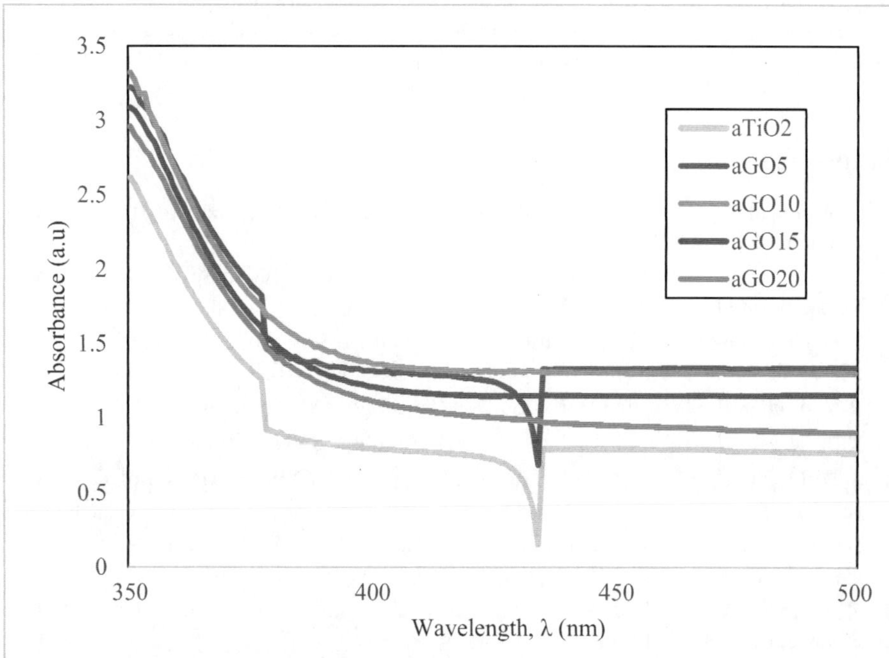

Fig. 4. Absorbance versus wavelength of GO/TiO_2 thin films.

The cut-off wavelength of annealed GO/TiO_2 thin film shifted from 380nm to 400 nm with increasing amount of GO. This indicates that the absorption range of thin film with the addition of GO is wider than that of $aTiO_2$ thin film. Wider cut-off wavelength in aGO_5 indicates aGO_5 is the optimum amount of GO addition to improve the performance of TiO_2 because it became narrower as more GO added. High absorption is due to low light scattering. Low light scattering is due to the increment of particle size and the decrease of nanocrystalline size distribution [17].

4 Conclusion

This work observes the effect of various GO amounts on morphological and optical properties of GO/TiO_2 thin films. The addition of various amount of GO have improved the formation of anatase phase of the thin films. It was revealed that, with small amount of GO addition, anatase phase of TiO_2 was formed in GO thin film and as the amount of GO increases the anatase TiO_2 decrease slightly and formed brookite TiO_2 again which means that the optimum amount of GO addition is 5mg. SEM image of aGO_5 was observed to have a large and unique surface area. Meanwhile, the optical properties of GO/TiO_2 thin film shows that the highest absorbance and most expended wavelength was found at aGO_5 with absorbance of 1.7(a.u) and cut-off wavelength at 400nm. Besides that, the energy gap of aGO_5 was also found to be the lowest which is 3.18eV.

The author would like to acknowledge the support from the Fundamental Research Grant Scheme (FRGS) under a grant number of FRGS/1/2017/TK07/UNIMAP/02/6 from the Ministry of Education Malaysia and Tin Solder Technology Research Grant (TSTRG) under grant number of 9002-00082 from Tin Industry (Research and Development) Board. The authors wish to thank the Center of Excellence Geopolymer & Green Technology (CEGeoGTech), School of Materials Engineering, Universiti Malaysia Perlis, UniMAP for their partial support.

References

1. S. Anandan, T. N. Rao, M. Sathish, D. Rangappa, I. Honma, and M. Miyauchi, ACS Appliled Mater. Interfaces, **5**, 1, (2012)
2. D. S. C. Halin, N. Mahmed, M. A. A. Mohd Salleh, A. N. Mohd Sakeri, and K. Abdul Razak, Solid State Phenom., **273**, (2018)
3. D. S. C. Halin, M. M. A. B. Abdullah, N. Mahmed, S. N. A. Abdul Malek, P. Vizureanu, and A. W. Azhari, *IOP Conf. Ser. Mater. Sci. Eng.*, **209**, (2017)
4. R. A. Rather, S. Singh, and B. Pal, *Sol. Energy Mater. Sol. Cells*, **160**, (2017)
5. K. A. Razak, D. S. C. Halin, and M. M. A. B. Abdullah, Solid State Phenom., **280**, (2018)
6. S. Prabhu, L. Cindrella, O. Joong, and K. Mohanraju, Sol. Energy Mater. Sol. Cells, **169**, (2017)
7. S.Bakardjieva, V. Stengl, L. Szatmary, J. Subrt, J. Lukac, N. Murafa, D. Niznansky, K. Cizek, J. Jirkovsky and N. Petrova , J. Mater. Chem., **16**, 18, (2006)
8. K. Fischer , A. Gawel, D. Rosen, M. Krause, A.A. Latif, J. Griebel, A. Prager and A. Schulze, Catalysts, **7**, (2017)
9. S. Liu, H. Sun, S. Liu, and S. Wang, Chem. Eng. J., **214**, (2013)
10. Sahbeni K., Sta I., Jlassi M., Kandyla M., Hajji M., Kompitsas M. and Dimassi W., J. Phys. Chem. Biophys., **7**, 3, (2017)
11. A. M. Ramli, M. Z. Razali, and N. A. Ludin, Malaysian J. Anal. Sci., **21**, 4, (2017)
12. A. Timoumi, S. N. Alamri, and H. Alamri, Results Phys., **11**, (2018)
13. Y. Ni, W. Wang, W. Huang, C. Lu, and Z. Xu, J. Colloid Interface Sci., **428**, (2014)
14. Y. Zhang, Z.-R. Tang, X. Fu, and Yi-Jun Xu, ASC Nano, **4**, 12, (2010)
15. M. Hashemi, B. Muralidharan, M. Omidi, J. Mohammadi, Y. Sefidbakht, E. S. Kima, H.D.C. Smyth, M. Shalbaf and T.E. Milner, J. Biomed. Opt., **23**, 8, (2018)
16. A. E. Athare, IJSRSET, **4**, 1, (2018)

17. J. Wen, X. Li, W. Liu, Y. Fang, J. Xie and Y.Xu, Chinese J. Catal., **36**, 12, (2015)

The influence of Copper and Bismuth element in Sn-Ag solders and ENIMAG substrate

Jaidi Zolhafizi[1] *and Osman.* Saliza Azlina[1,*]

[1]Faculty of Mechanical and Manufacturing Engineering, UTHM, 86400 Batu Pahat, Johor, Malaysia

Abstract. The addition of copper and bismuth elements in Sn-Ag solders on interfacial reaction during reflow soldering and isothermal aging of SA25 and SACB25051 was investigated. The growth of intermetallic compound (IMC) was characterized by using scanning electron microscopy (SEM), optical microscope (OM) and energy dispersive x-ray (EDX). The results show the irregular circle shapes of $(Cu,Ni)_6Sn_5$ IMC at the interface of solder joint for SA25 while rod-like shape for SACB25051 during both reflow soldering process and isothermal aging. The SACB25051 solder showed the thinner IMC thickness compared to SA25 solder for both conditions. The Cu and Bi element in Sn-Ag solder promotes a slower growth rate of IMC formation during interaction. The thickness and morphology of the IMC for both solders increase gradually with increasing of aging time.

1 Introduction

The expanding green technology concerns due to toxic element in Sn-Pb solders caused the development of many lead-free solders are growing rapidly in line with the Restriction of Hazardous Substances (RoHS) Regulation [1]. Recently, there are several promising candidates can replace the Sn-Pb including Sn-Ag-Cu [2,3], Sn-Ag [4,5] and Sn-Ag-Bi [2]. Even Sn-Ag-Cu (SAC) is one of the most popular lead-free solder system, it still have their own drawback where the melting-point of Sn–Ag–Cu system is higher than that of Sn–Pb solders. This can cause the soldering temperature were increased and slightly deteriorated the processing performance of the solder [2]. However, the additional of bismuth (Bi) element to tin based solders can reduce melting temperature [2,6] and improve processing performance of the solder [7].

Since several reports have shown that the doping element can improve the properties and performance [6,8], and also can reduce the growth rate of IMC layer [9]. Thus, the aim of this paper is to investigate the influence of copper and bismuth element in Sn-Ag solders and the ENIMAG substrate on intermetallic growth. The samples of isothermal aging also been investigated in order to compare with as-reflowed samples.

* Corresponding author: salizaz@uthm.edu.my

2 Materials and methods

The double type copper polymer (FR-4) was prepared as substrate materials with dimensions of 52 mm x 72 mm x 1.2 mm. The depositions of electroless nickel/immersion silver (ENImAg) were coated to the substrates. The plating baths for both substrates were maintained at 90°C for 1 hour and 40°C for 8 minutes, respectively. The Sn-2.5Ag (SA25) and Sn-2.5Ag-0.5Cu-1.0Bi (SACB25051) solder balls with a diameter of 500 μm was arranged on the substrates. The reflow process will undergo to 230°C and kept for 20 minutes. The sample then were subjected into the isothermal aging process at 150°C for 250 hours. Several characterization was carried out via cross-sectional and top surface examination to identify the type, thickness and morphology of IMC by using the scanning electron microscope (SEM) and an image analyzer.

3 Results and discussion

3.1 Top view examination

During the reflow soldering process, there are ternary compounds were detected at the interface between solder bump and Cu/ENIMAG substrate. Fig. 1 shows the images of Sn-2.5Ag (SA25) and Sn-2.5Ag-0.5Cu-1.0Bi (SACB25051) after exposed in the reflow soldering and isothermal aging process, respectively. The results show that only $(Cu,Ni)_6Sn_5$ IMC was observed at the interface for both solders during reflow soldering as well as isothermal aging. This is due to the diffusion/dissolution process between solder and substrate when the atoms moving in and out through a boundary, where Cu-Ni-Sn IMC layers might be nucleate [4]. The composition of $(Cu,Ni)_6Sn_5$ IMC was confirmed by energy dispersive x-ray (EDX) spectrum analysis as in Fig. 2. Meanwhile, a new $(Ni,Cu)_3Sn_4$ IMC layer was created at the interface during the solid state reaction. Even, there are obstacles existed at the boundary due to $(Cu,Ni)_6Sn_5$ formation, the Cu atoms are still diffuse from the substrate and solder at the slow and limited rate process. In terms of IMC morphology, the irregular circle shapes of IMC were observed at the solder joint for SA25 while rod-like shape for SACB25051 during both reflow soldering and isothermal aging process as shown in Fig. 1. The continuous heat exposure to the substrate during isothermal aging made the IMC grain size became bigger and denser. The continuous growth of IMC has been recorded until aging duration reached 250 hours. These results are consistent with previous research [10, 11].

3.2 Top view examination

Fig. 2 represents the cross-sectional examination during reflow soldering and isothermal aging for SA25 and SACB25051 solders with ENIMAG finish. After reflow soldering, the $(Cu,Ni)_6Sn_5$ layer of SACB25051 solder shows a thinner IMC than SA25 solder where the average thicknesses are 2.86 μm and 3.57 μm, respectively. Meanwhile, the total of IMC thickness for both SA25 and SACB25051 solders after exposing in isothermal aging shows a similar trend with reflow soldering where the IMC thickness recorded as 3.94 μm and 3.13 μm, respectively as illustrated in Fig. 4. From observation, the growth of IMC for both solders increase gradually with increasing the aging time. It was realized that Cu and Bi element in Sn-Ag solder managed to change the IMC thickness to be thinner than Sn-Ag solder. This explains that both Cu and Bi manage to promote a slow diffusion rate of the interaction between solder and ENIMAG, and then causes retardation of IMC growth. This statement also has a good agreement with previous researchers [12,13].

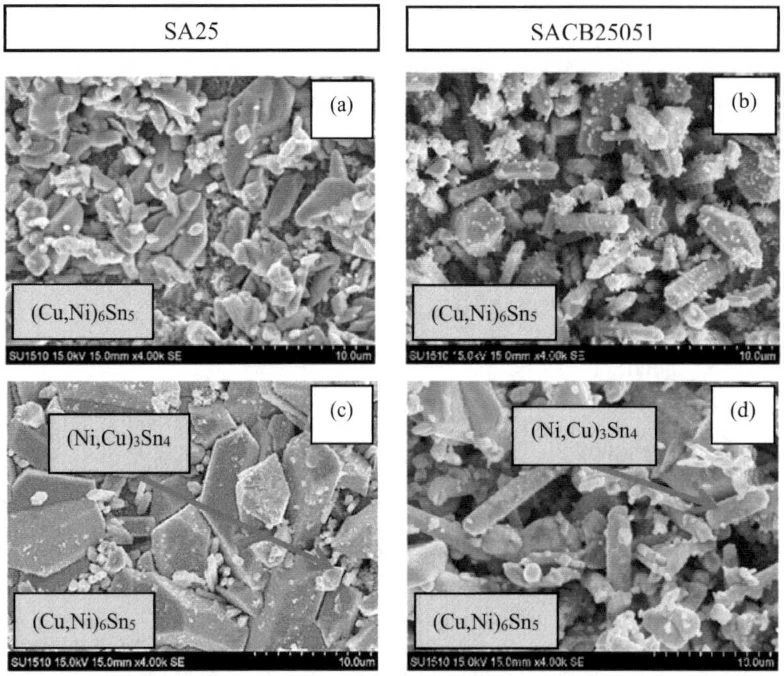

Fig. 1 SEM images of top surface morphology with magnification x4000 for SA25 and SACB25051. (a, b) reflow and (c, d) aging 250 hours.

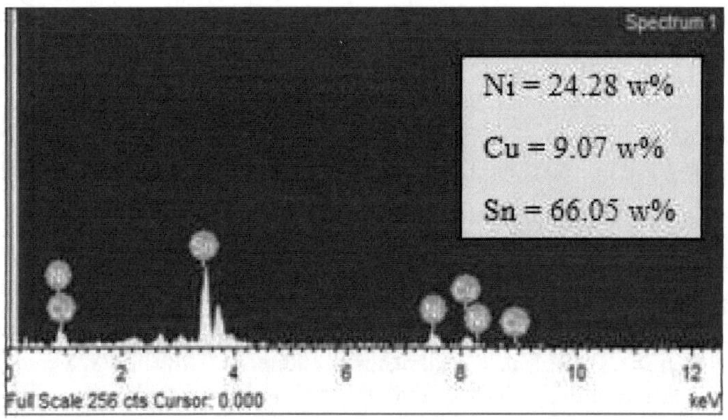

Fig. 2. EDX spectrum of $(Cu,Ni)_6Sn_5$ IMC formation.

IOP Conf. Series: Materials Science and Engineering **701** (2019) 012012 doi:10.1088/1757-899X/701/1/012012

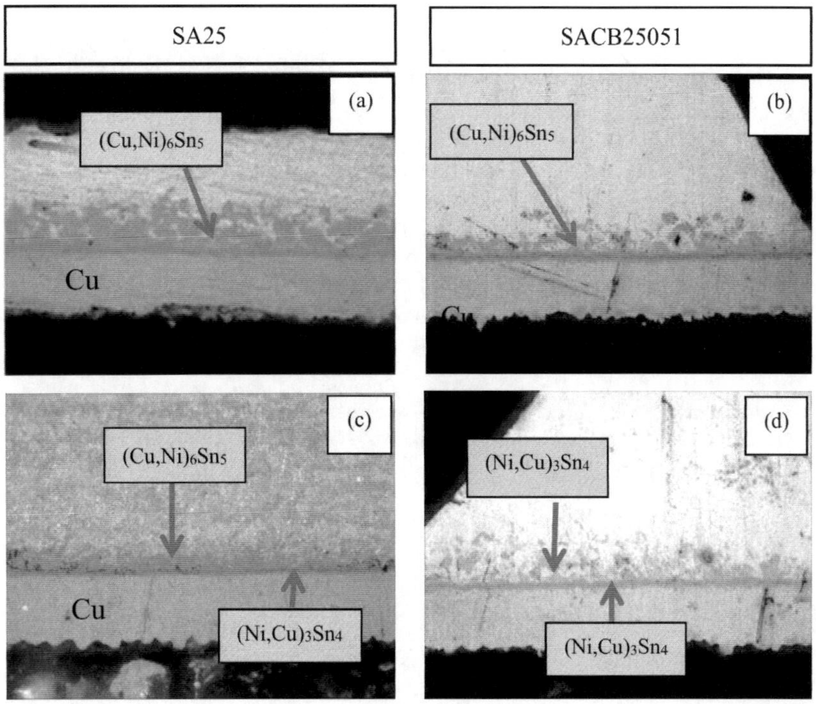

Fig. 3 Optical microscope images of cross sectional morphology with magnification x500 for SA25 and SACB25051. (a, b) reflow and (c, d) 250 hours.

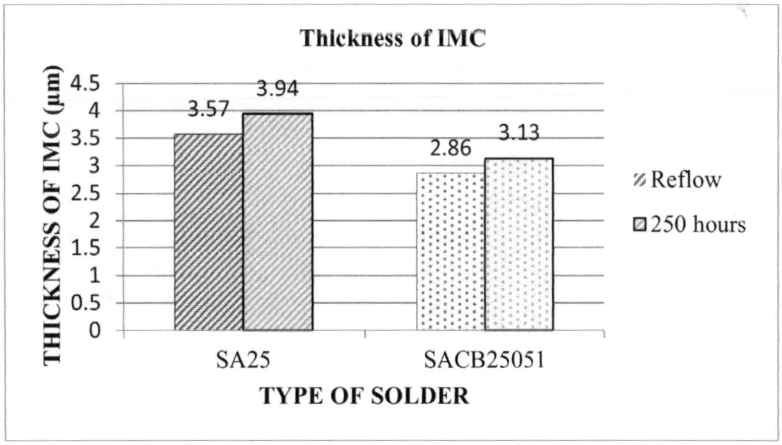

Fig. 4 The total thickness of IMC for SA25 and SACB25051 solders.

4 Conclusion

From the investigations, the following conclusions are successfully made:

1) As-reflow condition, the main intermetallic was formed between both types of lead-free solders with an ENIMAG surface finish are $(Cu,Ni)_6Sn_5$ IMC while $(Ni,Cu)_3Sn_4$ IMC was created beneath the (Cu,Ni)6Sn5 after the isothermal aging process.

2) The thickness of SACB25051 solder is thinner than SA25 solder for both reflow and aging condition. The Cu and Bi element in Sn-Ag solder promotes a slower growth rate of IMC formation.
3) The thickness and morphology of the IMC for both solders increase gradually with increasing of aging time

This work was financially supported by Universiti Tun Hussein Onn Malaysia Research Grant Scheme (Tier-1 H225), and facilities provided by Faculty of Mechanical and Manufacturing Engineering, Universiti Tun Hussein Onn Malaysia.

References

1. K. Kanlayasiri and R. Kongchayasukawat, *Trans. Nonferrous Met. Soc.*, **28**, (2018)
2. X.P. Zhang, C.B. Yu, Y.P. Zhang, S. Shrestha & L. Dorn. Journal of Materials Processing Technology, **192–193**, (2007)
3. Hiroshi Nishikawaa & Noriya Iwata, J. Mater. **215**, (2015)
4. C.E. Ho, W.Z. Hsieh, C.S. Liu & C.H. Yang, Thin Solid Films, **572**, (2014)
5. Wislei R. Osório, Daniel R. Leiva, Leandro C. Peixoto, Leonardo R. Garcia, Amauri Garcia, J Alloys Compd, **562**, (2013)
6. A.A.El-Daly,A.M.El-Taher, S.Gouda, J Alloys Compd, **627**, (2015)
7. X.P. Zhang, H.W. Wang, Y.W. Shi, J. Mater. Sci.: Mater. Electron, **15**, (2004)
8. A.A. El-Daly, A.E. Hammad, G.S. Al-Ganainy, M. Ragab, Mater. Sci. Eng., **608**, (2014)
9. I. Siti Rabiatul Aisha, A. Ourdjini, and O. Saliza Azlina, Int. J. Chem. Mol. Nucl. Mater. Metall. Eng., **10**, no. 1, (2016)
10. L. Guo-yuan and S. Xun-qing, (2006)
11. M.A. Azmah Hanim, A. Ourdjini, I. Siti Rabiatul Aisha, O. Saliza Azlina, **650**, (2013)
12. X. Hu, Y. Li, Y. Liu, and Z. Min, J. Alloys Compd., **625**, (2015)
13. M. H. Mahdavifard, M. F. M. Sabri, D. A. Shnawah, S. M. Said, I. A. Badruddin, and S. Rozali, Microelectron. Reliab., **55**, (2015)

Decolorization of reactive red 2 dye by using pva-polyaniline immobilized system

Mohd Fairul Sharin Abdul Razak[1,*], *Mohd Nazry* Salleh[1], *Azliza* Azani[1], *Nurul Izzati* Muhammad Zakir[1]

[1]School of Materials Engineering, Universiti Malaysia Perlis, (UniMAP), 02600 Jalan Kangar-Arau, Perlis.

Abstract. The immobilized of polyaniline (PANi) and polyvinyl alcohol (PVA) in bilayer system were studied base on several operational condition parameters such as loading and pH of the reactive red 2 dye solution. The PVA/PANi immobilized system was prepared by coating the PVA layer followed by deposition of PANi onto the surface of the glass plate. In this system, the addition of PVA particles works as binder between the PANi and the substrates. It was found that, PVA/PANi works as absorbent in this system and shown good response towards to the adsorption process of RR2 dye. This process can be seen by the decolorization of the dye. This decolorization was influenced by several factors which related to the interaction of dye with PVA/PANi composite and also during conditioning of dye as well. As a result, the adsorption process of RR2 dye in the immobilized system showed the best removal of dye under aerated system with the solution condition at pH 7. The percentage of dye removal was increased linearly with the amount of loading PVA/PANi. The rate of adsorption of dye was high at the beginning and after 30 minutes and onwards, all loading absorption rate has drop and the reading started to plateau indicates the surface areas of the system were nearly filled up by the dye particles. Thus, the highest adsorption was achieved was at 80 % removal after two hours. Indeed, by prolonging the time did not change the colour of the solution indicated to the maximum adsorption capacity of the process has been achieved.

1 Introduction

Anionic dye in textile industry is widely use around the world especially for more than 50% mainly for cotton dye which has been introduced by ICI in 1950s commercially, due to its bright colors, various range of colour and shades to be used, good washing fastness and are easy to apply to the textile and products [1]. Unfortunately, the usage of this type of dye has contributes to another problem that cause effluent water pollution as the dyes have oxygen demand, colour, and salt load [2].

It is more difficult to remove the presences of the dye using conventional bio-systems as the complex and stable chemical structure of the anionic dye make them resistant to bio-degradation [1]. The common conventional methods to remove the synthetic azo dyes are ultrafiltration, extraction,

* Corresponding author : fairulsharin@unimap.edu.my

absorption and oxidation with ozone and hydrogen peroxide. Alternatively, photocatalyst can be used to treat the polluted effluent water by using advanced oxidation processes (AOP).

During photocatalyst process, adsorption process also occurs spontaneously before photocatalysis process taking place. The advantages of adsorption process are for its simplicity in operation, inexpensive (compared to other separation processes) and without sludge formation. Textile industries are the major consumers of water and they impart a reasonable amount of color in the effluent [3-4].

2 Experimental

Soluble polyaniline was synthesis by using emulsion polymerization technique [5].While PVA solution was prepared based on the report [6] whereby the solubility of the PVA powder was improved by increasing the temperature of the solution. Thus, both layers were coated onto the glass plate with a dimension of 4.5 cm x 4 cm which form a bilayer PVA-PANi immobilized system. For the study of loading bilayer PVA/PANI, a series of different layers PVA/PANi were varied from 0.01, 0.03, 0.05, 0.07 and 0.09 g. The experiment to study the adsorption of the RR2 dye by PVA/PANi system was set up in the dark place. The RR2 dye solution was poured into the glass reactor cell, and then the glass plate was immersed into the reactor cell. The immersion of the PVA/PANi in the dye is set for 3 hours with interval every 30 minutes. The pH of the dye and the presence of aeration in the dye were also varied to study the efficiency of the absorption of dye by the PVA/PANi immobilized system.

3 Results and Discusssion

3.1 Effect of PVA/PANi loading

The loading of PVA/PANi was varied from 0.01, 0.03, 0.05, 0.07 and 0.09 g immersed into the 30 ppm RR2 dye solution. By manipulating the loading of the PVA/PANi, the surface area available for the RR2 to adhere to PVA/PANi also varies which the heavier the PVA/PANI loading, more surface area available for adsorption [7].From Fig.1. it shows that percentage of adsorption for set of loading has reached to 80% and more dye will be adsorbed removal after 90 minutes. At the first 30 minutes, the adsorption for all loading showed high rate of adsorption. This is due to the high surface area of the PANI available for RR2 dye to be absorbed. After 30 minutes and onwards, all loading adsorption rate has drop and the reading started to plateau.

As the time increasing, the surface area available for adsorption was reduced which make less dye being absorbed. 0.01g loading has the lowest adsorption compared to other. This is due to less area available for dye to be absorbed by the PVA/PANi, followed by 0.03 and 0.05 g loading respectively. However most of the samples showed the same trend of adsorption after one hour and it started to slow down. Based on K value for each samples, 0.05g loading slightly higher than 0.03g loading which is 0.414 and 0.396 K min^{-1}

IOP Conf. Series: Materials Science and Engineering **701** (2019) 012013 doi:10.1088/1757-899X/701/1/012013

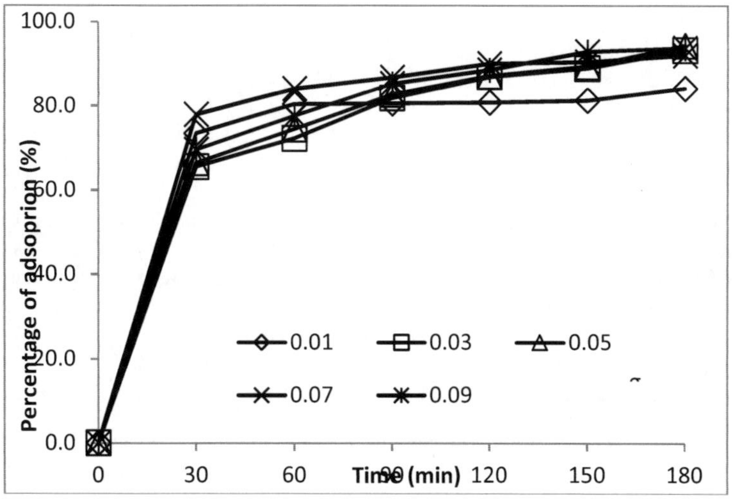

Fig. 1. The percentage of dye adsorption at different loading of PVA/PANi immobilized system.

3.2 Microstructure Analysis

Microstructures of PVA/PANi bilayer were analysed through scanning electron microscope (SEM) at 1000 and 5000 times magnification with 20kV energy. The SEM images of PVA/PANi immobilized system before immersion in the RR2 dye has shown in Fig. 2. In Fig. 2(a), it shows that PANi has interaction with the PVA as PVA act as a binder to PANI for hold the PANI coated onto glass substrate. PANI is presence in the agglomerated form and was dispersed on the surface of the PVA phase.

From the Fig. 2(b), it shows the structure of PVA/PANi immobilized film at x5000 magnifications. The agglomerated structure of the PANI on the PVA phase provide more surface area for the removal of the dye as more dye can be deposited or reacted with the PANi upon immersion [8].

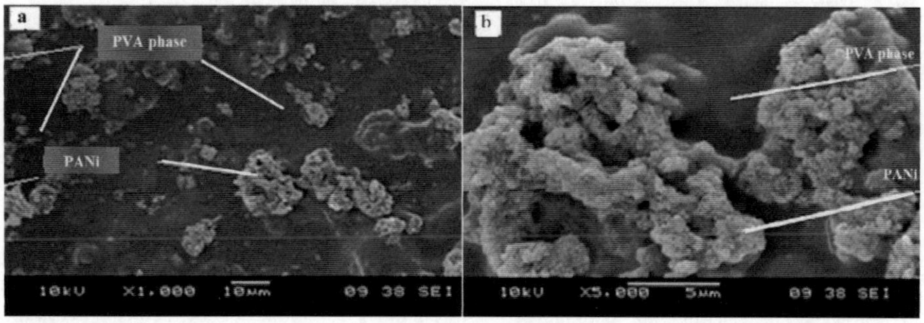

Fig. 2. SEM images for annealed PVA/PANi bilayer at (a) 1,000 and (b) 5,000 magnification.

3.3 The aeration study

The effect of aeration study of PVA/PANi immobilized system toward RR2 dye was compared with the aerated immobilized system under the same experimental conditions. Both systems are using the same composite loading 0.05g with the concentration of RR2 is 30 ppm.

From the Fig. 3, it shows that the percentages of adsorption for both systems had reached 20 % difference just at 30 minutes contact time between aerated system and non-aerated system. Obviously, the aerated system had improved the adsorption process tremendously with nearly reached at almost 95% dye removal for 180 minutes while the non-aerated system only reach maximum 44% dye removal at the same time.

The main purpose of the aeration process is to supply the oxygen to the RR2 dye and fasten the process of the dye removal and aid the adsorption process by the PVA/PANi immobilized system [5].

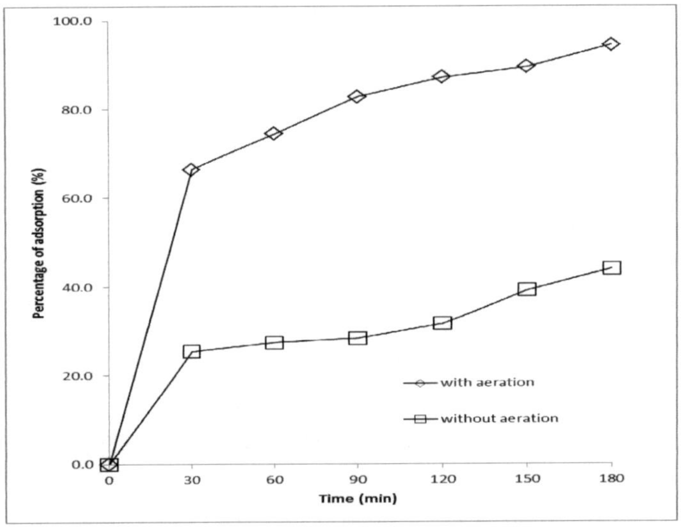

Fig. 3. The percentage of adsorption dye under aerated and non-aerated system.

3.4 Effect of Different pH solution of RR2 Dye

The pH of the dye can affect the efficiency of the adsorption rate of the dye onto PVA/PANi immobilized film [7]. Fig. 4 shows the adsorption process by PVA/PANi immobilized film work the best at pH 7 which it remove almost 54% of the dye, followed by pH 3, 9, 5 and 11 with percentage of dye removal at 45%, 40%, 36% and 23% respectively after 180 minutes. In acidic condition, the dye absorption works the best in pH 3 as compared to pH 5. With the addition of $H^{+,}$ it helps the absorption to be more effective. Meanwhile, in alkaline condition, with increasing OH^- in the dye solution from pH 9 to pH 11, the absorption become inefficient, as the OH^- species limit the interaction between the RR2 dye and the PVA/PANi immobilized film [9-11]. Therefore, the PVA/PANi immobilized film in neutral condition at pH 7 shows the highest adsorption of RR2 dye which is more than 50%. This shows that, without interfering with addition of OH^- and H^+, the RR2 dye adsorption by PVA/PANi immobilized system can occur in the optimum condition.

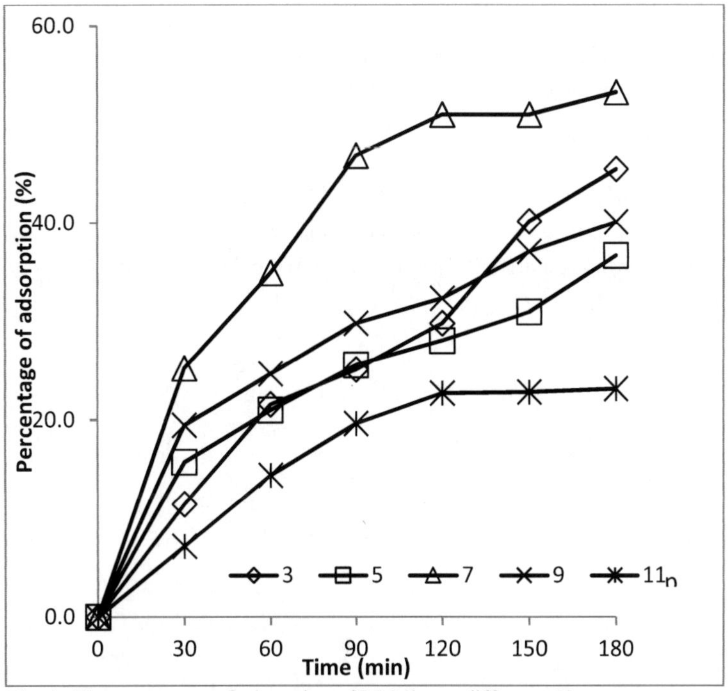

Fig. 4. The percentage of adsorption of RR2 dye at different pH.

4 Conclusion

The effect of loading of PVA/PANI bilayer system was optimized while the effect of aeration study and pH value of RR2 dye was observed towards the adsorption process by PVA/PANi bilayer system. The PVA/PANi immobilized system works the best as the adsorption agent for removing the RR2 dye at optimum condition for more than 80% dye removed from the original dye concentration. The aeration process contributes to more efficient dye removal and the loading of the PVA/PANi works the best at 0.05g loading. Moreover, at pH 7 showed the adsorption process works the best for removing of the dye. Lastly, SEM image revealed that PVA works as binder hold tight the particles of PANI and also improved the adsorption process due to better contact of surface area with the dye.

The authors wish to gratefully acknowledge the School of Materials Engineering, as well as Research and Innovation Department under Universiti Malaysia Perlis and Kementerian Pengajian Tinggi Malaysia (Ref: FRGS/1/2019/STG01/UNIMAP/02/2) for supporting this research project.

References

1. J. Tie, X. Fang, X. Wang, Y. Zhang, T. Gu, S. Deng, G. Li, D. Tang, Pol. J. Environ. Stud. **26**, (2017)
2. A. Khatri, M. H. Peerzada, M. Mohsin, and M. White, J. Clean. Prod., **87**, (2015)
3. B. Ohtani, J. Photochem. Photobiol. *C*, **11**, (2010)
4. B. Haspulat, A. Gülce, and H. Gülce, J. Hazard. Mater., **260**, (2013)

5. D. Sivakumar, and D. Shankar. International journal of environmental sciences 2, (2012)
6. C. I. Awuzie, Mater. Today Proc., **4**, (2017)
7. D. S. Patil, J. S. Shaikh, D. S. Dalavi, S. S. Kalagi, and P. S. Patil, Mater. Chem. Phys., **128**, (2011)
8. R. Maas and S. Chaudhari, Process Biochem., **40**, (2005)
9. K. M. Molapo *et al.*, Int. J. Electrochem. Sci., **7**, (2012)
10. S. K. Shukla, C. S. Kushwaha, and N. B. Singh, Mater. Today Proc., **4**, (2017)
11. Z. A. Boeva and V. G. Sergeyev, Polym. Sci. Ser. C, **56**, (2014)

IOP Conf. Series: Materials Science and Engineering **701** (2019) 012014 doi:10.1088/1757-899X/701/1/012014

Molecular dynamics simulation of the nano-reinforced lead-free solder at different reflow soldering process temperature

Intan Norshalina Sahrudin[1], **Mohd Shahrizal* Abdul Aziz[1], *Mohd Zulkifly* Abdullah[1], *Mohd Arif Anuar* Mohd Salleh[2]

[1]School of Mechanical Engineering, Universiti Sains Malaysia, 14300 Nibong Tebal, Penang, Malaysia
[2]Department of Mechanical Engineering Technology, Faculty of Engineering Technology, Universiti Malaysia Perlis (UniMAP), Sungai Chuchuh 02100, Padang Besar, Perlis

Abstract. Temperature plays an important role in the reflow soldering process. Reflow soldering process which involving four basic stages including preheating, soaking, reflow and cooling require a specific range of temperature at each stage and the highest temperature is the reflow stage. In this study, the movement of the reinforced nanoparticles in the lead-free solder at three different stage temperatures which are 30°C,150°C and 250°C will be shown via Large-scale Atomic/Molecular Massively Parallel Simulator (LAMMPS) software. The simulation model able to show the movement of nanoparticles in the solder at three different reflow soldering process stages. The most vigorous movement of nanoparticles is shown at the temperature of 250°C while the least vigorous movement is shown at room temperature of solder which is at 30°C.

1 Introduction

Most of today's electrical and electronic equipment which commercially manufactured are using surface mount technology (SMT) as it gives ample advantages to replace through-hole technology (THT). SMT offers efficient uses of printed circuit board (PCB) space as more electronic components can be packed onto the PCB which next increasing the possibility to build complex electronic circuits. In addition to that, SMT allowing the automated PCB assembly which brings to reduction of manufacturing cost and speed up the manufacturing process which then improving the reliability of electronic assembly.

The reflow soldering is the best method for SMT compared to wave soldering which is more complex as the board temperature and board time spends in the wave soldering machine needs to be carefully monitored. The mistake in maintaining the right wave soldering environment lead to the imperfection in soldering process. Reflow soldering process uses a solder paste out of powdered solder and flux, which then used that paste to attach the electronic components to the contact pads. Next, the entire assembly is being heated in a reflow oven to melt the solder and then connect the joint. Reflow soldering process undergoes four basic stages including preheat, thermal soak, reflow, and cooling.

*Corresponding: msharizal@usm.my

In the recent 10 years, the improvement in the solder technology which is the reinforcement of nanoparticles into the solder becomes more vigorously as researchers found out that by adding nanoparticles such as cobalt (1), copper (2), silver (3), nickel (4) diamond (5), iron nickel oxide (6), iron oxide (7), titanium dioxide (8), zinc oxide (9) and many more is proven to increase the solder properties in terms of either one, mechanical, thermal, wetting and intermetallic compound layer or the four of them.

Numerical approaches are introduced in the SMT field to reduce the cost and time-consuming experimental work. This approaches also make ease for showing the reaction of the solder in certain condition clearly and accurately. Over years, there are lots of numerical approaches to simulate the reflow soldering process environment including finite element method (FEM) (10), finite volume method (FVM) (11), fluid-structure interface (FSI) (12), lattice Boltzmann's method (LBM) (13), discrete phase method (DPM) (14) and molecular dynamic (MD) (15). This paper is focussing on the used of MD to simulate the nano-reinforcement lead-free solder during the reflow soldering process.

2 Methodology

2.1 Simulation tools

The problem is to simulate the solder atom with added nanoparticles in the atomic scale. A molecular dynamics (MD) software and other tools such as the Notepad++, command prompt and Ovito have been used to perform the simulation.

2.1.1 MD software

The Large-scale Atomic/Molecular Massively Parallel Simulator (LAMMPS) software by Sandia National Laboratories is used to simulate the nano-reinforced solder paste environment. LAMMPS is a free open source software that uses Message Passing Interface (MPI) for parallel communication and it uses neighbor lists (Verlet lists) to keep track nearby particles. LAMMPS has potentials for solid-state material, soft-matter and coarse-grained or mesocospic systems (16).

2.1.2 Other tools

LAMMPS is executed by reading the commands in the input script (text file) which is written in the Notepad++. To perform the LAMMPS simulation, the input script is read through the command prompt application which is the command line interpreter. Lastly, Ovito is used as the visualization tool to show the trajectory of the nanoparticles in the solder.

2.2 Input script

There are 4 major parts that must be considered in the input script which are initialization, atom definition, settings, and run a simulation. The first one is the initialization part, where the parameters for the simulation is set up. The command such as units, boundary, and atom_style is used to express the parameters. The units style for this simulation is set first with any of units style which lj, real, metal, si and others accordingly. The choice of the units style must consider the conversion of parameters unit following the units style. This simulation is set as real. Next, the boundary for the simulation can be set as periodic, fix, shrink-wrapped and shrink-wrapped with a minimum value at x,y and z direction. The boundary for this simulation is set as periodic in all directions. This shows the atoms can exit at one end and re-enter at the other end. For inter-atomic potential definition, the pair_style command is used to set up the force field style. There are variety of pair_style that can be chose. For this simulation, the lj/cut pair style is used to set the interaction between two different atoms. The command pair_coeff is used to set up the force field coefficient for each interaction of atoms.

Next part is the atom definition. The geometry and the atoms for the simulation model will be created. The commands such as lattice, region, create_box, and create_atoms are commonly used. The dimension of the simulation model can be set up as 2D or 3D and for this simulation it was set as 3D while the lattice is set to face-centered cubic (fcc). Then the region for the simulation model is set up by command region box follows by the required dimension to make a box shape model. To set up two different atoms in a simulation model, command create_box is used to differentiate another atom that is in the region box. The create_atoms command is used to create the solder atoms in the region box. The nanoparticles will be introduced in a sphere region. After the region for the nanoparticles has existed, command delete_atoms is used to delete the atoms in the sphere region so that the new atoms will be created in the sphere by command create_atoms.

The third part is the setting. The setting for the force field, simulation parameters and output options are set up after the solder atoms, and nanoparticles have existed. The command pair_coeff is used to set up the force field coefficient for each interaction of atoms. Next, the simulation parameters are set up with command neighbor followed by a distance beyond the cut off and the style which is binary. Another command used is the neighbour_modify which affected the building of solder paste environment. The min_style command is used to do minimization for the simulation. For this simulation, the minimization style used is the conjugated gradient. Then, the fix command is used to set the boundary conditions, time integration, and diagnostic options. For computation purpose, the command compute and variable are used in this simulation. Command compute is used to calculate the required value such as the temperature and energy and the variable command is the command to set up the temperature, volume, pressure and some other variables according to the reflow stage environment. The temperature will be set at different temperature to see the different nanoparticles trajectory in the solder. In this simulation, the temperature is set up as 30°C (room temperature), 150°C (ramping temperature), and 250°C (reflow temperature). For the output option, the thermo and dump command is used in this simulation model. This command is used to ensure, the simulation model can be shown in the visualization tools.

Last but not least, the run a simulation part, the run command is used to run the calculation for the simulation model via command prompt. Then, the new file has existed in the directory, which is the dump file. The dump file can show the simulation's image or video via visualization tools.

3 Results

3.1 Molecular dynamics simulation of nano-reinforced solder paste

The molecular dynamics simulation was applied to investigate the nanoparticles trajectory in the lead-free solder paste according to the reflow zone temperature. In this 3D simulation model, the nanoparticles are introduced in a spherical shape and as agglomerated atoms while the solder atoms are introduced as a bunch of single atoms that are freely moved. The color of solder and nanoparticles atoms also represented differently to show clearly the movement of each atom as in Fig.1. The agglomerated, purple color atoms show the Cu nanoparticles, the red color atoms show the solder atoms while the yellow color in Fig.2, Fig. 4. and Fig. 6. shows the atoms trajectory in the solder in 2D.

Before the reflow phase, which is the main phase in reflow soldering process, there are several stages that occurred first including the preheating, ramping and soaking which has increased the solder temperature from room temperature which is about 30°C, to the reflow temperature of about 250°C. The velocity is added to all atoms as to simulate the increment of the temperature and thus make sure the atoms are not in the static condition. This shows that the thermophoresis-effect occurred.

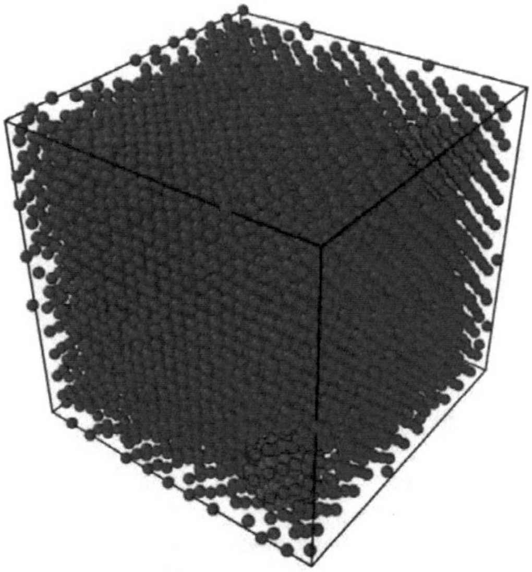

Fig. 1. Solder paste with doped nanoparticles.

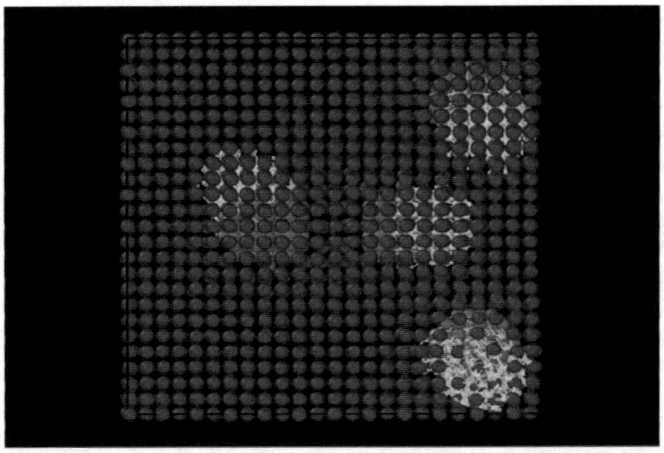

Fig. 2. Solder paste with doped nanoparticles with the mark of trajectory at temperature of 30°C.

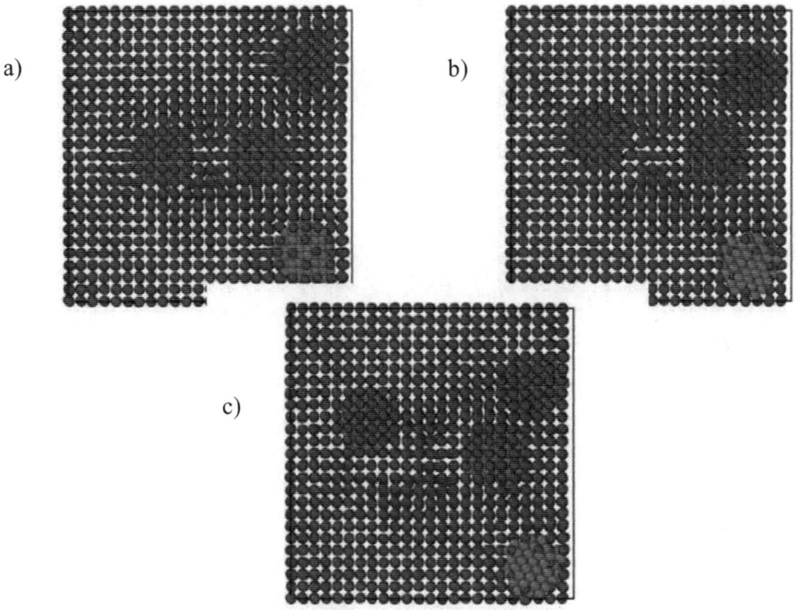

Fig. 3. Images of the position of the nanoparticles in the solder paste model at temperature of 30°C at a) 0 second, b) 10 seconds and c) 20 seconds in the x-y plane.

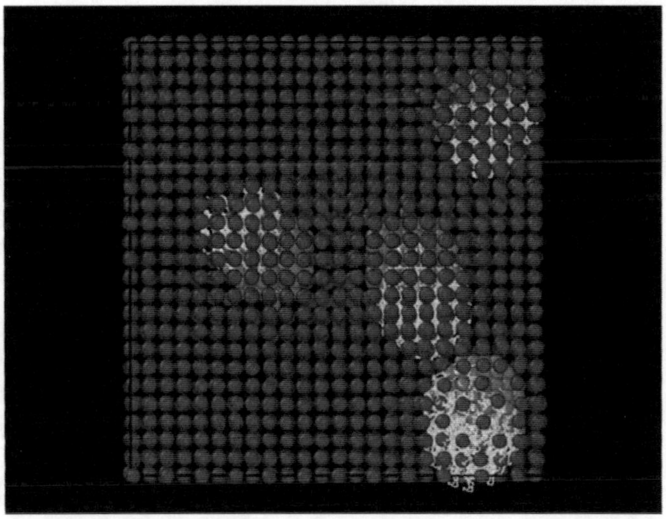

Fig. 4. Solder paste with doped nanoparticles with the mark of trajectory at temperature of 150°C.

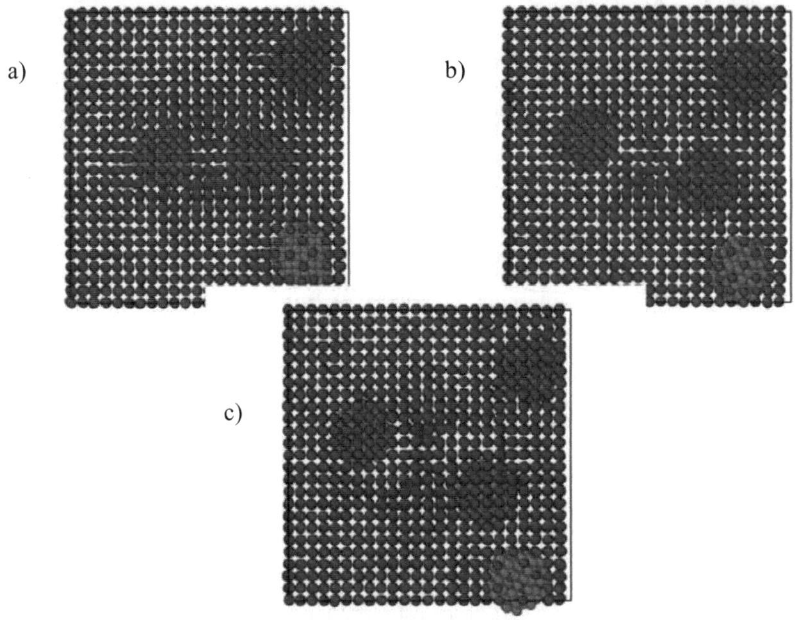

Fig. 5. Images of the position of the nanoparticles in the solder paste model at temperature of 150°C at a) 0 second, b) 10 seconds and c) 20 seconds in the x-y plane.

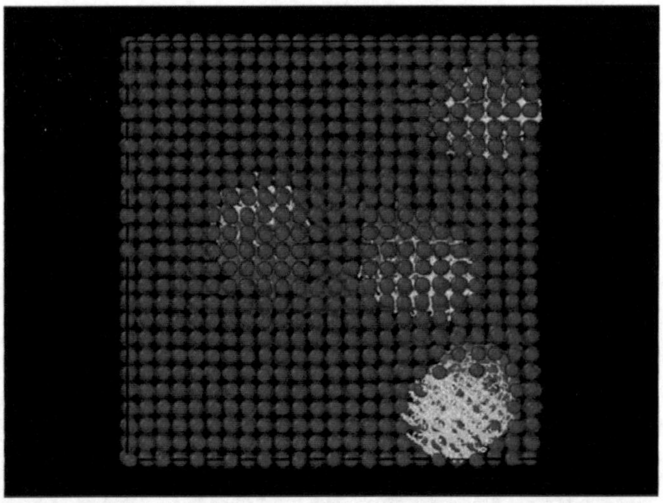

Fig. 6. Solder paste with doped nanoparticles with the mark of trajectory at temperature of 250°C.

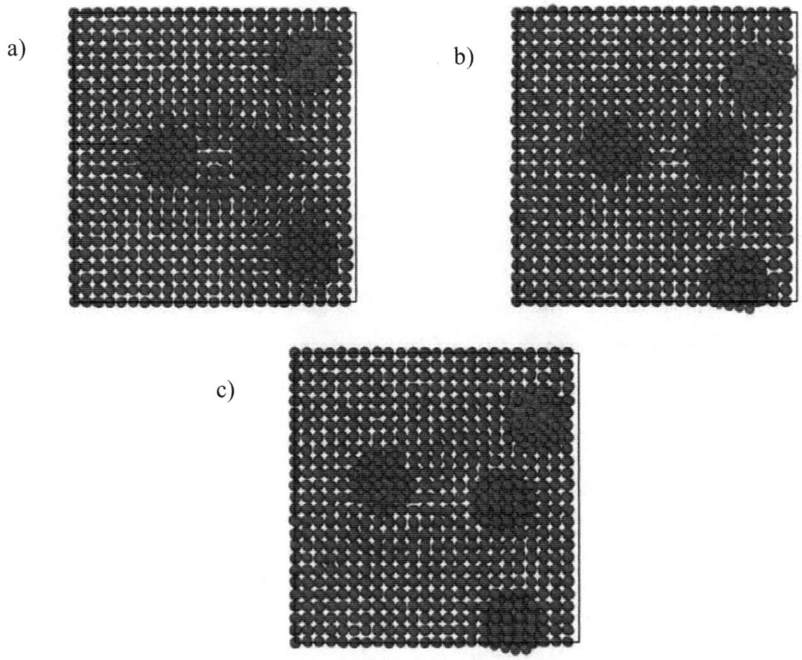

Fig. 7. Images of the position of the nanoparticles in the solder paste model at temperature of 250°C at a) 0 second, b) 10 seconds and c) 20 seconds in the x-y plane.

As shown in the Fig. 3., Fig. 5. and Fig. 7 a),b) and c) the position of the nanoparticles has changed compared to the original position at 0 second at all temperature. This shows that the nanoparticles has moved within the timestep. The collision of the fast moving atoms with the nanoparticles causing the random movement of the nanoparticles which called the Brownian motion.

4 Conclusion

The focus of this paper is to emphasize the steps to simulate the trajectory of the nanoparticles in the solder paste at different solder temperature and it is shown clearly with the use of LAMMPS software as the software focused on the simulation in the atomic scale. The nanoparticles moved according to the temperature in random directions. The most vigorous movement of nanoparticles is shown at the temperature of 250°C while the least vigorous movement is shown at room temperature of solder which is at 30°C.

References

1. Tay S. L., Haseeb A. S. M. A., Mohd Rafie J., *12th Electronics Packaging Technology Conference*, (2010)
2. Liu P., Gu X. T., *17th International Conference on Electronic Packaging Technology (ICEPT)*, (2016)
3. K. Bukat, M. Kos´cielski and J. Sitek, M. Jakubowska A. M. S., Solder Surf Mt Technol. **23(3),** (2012)

4 El-Daly AA, El-Taher A.M., Dalloul T.R., Mater Des. **55,** (2014)

5. Mui G.K., Wu X.-., Hu K.X, Yeh C., *Proceedings 47th Electronic Components and Technology Conference.*, (1997)

6. Chellvarajoo S., Abdullah M. Z., Samsudin Z., Mater Des (2017)

7. Gu Y., Zhao X., Li Y., Liu Y., Wang Y., Li Z., J Alloys Compd. **627**, (2015)

8. Li Z. L., Li G.Y., Li B., Cheng L.X., Huang J.H., Tang Y., J Alloys Compd, (2016)

9. Kanlayasiri K, Meesathien N. Adv. Mater. Sci. Eng., (2018).

10. Esfandyari A, Bachy B, Raithel S, Syed-Khaja A, Franke J., Procedia CIRP ,**62**, (2017)

11. Balázs I., Surf Mt Technol, (2014)

12. Lau C, Abdullah M.Z., Int J Comput Theory Eng. **5(4)**, (2013)

13. Abas A, Ishak M.H.H., Abdullah M.Z., Che Ani F, Khor S.F., Microelectron Reliab . **56**, (2016)

14. Haslinda M.S, Abas A, Che Ani F, Jalar A, Saad A.A., Abdullah M.Z. , Microelectron Reliab, **79**, (2017)

15. Dong H, Fan L, Moon K, Wong C.P., *Proc Electron Components Technol 2005 ECTC '05*, (2005)

16. LAMMPS Molecular Dynamics Simulator

The influence of graphite on conductivity, crystallinity and tensile properties of hydroxyethyl cellulose (hec) / graphite composite films

Mohd Pisal Mohd Hanif[1,*], *Abd. Jalil* Jalilah[1,2*], *Farah* Badrul[1] and *Syed Mahamud* Syarifah Nuraqmar[1]

[1]School of Materials Engineering, Universiti Malaysia Perlis (UniMAP), Kompleks Taman Muhibbah, Jejawi 2, 02600 Arau Perlis, Malaysia.
[2]Centre of Excellence Frontier Materials Research, Universiti Malaysia Perlis (UniMAP), Jalan Kangar - Alor Setar, Kampung Seriab, 01000 Kangar, Perlis, Malaysia.

Abstract. Conductive films based on natural polymers may find potential in food packaging and biomedical applications owing to their advantages of biodegradability and biocompatibility. This work demonstrated conductive composite films from hydroxyethyl cellulose (HEC) incorporated with graphite prepared by solution casting. Fourier Transform Infrared (FTIR) spectra confirmed the absence of covalent bond between the HEC matrix and graphite. The electrical conductivity of the composite films with various concentrations of graphite was investigated by a four-point probe. We observed that the conductivity of the HEC film increased remarkably from 10^{-8} S/cm to 10^{-5} S/cm upon addition of graphite. The highest conductivity of 9.44 x 10^{-5} S/cm was obtained for the film with 30 wt% of filler and further addition of filler reduced the conductivity. The correlation between conductivity and crystallinity was elucidated by X-ray diffraction (XRD) patterns. The difference in *d*-spacing values between HEC and HEC/Graphite films and the decrement of crystallinity percentage suggest that the rearrangement of ordered HEC crystalline structure was perturbed by the filler resulting in reduction of crystallinity phase. The tensile properties results clearly show that the incorporation of graphite into HEC films has reduced the tensile strength which is in agreement with the obtained electrical conductivity and crystallinity percentage.

1 Introduction

The development of conductive polymer composites (CPCs) based on natural polymers has generated considerable interest due to their advantages of renewability resources, biocompatibility and environmentally friendly. CPCs from non-conjugated polymers are made conductive by the incorporation of conductive fillers such as carbon black, carbon fibre and metallic particles [1]. With the advent of biosimilars in biomedical engineering, the

* Corresponding author: [1]hanifpisal@unimap.edu.my [2]jalilahjalil@unimap.edu.my

biopolymers are becoming an attractive alternative to replace the conventional thermoplastics that have limitation in biocompatibility. Conductive biopolymer composites may find importance in electrically induced drug delivery system, electrode coatings for neural sensing and stimulation, biosensors and etc. [2]. Besides that, conductive biopolymers may also be utilized as antistatic used in packaging films, containers and PET bottles to reduce attraction of dust [3].

Among natural polymers, cellulose is the most abundant and has excellent properties including flexibility and high mechanical strength [4]. Because of its lightweight and high thermal stability and also for being inexpensive, cellulose has also been a potential material for flexible sensors and electronic device applications [5]. Fortunato et al. [6] studied surface modification of a new flexible substrate based on hydroxypropyl cellulose for optoelectronic applications. Recently, carbon-based materials such as graphite, carbon nanotube, carbon fiber, and graphene that have high conductivity are mostly applied in the applications of lightweight and high-performance composite materials. Kim et al. [7] claimed that graphite is good as conductive filler for film because it has good thermal resistance besides its excellent electrical conductivity.

In this present work, we prepared conductive composite films from hydroxyethyl cellulose (HEC) incorporated with graphite as conductive filler by solution casting. The films were characterized by Fourier transform infrared (FTIR) spectroscopy. Electrical conductivity measurement for the composite films was carried out by four-point probe method at various graphite concentrations. The relationship between crystallinity and conductivity was elucidated by the crystallinity percentage obtained from X-ray diffraction (XRD) patterns. The influence of graphite on mechanical properties of the composite films was also investigated.

2 Experimental

2.1 Materials

HEC (cellulose ether = 92 %) was purchased from Merck KGaA (Darmstadt, Germany), graphite (carbon content = 80-98.5 %; particle size =63 μm or 250 microns), glycerol was supplied by HmbG Chemicals.

2.2 Samples Preparation

2 g of HEC was dissolved in 100 ml of distilled water. 20 wt% of glycerol was added as plasticizer followed by the addition of graphite powder. The mixture was stirred until homogeneous under magnetic stirring at room temperature for about 3 hours. Then, the solution mixture was cast into a glass mould and allowed to dry in oven for 6 hours at 50-60 °C. Thin films of 1 mm were obtained when the water evaporated. The amount of graphite was varied from 15-35 wt%. The formulations of HEC film and HEC/Graphite composite films with different loading of graphite are shown in Table 1.

Table 1. Formulations of HEC film and HEC/Graphite composite films with different loading of graphite.

Film Code	HEC (wt%)	Glycerol (wt%)	Graphite (wt%)
HEC Film	100	20	-
HEC/Graphite Composite Films	100	20	15, 20, 25, 30, 35

2.3 Measurements

FTIR Analysis Attenuated total reflectance (ATR-FTIR) spectra were recorded using Perkin Elmer Spectrum 400 FTIR with the wavenumber range of 650 – 4000 cm^{-1}.

Conductivity Conductivity measurement was carried out at room temperature by standard four-point probe method, using a Keithley Model 4200 Semiconductor Characterization System interfaced with a computer for automatic data collection. The samples were measured with voltage varied from 5 V to 10 V. The electrical resistivity ρ was determined using Equation 1,

$$\rho = R.\left(\frac{w \times t}{l}\right) \tag{1}$$

where; t is the sample thickness, R is the measured four-point resistance, w is the width of the sample and l is the length of the sample. The electrical conductivity σ was obtained by using Equation 2,

$$\sigma = \frac{1}{\rho} \tag{2}$$

Crystallinity X-ray diffraction (XRD) patterns were performed on a Bruker, D2 Phaser XRD Analyzer equipped with a monochromic CuKα ($\lambda = 1.5418$ Å) radiation. The data were collected at scan rate of 0.1°/min over an angular range from 10° to 60° 2θ in a continuous mode. The basal spacing (d) of the films was calculated based on crystal orientation ratio and the Bragg's Law formula as given in Equation 3 and 4,

$$I = {}^{I_2}/_{I_1} \tag{3}$$

where I is the crystal orientation ratio of the crystal planes, I_2 is the intensity peak number 2, and I_1 is the intensity peak number 1.

$$n\lambda = 2d \sin\theta \tag{4}$$

where is the distance between atoms, λ is the wavelength of the incident radiation, and θ is the incidence angle.

Mechanical Properties Tensile tests were carried out using Instron Universal Testing Machine Model 5569 based on ASTM D882 method to determine the tensile strength, percentage elongation at break and modulus of elasticity of the films. The crosshead speed was set at 5 mm/min. The measurements were replicated at least three times for each type of film.

3 Results and Discussion

Characterisation by FTIR Spectroscopy From the spectra of pure HEC shown in Fig. 1 (a), the strong and broad bands that correspond to hydroxyl group in the pyranose unit and C-O-C skeletal vibration appeared at 3334 and 1042 cm^{-1} [8,9], respectively. The band at 2886 cm^{-1} belongs to the C-H stretching due to the presence of the CH and CH_2 groups [10]. The absorption peak at 1643 cm^{-1} can be assigned to the bending mode of naturally absorbed water [8,9] and the peak at 1357 cm^{-1} is attributed to the O-H bending [8]. For the spectra of HEC film filled with graphite Fig. 1(b), the broad O-H stretching peak slightly shifted to 3341 cm^{-1} and broadened whereas the C-O-C band shifted to 1039 cm^{-1}. Other peaks were also slightly shifted and there was no any new peaks appeared. This indicates the absence of covalent or strong bond between graphite and the HEC matrix but the incorporation of graphite has somehow altered the environment of the composite films.

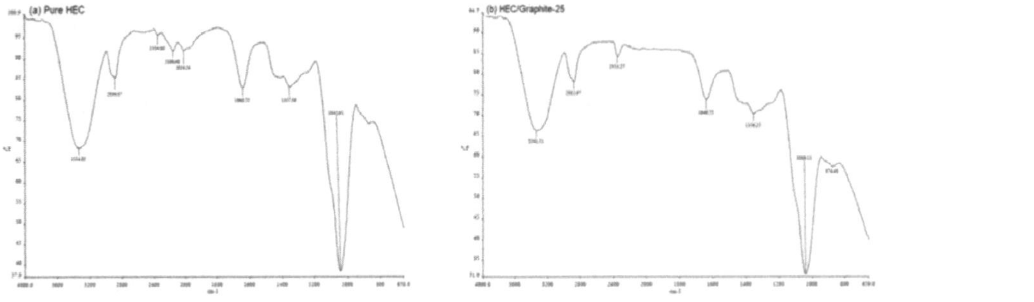

Fig 1. FTIR spectra of (a) Pure HEC film and (b) HEC/Graphite-25 composite film.

Conductivity The electrical conductivity calculated from the obtained resistivity is plotted in Fig. 2. Pure HEC was obviously insulative with the conductivity of 9.12 x 10^{-8} S/cm. From the graph, it is clearly seen that the conductivity increased two-order magnitude to 2.44 x 10^{-6} S/cm with the incorporation of graphite starting from 15 wt%. The value amplified at 30 wt% to 9.44 x 10^{-5} S/cm. Almost the same value was demonstrated by Hajar et al. from graphite filled polyethylene oxide/polyvinyl chloride polymer film at 25 wt% of graphite content [11]. However, further addition of graphite gave lower values. At lower graphite content, the composite film was dominated by the HEC matrix resulting in low electron mobility due to the absence of conductive interconnected network in the film. The interconnected network was built up as the amount of graphite reached the percolation threshold at 15 wt%. However, the amount of graphite reached saturation at 30 wt% and hence, the conductivity reduced above the saturation point.

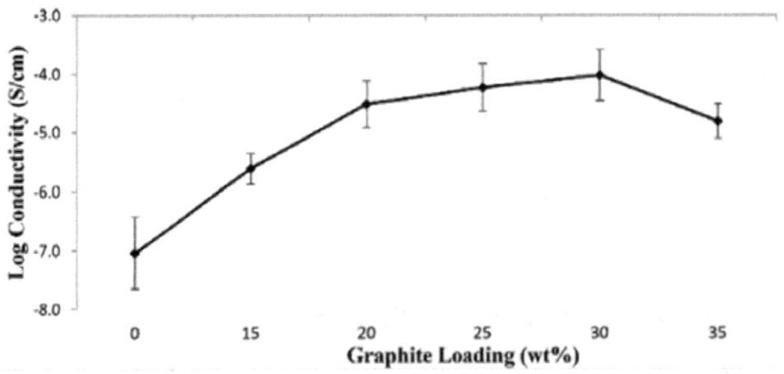

Fig. 2. Electrical conductivity with increasing concentration of graphite in HEC composite films.

X-Ray Diffractogram (XRD) Fig. 3 represents the XRD patterns of (a) pure HEC film and (b) HEC/graphite composite film with 30 wt% of graphite. The diffraction pattern of pure HEC film exhibits diffraction peaks at 2θ= 23.2°, 34.8° and 48.5°. The strong and sharp peak at 2θ= 23.2° depicts the semi-crytalline nature of HEC as reported by Gupta and Varshney [12]. For the XRD pattern of HEC/Graphite-30 film, a high intensity peak around 26.7° corresponding to the graphitic structure (002) [13] appeared in between the first and second peak of pure HEC. The shape of the HEC peaks remained unchanged but the intensity was slightly reduced. This explains that there was no strong interaction between the matrix and the filler but suggests that the ordered structure of HEC was slightly perturbed. The calculated *d*-spacing values and crystallinity percentage for pure HEC and HEC/Graphite-30 composite film are shown in Table 2. The change in *d*-spacing value when graphite is incorporated into HEC suggesting a rearrangement occurred in the composite film. Suppiah et al. [14] reported a reduction in *d*-spacing when halloysite nanotube (HNT) filler was increased in carboxymethyl cellulose (CMC). The decrement in crystallinity percentage

IOP Conf. Series: Materials Science and Engineering **701** (2019) 012015 doi:10.1088/1757-899X/701/1/012015

shows that the crytallinity reduced by the addition of graphite. The results also confirmed that the graphite filler was insoluble in the matrix.

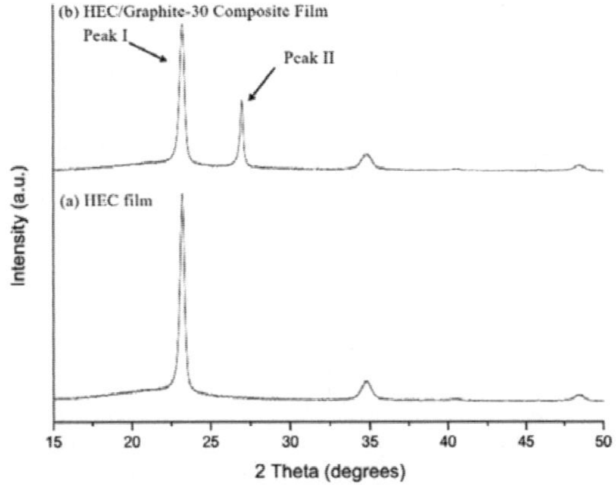

Fig. 3. XRD diffractograms of (a) HEC film and (b) HEC/Graphite-30 composite film.

Table 2. Crystal orientation ratio, interparticle spacing and crystallinity percentage (%) of HEC film and HEC/Graphite-30 composite film.

Composite code	Intensity (counts)			Interparticle spacing, *d* (nm)	Crystallinity percentage (%)
	I_1	I_2	I_2/I_1		
HEC	13,164	606	0.05	4.15	66.8
HEC/Graphite-30	9,424	4,747	0.50	3.15	64.7

Tensile Properties The tensile properties of the pure HEC and HEC/Graphite films are displayed in Fig 4. It can be seen that the tensile strength of the films were very much affected by the addition of graphite Fig. 4(a). Pure HEC shows tensile strength at around 6 MPa and the value decreased into half upon addition of graphite at 15 wt%. The tensile strength values gradually decreased with the increasing amount of graphite up to 35 wt%. This could be due to the interruption caused by graphite particles on the intermolecular interactions, namely, hydrogen bonding between the polymer chains in the HEC. The excellent mechanical strength in cellulose is attributed to the 3D structure formed by hydrogen bonding between cellulose chains [4]. As demonstrated by the FTIR spectra, there was no evident of covalent bonding formed between HEC and graphite. For the elongation at break as shown in Fig. 4 (b), the value increased from 33% to 47% in the presence of graphite at 15 wt%. However, the flexibility of the films filled with graphite reduced with the increment of graphite loading. This is because of the graphite characteristic which has high stiffness with high aspect ratio as compared to HEC matrix [15]. Nonetheless, the modulus of elasticity increased as the amount of graphite was increased. Among the five varied concentrations, the highest tensile modulus was seen for the film with 35 wt% of graphite. The correlation of modulus and crystallinity was not seen probably because the distribution may not be homogeneous as graphite content increases.

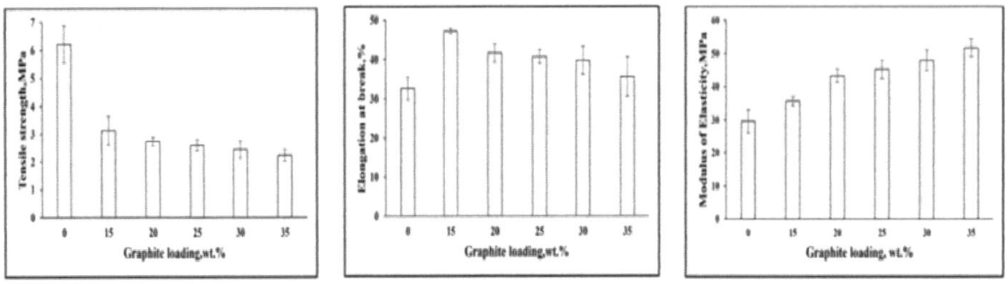

(a)Tensile strength (b)Elongation at break (c)Modulus of elasticity

Fig. 4. Tensile properties with increasing loading of graphite in HEC composite films.

4 Conclusions

We have successfully prepared free-standing films of HEC incorporated with graphite powder by solution casting. Strong interaction was not evident between the HEC matrix and graphite as there was no trace of new peak appeared nor any peaks significantly distorted in the FTIR profiles. The conductivity increased three-fold from 10^{-8} S/cm to 10^{-5} S/cm upon addition of graphite and optimized at 30 wt% of graphite content. The highest conductivity obtained was 9.44 x 10^{-5} S/cm. From the XRD results, we conclude that the incorporation of graphite has slightly altered the arrangement of highly-ordered crystalline structures of the HEC resulting in reduction of the crystallinity which promotes conductivity. The addition of graphite considerably affected the mechanical properties probably because the strong intermolecular interactions in the HEC matrix were destroyed by the graphite filler. Lower in tensile strength upon addition of graphite is associated with reduced crystallinity as proven by the XRD results. This research provides an environment friendly method to prepare conductive composite films. The selection of filler may boost the conductivity and may find applications that crucially require biocompatibility.

The authors gratefully acknowledge Mr. Faisal from the Centre of Excellence (COE) Frontier Materials Research of Universiti Malaysia Perlis (UniMAP) for the XRD measurement. We also thank everyone who directly or indirectly contributed to this research especially the lecturers and staffs of the School of Materials Engineering, UniMAP for completing this work.

References

1. J. Stejskal, *Chemical Papers,* **67**, (2013)
2. J. Jagur-Grodzinski, *e-Polymers,* **12**, (2012)
3. M. J. Kirwan and J. W. Strawbridge, *Food packaging technology,* **1**, (2003)
4. K. Gardner and J. Blackwell, *Biochimica et Biophysica Acta (BBA)-General Subjects,* **343**, (1974)
5. N. Yamazoe and Y. Shimizu, *Sensors and Actuators*, (1986)
6. E. Fortunato *et al.*, *Thin Solid Films,* **442**, (2003)
7. S. Kim *et al.*, *International Journal of Polymer Science*, (2019)
8. B. Orhan, C. A. Ziba, M. H. Morcali, and M. Dolaz, *Sustainable Environment Research,* **28**, (2018)
9. J. Sun, X. Sun, H. Zhao, and R. Sun, *Polymer degradation and stability,* **84**, (2004)
10. R. L. Oliveira *et al.*, *Journal of the Brazilian Chemical Society,* **26**, (2015)
11. M. Hajar, A. Supri, M. Hanif, and M. Yazid, in *Journal of Physics: Conference Series,* **908**, (2017)

12. S. Gupta and P. K. Varshney, *Ionics,***23**, (2017)
13. K. Mishra and S. Ramaprabhu, *AIP Advances,* **1**, (2011)
14. K. Suppiah, P. L. Teh, S. Husseinsyah, and R. Rahman, *Polymer Bulletin,***76**, (2019)
15. B. A. Alshammari, F. S. Al-Mubaddel, M. R. Karim, M. Hossain, A. S. Al-Mutairi, and A. N. Wilkinson, *Polymers,***11,** (2019)

Characterization of geopolymer ceramic reinforced Sn-0.7Cu composite solder: Effect of milling time and speed.

Marliza Mostapha[1,2*], *Mohd Arif Anuar* Mohd Salleh[1,2], *Mohd Mustafa Al Bakri* Abdullah[1,2], *Maiteli* Thiaghu[2], *Nur Syahirah* Mohamad Zaimi[1,2] and *Romisuhani* Ahmad[1,3]

[1]Centre of Excellence Geopolymer and Green Technology, School of Materials Engineering, Universiti Malaysia Perlis (UniMAP), Taman Muhibbah 02600, Jejawi, Arau, Perlis, Malaysia
[2]School of Materials Engineering, Universiti Malaysia Perlis (UniMAP), 01000 Jejawi, Perlis, Malaysia
[3]Faculty of Engineering Technology, Uniciti Alam Campus, Universiti Malaysia Perlis (UniMAP), Perlis, Malaysia

Abstract. The effect on the addition of optimally ball milled kaolin and fly ash geopolymer ceramic as reinforcement on the morphology and electrical properties of Sn-0.7Cu composite solder were explored. Geopolymer ceramics from fly ash class F and kaolin were prepared using geopolymer technology, milled at various speed and time. 1.0 wt. % of each sample were used to form composite solder via microwave sintered through powder metallurgy method. Structural characterization via SEM reveals that kaolin geopolymer ceramics has nano-sized subangular powder particles with larger amount of open porosity compared to fly ash geopolymer ceramics when milled at optimum speed and time. Four Point Probe test results showed a decreasing trend of electrical resistivity for kaolin geopolymer ceramics as the milling speeds and times increased. Overall, the results compared to electrical resistivity of other composite solder with various typical ceramic reinforcement additions, proves that kaolin geopolymer ceramics reinforcement to be the best option so far in term of morphology, electrical properties and its sustainable manufacturability.

1 Introduction

Geopolymers is an inorganic polymer that formed at low temperature, normally under 100 °C [1,2] and contain amorphous to semi-crystalline crystal structure which can be converted to crystalline ceramic phase through sintering [3]. Compared to conventional ceramic production, via sol gel or hydrothermal techniques, this method of producing ceramic through geopolymer starting material has improvised the material strength [4]. In geopolymer process, there is no release of bounded carbon dioxide since this process does not involve any calcination of calcium carbonate and the process does not need for facilities

* Corresponding author: marliza@unimap.edu.my

such as extreme high temperature kilns, with large outlay on fuel, plants and equipment. The mechanical properties of this geopolymer ceramic are similar to those of industrial ceramics. The availability of suitable raw materials from appropriate geological resources also utilization from industrial waste product as raw material can be considered as an advantage of this new ceramic production technology [5,6].

Fly ash and kaolin geopolymer besides its plus point of vast availability, its good processing workability, and improved mechanical properties in final product is the cause why is being studied for its potential application as ceramic precursor [4]. Nano-geopolymer on the other hand portrays quality of forming thermally-stable high-strength and close to final shape structures at room temperature. Also, nanometres sized crystallite novel amorphous and crystalline materials are produced. High-energy ball milling becomes handy for particle size reduction (comminution) down to the nanometres scale which corresponds to increased reactivity of nanostructured powders promoted by higher surface area [7]. Among high-energy ball mills; the planetary ball mill is a mechanically simple, low cost, versatile device for efficient grinding and suitable for any class of materials, and a larger quantity of powder can be produced easily [8]. According to several reference papers, particle size reduction prior to milling not just increases the reactivity, but also proved to contribute increases in compressive strength of the geopolymers [9–13]. Therefore, the feasibility of forming nanostructured ceramics from geopolymer and its effect on mechanical properties has been widely demonstrated, hitherto there are limited literatures available on incorporation of nano-sized kaolin and fly ash geopolymer ceramics into composite solder [14-17]. Little work has been conducted to investigate the comparative effect of addition of these ceramic reinforcements to the electrical properties of solder.

In this paper, two types of geopolymer ceramics which from fly ash and kaolin were milled for optimal time and speed. Both of this geopolymer ceramics were characterized in aspect of particle size and surface morphologies. The electrical resistivity of composite solder incorporated with two different types of geopolymer ceramics has been investigated by mechanical mixing of 1wt% of respective ceramic reinforcement with Sn-0.7Cu solder, compacted via powder metallurgy technique, microwave sintered and tested via four-point probe method. The SEM analysis is used to correlate the particles morphology, shape, size and porosity presence with the electrical resistivity of composite solder.

2 Experimental procedure

In this research, fly ash and kaolin were used as raw materials to form geopolymer ceramic. Fly ash supplied by Manjung Power Plant, Perak and kaolin supplied by Associated Kaolin Industries Sdn Bhd, Malaysia. The element composition of as-received fly ash and kaolin were given in Table 1. Sodium hydroxide (NaOH) pellet with 99% purity supplied by Formsoda Plastic Corporation, Taiwan and Sodium Silicate (Na_2SiO_3) solution supplied by South Pacific Chemical Industries Sdn Bhd, Malaysia. Distilled water used throughout the research. Sn0.7Cu were used as base lead-free solder supplied by Nihon Superior Co. Ltd., Japan.

Table 1. Elemental composition of as-received fly ash and kaolin.

Element (%)	Si	Al	Fe	Ca	K	Ti	Loss on ignition
Fly ash	25.8	14.6	5.3	2.9	1.4	1.3	0.6
Kaolin	26.23	14.71	4.60	-	5.03	1.36	0.99

NaOH solution with 12 M were used to prepare alkaline activator with Na_2SiO_3 /NaOH ratio of 0.24:1. The mixture of kaolin or fly ash, and alkaline activator with ratio of 1:1 were mixed together for a few minutes and cured at 80 ˚C for 24 hours. Kaolin and fly ash geopolymer were crushed and sieved into 150 μm and compacted at 5 tonnes for 5 minutes. The sample were heated at 1200 ˚C for 3 hours soaking time. Then, sample were crushed and sieved into 43 μm before ball milling process according to the parameter set up showed in Table 2 using stainless steel balls at 10:1 ball-to-powder ratio in dry environment. Monolithic Sn0.7Cu solder and its composites were prepared by powder metallurgy method through microwave sintered by mixed 1 wt.% of milled geopolymer ceramics with Sn0.7Cu solder in airtight container via planetary mill for 1 hour at 200 rpm. The sample were compacted in 12 mm diameter mould at 4.5 tonnes for 5 minutes and sintered through microwave sintering at 185 ˚C under ambient condition in 800W, 50Hz Panasonic microwave oven. The particle size of fly ash and kaolin geopolymer ceramic before and after milling were obtained using a Malvern particle size analyser. Scanning electron microscope (SEM) were used to study the changes on surface morphology before and after milled and four-point probe were used to analysed the electrical resistivity of monolithic Sn 0.7Cu and its composites.

Table 2. The ball milling parameter.

No.	Raw materials	Milling time (hours)	Milling speed (rpm)
1	Fly-ash geopolymer ceramic (FAGC)	5	0, 100, 300, 500
2	Kaolin geopolymer ceramic (KGC)		
3	Fly-ash geopolymer ceramic (FAGC)	0, 5, 7, 10	*optimum speed
4	Kaolin geopolymer ceramic (KGC)		

* Depend on the optimum speed that produced the smallest particle size.

3 Results and Discussions

3.1 Particle size and surface morphologies

The surface morphologies of fly-ash geopolymer ceramic (FGC) before and after milling are shown in Fig. 1 (a) and Fig. 2 (a, b, c, d), and kaolin geopolymer ceramics (KGC) before and after milling are shown in Fig. 1 (b) and Fig. 2 (e, f, g, h), respectively. It can be seen that FGC surface contain large amount of open micropores, irregular and nearly rounded shape whereas KGC surface contain small amount of open micropores that actually larger in size compare to FGC, irregular and angular in shape. The presence of many large particles in FGC powder before milling are shown in Fig. 2 (a). Once the milling speed increases, the reduction of FGC particle sizes also happens from average size of 43 μm down to 7.5 μm. The irregular and nearly rounded shape of FGC starts to become more angular, smaller and nearly in the same size.

Similarly, the same pattern was observed in KGC powder where the powder starts to decrease in size from average size of 43 μm down to 7 μm. The process of particle size reduction causes the morphological changes of FGC and KGC and this can be clearly observed in the SEM micrograph shown in Fig. 2. The plastic deformation started to increase as the milling speed increases and in turn increased the changes in particle morphologies. Further increase in milling speed lowered the ability of powder to withstand the deformation without fracturing [18,19]. Consequently, smaller particle size was formed as tabulated in Table 3. Thus, with the optimal speed of 500rpm that was observed and found to have the lowest particle size, the geopolymer ceramic powder further milled for 7 and 10 h at 500rpm to investigate the optimal milling time in particle size reduction.

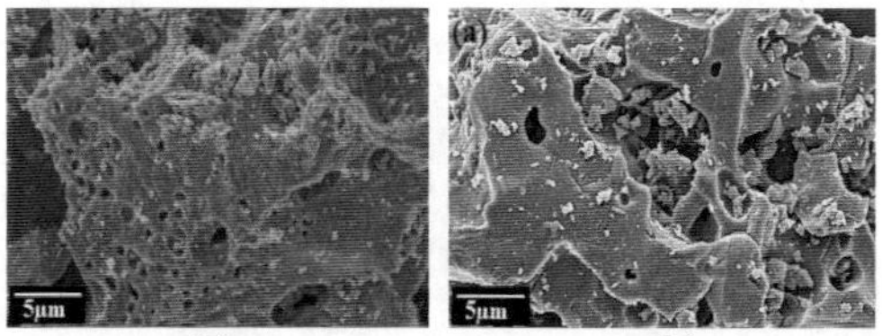

(a) **(b)**

Fig. 1. Surface morphologies of **(a)** FGC before milling and **(b)** KGC before milling.

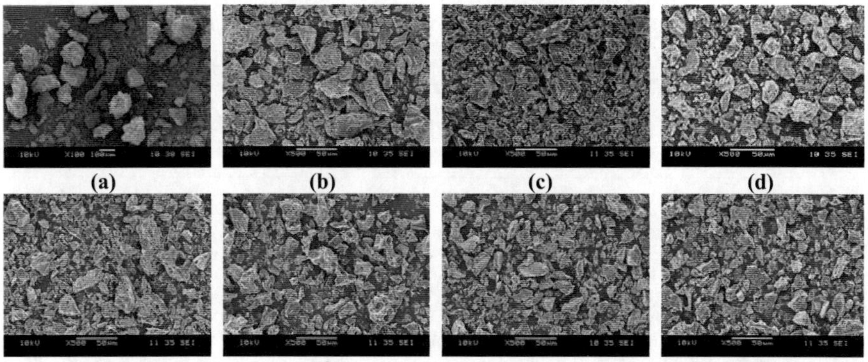

(a) **(b)** **(c)** **(d)**

(e) **(f)** **(g)** **(h)**

Fig. 2.Surface morphologies of geopolymer ceramic powders milled at 5 hours for different milling speed **(a)** FGC at 0 rpm, **(b)** FGC at 100 rpm, **(c)** FGC at 300 rpm, **(d)** FGC at 500 rpm, **(e)** KGC at 0 rpm, **(f)** KGC at 100 rpm, **(g)** KGC at 300 rpm and **(h)** KGC at 500 rpm.

Table 3. Particle size of KGC and FGC with different milling speed for 5 hours.

Speed of milling (rpm)	Particle size (KGC)	Particle size (FGC)
0	43 μm	43 μm
100	23 μm	29 μm
300	13 μm	16 μm
500	7 μm	7.5 μm

The effect of milling time on the particle size of FGC and KGC powder at 500 rpm was tabulated in Table 4. Increase of ball milling duration contributes to particle fineness indicating the extent of breakdown of structure [20]. When the milling time was increased up to 7 hours, particle size of FGC powder was drastically decreased while particle size of KGC was slightly decreased compared to FGC. This shows that FGC experienced major deformation and lost their strength to withstand the fracture effect that took place after 5 hours of milling. Fig. 5. (a) shows that the particle size become smaller and more angular for KGC powder while Fig. 5 (c) shows that there were small changes in size and shape of KGC powder compared to KGC powder milled for 5 hours. Further increased in milling time to 10 hours, particle size of FGC was increased significantly. This aligned with the increased in particle size shown in Fig. 5 (b). The FGC particles happen to be in cold-welding stage when milled for 10 hours.

During the process of milling, the geopolymer ceramic powder is subjected to three important process that are severe plastic deformation that leads to change in the size of particles, fracturing effect that cause the breaks of particles into smaller size and cold-welding process where this process cause the re-joining of the particles and thus increases the size of the particles [19]. For independent ceramic powder, the average particle size tends to be increased upon attaining steady-state equilibrium when balance between welding rates is achieved after milling for certain time interval [8]. At this point, the smaller FGC particles can withstand the deformation without further fracture and re-welded into larger particles [9].

In contrast, particle size of KGC was continue decreased. Reduction in size of particles larger than average and propagation through smaller particle cluster of trashes smaller than average results in finer particle size of KGC at this stage in agreement with Rao et al.'s work [20]. After 10 hours of milling time at 500 rpm, the KGC powder became sub angularly shaped with coarse surface morphology as shown in Fig. 5 (d) .

Table 4. Particle size of KGC and FGC with different milling time at 500 rpm.

Milling time (hours)	Particle size (KGC)	Particle size (FGC)
0	43 μm	43 μm
5	7 μm	7.5 μm
7	5.5 μm	0.97 μm
10	0.88 μm	6 μm

Fig. 3.Surface morphologies of geopolymer ceramic powders milled at 500 rpm for different milling time **(a)** FGC at 7 hours, **(b)** FGC at 10 hours, **(c)** KGC at 7 hours, **(d)** KGC at 10 hours.

3.2 Electrical resistivity of Sn0.7Cu solder composites

Good electrical conductivity is a key criterion for a solder which serves its function as electrical interconnection which allows current conduction through it. Thus, a low electrical resistivity of the material is expected for effective functioning of an electronic device. Presence, shape, volume, size and type of reinforcement and matrix are found to be a few factors that affect electrical resistivity of a composite material [21-26]. In this research, three of the factors; presence, type and size of reinforcement definitely affected the electrical resistivity of the composite solder.

Fig. 4 shows the effect on electrical resistivity of the composites solder when FGC and KGC prepared by milling at different time and speed were used as reinforcement material to form composite solder. Base on Fig. 4. (a), the lowest electrical resistivity was from composite solder incorporated with 1wt% of KGC milled at 500 rpm. A slightly similar low electrical resistivity value of 8.82 μΩ.cm obtained for composite solder incorporated with 1wt% of FGC powder milled at 500 rpm indicating 500 rpm as the optimal milling speed for the reinforcement material. Preceding the study in finding the optimal milling time, the result as shown in on Fig. 4. (b) suggests, 7 hours as the optimal milling time for FGC powder as the resistivity value tend to increase when the powder milled for period of time longer than 7 hours, whereas 10 hours is found to be the optimal milling time for KGC powder upon achieving the lowest electrical resistivity value of 2.78 μΩ.cm. These results

IOP Conf. Series: Materials Science and Engineering **701** (2019) 012016 doi:10.1088/1757-899X/701/1/012016

do not just directly relate to the size of the reinforcement, but also the volume of porosity in the reinforcement material itself [22].

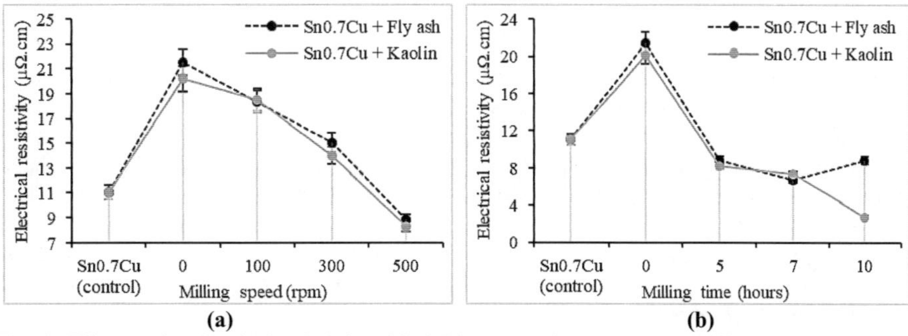

Fig. 4. Effect on the electrical resistivity of Sn0.7Cu composites solder with different **(a)** milling speed and **(b)** milling time.

Fig. 5 shows the comparison between presence of porosity in KGC particles and FGC particles. This micro porosity present in respective geopolymer ceramic particles largely influence the homogeneity of matrix and reinforcement as the porosity allows better bonding between reinforcing particles and solder thus enhancing the overall uniform distribution [27].

Based on Fig. 5, the obvious effect of porosity to electrical resistivity can be understood. Although practically in most cases, porosity is found to have very little effect on conductivity, but theoretically presence of open porosity found to increase the conductivity. This is possible if the electronically conducting species are adsorbed onto the surface of pores, proving additional conduction paths also said to act as active site provider [28]. Also, as per Hall Petch relationship, grain size increase causes removal of high resistance grain boundaries. This is the factor which contributes to electronic conduction of ceramics. KGC surface morphology as in Fig. 5. (b) shows a large number of open pores compared to FGC where this in turn became a path for above phenomenon to happened and consequently decreased the electrical resistivity of the composites solder. This finding is aligned with the trend that can be observed in Fig. 4. (b).

Due to above mentioned factor, many nano-sized porous ceramics have been used as reinforcement material in composite solder. Even so, addition of kaolin and fly ash geopolymer ceramic is found to be most effective not just in term of electrical resistivity, but also in term of sustainability. Production of geopolymer ceramic is a simple, low cost process that requires very low temperature sintering for ceramic formation compared to production of typical ceramics.

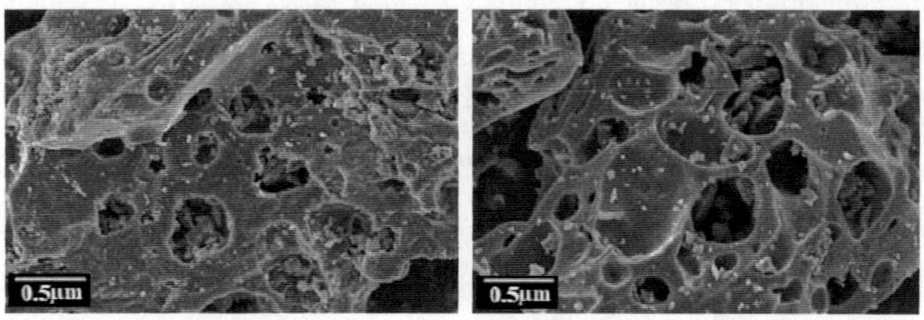

(a) (b)

Fig. 5. **(a)** FGC at 7 hours and **(b)** KGC milled at 10 hours.

4 Conclusion

In conclusions, the milling speed and the milling times are two predominant parameters that affects the size of particles and morphological changes in geopolymer ceramics. The optimal milling speed to produce nano-sized powder particles for both KGC and FGC is found to be 500 rpm, with different optimal milling time of 10 hours for KGC and 7 hours for FGC at the optimal milling speed. KGC is found to be composed of sub angular structured, irregularly shaped particles with larger amount of open porosity initiating from its plate like structure compared to FGC that have lesser particle size reduction, shape irregularity and open porosity which is the factors that support the resulting larger electrical resistivity of FGC compared to KGC in composite solder. Overall, comparing and considering the performance of KGC and FGC also other typical ceramics in aspect of morphology, electrical properties and sustainability, KGC is found be the best ceramic reinforcement for composite solder.

Acknowledgements

This work was supported by an Institutional Links grant, ID 332397914, under the Newton-Ungku Omar Fund partnership. The grant is funded by the UK Department of Business, Energy and Industrial Strategy (BEIS) and Malaysia and delivered by the British Council. The authors acknowledge the financial support through the Ministry of Education (MoE), Malaysia, under the Fundamental Research Grant Scheme (9003-00612) (FRGS/1/2017/TK10/UNIMAP/03/2).

References

1. Davidovits, J. J. Ceram. Sci. Technol. 8, 335–350 (2017).
2. Azimi. E.A, M. M. A. B. Abdullah, Y.M. Liew, C.Y. Heah., K. Hussin, I.H. Aziz, Matec Web of Conferences, 78, (2016)
3. Jaya, N. A. et al. Key Eng. Mater. 700, 3–11 (2016).
4. Jaya, N. A. et al. Appl. Mech. Mater. 754–755, 152–156 (2015).
5. Al Bakri, A. M. M. et al. Key Eng. Mater. 594–595, 575–580 (2013).
6. I. H. Aziz, M. M. A. B. Abdullah, C.Y. Heah, and Y.M. Liew, D.Panias, K.Sakkas. IOP Conference Series : Materials Science and Engineering, 209,(2017).
7. Abdullah, M. M. A. B., Ming, L. Y., Yong, H. C. & Tahir, M. F. M. Clay-Based Materials in Geopolymer Technology. in Cement Based Materials (InTech, 2018). doi:10.5772/intechopen.74438
8. Broseghini, M. et al. J. Eur. Ceram. Soc. 36, 2205–2212 (2016).
9. Sharma, P., Sharma, S. & Khanduja, D. Mater. Manuf. Process. 30, 1370–1376 (2015).
10. Assaedi, H., Shaikh, F. U. A. & Low, I. M. J. Asian Ceram. Soc. 4, 19–28 (2016).
11. Assi, L. N., Eddie Deaver, E. & Ziehl, P. Constr. Build. Mater. 167, 372–380 (2018).
12. Zhang, P. *et al. Appl. Sci.* **9**, (2019).
13. Zhang, L. & Tu, K. N. *Mater. Sci. Eng. R Reports* **82**, 1–32 (2014).
14. N.M. Nasir, N. Saud, M.N. Deraman, A.A.M. Salleh, M.I.I. Ramli, and R.M. Said, Materials Science Forum. **803** (2015)
15. M.I.I. Ramli, M.S.S. Yusof, M.A.A.M. Salleh, R.M. Said, and K. Nogita, Solid State Phenomena. **273** (2018)

16. F. Somidin, H. Maeno, M.M. Salleh, X.Q. Tran, S.D. McDonald, S. Matsumura, and K. Nogita, Materials Characterization. **138** (2018)
17. M.I.I. Ramli, M.A.A.M. Salleh, F.A.M. Sobri, P. Narayanan, K. Sweatman, and K. Nogita, Journal of Materials Science: Materials in Electronics. **30**(4) (2019)
18. Biyik, S. & Aydin, M. Acta Phys. Pol. A 127, 1255–1260 (2015).
19. Indexed, S., Vasamsetti, S., Dumpala, L. & Subbarao, V. V. 9, 1579–1589 (2018).
20. Rao, J., Narayanaswami, P. & Prasad, K. Int. J. Eng. Sci. Technol. 2, 284–299 (2010).
21. Allemang, R., De Clerck, J., Niezrecki, C. & Wicks, Conf. Proc. Soc. Exp. Mech. Ser. 45, 577–617 (2014).
22. M.A.A.M. Salleh, M.H. Hazizi, M.A.B.M. Abdullah, N. Noriman, R. Mayapan, and Z.A. Ahmad, Advanced Materials Research. **626** (2012)
23. K. Nogita, M.A.A.M. Salleh, S. Smith, Y. Q Wu, S.D. McDonald, A.G.A. Razak, T. Akaiwa, and T. Nishimura, 2017 International Conference on Electronics Packaging (ICEP). (2017)
24. K. Nogita, M.A.A.M. Salleh, E. Tanaka, G. Zeng, S.D. McDonald, and S. Matsumura, JOM. **68**(11) (2016)
25. M.A.A.M. Salleh, A.M.M.A. Bakri, F. Somidin, and H. Kamarudin, International Review of Mechanical Engineering. (2013)
26. M.A.A.M. Salleh, M.M.A.B. Abdullah, F. Somidin, A.V. Sandu, N. Saud, K. Hussin, S.D. McDonald, and K. Nogita, Revista de Chimie **64**(7) (2013)
27. Babaghorbani, P., Nai, S. M. L. & Gupta, M. J. Mater. Sci. Mater. Electron. 20, 571–576 (2009).
28. Dorey, R. A. (William Andrew, 2011).

IOP Conf. Series: Materials Science and Engineering **701** (2019) 012017 doi:10.1088/1757-899X/701/1/012017

Magnetic properties of TPU/NR blends: low loading NiZn and LNR added

Nor Azwin Ahad[1] *and Sahrim* Hj Ahmad[2]

[1]School of Materials Engineering, Kompleks Pusat Pengajian Jejawi 2, Universiti Malaysia Perlis 02600 Jejawi, Perlis, Malaysia.
[2]School of Applied Physic, Faculty of Science and Technology, Universiti Kebangsaan Malaysia, 43600 Bangi, Selangor, Malaysia.

Abstract. Thermoplastic polyurethane and natural rubber blending are able to change the origin properties polymer material. Then, this blend was added with magnetic filler, NiZn at low loading of 1wt% to 5wt%. Other than that, liquid natural rubber (LNR) was also added as compatibilizer. The previous study presented, NiZn ferrite at 3wt% in TPU/NR blend is the optimum filler loading, but LNR was not unbeaten in its role as compatibilizer. The addition of this magnetic filler enhanced the magnetic properties on saturation magnetization and remanence magnetization but decreased the value of coercive force.

1 Introduction

Modification in the behaviour and the improvement of the polymer properties has made it an ideal material for use in aerospace, transportation, packaging and food industries. The nature of the origin of the polymeric material is an electric resistor and nonmagnetic material. However, these properties may be changed by adding magnetic materials. For that purpose, NiZn were used in this study, which are blended together with TPU/NR. TPU are easily processing and make it possible to be blended with NR which provides low compounding temperature and good nature as polymeric material. Previous studies have been carried out, regarding TPU / NR formulations with NiZn and demonstrated on thermomechanical [1], DMA [2] and others. Besides, we studied also about tensile properties and swelling behaviour [3]. In this paper, we investigated the effect of adding low loading NiZn as filler and LNR as compatibilizer in the blends, to explore on their magnetic properties.

2 Methodology

Thermoplastic polyurethane (TPU) is a type of polyester-based 11T80 PEARLTHANE with A82 Shore hardness obtained from MERQUINSA Company. Natural rubber (NR) grade SMR-L with a density of 0.92g cm3 supplied by the Rubber Institute of Malaysia (RRIM). Liquid natural rubber (LNR) is available in research laboratories with natural rubber photo oxidation degradation method [3]. NiZn ferrite was used as filler supplied by Nanostructures & Amorphous Materials, Inc. USA, where its purity is 98.5% and particle size of 3 -10 nm.

IOP Conf. Series: Materials Science and Engineering **701** (2019) 012017 doi:10.1088/1757-899X/701/1/012017

As this study was carried out from the previous [1, 3], the samples were prepared by the optimal composition for TPU / NR and LNR i.e. 85TPU/15NR and 85TPU/5NR/10LNR [21], as well as a filler content of 1 to 5% by eight. The TPU/NR blends were prepared by melt blending technique using laboratory internal mixer (model RheoHaake 600p). Blending was carried out at temperature 180°C, with a mixing speed of 60rpm for 10 min and then pressed at 185°C under 45MPa of pressure for about 2 min using hot press (Carver Laboratory Press) into thin sheets.

The magnetic properties were measured using vibrating sample magnetometer (VSM model LDJ 9600) at room temperature. The measurements were carried out in a maximum field of 12kOe. Magnetic parameters such as

saturation magnetization (Ms), remanence (Mr) and coercivity (Hc) were determined.

3 Results and Discussion

The investigation of the magnetic properties of a material is important to understand the usefulness of the material. The sample is tested to reach a saturation state in which the material becomes a single domain in that state. The magnetic properties of TPU / NR and TPU / NR / LNR composites containing NiZn ferrite are explained using magnetic hysteresis loops versus magnetic fields. Fig. 1and Fig. 2 shows a hysteresis loop for magnetizing against the maximum magnetic field, which is applied in the magnetic field range of -12 kOe to 12 kOe at room temperature. Practically, all the samples have the same form of hysterisis loop. So, it can be concluded that the composite is soft magnetic at room temperature, due to the resulting hysterisis loop being narrow with little coercive force. Saturation magnetization (Ms), remenance magnetization (Mr), saturated magnetic field and coercive force (Hc) are magnetic properties parameters that can be obtained from magnetic hysterisis loop.

For all the samples, it was showed that the alignment of the magnet moment increases with an increase in the magnetic field. The magnetic moments in the sample align with the direction of the applied magnet field. The magnetization increases rapidly in a low magnetic field region before reaching a constant value held by the magnetizing magnet (Ms). When the magnetic field is amplified, Ms is found to increase with an increase in the filler in the matrix. Composite samples tend to follow the nature of ferrite if the filler content increases, because the magnetic nature of the filler will influence the composites.

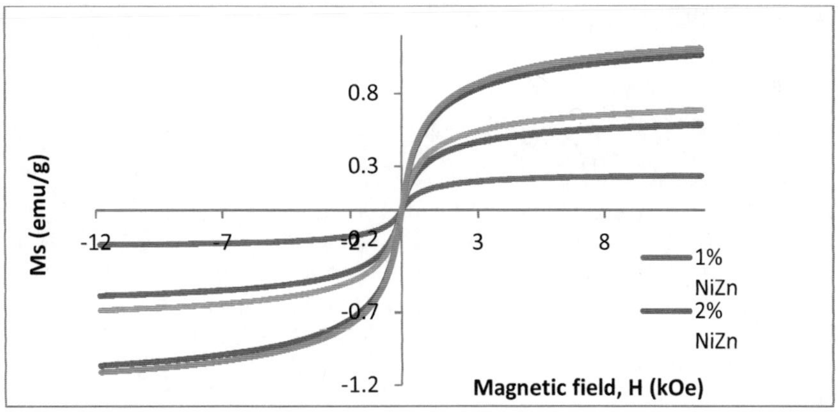

Fig. 1 The room temperature magnetization curves of TPU/NR blends.

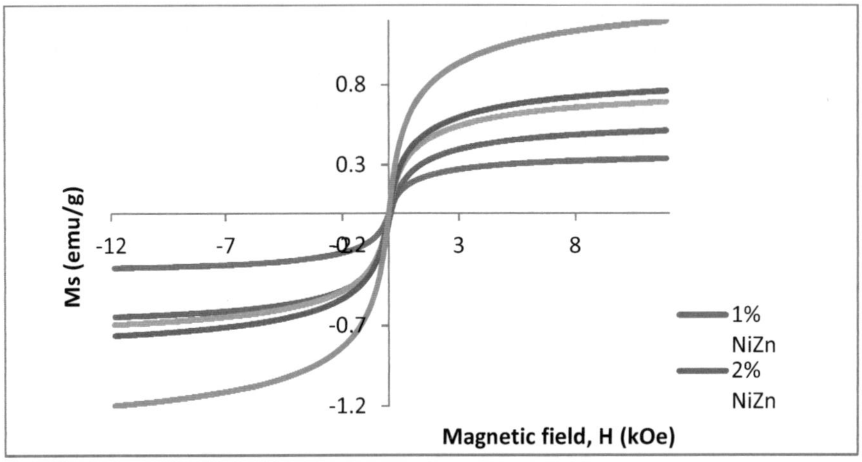

Fig. 2 The room temperature magnetization curves of TPU/NR/LNR blends.

Saturation magnetization (Mr) refers to magnetizing that remains in the magnetic material when the magnetic field is removed. It is a measure of the degree of magnetic resistance for a material, where saturation magnetization is the magnetic direction remaining in the magnetic material when the external magnetic field is removed. Increased saturation magnetization is characterized as a change in interaction between particles. The value of Ms and Mr for polymer magnetic composites increased with an increase in the magnetic powder loading in the polymer [4, 5]. The proportion of TPU / NR and TPU / NR / LNR composites increases linearly with the increase in Ms as shown in Fig. 3 and Fig. 4.

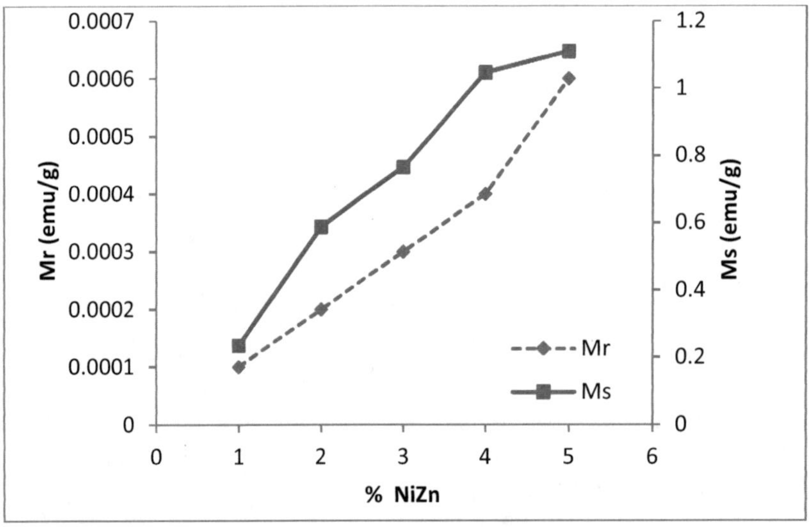

Fig. 3 Saturation magnetization and remenance magnetization of TPU/NR blends.

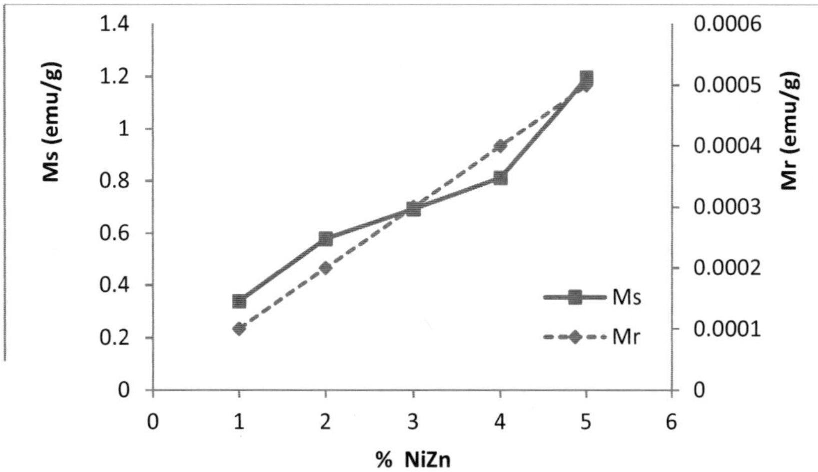

Fig. 4 Saturation magnetization and remenance magnetization of TPU/NR/LNR blends.

Ms value is very dependent on the effect of limited size. This impression includes changes in the average coordination number and the presence of an unbalanced spin rather than a break in symmetry at the border. Other characteristics are such as violence, relaxation and structural irregularities on the upper surface which increase the complexity of the filler system [6].

Saturation magnetization, remenance magnetization and magnetic field will be increases, when coercive force decreases with an increase in ferrite fillers [7, 8]. Coercive force is a forced force to return magnetizing of the material to zero. The coercive force can exhibit the sphere of ease of spin or magnetic moment for the filler to experience its original orientation towards the magnetic field in the matrix. If the saturation magnetization of the sample is retained, the magnetic moment will be forced to orient everything with a magnetic field. So that, the net magnetization becomes zero [6]. The orientation of a spin or magnetic fill moment is easy if the interaction of the magnetic moment between the ferrite filler is strong. This means, at a low fill composition, the initial orientation of the magnetic moment is predicted to be difficult because of the high obstacle or polymer matrix obstructions to the fill particle rotation.

According to Youyi [5] and El-Nashar [8], the value of Hc decreased with the increase in filler loading. This pattern may also be seen previously in figure above. They stated that, the decrease in Hc value when the load of NiZn ferrite fillers was added from 2% to 5% by weight was due to the increased magnetic interaction between NiZn ferrite. Yu [9] also stated in their study of natural rubber filled NiZn, that the value of Hc would be reduced at high filler load due to the increased magnetic interaction between in the matrix. However, he emphasized the high filler load when it exceeds 100% by weight. This phenomenon is also explained in the study of Ing [6] and Yu [9], where a uniform and random dispersion of fillers will reduce the average distance between the fillers in the polymer matrix and will add magnetic interactions.

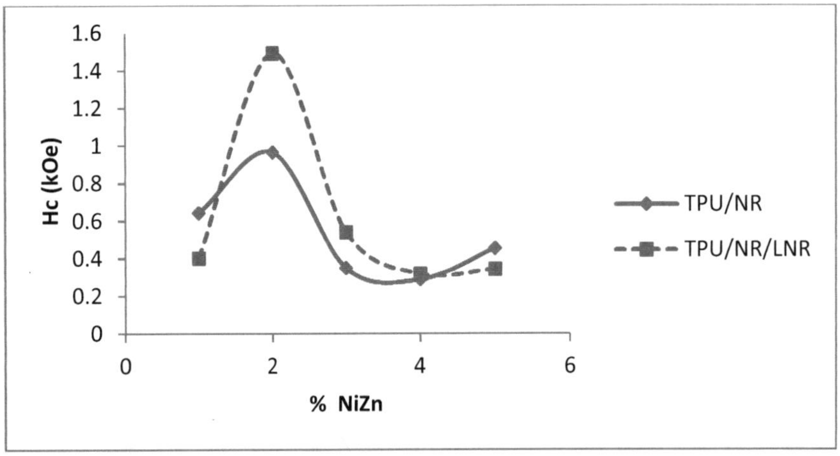

Fig. 5 Coercive forece of TPU/NR and TPU/NR/LNR blends.

If a comparison is made between the two types of composites in this study, it is found that both show the same graph plot style, but the graph plot for TPU / NR composites is lower. This shows, the filler scatter in this composite is better than the TPU / NR / LNR composite even though it can be seen, at 4% by weight of NiZn ferrite, the Hc values are almost the same and at 5% by weight, the TPU / NR / LNR composite is found to have lower Hc value. This may be due to the presence of LNR which can help in the spread of NiZn ferrite better than the matrix without LNR.

4 Conclusions

The saturation magnetization, Mr and remenance magnetization, Mr has increased. The other hand reduced the coercive force, Hc as parallel to the increasing of NiZn content. The magnetic filler should be well dispersed in the polymer matrix, thus will improve the properties of the blends. However, if the magnetic filler is more than the optimum percent, that can be represented by a matrix the filler is not well dispersed and tend to form agglomerates. The filler particles itself will tend to agglomerate due to the magnetic attraction between one another.

The authors would like to thank the Ministry of Higher Education Malaysia (MOHE) for the research fund under the Fundamental research Grant Scheme (FRGS) of the Project Code: FRGS/1/2018/TK05/UNIMAP/02/3).

References

1. N.A Ahad, & S.H Ahmad, J. of Sc. & Eng. Rsrch, 3(**4**) (2016)
2. N.A Ahad, N Parimin, & S.H. Ahmad, *Pro. Of Mech Eng Rsrch Day MERD2018* (2018)
3. N.A Ahad, S.H.Ahmad, & N.M. Zain, Adv Mat Rsrch, **795** (2013)
4. M.S Boon, W.P Șerena & M.Mariatti, J. of Mgntsm and Mgntc Mat., **5** (2012)
5. Y. Sun, X. Zhou, Y. Liu, G. Zhao & Y. Jiang, Mat Rsrch Bul., **45** (2010)

6. I. Kong, S.H Ahmad M.H Abdullah, D.Hui, N. Yusoff, & D.Purniyati, J. of Mgntsm and Mgntc Mat, **21** (2010)
7. N.A Nassim, Y.Atassi, A. Solloum, A. Charba & A. Maliki, Mat Chem & Phy, **211** (2018)
8. D.E El-Nashar, Mater Des., **52** (2013)
9. L.J. Yu, S.H Ahmad, I.Kong, M.A Tarawneh, S.B.B Abd Razak, E. Natarajah & C.K Ang, *De Tech* (2019)

IOP Conf. Series: Materials Science and Engineering **701** (2019) 012018 doi:10.1088/1757-899X/701/1/012018

Effect of solution concentration and applied voltage on electrospun polyacrylonitrile fibers

Sharifah Shahnaz Syed Bakar[1,*], *Kar Mun* Foong[1], *Norzilah* Abdul Halif[1] and *Shuhaida* Yahud[2]

[1]School of Materials Engineering, Universiti Malaysia Perlis, Pusat Pengajian Jejawi 2, Taman Muhibbah, 02600 Jejawi Perlis.
[2]School of Mechatronic Engineering, Universiti Malaysia Perlis, 02600 Kangar, Perlis, Malaysia

Abstract. Electrospinning is a simple and convenient method for fabrication of polymeric nanofibers from polymeric solution. In this paper, polyacrylonitrile (PAN) fibers are fabricated from different PAN solution concentration and applied voltages by electrospinning process. The fibers were then analyzed with Scanning Electron Microscopy (SEM), X-Ray Diffraction (XRD) and four-point probe to study the effect of PAN solution concentration and applied voltage on electrospun PAN fibers in terms of fiber diameters and morphologies, crystallinity and electrical properties. SEM showed that the electrospun PAN fibers with fiber diameter ranging from 0.5 to 2.7 μm. The diameter of fibers were increased with increasing PAN concentration and voltage applied. The increasing voltage also has increased the bead formation on the fibers. The amorphous structure of electrospun PAN fibers was exhibited by XRD. Electrical conductivity of electrospun PAN fibers were increased with increasing PAN concentration but somehow reduced with increasing voltage of the electrospinning.

1 Introduction

Nanofibers are nanostructured polymer fibers with fibers diameter ranging from micrometers to nanometers [1, 2]. Electrospinning was mostly employed to fabricate nanofibers with light weight, small diameters, long lengths, large surface area per volume, high porosity and tiny pore size [3, 4]. The properties and morphologies of electrospun nanofibers are significantly being influenced by numerous electrospinning processing parameters [4]. For instance, polymer (types, molecular weight, concentration), solvent (types, vapour pressure, diffusivity in air), additives (surfactants, salts), solution properties (surface tension, viscosity, rheological behavior, relaxation time, electrical conductivity, dielectric permittivity), applied voltage, volumetric flow rate, electrospinning time, spinneret orifice diameter, distance between spinneret and collector and ambient conditions (relative humidity, temperature) [3, 4].

Based on this study, polyacrylonitrile (PAN) is chosen as a raw material to produce the nanofibers via electrospinning since PAN is the most broadly used in electrospinning because of its superior properties [5]. PAN fibers are white in colour and non-conductive. It has excellent mechanical properties and always played an important role for biomaterials due to its outstanding biocompatibility.

*Corresponding author: shahnaz@unimap.edu.my

PAN fibers is a copolymer which consist of first monomer of acrylonitrile and the second monomer of vinyl acetate. In addition, PAN fiber possess excellent stretching strength and also high stretching strain [6]. PAN is the most ordinary among the various precursors for fabricating carbon nanofibers. The main reasons are high carbon yield and flexibility for tailoring the structure of the finishing carbon nanofiber products and the ease of achieving the stabilized products by the formation of a ladder structure through nitrile polymerization. Apart from this, PAN nanofibers owned desired properties such as high ionic transport, low diffusion resistance, high electrolyte uptake, good compatibility with electrodes, excellent chemical, thermal, mechanical stability and long life cycle [7].

Electrospinning also known as electrostatic fiber spinning is a simple and convenient method for the fabrication of polymeric nanofibers and nanocomposites from both natural and synthetic polymers [4, 8]. The main principle of electrospinning is to generate a charged jet of polymeric solution by applying an electric field [8]. A polymeric solution is accelerated in an electric field between a charged spinneret and a grounded collector since electrospinning is a fiber production method which uses electric force to form fibers from polymeric solution [4, 8]. When the applied electric field overcomes the surface tension of the droplet, a charged jet of polymeric solution is ejected. As the jet travels in the air, the solvent evaporates and a charged fiber is left behind which can be able to collect on a grounded collector. The jet grows longer and thinner and extends towards the grounded collector due to bending instability until it solidifies on the grounded collector [9].

Thus, electrospinning has been chosen for fabrication of nanofiber in this study. The effect of solution (in term of materials) and processing parameter via applied voltage during electrospinning on the fiber yield diameters and morphologies, crystallinity and electrical properties will be investigated.

2 Methodology

The experimental approach involved the raw materials, sample preparation, electrospinning set-up and characterization of electrospun PAN fibers. The characterization of electrospun PAN fibers were determined by using Scanning Electron Microscopy (SEM), X-Ray Diffraction (XRD) and four-point probe.

2.1 Raw materials

Polyacrylonitrile (PAN) and N, N-Dimethylformamide (DMF) are the raw materials that were used to prepare PAN solution. Polyacrylonitrile (PAN) with an average molecular weight of 150000 was purchased from Sigma-Aldrich and used without further purification. 99.8% of N, N-Dimethylformamide (DMF) was purchased from Avantor Performance Materials, was chosen as a solvent.

2.2 Sample preparation

PAN powder was dissolved in DMF to prepare PAN solution with different concentrations of 8 wt%, 10 wt% and 12 wt%. Then, the mixtures were stirred under constant magnetic stirring for 24 hours at room temperature prior electrospinning in order to obtain homogeneous solution.

5ml of syringe filled with PAN solution was placed horizontally by varying the concentrations (8 wt%, 10 wt% and 12 wt%). The syringe with a spinneret nozzle of 0.6mm in diameter was held with a distance of 9cm from spinneret to collector. In this experiment, the flow rate of PAN solution is fixed at 20 μL/min. A high voltage power supply which can generate up to 50 kV was connected to the syringe. A flat metal plate covered with aluminum foil was placed vertically serving as a grounded counter electrode. An electric field is formed between the spinneret nozzle and the grounded collector by applying voltage of 10 kV, 15 kV and 20 kV. The fibers that were charged onto the plate were then collected by using a pair of tweezers. The detailed information on electrospinning parameters of this experiment are shown in Table 1.

Table 1: The parameters of electrospinning process

PAN solution concentration (wt %)	Applied voltage (kV)	Flow rate of PAN solution (μL/min)	Collector distance (cm)
8	10	20	9
	15		
	20		
10	10		
	15		
	20		
12	10		
	15		
	20		

2.3 Characterization of electrospun PAN fibers

2.3.1 Scanning electron microscopy (SEM)

Scanning Electron Microscopy (SEM) (Model: JEOL JSM-6460LA) is employed to analyze the characterization of electrospun PAN fibers (sample). Firstly, electrospun PAN fibers were kept dried in a vacuum for 24 hours before it undergoes the characterization since the sample has to be prepared to withstand the vacuum condition and high energy electron beam. After that, the sample was placed on the sample holder and taped with double sided tape. The sample was coated with a thin gold layer by using sputter coater (Model: JFC 1600). The sample sputter-coated with gold having been prepared for viewing with an SEM. Then, high resolution images of electrospun PAN fibers surface were evaluated by an image analyzer (Sem Afore 5.0, JEOL).

2.3.2 X-Ray diffraction (XRD)

The crystalline phase and composition of electrospun PAN fibers are analyzed by using X-Ray Diffraction (XRD) (Model: XRD-6000, Shimadzu, Tokyo, Japan). The diffraction patterns were recorded using Cu Kα radiation over a 2θ range of 15° to 60°, a position-sensitive detector using a step size of 0.02° and scan rate of 10°/min. The sample holder was cleaned before put it into the chamber. The specimen was well-compacted and placed on the sample holder to obtain an optimal result. The analyzed data is then interpreted by using a software called HighScore Plus.

2.3.3 Four point probe test

The electrical conductivities of electrospun PAN fibers was measured with four-point probe. The four-point probe was typically used to measure the resistivity of sample in nanofabrication. The four equally spaced probes were linked with the sample of unknown resistivity. A high impedance current source is applied by passing a current through the two

IOP Conf. Series: Materials Science and Engineering **701** (2019) 012018 doi:10.1088/1757-899X/701/1/012018

outer probes while the voltage across the two inner probes is measured by a voltmeter to measure the sample resistivity.

3 RESULTS AND DISCUSSION

The results of the sample characterization including Scanning Electron Microscopy (SEM), X-Ray Diffraction (XRD) and Four Point Probe are discussed in the following section.

3.1 Scanning Electron Microscopy (SEM)

3.1.1 Effect of PAN Solution Concentration

Based on the Figure 1 (a) to (c), there was an increase in fiber diameter with higher polymer concentration. This statement was in agreement with the finding by Shao-Hua Wu et al. since the electrospun fibers produced were same with the conventional electrospinning results [11]. As described by Wang et al., the fiber diameter and flux of electrospun PAN nanofibers has been increased by elevating the PAN solution concentration [12]. As the concentration and viscosity of the electrospinning solution increases, the number of macromolecular chains as well as chain entanglements in the electrospinning solution also increase, while the size of Taylor cone keeps relatively constant. Therefore, the charged jet becomes thicker and longer, due to the higher viscosity resistance leading to the increase of fiber diameter [13,14]. The tendency of bead formation (marked with red circle) in Figure 1(a) were reduced when the PAN solution concentration increased as can be seen in Figure 1(b) and (c) where there are no bead was can be observed in these figures. This phenomenon is owing to the increase in the concentration that has resulting in larger fiber which then regulated the flow, thus reduced the formation of beads.

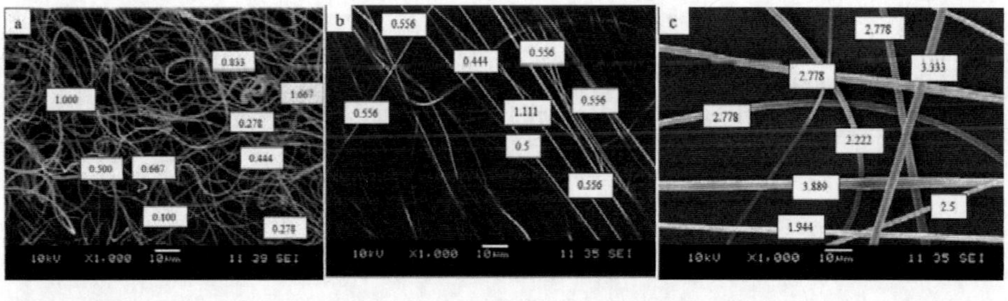

Fig. 1: SEM images of electrospun PAN fibers at 10 kV with different PAN solution concentration (a) 8 wt%, (b) 10 wt% and (c) 12 wt% at 1k magnification

3.1.2 Effect of Applied Voltage

Some researchers found that the PAN fiber diameter become larger with increasing applied voltage. The optimal charge density was 2.5 kV/cm and then further increasing the charge density would increase the diameter of electrospun PAN fibers has been reported by Ashraf and Hamid [15].

Normally, more fluid will be ejected in a jet at a higher applied voltage, thus resulting in a larger fiber diameter [16]. This statement has been agreed by Zhang and Demir et al. since they have proposed that there is more ejection of polymer when higher voltages were applied, causing the formation of fiber with larger diameter [1]. However, the higher the applied voltage, the higher the probability of bead formation. According to the Figure 2 (a) to (c), the beaded fibers (marked with red circle) were electrospun from 8 wt% of PAN solution with applied voltage increased from 10 to 20 kV. As can be seen, higher bead formation can be observed when as the applied voltage were increased. This was due to the increase the electrostatic repulsive force on the liquid jet which then increased the formation of beads on fibers.

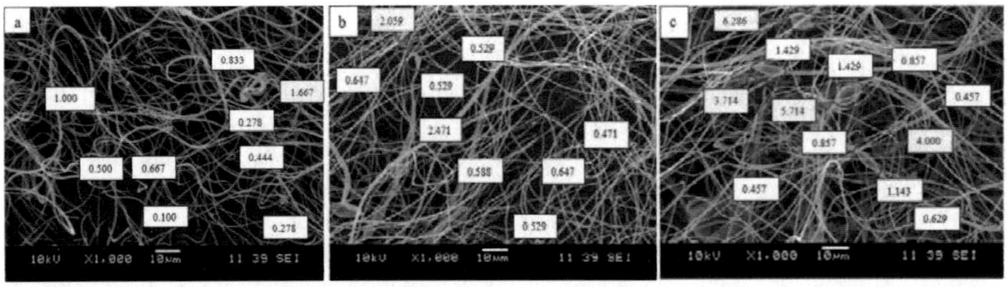

Fig. 2: SEM images of electrospun PAN fibers with 8 wt% of PAN solution at different applied voltage (a) 10 kV, (d) 15 kV and (g) 20 kV at 1k magnification

3.2 X-Ray Diffraction (XRD)

Figure 3 shows the XRD diffraction patterns of electrospun PAN fibers at 10kV with different concentrations. PAN is a semi-crystalline polymer [17]. All samples showed two equatorial peaks with a diffuse meridian peak (broad peak) that represents rather random molecular orientation. These data exhibited that the molecular arrangement in electrospun PAN fibers were mostly similar across all samples. Thus, equatorial peaks are normally found in PAN fibers. In fact, the molecules of PAN have a very short crystallization time and corresponding PAN in all the fibers mainly existed in an amorphous state with a small difference between the fibers [18, 19].

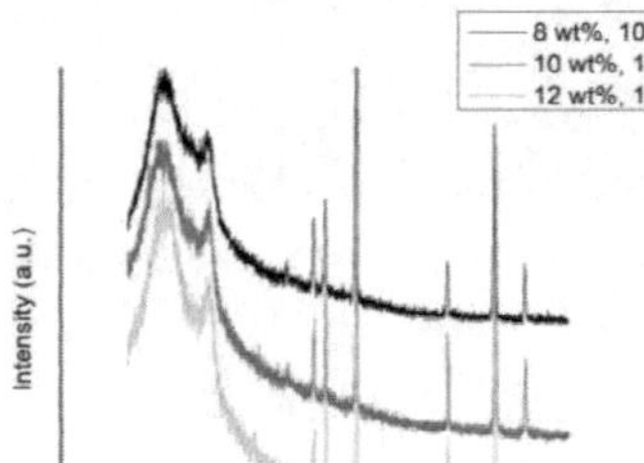

Fig. 3: XRD patterns of electrospun PAN fibers at 10 kV with different PAN concentration (a) 8 wt%, (b) 10 wt% and (c) 12 wt%

3.3 Four Point Probe Test

The electrical conductivity of electrospun PAN fibers with different parameters was shown in Table 2. The four-point probe test which is used for electrical conductivity was carried out at room temperature demonstrated that all electrospun PAN fibers formed have different conductivity values with different parameters. Based on the Figure 4, there was an increasing trend of electrical conductivity for the electrospun PAN fibers produced at applied voltage of 10 kV with different concentrations of 8 wt%, 10 wt% and 12 wt%.

Table 2: Effect of PAN concentration and voltage of electrospinning on electrical conductivity of electrospun PAN fibers

PAN Concentration (wt%)	Voltage (kV)	Electrical conductivity (S/mm)
8	10	0.00165
	15	0.00172
	20	0.00181
10	10	0.00181
	15	0.00175
	20	0.00174
12	10	0.00200
	15	0.00180
	20	0.00203

The electrical conductivity of electrospun PAN fibers increased from 0.00165 S/mm, 0.00181 S/mm to 0.002 S/mm when the PAN solution concentration increased from 8 wt%, 10 wt%, to 12 wt% while maintaining the applied voltage at 10 kV. The electrospun PAN fibers fabricated at applied voltage of 15 kV with increasing the concentration from 8 wt%, 10 wt% to 12 wt% was showed an increase in electrical conductivity. The electrical conductivity of electrospun PAN fibers increased from 0.00172 S/mm, 0.00175 S/mm, to 0.0018 S/mm when the PAN solution concentration increased from 8 wt%, 10 wt%, to 12 wt% while maintaining the applied voltage at 15 kV. Initially, there was a decreasing trend of electrical conductivity for the electrospun PAN fibers produced at applied voltage of 20 kV with 8 wt% and 10 wt% of PAN solution concentration, then it is start to increase gradually from 10 wt% to 12 wt%. The electrical conductivity of electrospun PAN fibers decreased from 0.00181 S/mm to 0.00174 S/mm and then increasing to 0.00203 S/mm when the PAN solution concentration increased from 8 wt%, 10 wt%, to 12 wt% while maintaining the applied voltage at 20 kV. The electrical conductivity of electrospun PAN fibers did not varies much. The small increased in conductivity of the yield fibers was due to the increasing fiber diameter with increasing PAN concentration and voltage applied during electrospinning process.

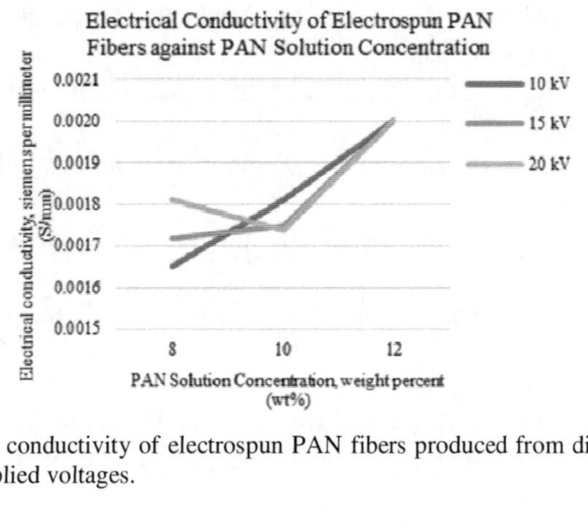

Fig. 4: The electrical conductivity of electrospun PAN fibers produced from different PAN solution concentration and applied voltages.

Conclusion

In this work, the effect of PAN solution concentration and applied voltage on electrospun PAN fibers in terms of the morphologies, crystallinity and electrical properties has been explored. It was found that the properties of electrospun PAN fibers fabricated is critically depending on the electrospinning parameters such as PAN solution concentration and applied voltage. With decreasing the PAN solution concentration, the fiber diameter decreases. When applied voltage increases, there is increasing in bead formation on electrospun PAN fibers has been observed. As analyzed by XRD, all electrospun PAN fibers mainly existed in an amorphous state without any changes with increasing PAN concentration or voltages. Electrical conductivity of electrospun PAN fibers were increased with increasing PAN concentration and voltage of the electrospinning.

The author would like to acknowledge and convey appreciation towards FRGS grant 9003-00400 Derivation of Carbon Nanowires from Electrospinning of Photoresist Polymer for supporting this work.

References

1. N. Bhardwaj, S.C. Kundu, Biotech. Adv. E **28**, 3 (2010)
2. K. Balagangadharan, S. Dhivya, N. Selvamurugan, Int. J. Bio. Macro. E **104** (2017)
3. C.J. Thompson, G.G. Chase, A.L. Yarin, D.H. Reneker, Polymer. E **48**, 23 (2007)
4. N. Ismail, F. Junior, Maksoud, N. Ghaddar, K. Ghali, A. Tehrani-Bagha, J. Aerosol Sc. E **113** (2017)
5. H. Ji-Huan, W. Yu-Qin, Y. Jian-Yong, Fibers and Polymers. E **9**, 2 (2008)
6. Y. Xia, Y. Lu, Polymer Composites. E **31,** 2 (2009)
7. X. Maa, P. Kollab, R. Yang, Z. Wang, Y. Zhao, A.L.Smirnova, H. Fong, Electrochimica Acta. E **236** (2017)
8. O.S. Yördem, M. Papila, Y.Z. Menceloğlu, Materials & Design. E **29**, 1 (2008).
9. S.Y. Gu, J. Ren, G.J. Vancso, Eur. Polymer J. E **41**, 11 (2005)
10. S.-H. Wu, X.-H. Qin, Materials Letters. E **106** (2013)
11. K. Yoon, K. Kim Xuefen, W. Dufei Fang, B.S. Hsiao, B. Chu, Polymer. E **47**, 7 (2006)
12. A. Shahram, T. Arash, M. Hamideh, C.A. Dominique, N. Babak, N. Parviz, Polymer Deg & Stab. E **142** (2017)
13. J. Sutasinpromprae, S. Jitjaicham, M. Nithitanakul, C. Meechaisue, P. Supapho, Polymer Int. E **55,** 8 (2006)
14. A. Eleyas, S. S. S. Bakar, N. Abdul Halif, S. Yahud, J. Built Environment, Tech. and Eng. E **2** (2017)
15. Z.-M. Huang, Y.-Z. Zhang, M. Kotaki, S. Ramakrishna, Composites Sc. and Tech. E **63,** 15 (2003)
16. S. Almuhamed, M. Bonne, N. Khenoussi, J. Brendle, L. Schacher, B. Lebeau, D. C. Adolphe, J. Ind. and Eng. Chem. E **35** (2016)
17. D.-G Yu, G.R. Williams, Li.-D. Gao, S.W. Annie Bligh, J.-He Yang, X. Wang, Colloids and Surfaces A: Physicochemical and Eng. Aspects. E **396** (2012)
18. E. Zussman, X. Chen, W.Ding, L. Calabri, D. A. Dikin, J. P. Quintana, R. S. Ruoff, Carbon. E **43**, 10 (2005)

IOP Conf. Series: Materials Science and Engineering **701** (2019) 012019 doi:10.1088/1757-899X/701/1/012019

Recovery of tin from copper dross by acid leaching

Nurul Ain Jabit[1,*], *Suhaina* Ismail[1], and *Muhammad Alif* Noor Azhar[1]

[1] Strategic Mineral Niche Area Group, School of Materials and Mineral Resources Engineering, Universiti Sains Malaysia, 14300 Penang, Malaysia

Abstract. Copper dross(CD) is a waste product formed during copper doping where sulphur was added to the tin metal during refining process to remove copper in the form of copper sulphide. CD was found to contain significant amount of tin and other valuable metal such as copper that could be trapped during copper doping process. The main objective of this work is to assess the potential of CD as a secondary resource for tin product and the applicability of hydrometallurgical process to extract tin and copper. Two main parts involves in this work are CD characterization and recovery of tin by acid leaching. From elemental composition analysis by XRF, in CD sample, amount of SnO_2 present is about 63% followed by SO_3 with 14%, Fe_2O_3 is 9% and CuO with 2.5%. Cassiterite and magnetite is the main phases found by XRD analysis. From the leaching study, it was found that almost 60% of tin has been dissolved in acid solution at condition of 80 °C, 2 M HCl with 8 hours leaching time.

1 Introduction

The increasing demand and resource depletion for tin in the world has required intensive studies for the recovery and extraction of this metal from different source. There are more than 50 kinds of tin bearing minerals on Earth; among them, cassiterite (SnO_2) is the only natural mineral from which tin metal can be economically extracted. According to the Minerals Commodity Summaries 2017 issued by Geological Survey of the United States, the proven global reserve of tin resources are 4.7 Mts, and they are mainly concentrated in China (110 kt), Indonesia (80 kt), Brazil (70 kt) and Malaysia (25 kt). Tin reserves was reported to decrease sharply from 9.6 Mts down to 4.7 Mts from the year 1999 to 2016. Since tin mine production cannot meet the demand for global tin consumption, tin-bearing secondary resources become important alternative sources for tin production. During the traditional tin extraction process, tin bearing by-products are generated with different tin contents which includes tin-bearing tailing, tin middlings, tin slag, hardhead and tin anode slime. In addition tin bearing solid waste, such as tin plates, tin alloys and electronic waste (E-waste) are also recyclable [1].

In this case, copper dross (CD) is the example of by-products from tin processing that is underutilize as when the copper dross were produced from the copper doping in refining process, it is treated as a waste or by-product. The product of CD has become a problem for the industry to store it and it also does not yield a convincing profit. Some of other industries

* Corresponding author: ainjabit@usm.my

will be interested to buy the copper dross at very low price. For that reason, the industry has made the effort to utilize the presence of copper dross but could not made a clear methodology in searching for the suitable parameters to extract the copper and also tin from the copper dross. There was also still lack of research or study on tin recovery from CD that has been reported particularly.

Apart from that, there are few published works on extraction of copper from slag [2], recovery of tin from tin soldering dross [3] and extraction of zinc and copper from brass slag [4]. All of these published works were using either hydrometallurgical or pyrometallurgical process for metal extraction from slag or dross. However, there are limitation where some of the metal oxides that formed during roasting or refining process will have resistant to the acid/lixiviants that most commonly used for hydrometallurgical treatment [2]. For that reason, an extensive work has to be done on development of tin extraction process from secondary resources that has been produced by tin industry particularly in Malaysia. In order to study the feasibility of tin recovery from CD, a few stage of research work has to be implemented such as characterization of CD, statistical evaluation for process optimization and development of mineral/metal processing circuit. In this work, the preliminary work has been initiated with characterization and leaching studied.

2 Experimental Work

2.1 Sample Preparation

10 kg of copper dross (CD) received from Malaysian Smelting Corporation (MSC) were used in this work. The cone and quartering method was performed on received sample until achieved the representative sample size. This representative sample then were crushed and ground to reduce the particle size until achieved 80% passing 75μm. Then the samples were divided into few sub sample for characterization analysis and leaching experiment.

2.2 Characterization Studies

The representative samples have been undergo characterization analysis such as X-Ray Diffraction (XRD) and X-Ray Fluorescence (XRF). The objective of characterization is to understand the composition of sample, percentage of elements, physical and chemical characteristics of the sample.

2.3 Leaching study

For leaching study, hydrochloric acid (HCl) as leaching reagents and millipore water was used throughout the experiments. Batch leaching test was carried out using 500 cm^3 glass reactor provided with a reflux condenser and magnetic stirrer. The desired volume of HCl of the required concentration was poured in the reactor and about 10 g of CD was added. The reactor was heated to the desired temperature using a thermostatically controlled water bath. The leachate were taken from the reactor using syringe at various time intervals and filtered through Whatman Cellulose Nitrate membrane filters with pore size of 0.45 μm. Leachate samples was analysed using Inductively coupled plasma atomic emission spectroscopy (ICP-AES) to determine tin content in solution.

IOP Conf. Series: Materials Science and Engineering **701** (2019) 012019 doi:10.1088/1757-899X/701/1/012019

3 Results and Discussion

3.1 Chemical compositions and Phases

Chemical analysis and X-ray diffraction data for CD sample are shown in Table 1 and Fig. 1 respectively. From Table 1 it is revealed that SnO_2 is the major component of the dross (62.7%) followed by SO_3 (14.3%), Fe_2O_3 (9.4%), As_2O_3 (6.01%) and CuO (2.5%). XRD characterization of CD indicated that the major crystalline phases are cassiterite (SnO_2), quarzt (SiO_2), corundum (Al_2O_3), chalcocite (CuS) and magnetite (Fe_3O_4). Due to complexity of the materials, thorough XRD analysis should be done to identify each of mineral phases that present in the materials.

Table 1. Chemical composition of CD sample at -75 μm before leaching.

Composition	Mass (%)
SnO_2	62.7
SO_3	14.3
Fe_2O_3	9.4
As_2O_3	6.01
CuO	2.5
SiO_2	2.1
Al_2O_3	1.7
MnO	0.02
Co_2O_3	0.32
NiO	0.033
ZnO	0.12

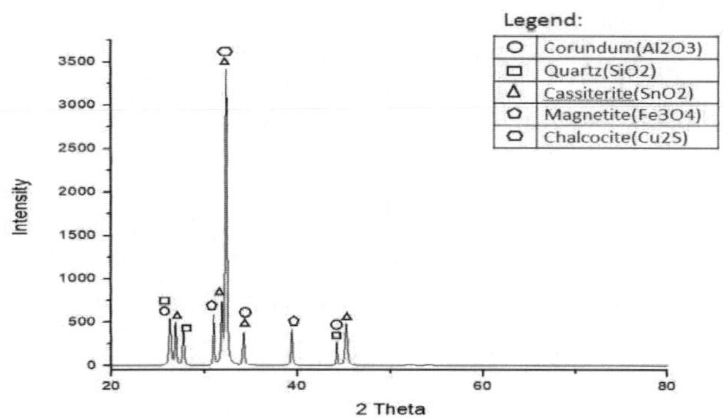

Fig. 1. X-ray diffractogram for CD sample.

The morphological surface of CD was also observed under SEM and analyzed using EDX as shown in Fig. 2. Different morphologies were observed on the sample revealed that the sample contain different types of minerals due to different shapes and variation of brightness. EDX analysis indicated that Sn and sulphur was the main elements consists in the samples.

IOP Conf. Series: Materials Science and Engineering **701** (2019) 012019 doi:10.1088/1757-899X/701/1/012019

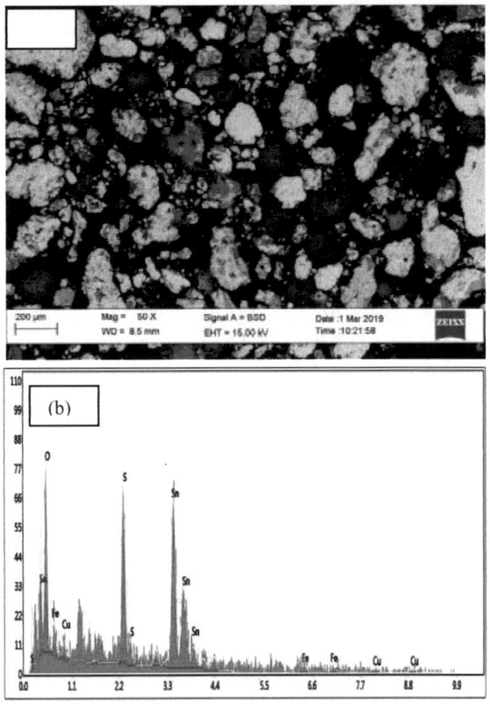

Fig. 2. SEM Micrograph (a) and EDX spectrum of CD sample (b).

3.2 Leaching

For the tin recovery from CD sample, leaching method using HCl as a leaching reagent has been accomplished. Parameters such as temperature (60 – 80°C), acid concentration (1.0 M -2M), stirring rate (350rpm) and leaching time (8 hours) was studied. It can be observed from the results obtained in Table 2, the recovery of tin depends strongly on acid concentration and leaching temperature. The highest recovery of tin (63%) achieved at 80°C in 2.0 M HCl. SnO_2 is reported to be refractory materials, therefore it is not soluble in acid or in bases [3]. This is explained why the recovery of Sn was low even at high leaching temperature. Therefore, CD maybe need to undergo pre-treatment process to convert insoluble SnO_2 to soluble Sn.

Table 2. Sn recovery (%) from selected leaching experiments.

Exp	T/(°C)	[HCl]/(M)	Stirring rate(rpm)	Leaching Time (hours)	Sn Recovery (%)
1	60	1.0	350	8	13.56
2	60	2.0	350	8	50.20
3	70	1.5	350	8	50.21
4	80	1.0	350	8	18.08
5	80	2.0	350	8	63.03

3.3 Proposed process flowsheet

A simplified process flowsheet for the recovery of tin from copper dross using roasting and acid leaching process is shown in Fig. 3. Copper dross sample can produced two different types of product which are tin oxide powder and tin metal through precipitation and electrowinning process. This work need further studies on solubility of Sn from CD in acid solution and also separation process.

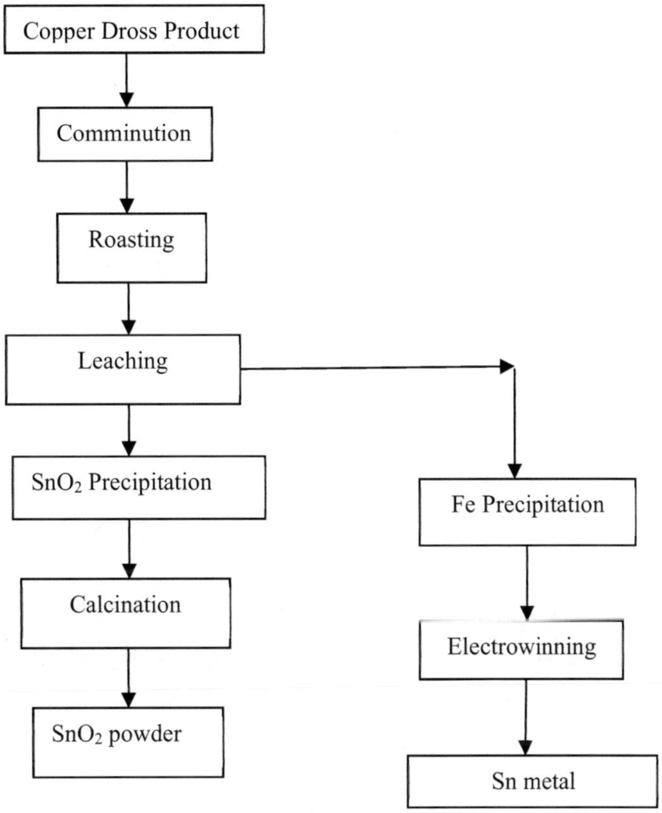

Fig.3. Proposed process flowsheet for SnO_2 and Sn metal product.

4 Conclusion

Copper dross is highly potential to become a secondary resources for tin industry. With significant amount of tin in CD samples it is crucial to recover tin from the waste product. It is also promote sustainable route for production of tin at the same time reduce the waste stockpiles. Successful extraction of tin will upgrade the application of this by-product and value added to the industry.

Acknowledgement- The authors are greatly appreciate to Tin Industry (R&D) Board, Ministry of Water, Land and Natural Resources and Malaysia Smelting Corporation (MSC) for providing materials and for the financial support of this project.

References

1. Z. Su, Y. Zhang, B. Liu. M. Lu, G. Li, T. Jiang, JOM 69, **11** (2017)
2. A. Mecke, I. Lee, J.R. Baker jr., M.M. Banaszak Holl, B.G. Orr, Eur. Phys. J. E **14**, 7 (2004)
3. M.D. Dimitrijevic, D. M. Urosevic, Z. D. Jankovic, Physicochem.Probl.Miner.Process ***52***. (2016)
4. Z. Harangi, T. Kékesi, Mater.Sci. Eng. **39**, (2014)

IOP Conf. Series: Materials Science and Engineering **701** (2019) 012020 doi:10.1088/1757-899X/701/1/012020

The appearance of machining burr as a cracking failure on twistlock pin

Wahid Mohamad Fitri Mohd[1*], *Shamsudin* Saiful Rizam[1], *Sanusi* Mohamad Syazwan, *Wardan* Rajaselan[1], *Rahmat* Azmi[1]

[1]Center of Excellence Geopolymer and Green Technology (CEGeoGTech), School of Materials Engineering, Universiti Malaysia Perlis, 02600, Arau, Perlis, Malaysia

Abstract. The failure analysis of the twist lock pin made of alloy steel is presented. The pin was visually observed to have a cracking character at curvature area after 45,000 cycles. Some non-destructive inspections were previously done. However it failed to detect any crack defect. The twist lock pin sample was further macro and microscopic analysis. The analysis found that the surface discontinuities that were previously presumed as a cracking morphology was actually a machining burr and improper machining marks. There were some micro- and nano-sized pinholes dispersed over the metal matrix of the twistlock pin sample. The microstructure of the twistlock pin was tempered martensite structure.

1 Introduction

Twistlock is a standard fixing device, which locates in a corner casting, for securing container or a swap body to a road vehicle, rail wagon or attached to spreader for top lifting of containers. Generally, the semi-automatic twistlocks are comprised of a single-piece steel shaft (twistlock pin) encased in a steel housing. The shaft has an upper and a lower cone that are oriented in different directions and fit inside the corner fittings.

Twist lock pin is generally made of 36CrNiMo6 steel, which is equivalent to AISI 4340 alloy steel. This alloy is known for its toughness, ability to attain high strengths in the heat-treated condition and has very good fatigue resistance. Air-cooling from autenitizing temperature produces a bainitic structure. Oil quenching from autenitizing produces a martensitic structure with some possible retained austenite. The martensite of this steel consists of many laths of about the same orientation within packets.

Twistlock and its pin should have high tensile and shear strengths, high fatigue resistance. It carries concentrated load either in transverse or longitudinal directions. Proper maintenance and periodically check on the twistlock are important to avoid any sudden accident.

In July 2015, an incident occurred where several rail wagons were collided from container freight train. Investigation found that the twistlocks were in poor condition to be able for service [1]. Almost a year ago, 5-containers stack collapsed and crashed down from container ship onto deck of bunker tanker, moored alongside container ship.

* Corresponding author: fitri@unimap.edu.my

IOP Conf. Series: Materials Science and Engineering **701** (2019) 012020 doi:10.1088/1757-899X/701/1/012020

Twistlock failure is believed to be the cause of an accident, 40-foot container was to be lifted from container ship, but it didn't go off smoothly and destabilized container below [2]. Burr forms during machining process. It involves large plastic deformation of work material. The type of burrs and their characteristics depend on the type of machining process, the process parameters, tool property, tool geometry, tool edge configuration, coolant, and workpiece material properties [3].

This paper presents the investigation of twistlock pin, which visually observed to have a cracking characteristic after 45,000 cycles.

2 Materials and method

As-received sample was cleaned to remove rust and some residual coating on the sample surface. The visual inspections were documented by D3100 Nikon digital DSLR camera to record the original condition of the sample (Figure 1). The sample was sectioned into 2 workpieces for chemical composition and metallographic analysis, respectively. The chemical composition was analyzed by using optical emission spectrometer, OES (Q8 Magellan, Bruker).

Fig. 1. As-reveiced twistlock .

Prior to the analysis, the metal surface was ground using 80 grit SiC paper, and the chemical composition was examined at 5 different spots on the metal surface. The macroscopic analysis was done by stereomicroscope (SZX 16, Olympus). The samples were ground using 180, 240, 320, 440, 800 & 1200 SiC grit papers. Then, the samples were polished using 6- and 1-micron diamond solution. As-polished surface were examined to study the presence of defect or surface discontinuity such as cracks and pinholes. The as-polished samples were etched using 15 ml HNO_3 + 5 ml methanol solution to reveal the microstructural pattern on the samples. All the metallographic analyses were carried out by metallurgical microscope (BX41, Olympus).

3 Results and discussion

Quantitative analysis was done to inspect whether the material meets the specified composition limits of 36CrNiMo6 steel (Table 1). The nickel (Ni) element was found slightly below the range, thus the twist lock pin materials are not in compliance with 34CrNiMo6 steel. The lack of Ni content relatively reduces its special properties such as elastic limits, hardenability, fatigue and impact resistance.

Table 1 Comparison between standard composition [4] and twistlock composition.

Element	36CrNiMo6 steel (%)	Twist lock pin (%)
C	0.30 - 0.38	0.371

IOP Conf. Series: Materials Science and Engineering **701** (2019) 012020 doi:10.1088/1757-899X/701/1/012020

Si	≤ 0.40	0.298
Mn	0.50 - 0.80	0.596
P	≤ 0.025	0.015
S	≤ 0.035	0.024
Cr	1.30 – 1.70	1.538
Mo	0.15 – 0.30	0.153
Ni	1.30 – 1.70	1.148
Fe	bal.	bal.

B

(a) (b)

A

Fig. 2. Camera photographs of (a) as-cleaned sample, show the focusing spot; (b) along curvatures area A-B.

From visual inspection performed on sample, it is observed that the line morphology along the curvature area A seems to have a cracking character on it (Fig. 2(b)). The cracking characters clearly appear to occur along the curved area as depicted by the red marked arrows. The visual appearance of such morphology like-crack creates anxiety to the user as it is inevitable will cause a catastrophic accident to happen.

The observation is focused on the cracking character adjacent to the curvature area, as the stress concentration factor is known geometrically increases here [5,6]. The line morphology closes to the curvature surfaces of A (Fig. 2(b)) shows the surface discontinuities as indicated by red arrow in Fig. 3. The surface discontinuities were clearly observed. However, macroscopically observation on B (not shown here) is believed not a cracking morphology. It is noticed that it just

Fig. 3. Stereomicrograph image at curvature surface.

a scratchy surface affected by the operation service condition. Therefore, the deficiencies observed in the curvature of area B are non-cracking morphology. The preliminary assumption by the user can be ruled out. The above stereomicroscope image, still inadequate to complete the investigation, therefore this twist lock was cross-sectioned to detect any crack presence in the sample. The suspected surface discontinuities were detected as shown by the red arrow bar (Fig. 4). There was no indication of crack detected at both cross-sectional curvatures areas. As-polished surface from the discontinuities that were previously presumed as a cracking morphology were actually a machining burr and improper machining mark as shown in Fig. 4. It was also observed that several pinholes scattered in the metal matrix near the curvature area.

As-polished metallographic images for metal matrix are shown in Fig. 5. There were some micro- and nano-sized pinholes dispersed over the metal matrix of the twisted pin sample. The presence of pinholes imperfection was due to many factors, some of them were attributed from the gas dissolved and/ or not properly degassed during steel making processes. The presence of pinholes is probably contributed to weaken the overall mechanical properties of steel.

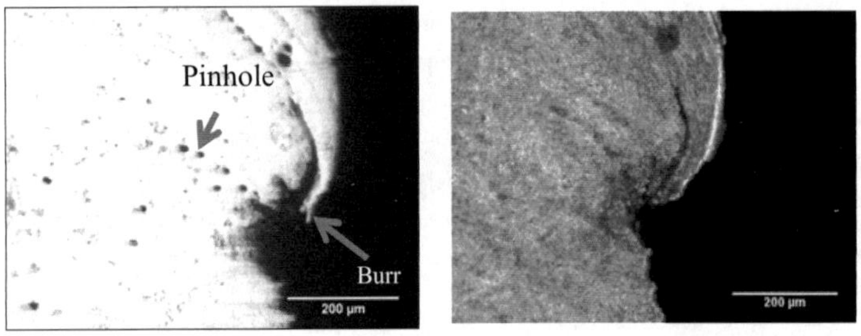

(a) (b)

Fig. 4. Metallographs of surface discontinuities of (a) as-polished (b) etched curvature area.

As-polished samples were etched using HNO_3^+ methanol to reveal its microstructure as shown in Fig. 6. Generally, the cross-sectional area that is close to the surface has a fully fine tempered martensite structure while forming bainite away from the surface.

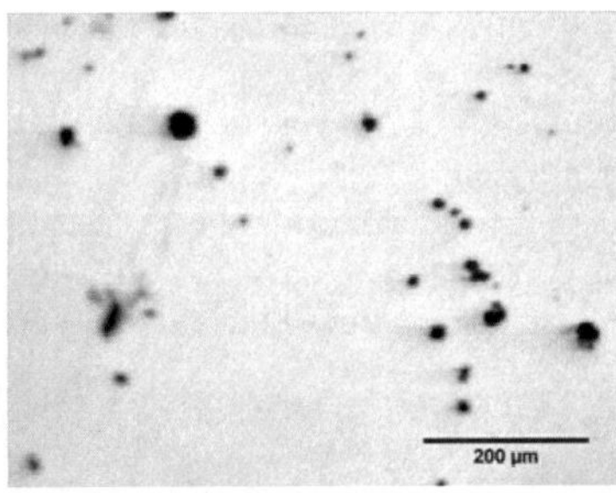

Fig. 5. As-polished micrograph image 10 mm from the curvature area.

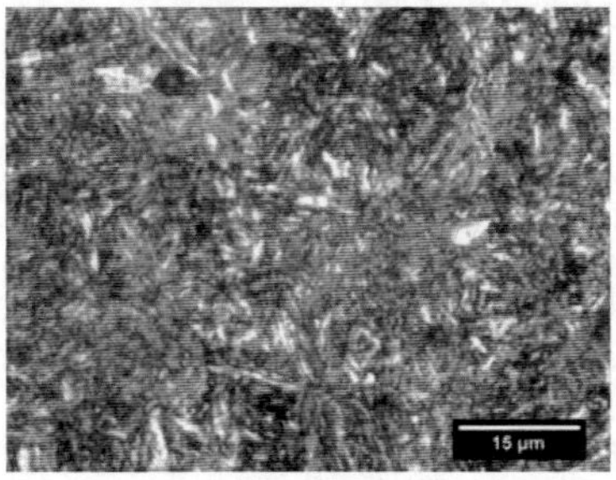

Fig. 6. Optical micrograph 10 mm from curvature surface.

4 Conclusion

The twist lock pins material is not in compliance to 34CrNiMo6 steel as Ni element was found slightly below the standard range of chemical composition. This is relatively reducing its special properties such as elastic limits, hardenability, impact and fatigue resistance. Upon observation there was no indication of cracking on the twist lock pins sample. There were surface discontinuities, especially on curvature area A. However, the surface discontinuities that were previously presumed as a cracking morphology were actually a machining burr. There are some micro- and nano-sized pinholes dispersed over the metal matrix of the twisted pin sample. The imperfection was originated during the steel making processes. The microstructure of twist lock pins was a tempered martensite structure.

The authors would like to thank School of Materials Engineering, Universiti Malaysia Perlis for giving the opportunity to perform field investigation on the above matter.

References

1. Australia Transport Safety Bureau, ATSB Transport Safety Report, Rail Occurrence Investigation, (2016)
2. https://www.heavyliftnews.com/containers-from-cscl-container-ship-fell-onto-tanker/
3. I. A. Choudhury, S. A. Lawal, Comprehensive Mater. Process., **11**, (2014)
4. ASM International: Properties and Selection: Irons, Steels, and High-Performance Alloys, ASM Handbook, **1**, (1990)
5. Mari Åman, Yuzo Tanaka, Yukitaka Murakami, Heikki Remes, Gary Marquis, Procedia Structural Integrity, **7**, (2017)
6. C. W. Springfield, H. Y. Jung, Engng. Fracture Mech., **31**, (1988)

Removal of iron in rice husk via oxalic acid leaching process

Faizul Che Pa[1,*], and *Wong* Kok Kein[1]

[1] Centre of Excellence Geopolymer & Green Technology, School of Material Engineering, Universiti Malaysia Perlis, 02600 Jejawi, Perlis, Malaysia

Abstract. Rice husk (RH), an agricultural waste which has been used as an alternative source for silica (SiO_2). In order to produce high purity amorphous SiO_2 from RH, acid leaching pre-treatment was required to remove the metallic impurities, followed by calcination at 800 °C to obtain the ashes. This research was studied about using oxalic acid as the leaching agent to remove iron (Fe) content in RH. The experimental parameters used in this study were acid concentration, reaction temperature, and leaching time. Based on the chemical composition analysis, the highest SiO_2 content of 99.30% and the lowest Fe_2O_3 content of 0.43% were to be found in RHA which had treated under optimum parameters of 1.0 M of oxalic acid concentration, 90 °C of reaction temperature and 3 hours of leaching time. In addition, the particles of TRHA were observed with a tubular porous structure and irregular shape due to the leaching effect. Phase analysis by x-ray diffraction show that the amorphous SiO_2 was confirmed as a broad diffused peak. It's been found in the 2θ angle range of 18° to 30° with no crystalline peak observed.

1 Introduction

Rice husk was the by-product from paddy field and it was yielded about 20% from the rice production [1]. Although the rice production industry was started for the past centuries, the problems of rice husk overproduction remain unsolved. In particular, the disposal of rice husk and rice husk ash (RHA) were to give risen to the environmental pollution and human health issues [2]. In the past decades, the researchers found out this agriculture wastes could be utilised into useful application. The most common application to be found was replacing the conventional silica (SiO_2) source [1].

As an alternative silica source, the purity of SiO_2 in RHA is crucial factor in order to produce high quality of ceramic-based end product. The metal impurities to be found in the rice husk may contaminate the end product. In particular, iron (Fe) affects the degree of transparency, purity, and functionality of glass based and ceramic-based end products due to its oxidation state nature [3]. Different oxidation states give different colours to the iron

* Corresponding author: faizul@unimap.edu.my

compounds. Among the industries, acid leaching process is the most common technique to remove metal impurities.

Several researchers had studied the usage of strong inorganic acids to remove the metal impurities in agricultural wastes by using acid leaching process in order to extract high purity of SiO_2. However, these strong acids were difficult to dispose and destructive to the human health and environment [4]. Weak organic acids are having moderate acidity, less processing cost, and remarkable leaching effect which are suitable to replace the strong inorganic acids [5]. In this study, the effect of oxalic acid in removing Fe content and yielding high purity of SiO_2 in rice husk was determined by manipulating the parameters which are acid concentration, reaction temperature, and leaching time. The chemical composition, surface morphology, and phase identification of RHA were characterised as well.

2 Experimental

2.1 Material used and sample preparation

The raw RHs used in this study were originated from the disposing field in Kuala Kedah, Kedah, Malaysia. It washed with tap water to remove dusts and contaminants followed by drying under the hot sun for a day. Next, the dried husks were pulverised into powder by using ring milling machine. The coarse powder was later sieved into fine powder (< 63μm) using sieve shaker to increase leaching efficiency.

2.2 Leaching process

All the experiments were conducted under ambient condition. Oxalic acid dehydrate powders were dissolved in distilled water to prepare oxalic acid solution. 10 g of RHs was put into a beaker filled with 250 ml of oxalic acid solution. The chosen experimental parameters were acid concentration, reaction temperature, and leaching time. Acid concentration was carried out with 0.1 M, 0.5 M, and 1.0 M; reaction temperature was selected with 70 °C, 80 °C, and 90 °C; three leaching durations (1 hour- 3 hours) were chosen. The mixture was put on a hot plate magnetic stirrer with variation among the parameters. After leaching process, the leached RHs were filtered and rinsed with distilled water to remove excess oxalic acid until the filtered water turned clear. The materials were then dried in the oven at 100 °C for 3 hours.

2.3 Calcination process

The dried RHs were calcined in the muffle furnace at controlled temperature of 800 °C for 3 hours. This calcination process allowed the RHs undergone chemical reaction to become RHA with high purity amorphous SiO_2.

2.4 Characterisation analysis

The chemical composition of untreated rice husk ash (URHA) and treated rice husk ash (TRHA) were analysed by X-ray fluorescence spectrometer (XRF). Scanning electron microscopy (SEM) was used to study the morphology of URHA and TRHA. Since the RHAs are non-conductive materials, they were sputter coated a layer of platinum to enhance the analysis. For phase identification analysis in both URHA and TRHA, X-ray diffractometer (XRD) was employed. The diffraction angle, 2θ was scanned from 10° to 80° over the samples with 0.2° step size and 2°/min scan rate.

IOP Conf. Series: Materials Science and Engineering **701** (2019) 012021 doi:10.1088/1757-899X/701/1/012021

3 Results and discussion

3.1 Material used and sample preparation

The results of chemical composition of URHA and TRHA obtained from XRF analysis were tabulated as shown in Table 1, Table 2, and Table 3 according to the manipulation of acid concentration, reaction temperature, and leaching time, respectively.

3.1.1 Oxalic acid concentration

The effect of oxalic acid concentration was studied by conducting experiment at constant 90 °C and 3 hours of leaching time. As Table 1 shown, the SiO_2 content increased whereas the Fe_2O_3 content decreased as the acid concentration used increased. The SiO_2 was obtained with the highest amount of 99.30 % while the Fe_2O_3 was obtained with the lowest amount of 0.43 % when 1.0 M of acid concentration used in leaching rice husk. It was also observed the metal oxide impurities such as K_2O and CaO were significantly decreased as the acid concentration increased.

The complete elimination of CaO occurred at 1.0 M of acid concentration used. It happened the same on K_2O with acid concentration of 0.5 M and above was used. It has proven that the chelating reaction between carboxylic (-COOH) groups and the metal elements was the driving force to remove metal elements in rice husk by forming the metal complexes [6]. The relative amount of SiO_2 was increased due to the reduction of carbonaceous elements and thermal decomposition of organic matters in rice husk during the calcination process [7]. The optimum acid concentration was 1.0 M and it was constraint for the next parameter.

Table 1. Chemical composition of URHA and TRHA with different oxalic acid concentration.

Oxalic Acid Concentration (M)	Chemical Compound (wt %)				
	SiO_2	Fe_2O_3	K_2O	CaO	Others
URHA	92.40	2.26	2.06	1.96	1.32
1.0	99.30	0.43	-	-	0.27
0.5	98.00	1.40	-	0.04	0.56
0.1	96.60	1.94	0.18	0.20	1.08

3.1.2 Reaction temperature

Table 2 showed the chemical composition of URHA and TRHA at 80 °C and 90 °C under controlled variables which are 1.0 M of acid concentration and 3 hours of leaching time. Based on the chemical compounds content in both TRHA, the leaching performance at 90 °C was better than at 80 °C as the SiO_2 content in RHA was increased from 92.40% to 99.30% and the Fe_2O_3 content was decreased from 2.26% to 0.43% at 90 °C. The acid molecules gained extra kinetic energy from the raised temperature, the movement of molecules greatly increased results in chelating rate increased. In addition, the particle size of husk powder also contributed in increasing leaching rate as the smaller the particle size, the more the surface area exposed, the easier the heat and mass transfer to the rice husk [8]. Temperature and particle size were able to enhance leaching effect.

Table 2. Chemical composition of URHA and TRHA with different oxalic acid concentration.

Reaction Temperature (°C)	Chemical Compound (wt %)				
	SiO_2	Fe_2O_3	K_2O	CaO	Others
URHA	92.40	2.26	2.06	1.96	1.32
90	99.30	0.43	-	-	0.27
80	97.40	1.67	-	0.03	0.90

3.1.3 Leaching time

The leaching time had the least effect in this experiment while the acid concentration and the reaction temperature were controlled at optimum condition (1.0 M, and 90 °C, respectively). Based on Table 3, it showed that the highest SiO_2 content (99.30%) yielded in RHA was obtained at 3 hours of leaching time as well as the lowest Fe_2O_3 content (0.43%). In contrast, the experiment conducted for 2 hours was not effective in removing the metal impurities, especially Fe. There was a small amount of Fe content removed from rice husk as there was 2.02% of Fe_2O_3 remained in RHA after leaching. In general, the leaching effect is better with longer leaching time with the addition of other variable.

Table 3. Chemical composition of URHA and TRHA with different leaching time.

Leaching Time (hours)	Chemical Compound (wt %)				
	SiO_2	Fe_2O_3	K_2O	CaO	Others
URHA	92.40	2.26	2.06	1.96	1.32
3	99.30	0.43	-	-	0.27
2	96.50	2.02	0.28	0.12	1.08

3.2 Surface morphology analysis of RHA

In URHA, the particles were irregular size and shapes and distributed evenly as shown in Figure 1(a). This might cause by the grinding action on the RH to obtain a mean particle size of 63μm for better leaching efficiency. However, there were some agglomeration observed (showed by the yellow circles) and it showed a clearer in Figure 1(b) whereby the particles were not decomposed but agglomerated together instead. This phenomena might be happened due to the sample was not undergoing acid leaching treatment [4]. On the other hand, the particles in TRHA were decomposed but they were remained in irregular size and tubular shape as shown in Figure 1(c). In Figure 1(d), the surface condition on the TRHA particle was clearly observed with the porous structure due to the leaching effect and calcination on organic matters [9].

Fig. 1. SEM images for both URHA (a & b) and TRHA (c & d) under different magnification at 500x and 5000x.

3.3. Phase Identification Analysis of RHA

The crystallinity of SiO_2 in RHA was strongly affected by the combustion temperature and time [4]. Both URHA and TRHA were undergone calcination at 800 °C for 3 hours and showed the same broad diffused pattern at the 2θ angles range of 18° to 30° as shown in Figure 2. This phase analysis was similar to other researcher's result whereby the broad diffused peak was indicating SiO_2. No sharp peak was observed. As a result, the identification of amorphous SiO_2 was confirmed in RHA [10].

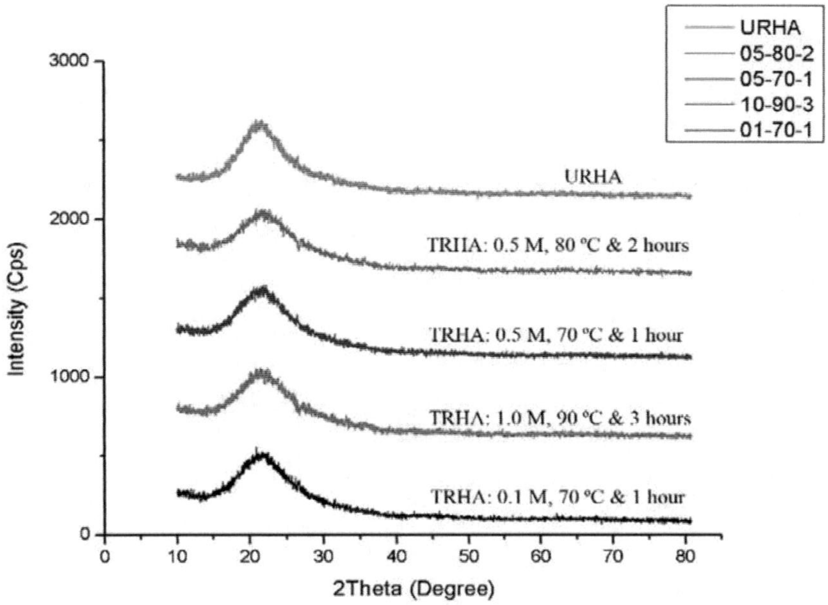

Fig. 2. XRD results of URHA and TRHA with various experimental parameters.

4 Conclusion

Acid leaching using oxalic acid was effective and efficient in removing metal elements especially iron (Fe) in rice husk. Removal of metal impurities in rice husk was greatly enhanced the purification of silica (SiO_2) yielded after calcination at 800 °C. The optimum parameters in leaching rice husk were 1.0 M of oxalic acid concentration, 90 °C of reaction temperature, and 3 hours of leaching time. Particles of TRHA showed porous structure due to elimination of organic matters whereas particles of URHA were not decomposed but agglomerated under morphology analysis. Phase identification analysis between both URHA and TRHA was showing a board diffused peak at 2θ angles range of 18° to 30°, amorphous SiO_2 observed.

The authors would like to thank the Center of Excellence Geopolymer & Green Technology and School of Materials Engineering, Universiti Malaysia Perlis Malaysian Government for giving us facilities to run this research project.

References

1. K. Mohanta, D. Kumar, & O. Parkash,. Int. J. Emerging Technol., **2**, (2012)
2. Uddin & Fazul Rahaman. *Inorganic Pollutant in Wastewater: Method of Analysis, Removal and Treatment*, **16**, (2018)
3. F. Vegliò, B. Passariello, M. Barbaro, P. Plescia, & A. M. Marabini (1998). Inter. J. Miner. Procs. , **54**, (1998)
4. N. H. M. Muzni, N. Hidayu Jamil, & C. K. Yeoh. *Materials Science Forum*, **857**, (2016)
5. V. R. Ambikadevi, & M. Lalithambika,. Appl. Clay Sci., **16**, (2000)
6. N. Syahirah, M. Zarib, & S. Abdullah. Int. J. Eng. Tech. **7**, (2018)
7. E. E. Mon. Inter. J Sci. Eng. Res. **9**, (2018)
8. F. C. Pa, A. Chik, & M. F. Bari. *MATEC Web of Conferences*, **78**, (2016)
9. C. P. Faizul, C. Abdullah, & B. Fazlul. Adv. Env. Bio., **7**, (2013)
10. O. Olamide. *Genetic Algorithm Approach For Optimisation Of Silica Extraction For Micro-Crystalline Silicon Production*, (Olawale B. Tech (Lautech), M.Sc . Industrial and Production Engineering (Ibadan) Matric Number : 130418 A Thesis in the Department, 2016)

Oxide scale growth on Fe-40Ni-24Cr alloy at high temperature oxidation

Noraziana Parimin[1,*], and *Esah* Hamzah[2]

[1]School of Materials Engineering, Universiti Malaysia Perlis, 02600 Arau, Perlis, Malaysia
[2]Faculty of Mechanical Engineering, Universiti Teknologi Malaysia, 81310 Skudai, Johor, Malaysia

Abstract. Fe-40Ni-24Cr alloy is a Ni-based alloy used at high temperature application. In this study, Fe-40Ni-24Cr alloy had undergo a solution treatment at two different temperatures, namely 1050°C and 1150°C, for 3 hours soaking times followed by water quench. A solution-treated Fe-40Ni-24Cr alloy was experienced a high temperature isothermal oxidation test at 500°C for 500 hours exposures duration. The growth of oxide scale was characterized in terms of oxidation kinetics and surface morphology using field emission scanning electron microscope equipped with energy dispersive x-ray spectroscopy. The oxidation kinetics of all samples shows an increasing weight gain pattern as the exposure durations increases. The oxidation kinetics was obeyed a parabolic rate law, indicating the oxide growth rate was followed diffusion-controlled mechanism. The surface morphology of oxidized solution-treated Fe-40Ni-24Cr alloy displayed a formation of continuous oxide scale with evidence of overgrown Nb-rich oxide particles distributed along the oxidized samples surface. Oxidized Fe-40Ni-24Cr alloy which undergo solution treatment at 1150°C exhibited the formation of oxide spallation. The spallation has a tendency to occur around the area of overgrown Nb-rich oxide particles.

1 Introduction

Ni-based alloy is a family of superalloy which is used at high temperature oxidation due to the excellent high temperature strength, oxidation resistance and creep resistance [1-3]. Fe-40Ni-24Cr alloy is a type of Ni-based alloy with good oxidation resistance and excellent strength at elevated temperature. It is used for heat treating equipment, waste incinerators, radiant tubes, muffles, retorts and furnace component which involved in the exposure of the component at high temperature condition and longer service duration. Therefore, oxidation behaviour has become one of the most important factors to assess the properties of this alloy. It is important to evaluate them through oxidation kinetics and the nature of oxide scale formed by high temperature oxidation.

Ni-based alloy is characterized by the high phase stability of the face centre cubic austenite matrix. The Fe-Ni-Cr alloy form a protective surface of Cr_2O_3, which have excellent oxidation resistance in many severe environments [4]. The high Ni content maintains an austenitic structure. The Fe content provides resistance to internal oxidation. The high Cr

* Corresponding author: noraziana@unimap.edu.my

content in this alloy has a capability to form a protective Cr O oxide which developed a protective layer on the component. Oxidation resistance is provided by the formation of a protective oxide layer which is formed when the metal is exposed to oxygen at high temperature and encapsulates the materials, and thus protecting the rest of the component. Many Ni-based alloy contains a combined of several alloying element, including Nb. Nb as alloying element has been extensively used in superalloy for many purposes for improvement of both mechanical properties and oxidation resistance [5-6]. In order to control oxidation, the favourable of Nb have been studied. The effect of Nb in Ni-based alloy is of great importance. Therefore, it needs further and deeper investigations for the high temperature application of Ni-base alloy.

Generally, the oxidation resistance of alloys depends on the protective properties of oxide layer formed on the surfaces in the oxidizing atmospheres. However, the formation of the oxide layer significantly varies with the chemical composition, microstructure and surface condition of the alloy. Therefore, various methods have been proposed to enhance the oxidation resistance of alloys either through altering the chemical composition of the materials or changing the structure of the alloys. Among these methods, grain refinement has been reported as an efficient way to enhance oxidation resistance of the alloys [7-9]. In this study, the effect of grain alteration by solution treatment has been investigated on the oxidation behaviour of Fe-40Ni-24Cr alloy which focused on the oxidation kinetics and oxide scale formation on the alloy surface.

2 Experimental

Fe-40Ni-24Cr alloy used in this study was commercial HAYNES® HR-120® alloy supplied by Haynes International, Inc. The chemical composition was measured using optical emission spectrometer (OES), which are: 40.45 Ni, 24.11 Cr, 0.05 C, 0.08 Al, 0.03 Ti, 0.44 Si, 0.7 Mn, 0.01 P, 0.11 Cu, 0.25 Mo, 0.17 Co, 0.44 Nb, 0.05 W and balance Fe. The samples were cut using wire electrical discharge machine to dimension of approximately 10 mm x 10 mm x 3 mm. The cut samples were undergoing a solution treatment process at 1050°C and 1150°C for 3 hours soaking time, with 10°C/minute heating rate, followed by water quench. These samples were denoted as ST1050 and ST1150 for samples solution-treated at 1050°C and 1150°C, respectively. The solution-treated samples were ground using silicon carbide grinding paper up to P600 grit finished. The samples dimension was measured prior the high temperature isothermal oxidation test. The samples weight was measured before and after the oxidation test. The solution-treated samples were then experienced a discontinuous isothermal oxidation test at 500°C for 500 hours. The oxidized solution-treated samples were characterized in terms of oxidation kinetics and surface morphology using field emission scanning electron microscope (FESEM) equipped with energy dispersive x-ray (EDX).

3 Results and Discussion

3.1 Solution Treatment

The results of solution treatment on Fe-40Ni-24Cr alloys displayed an increasing grain size of alloy as the solution treatment temperature increases. Solution treatment at 1050°C recorded a fine grain size which is 32.85 μm. Whereas, solution treatment at 1150°C recorded a coarse grain size which is 37.13 μm.

3.2 Oxidation Kinetics

The weight change per surface area (w/A) as a function of time (t) of Fe-40Ni-24Cr alloy oxidized at 500ºC is presented in Fig. 1. The oxidation rate law was identified using a double log plots as plotted in Fig. 2 (a). The equation (1) was use to fit the data where x is weight change per surface area, t is time and m is constant. The m value with 1, 2 and 3 represent linear, parabolic and cubic oxidation rate laws, which can be determined from the double log plots. Overall, all samples produced a good fitting results with m ranging from 1.83 to 2.37, representing the parabolic rate law ($m = 2$). The parabolic rate constant was determined from parabolic rate law equation (2), where K_p denote parabolic rate constant. The square of weight change as a function of time was plotted, and the K_p value is calculated as shown in Fig. 2 (b), and all data shows a good fitting result.

$$log\, x = 1/m\, log\, t + C \quad (1)$$

$$x^2 = K_p\, t + C \quad (2)$$

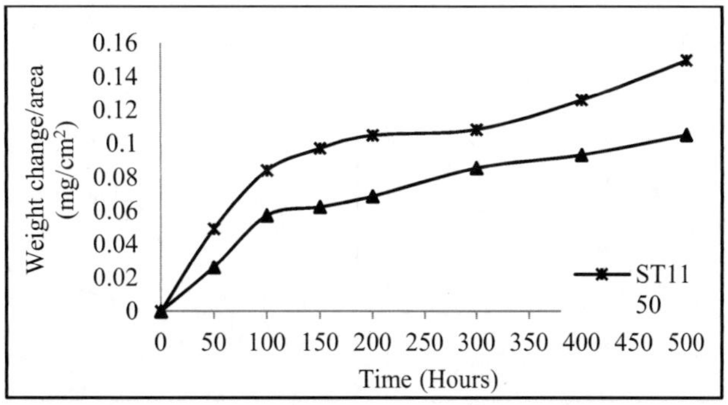

Fig. 1. Oxidation kinetics of solution-treated Fe-40Ni-24Cr alloy at 500°C.

The oxidation kinetics for fine grain ST1050 sample shows low oxidation rate. Whereas, ST1150 sample shows a slightly high oxidation rate compared to ST1050 sample. The weight gain of both samples shows a sharp increasing at the beginning up to 100 hours, then constantly gaining weight until 500 hours. The significant different in between these samples is, ST1150 recorded more weight gain compared to ST1050 which indicating the thick oxide layer was formed on the sample surface. The high oxidation rate was recorded for coarse grain ST1150 sample. The K_p value for ST1050 was smaller than ST1150 sample, which are 5.49 x 10^{-9} and 10.46 x 10^{-9} $mg^2cm^{-4}s^{-1}$, respectively, indicating that ST1050 sample has a low parabolic rate compared to ST1150. In parabolic rate law, the growth of oxide occurs with a continuing declining rate of oxidation. The reaction rate is inversely proportional to the weight of oxide formed or the scale thickness.

The low oxidation rate was determined from rapid initial oxide formation for fine grain sample, generates a rapid protective oxide scale. As the oxide scales thicken, the diffusion rate decrease, recorded a small weight gain. The mechanism of initial oxide formation for fine grain sample was begin with the formation of oxide nuclei when this alloy was exposed at high temperature condition. Oxygen ion from the atmosphere was absorb into the alloy surface, react at the metal-gas interface forming an oxide nuclei. As the exposure time increase, the oxide nuclei were growth laterally and the size was increase as the reaction rate increase. The oxide nuclei then start to impinge on each other to form a continuous oxide layer and will covered the alloy surface. This mechanism has decreased the reaction rate. When the continuous oxide layer completely formed on the alloy surface, the oxide scale will continue to growth vertically to further increase the oxide scale thickness. This phenomenon has developed a protective oxide scale on the alloy surface and governed by diffusion-controlled growth rate, which is, as the oxide scale thickness increase, the diffusion rate was decreases.

The formation of initial oxide scale on the coarse grain size was slightly comparable with fine grain size except for the impingement process of oxide nuclei. The oxide nuclei were initiate by the rapid outward metal ion diffusion through grain boundary area. The fine grain size has higher grain boundary area compared to coarse grain size. Therefore, the impingement process of fine grain size was occur faster compared to coarse grain size. The quick impingement of fine grain size due to close area of adjacent oxide nuclei. The delay of the impingement process of coarse grain size due to the faraway location of adjacent oxide nuclei to impinge on each other during lateral oxide growth. Consequently, the growth of the oxide nuclei will continue until all nuclei was impinge. On the other hand, the oxide formation also governs by the vertical oxide growth. As the exposure time increase, the oxide thickens vertically. The limited ion diffusion in coarse grain sample is most probably due to the lesser grain boundary area which provides a rapid diffusion path for outward ion movement toward metal-gas interface.

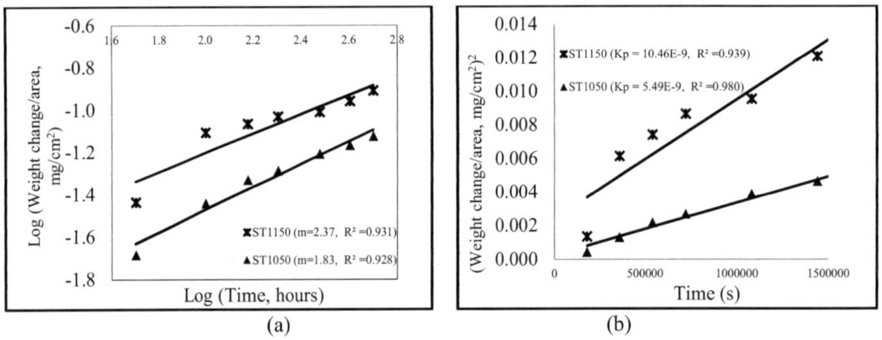

Fig. 2. Weight change as a function of exposure time of solution-treated Fe-40Ni-24Cr alloy at 500°C; (a) double log plots indicating a parabolic rate law and (b) square of weight change indicating parabolic rate constant, K_p value. The fitting parameter R^2 with a value approaching one denotes the increasing consistence of the data with the fitting.

3.3 Surface Morphology

Fig. 3 shows FESEM images of ST1050 sample at different magnifications. Low magnification image in Fig. 3 (a) displayed a continuous oxide layer was formed on the sample surface with formation of overgrown oxide particles. The close-up image of the overgrown oxide particle in Fig. 3 (b) demonstrated a protruding structure with some fine crack formation around the oxide particle. An elemental EDX analysis at the upmost area (A) on the oxide particle detected the presence of O and Nb, indicating the formation of NbO_2 oxide phase. This oxide phase was identified using XRD analysis which is not discussed in this paper. Fig. 3 (c) shows high magnification image (25,000 x) on the continuous oxide layer, demonstrated a uniform oxide structure formed at the alloy surface.

Nb alloying element was added to the Fe-40Ni-24Cr alloy system to enhance their properties, but significantly added to the development of Nb rich precipitates which are NbC and (Nb,Ti)C that will strengthen the alloy [10-12]. However, the exposure of this material to high temperature and longer time will increase the tendency to formed Nb-rich oxide precipitate which tend to grow faster with protruding structure that will served as a preferred site for crack formation and further deteriorate the materials into the certain extend. The main

IOP Conf. Series: Materials Science and Engineering **701** (2019) 012022 doi:10.1088/1757-899X/701/1/012022

feature of oxidized Fe-40Ni-24Cr alloy in this study is the formation of Nb-rich oxide particle which tend to form an overgrown large particle on the surface. Besides, this alloy formed a Cr-rich oxide which serve as a protective barrier between metal and environment. This was characterized by EDX analysis on the sample surface at Fig. 3 (c), recorded the enrichment of element Cr and O, which indicating the formation of Cr-rich oxide. Other researches have also revealed the similar results, that the Nb-rich precipitates encourages the occurance of pitting [13]. In addition, the pitting is due to the massive compositional differences between the matrix and precipitate. The extensive growth of oxide scale tend to create a small difference in volume thermal expansion coefficients between the oxides and metals [14].

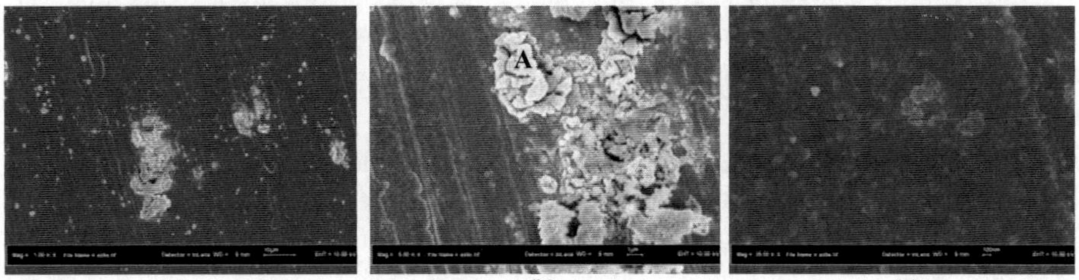

(a) (b) (c)

Fig. 3. FESEM images of ST1050 sample at 500°C at different magnification: (a) 1000x, (b) 5,000x and (c) 25,000x.

Fig. 4 (a) shows FESEM image of ST1150 sample exposed for 500 hours, displayed the formation of continuous oxide layer with occurrence of overgrown oxide particle distributed along the alloy surface. Fig. 4 (b) indicating the evidence of oxide spallation at the isolated oxide particle area. An elemental EDX analysis at this spallation area (B) recorded the presence of major element of O, Nb, Cr, Ti and C. This analysis indicates that the former oxide formed at this region composes of Nb-rich, Cr-Ti oxide, Cr-rich and/or Ti-rich oxide. Carbon element was detected from the carbides of NbC, (Nb,Ti)C and/or TiC. The formation of Nb-rich and Ti-rich oxide precipitates was formed on the alloy surface due to the presence of Nb-based and Ti-based precipitates in the alloy namely NbC, (Nb,Ti)C and/or TiC.

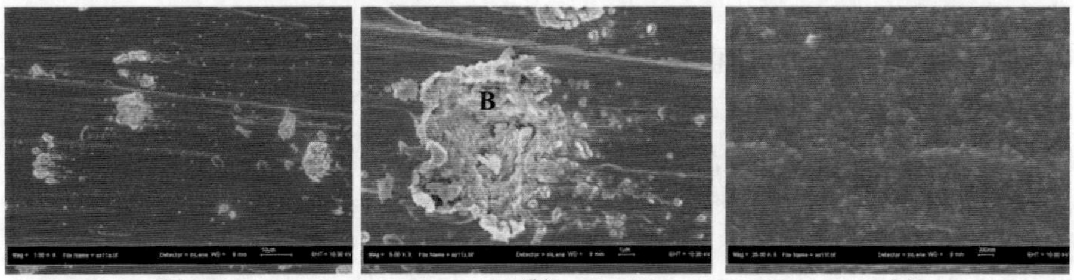

(a) (b) (c)

Fig. 4. FESEM images of ST1150 sample at 500°C at different magnification: (a) 1000x, (b) 5,000x and (c) 25,000x.

During isothermal oxidation test, the oxide nuclei start to develop at the alloy surface due to the absorption of the oxygen gas at the metal surface, interact with metal ion on the surface. The outward metal diffusion was induced by the outward metal ion diffusion to the metal-gas interface. The readily existence of Nb-based and/or Ti-based precipitates also contributed to the outward diffusion of Nb ion and/or Ti ion to the surface which developed a Nb and/or Ti containing oxide particle at the alloy surface. Since Nb and/or Ti precipitates are present at the isolated area in the alloy, the formation of Nb-rich and/or Ti-rich oxide was also developed at the isolated area which is distributed along the alloy surface. Fig. 4 (c) shows close-up image of the oxidized sample on the continuous oxide area exhibited the uniform and protective oxide scale was formed. The dense oxide scale was formed indicating a good protection layer.

4 Conclusion

Isothermal oxidation kinetics of Fe-40Ni-24Cr alloy obeyed parabolic rate law. Fine grained of solution-treated sample ST1050 exhibited low weight gain and low parabolic rate constant value, hence low oxidation rate compared to coarse-grained solution-treated sample ST1150. The surface morphology of both samples exhibited the formation of uniform oxide scale with overgrown Nb-rich oxide particles formed on the alloy surface. The main difference of these oxidized alloys is the formation of Nb-rich oxide particle on the ST1150 sample, tend to exfoliate due the faster growth rate of this overgrow oxide particles on the surface.

The authors would like to thank the Ministry of Education Malaysia for the research fund under the Fundamental Research Grant Scheme (FRGS) (Project Code. FRGS/1/2016/TK05/UNIMAP/02/4).

References

1. Z. P. Hao, R. R. Cui, Y. H. Fan, J. Q. Lin, Int. J. Mech. Sci. **150** (2019)
2. J. Jiang, G. Xiao, Y. Wang, Y. Liu, J. Alloys & Comp. **784** (2019)
3. C. N. Athreya, K. Deepak, D. Kim, B, de Boer, S. Mandal, V. S. Sarma, J. Alloys & Comp. **778** (2019)
4. F. C. Campbell, *Elements of Metallurgy and Engineering Alloys* (Ohio: ASM Int., 2008)
5. Y. X. Xu, J. T. Lu, W. Y. Li, X. W. Yang, Cor. Sci. **140** (2018)
6. Y. X. Xu, J. T. Lu, X. W. Yang, J. B. Yan, W. Y. Li, Cor. Sci. **127** (2017)
7. X. Wang, J. A. Szpunar, J. Alloys & Comp. **752** (2018)
8. T. S. Jo, S. H. Kim, D. G. Kim, J. Y. Park, Y. D. Kim, J. Nuc. Mat. **402** (2-3) (2010)
9. H. Z. Zheng, S. Q. Lu, Y. Huang, Cor. Sci. **51** (2) (2009)
10. I. A. Sustaita-Torres, S. H. Rodriguez, M. P. Guerrero-Mata, M. Garza, E. Valdes, F. Deschaux-Beaume, R. Colas, Mat. Chem. & Phys. **133** (2-3) (2012)
11. S. Borjali, R. Saeed, Allahkaram, K. Hamed, Mat. & Des. **34** (2012)
12. B. Piekarski, Mat. Char. **61** (9) (2010)
13. L. Tan, Cor. Sci. **50** (11) (2008).
14. L. Tan, X. Ren, K. Sridharan, T. R. Allen, Cor. Sci. **50** (7) (2008)

IOP Conf. Series: Materials Science and Engineering **701** (2019) 012023 doi:10.1088/1757-899X/701/1/012023

Thermal behaviour and microstructural analysis of Sn-0.7Cu alloy and Sn-0.7Cu soldered on electroless nickel/immersion gold

Ai Ling Teoh[1,*], *Mohd Arif Anuar* Mohd Salleh[1], *Dewi Suriyani* Che Halin[1], *Kai Loong* Foo[2], and *Juyana* A. Wahab[1]

[1]School of Materials Engineering, Universiti Malaysia Perlis (UniMAP), Taman Muhibbah, 02600 Jejawi, Arau, Perlis, Malaysia
[2]Nano Biochip Research Group, Institute of Nano Electronic Engineering (INEE), Universiti Malaysia Perlis (UniMAP), 01000 Kangar, Perlis, Malaysia

Abstract. Lead-free solder has been developing since decades due to the environmental and health harmful of lead solder. To compete with the lead solder, the behaviour and properties of lead-free solder have been continually researched and improved. Sn-0.7Cu is one of the most common solder in application and suitable to be used as a control in research. The thermal behaviour by differential scanning calorimeter (DSC) and microstructure formation with elemental analysis by scanning electron microscope (SEM) of Sn-0.7Cu solder alloy was compared to Sn-0.7Cu soldered on electroless nickel/immersion gold (ENIG) substrate in this study. The result showed that the pasty range of Sn-0.7Cu was lower than Sn-0.7Cu/ENIG while the undercooling was higher. Primary and interfacial $(Cu,Ni)_6Sn_5$ were formed in Sn-0.7Cu/ENIG due to the diffusion and dissolution of elements from ENIG substrate. The β-Sn dendrites and eutectic phase of Sn-0.7Cu/ENIG was coarser than Sn-0.7Cu due to the decreasing of undercooling.

1 Introduction

The Sn-Pb properties of eutectic composition with low cost, good wettability with low melting point and good in mechanical properties lead to the most preferable material for interconnect in electronic devices. However, lead is one of the toxic material that harmful to the environment and health [1-3]. This reason contributes to the development of lead-free solders that is friendly and green to the environment, where the directives on the restriction of hazardous substance (RoHS) and Waste of Electrical and Electronic Equipment (WEEE) play an important role in driving it. In the past studies and researches, it is found that the eutectic Sn-0.7Cu solder is one of the most trusted lead-free solder in replacing the lead solder [4-8].

In last decades, many researches have been done in order to study the thermal behaviour of the lead-free solder and its resulted microstructural after solidification process.

* Corresponding author: akirateoh@gmail.com

Schindler et al. studied on the undercooling of Sn-Cu solder alloys with different Cu content from 0 to 3 wt.% and found that only little influence by Cu composition in the degree of undercooling . As different microstructural formation gives rise to a different mechanical properties of solder joint, Ventura et al. focused on the morphologies forming during the eutectic growth in binary Sn-Cu solder alloys with different growth rate of directionally solidification [9, 10]. In order to improve the solder consistency in application, Silva et al. investigated the effects of Ag and Cu minor addition into the Sn-Bi solder by directional solidification, resulted in the different microstructural formation and changing in the tensile and ductility properties [11].

In the application, solder is always related to its soldering interface since the reactions between the solder and substrate are occurred during solidification with a degree of undercooling, affecting the microstructural formation of solder joint that contributes to a different performance of the electronic devices ; while most of the studies were done on copper substrate. Therefore, the purpose of this study is to investigate the influence of ENIG substrate to the Sn-0.7Cu solder by thermal behaviour and microstructural change where this Sn-0.7Cu solder will act as a control in the whole research.

2 Experimental Procedures

2.1 Sample preparation

The pure metal ingot of Sn-0.7Cu was supplied by Nihon Superior Co. Ltd., Japan. The Sn-0.7Cu solder alloy was first melted in the electric resistance furnace at 350 °C for 1 hour and then casted onto a stainless-steel mould. A several casted metal sheets were remained for the analysis and observation for solder, while the others were rolled into 50 μm thickness of foils and then punched with 3 mm diameter round shape puncher. Next, the 3 mm punched metal foils were put on the Pyrex sheet inside the reflow oven after applying the rosin mildly activated (RMA) flux on it for the purpose of 900 μm solder ball formation in a sphere shape by cause of surface tension. A several solder balls were remained for the analysis, while the others were reflowed on the substrate with ENIG surface finish after applying the RMA flux.

2.2 Thermal analysis

By using thermal analysis instruments which is differential scanning calorimeter (DSC) model Q10, 5 mg of Sn-0.7Cu metal sheet was analysed to obtain the thermal behaviour of solder. To study the thermal behaviour of Sn-0.7Cu/ENIG, 3.2 mg of solder ball with RMA flux was put on the 0.8 mg of ENIG substrate as the test sample. With the heating and cooling rate of 10 °C/min during the test, the sample was heat up to 270 °C and then cooled down to 30 °C in an inert nitrogen gas chamber to collect the endothermic and exothermic curves for analysis.

2.3 Microstructural analysis

The Sn-0.7Cu casted metal sheet and the reflowed Sn-0.7Cu/ENIG were polished and etched with the organic solderability preservative (OSP) for the microstructural observation and elemental analysis by using scanning electron microscope (SEM) model JEOL JSM 6460LA equipped with Energy Dispersive X-ray (EDX).

3 Result and Discussions

3.1 Thermal behaviour

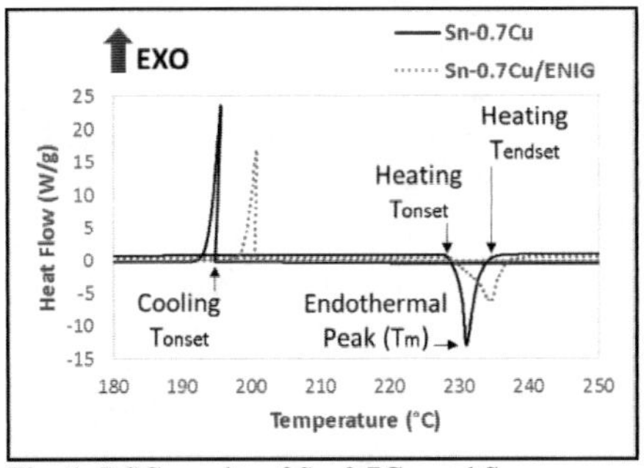

Fig. 1. DSC results of Sn-0.7Cu and Sn-0.7Cu/ENIG during heating (endothermal) and cooling (exothermal).

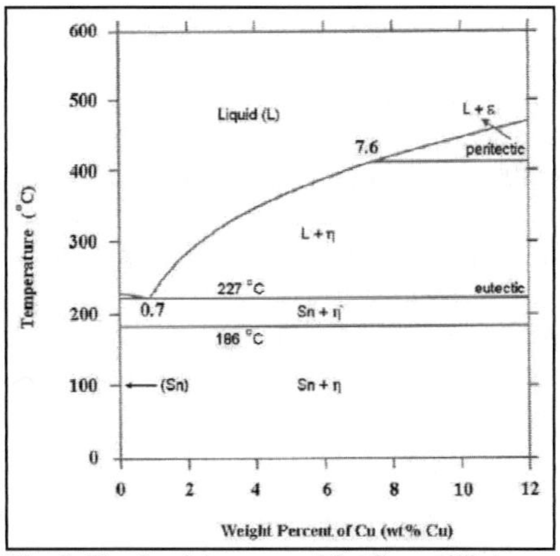

Fig. 2. The Sn-rich corner of Sn-Cu equilibrium diagram [12].

The melting and cooling characterization of solder plays an important role in affecting the solder joint properties contributes to the electronic packaging performance. The thermal behaviour of the samples was carried out by using DSC analysis at the heating and cooling rate of 10 °C/min to mimic the application of ball-grid array (BGA) [13]. The DSC results are displayed in the Fig. 1, Table 1 and Table 2. The thermal behaviour of Sn-0.7Cu solder on the substrate with ENIG surface coating was tested besides the thermal behaviour of Sn-0.7Cu solder itself. Sn-0.7Cu and Sn0.7Cu/ENIG are heated up and the endothermic reaction is started at 228.00 and 228.26 °C respectively as the heating T_{onset}. The heating curves are then reaching the peaks at 231.59 and 232.61 °C for Sn-0.7Cu and Sn0.7Cu/ENIG respectively as Tm. The Sn-Cu binary phase diagram is illustrated in Fig. 2. As can be viewed in the Sn-Cu equilibrium diagram, the T_{onset} and Tm of the DSC curves are resulted from the phase changing of (Sn+η') to liquid at the temperature of 227 °C. These DSC results are agreed by the study of Cho et al. [14], where the similar findings are reported.

On the other hand, the pasty range is formulated as:

$$\text{Pasty range} = \text{Heating } T_{endset} - \text{Heating } T_{onset} \quad (1)$$

where heating T_{endset} is the liquidus temperature and T_{onset} is the solidus temperature when heating [13, 15]. From Table 1, the calculation shows that the pasty range for Sn-0.7Cu and Sn-0.7Cu/ENIG is 10.27 and 11.21 °C. With the existing of substrate, the pasty range is slightly increased compared to solder itself since soldering process has taken part between the solder and substrate when heat applied. Besides the pasty range, the degree of undercooling is another main factor influencing the solder joint reliability that determines the solidification behaviour of the solder. The undercooling is defined as the difference between the beginning point of the melting (heating T_{onset}) in heating curve and the beginning point of solidification (cooling T_{onset}) in cooling curve [16]; formulated as:

IOP Conf. Series: Materials Science and Engineering **701** (2019) 012023 doi:10.1088/1757-899X/701/1/012023

$$\text{Undercooling, } \Delta T = \text{Heating } T_{\text{ondset}} - \text{Cooling } T_{\text{onset}} \quad (2)$$

The undercooling degree of Sn-0.7Cu and Sn-0.7Cu/ENIG is 32.19 and 31.45 °C respectively. With the existing of substrate, the degree of undercooling is slightly decreased compared to solder itself. Cho et al. demonstrated the same trend of results that the undercooling of Sn-rich solder alloys is reduced when it reacted with various under bump metallurgies (UBMs). The wettable surface of UBMs facilitates a heterogeneous nucleation of β-Sn phase during solidification [14]. In this study, the substrate with the surface coating of ENIG has provided a wettable surface to the solder for the soldering process of elemental diffusion and dissolution from substrate and the β-Sn nucleation in solder during solidification, causing the reduction of undercooling degree.

Table 1. Heating peak (T_m), solidus temperature (heating T_{onset}), liquidus temperature (heating T_{endset}) and pasty range for Sn-0.7Cu and Sn-0.7Cu/ENIG.

Sample	Heating Peak T_m (°C)	Heating T_{onset} (°C)	Heating T_{endset} (°C)	Pasty Range $T_{onset} - T_{endset}$ (°C)
Sn-0.7Cu	231.59	228.00	238.27	10.27
Sn-0.7Cu/ENIG	232.61	228.26	239.46	11.21

Table 2. Starting point of solidification (cooling T_{onset}) and degree of undercooling (ΔT) for Sn-0.7Cu and Sn-0.7Cu/ENIG.

Sample	Cooling T_{onset} (°C)	Undercooling ΔT (°C)
Sn-0.7Cu	195.81	32.19
Sn-0.7Cu/ENIG	196.81	31.45

3.2 Microstructure observation

The SEM image in Fig. 3 shows that a network band is formed among the matrix in Sn-0.7Cu solder alloy. The research by Zeng et al. states that this network band is the eutectic mixture consists of Sn and a small fraction of faceted intermetallic compound Cu_6Sn_5 as shown in Fig. 4; while its white color area is the primary β-Sn dendrites [17]; as agreed in other studies too [18]. As a result, Sn-0.7Cu solder only contains the elements of Sn and Cu in its microstructure.

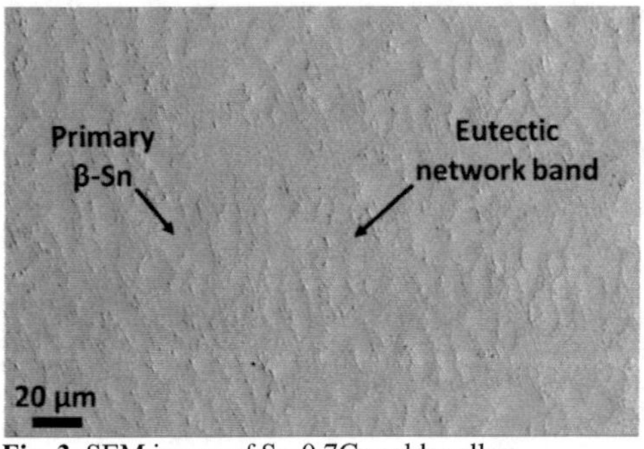

Fig. 3. SEM image of Sn-0.7Cu solder alloy.

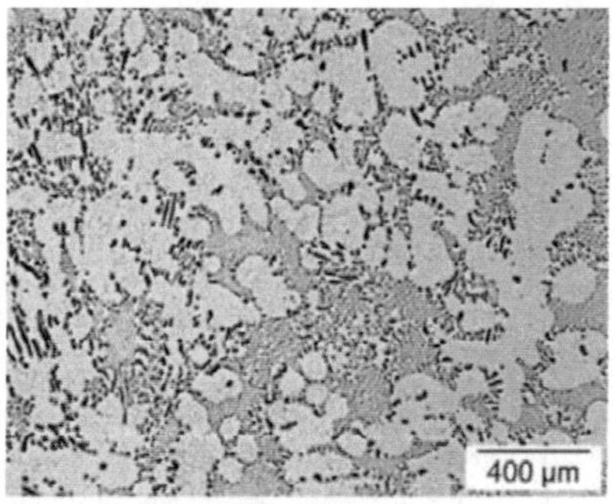

Fig. 4. Fully solidified microstructure of Sn-0.7Cu [17].

IOP Conf. Series: Materials Science and Engineering **701** (2019) 012023 doi:10.1088/1757-899X/701/1/012023

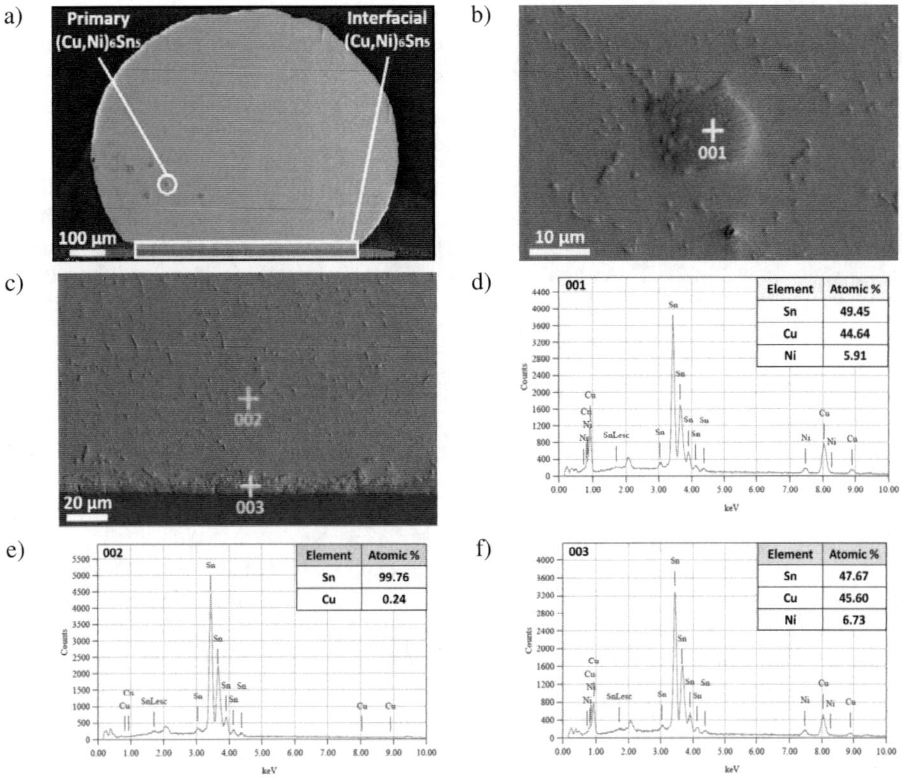

Fig. 5. SEM results of Sn-0.7Cu soldered on ENIG substrate. (a) Sn-0.7Cu/ENIG (b) Primary $(Cu,Ni)_6Sn_5$ (c) Interfacial $(Cu,Ni)_6Sn_5$ (d) EDX on primary $(Cu,Ni)_6Sn_5$ (e) EDX on β-Sn (f) EDX on interfacial $(Cu,Ni)_6Sn_5$.

The microstructural of Sn-0.7Cu solder has changed when it is soldered on a substrate as displayed in Fig. 5, and hence EDX analysis is carried out to understand the elemental in the microstructure. In Fig. 5a, it reveals that the formation of crystals (labelled as primary and interfacial $(Cu,Ni)_6Sn_5$ in figure) in the solder ball and at the interfacial between the substrate and the solder. From the EDX result of the crystal (Fig. 5b) inside the solder ball in Fig. 5d, the crystal is having the elements of Sn, Cu and Ni with the atomic percentage that forming the compound of $(Cu,Ni)_6Sn_5$. These crystals formed inside the solder ball is termed as the primary intermetallic compound (IMC) [19]; and the Ni elements is precipitated from the diffusion and dissolution from the substrate during the interfacial reaction [14]. Thus, the crystal formed inside solder ball after soldering in Fig. 5b is the primary $(Cu,Ni)_6Sn_5$ IMC due to the elemental diffusion and dissolution from the substrate.

Similar to Sn-0.7Cu solder, the Sn-0.7Cu/ENIG consists of the network band among the matrix as shown in the magnified Fig. 5c; and the EDX result in Fig. 5e explains that the area without the network is the β-Sn where it is the Sn-rich area. However, it can be seen that the β-Sn dendrites and the eutectic phase are coarser than the Sn-0.7Cu solder without the substrate, where the eutectic phase forms a better continuous network. This microstructural changes correlated to the interfacial reactions during soldering is due to the reduction of undercooling degree. According to the study by Cho et al., the temperature of nucleation and growth during the solidification state was higher than the solder without substrate due to the interfacial reactions resulted in the decreasing of the undercooling degree; and hence the solidifying grains were grow faster when this temperature raises leading to the formation of larger grain structure [14].

Furthermore, due to the interfacial reaction between the solder and substrate during soldering process, there is the growth of a crystal layer at the interfacial area of the solder and substrate as shown in Fig. 5c. Similar to the EDX result of primary IMC, Fig. 5f exhibits the interfacial crystals are formed by the Sn, Cu and Ni as the $(Cu,Ni)_6Sn_5$ IMC. The elements from substrate are diffused and dissolved into the solder during the interfacial reactions when heat is applied and resulting in the formation of IMC as reported by other researchers [14, 20-22]. The addition of other elements into the solder has reduced the degree of undercooling since the nucleation of new IMC precipitates has been started during the solidification process [18]. Other researches have also revealed the similar results that the heterogeneous nucleation sites for β-Sn dendrites are provided by the primary IMCs causing the reduction of undercooling degree [23, 24].

4 Conclusion

The pasty range of Sn-0.7Cu/ENIG is higher than Sn-0.7Cu while its degree of undercooling is lower due to the diffusion and dissolution of elements from substrate into the solder and the β-Sn nucleation when solder solidifies during the interfacial reactions. Both microstructures of Sn-0.7Cu and Sn-0.7Cu/ENIG consist of primary β-Sn and eutectic Cu_6Sn_5 while Sn-0.7Cu/ENIG has another extra 2 features which are the primary $(Cu,Ni)_6Sn_5$ and interfacial $(Cu,Ni)_6Sn_5$ due to the elemental diffusion and dissolution from substrate. The β-Sn dendrites and the eutectic phase of Sn-0.7Cu/ENIG are coarser than the Sn-0.7Cu due to the reduction of undercooling degree.

The authors thankfully acknowledge Nihon Superior as supporter in research and the Fundamental Research Grant Scheme (FRGS) under a grant number of FRGS/1/2017/TK05/UNIMAP/03/3 (No : 9003-00621) for the financial and materials support, but not least to School of Materials Engineering, Universiti Malaysia Perlis (UniMAP) for facilities supporting.

References

1. N.M. Nasir, N. Saud, M.N. Deraman, A.A.M. Salleh, M.I.I. Ramli, and R.M. Said, Materials Science Forum. **803** (2015)
2. M.I.I. Ramli, M.S.S. Yusof, M.A.A.M. Salleh, R.M. Said, and K. Nogita, Solid State Phenomena. **273** (2018)
3. F. Somidin, H. Maeno, M.A.A.M. Salleh, X.Q. Tran, S.D. McDonald, S. Matsumura, and K. Nogita, Materials Characterization. **138** (2018)
4. W.-T. Chen, C. Ho, and C. Kao, Journal of Materials Research. **17**(2) (2002)
5. A. El-Daly, A. El-Taher, and S. Gouda, Materials & Design (1980-2015). **65** (2015)
6. M. Fazal, N. Liyana, S. Rubaiee, and A. Anas, Measurement. (2018)
7. H.A. Jaffery, M.F.M. Sabri, S.M. Said, S.W. Hasan, I.H. Sajid, N.I.M. Nordin, M.M.I.M. Hasnan, D.A. Shnawah, and C.V. Moorthy, Journal of alloys and compounds. (2019)
8. M.I.I. Ramli, M.A.A.M. Salleh, F.A.M. Sobri, P. Narayanan, K. Sweatman, and K. Nogita, Journal of Materials Science: Materials in Electronics. **30**(4) (2019)

9. T. Ventura, S. Terzi, M. Rappaz, and A.K. Dahle, Acta Materialia. **59**(4) (2011)
10. M.A.A.M. Salleh, M.H. Hazizi, M.A.B.M. Abdullah, N. Noriman, R. Mayapan, and Z.A. Ahmad, Advanced Materials Research. **626** (2012)
11. B.L. Silva, M.G. Xavier, A. Garcia, and J.E. Spinelli, Materials Science and Engineering: A. **705** (2017)
12. A.A. El-Rehim and H. Zahran, Journal of alloys and compounds. **695** (2017)
13. A. El-Daly, W. Desoky, A. Saad, N. Mansor, E. Lotfy, H. Abd-Elmoniem, and H. Hashem, Materials & Design. **80** (2015)
14. M.G. Cho, S.K. Kang, and H.M. Lee, Journal of Materials Research. **23**(4) (2008)
15. R. Sayyadi and H. Naffakh-Moosavy, Materials Science and Engineering: A. **735** (2018)
16. J.W. Elmer, E.D. Specht, and M. Kumar, Journal of Electronic Materials. **39**(3) (2010)
17. G. Zeng, S.D. McDonald, Q. Gu, Y. Terada, K. Uesugi, H. Yasuda, and K. Nogita, Acta Materialia. **83** (2015)
18. T. Lu, D. Yi, H. Wang, X. Tu, and B. Wang, Journal of alloys and compounds. **781** (2019)
19. M.M. Salleh, C. Gourlay, J. Xian, S. Belyakov, H. Yasuda, S. McDonald, and K. Nogita, Scientific reports. **7** (2017)
20. K. Nogita, M.A.A.M. Salleh, E. Tanaka, G. Zeng, S.D. McDonald, and S. Matsumura, JOM. **68**(11) (2016)
21. M.A.A.M. Salleh, A.M.M.A. Bakri, F. Somidin, and H. Kamarudin, International Review of Mechanical Engineering. (2013)
22. M.A.A.M. Salleh, M.M.A.B. Abdullah, F. Somidin, A.V. Sandu, N. Saud, K. Hussin, S.D. McDonald, and K. Nogita, Revista de Chimie **64**(7) (2013)
23. H. Shang, Z. Ma, S. Belyakov, and C. Gourlay, Journal of alloys and compounds. **715** (2017)
24. X. Tu, D. Yi, J. Wu, and B. Wang, Journal of alloys and compounds. **698** (2017)

Wettability and thermal properties of Sn-0.7Cu-0.05Ni-xZn solder alloy

Mohd Izrul Izwan Ramli[1,*], *Mohd Arif Anuar* Mohd Salleh[1], *Nurul Ashikin* Saleh[1] and *Siti Farahnabilah* Muhd Amli[1]

[1]Center of Excellence Geopolymer & Green Technology (CeGeoTech), School of Materials Engineering, Universiti Malaysia Perlis (UniMAP), Taman Muhibbah, 02600 Jejawi, Perlis, Malaysia

Abstract. This study investigate the effect of Zn additions on wettability and thermal properties of Sn-0.7Cu-0.05Ni lead-free solder alloy. The additional of alloying element have been developed by using casting technique with different percentage of Zn which is 0.5, 1.0 and 1.5 wt.%. The Sn-0.7Cu-0.05Ni-xZn solder alloy samples were then analyzed by using the optical microscope (OM) to observerse the wettability perfomance by contact angle. From this study, it was found that different composition of Zn affected the wettability and thermal properties of Sn-0.7Cu-0.05Ni solder alloys. The addition of 0.5 wt.% has slightly increase the melting point while the addition of higher than 1.0 wt.% Zn can decrease the melting temperature. The addition of 0.5 wt.% of Zn also can reduce the undercooling of solder alloy. Thus, the minor addition of Zn for Sn-0.7Cu-0.05Ni can influence the wettability and thermal properties of solder alloy.

1 Introduction

Soldering is a process in which two or more metal items is joined by melting and flowing a filler metal (solder) into the joint. As a joining interaction, solder plays a critical role and has been extensively used in the electronic packaging industry. For years, Tin-Lead (Sn-Pb) solder has been widely used as interconnect materials. However, application of Pb in electronic product is being discontinued due to environmental and health concerns [1-4].

With the advancement of micro/nano systems technology towards faster speed, increasing service performance and miniaturization, packaging technology is facing its limitations. Therefore, it is an imperative issue for the electronic business to improve a viable alternative solder for electronic assemblies. Thus, to fulfil the ever-stricter service requirements, new interconnection materials have to be developed. Eutectic alloy of Sn-Cu that has recently gained researchers attention due to its availability and low production cost. However, Sn-0.7Cu has been reported to have poorer mechanical properties and wettability compared to lead free solder alloys containing silver [1, 5-7]. With a small amount of nickel (Ni) were added into Sn-Cu solder, it can and improve the fluidity of Sn-0.7Cu. So, Sn-0.7Cu-0.05Ni is most commonly used as lead free solders alloy that eco-friendly with the

* Corresponding author: mohdizrulizwan@gmail.com

IOP Conf. Series: Materials Science and Engineering **701** (2019) 012024 doi:10.1088/1757-899X/701/1/012024

environment [8-10]. The properties of Sn-0.7Cu-0.05Ni solder alloy are good on fluidity and give uniformity to the thickness in copper substrate. The addition of Zn can reduce the low melting temperature of the solders [11] and also good in improvement the wettability which is giving the ability to molten solder spread over the copper pad [11].

In this study, the development of new solder alloys which is Sn-0.7Cu-0.05Ni-xZn has been investigated. The result discusses the understanding of Zn addition on wettability and thermal properties of the solder alloys.

2 Experimental Procedure

The Sn-0.7Cu-0.05Ni based alloy was supplied by Nihon Superior Co Ltd., Japan. The solder alloy was mixed with different composition (wt.%) of Zn by using casting technique. The solder alloys were mixed with three different composition which is 0.5, 1.0 and 1.5 wt.% of Zn. The Sn-0.7Cu-0.05Ni solder alloy and Zn element was carefully weighing and put into graphite crucible. The graphite crucible was then put into the furnace at 600 °C. The liquid alloy was held at 600 °C for 1 h and was stirred sufficiently each 10 min, and then the liquid alloy was casted in the atmospheric environment and then was cooled to the ambient temperature.

For wettability analysis, the samples were then reflowed onto Cu-substrates in a F4N reflow oven. Some activated rosin flux was applied during reflow soldering to act as a cleaning agent to remove and prevent the formation of oxides. The reflowed sample were mounted with epoxy resin, grinded, polished. The wettability was carried out by optical microscope (OM). This analysis was investigated by measured the contact angle formed in the copper substrate solder as shown in Fig.1. For thermal analysis sample, the sample are in solder ball shaped. To making a solder ball shaped, the bulk sample must through a rolling process which making solder alloys thickness become 0.2 mm. Therefore, the piece of sheet solder alloy was cutting into 0.002 gram and reflow it. The solder ball is prepared on the Al pan with piece of sheet of Cu and some flux as shown in Fig. 2. In this study, differential scanning calorimetry (DSC) is used to determine the melting point of the material. The solder alloys were heated at temperature of 250 °C by using a heating rate of 10 °C/min under the nitrogen atmosphere.

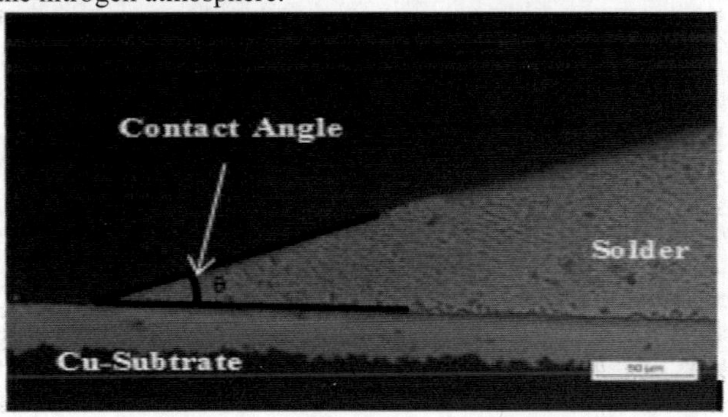

Fig. 1. Contact angle of the solder alloy.

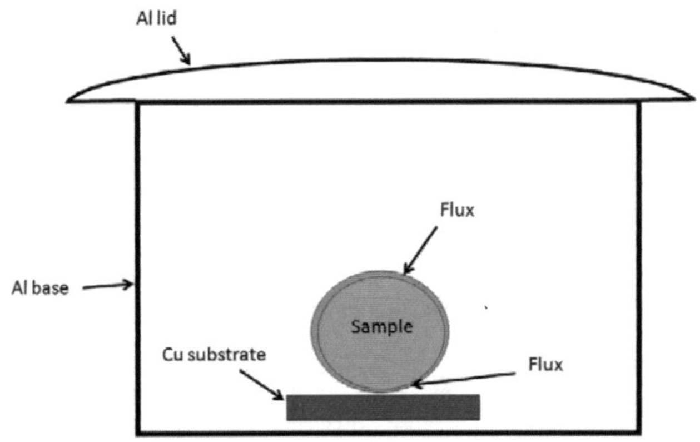

Fig. 2. Sample preparation for Differential Scanning Calorimetry (DSC).

3 Results and Discussion

3.1 Wettability Analysis

Wettability is described as the capability of the molten solder to spread over on a substrate during the reflow process and was determined with respect to the contact angle. Wettability depends on several factors such as surface properties and also solder alloys compositions [4, 12, 13]. By minimize the value of the contact angle, the wettability can be optimized that corresponds to lower surface-interfacial energy. Based on the graph in Fig. 3, the contact angle for pure Sn-0.7Cu-0.05Ni solder alloy sample is 23.61° which is higher than 0.5 wt.% Zn but lower than 1.0 and 1.5 wt.% addition of Zn. Shen et al. [14] also discovered that the addition of 0.01 and 0.03 wt.% Ni can decrease the wetting force and the wetting time of Sn-Cu solder. Based on the review, the minor addition of Ni can reduce the contact angle of Sn-0.7Cu-0.05Ni solder alloy. The similar observation can be seen in this study, with the addition of minor 0.5 wt.% Zn in the solder system make the contact angle decrease to 17.32 °.

However, when 1.0 and 1.5 wt.% Zn was added into the solder alloys, the contact angle slightly increases to 26.78 ° and 28.48 ° respectively. L. Zang et al. [15] reported that improvement of the wettability on the Cu substrate with molten solder alloy are related with the formation of intermetallic. By the formation of intermetallic that have been investigate for the addition of 1.0 and 1.5 wt.% of Zn, the higher thickness has been occurred and formed a higher wetting angle. This was reported by Ramli et al. [16] when they added in range 1.0-1.5 wt.% of Zn into Sn-0.7Cu solder alloy. In addition, alloying of Zn element into the Sn-based solder is concerned because Zn is very active and may deteriorate wetting performance of lead-free solder at molten state [17]. This is show that the Ni and Zn have exactly same properties to lowered the wetting angle by addition of small amount of element. However, with the higher amount of Zn can increase the contact angle value due to the oxidation and high activity of Zn in Sn alloy [18].

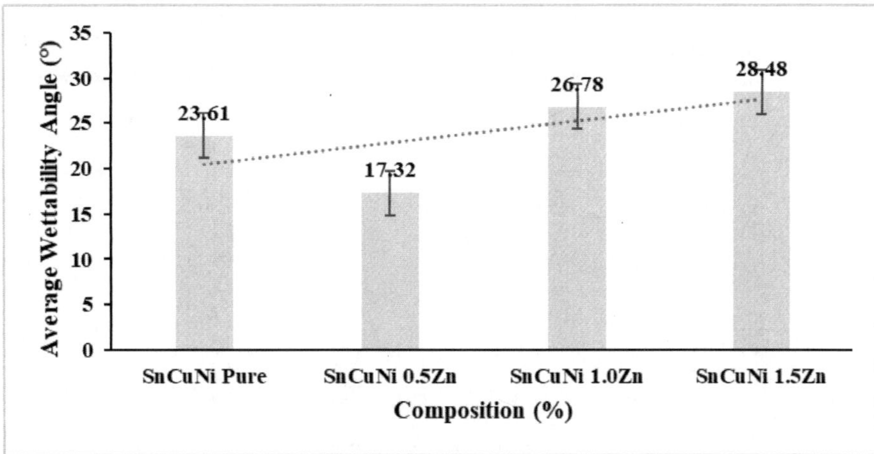

Fig. 3. The average wettability angle of Sn-0.7Cu-0.05Ni solder alloy with addition of 0.5, 1.0 and 1.5 wt.% Zn.

3.2 Thermal Properties

Thermal properties such as melting temperature is important factors in soldering industry where it can determines the operating system temperature. Fig. 4 shows the melting point of Sn-0.7Cu-0.05Ni solder alloy with addition of 0.5, 1.0 and 1.5 wt.% Zn. The Sn-0.7Cu-0.05Ni solder alloy with the addition of 0.5 wt.% has slightly increase the melting point with 230.93 °C while the addition of 1.0 wt.% Zn can decrease the melting temperature with 227.63 °C. This process is known as endothermic process because the process towards to heating process. Bao et al. [19] found interesting behavior in the phase diagram by using simple regular model which is claims when the positive interaction of solid and negative liquid occur, all the solidus and liquidous with the eutectic point move at the lower melting point component. Liu et al. [20] also report that the addition of Ni increase the melting temperature of the low-Ag solder.

For thermal analysis, it can be concluded that the low melting point achieved when 1.5 wt.% Zn are added. The range of the melting point also have low value when 1.5 wt.% Zn are added into Sn-0.7Cu-0.05Ni solder alloy than others because the composition just need a short time to fully melt. Solder alloys must have low melting temperature with narrow melting range [11]. However, there are no slightly different value of melting point temperature and melting range temperature for all samples. The melting point temperature are still around with eutectic or near-eutectic composition. For undercooling temperature, it shows that the addition of 0.5 wt.% of Zn has a short time to transfer from liquid to solid. According to Fig. 5, the onset and endset heating point was calculated the range melting point of solder alloys. The addition of 0.5 wt.% Zn give the higher value of melting range with 9.27 °C than others because the composition totally melting at high temperature which is 238.42 °C. In addition, Liu et al. [20] explain that the higher addition of Bi in the solder alloys could increase the melting range and can lead to the initiation of solidification crack of the solder joints.

Fig. 6 show the undercooling of Sn-0.7Cu-0.05Ni-xZn solder alloy. Undercooling range temperature refer to solder alloys transfer from liquid to solid. The undercooling value can be calculated by minus the onset heating point with onset cooling point. The graph shows the higher undercooling of Sn-0.7Cu-0.05Ni solder alloy is 63.84 °C compared to addition of 0.5 wt.% of Zn which have lower undercooling temperature. This is show that the optimum value for undercooling temperature is 0.5 wt.% addition of Zn because have low undercooling compare with addition different up to 1.0 and 1.5 wt.% Zn.

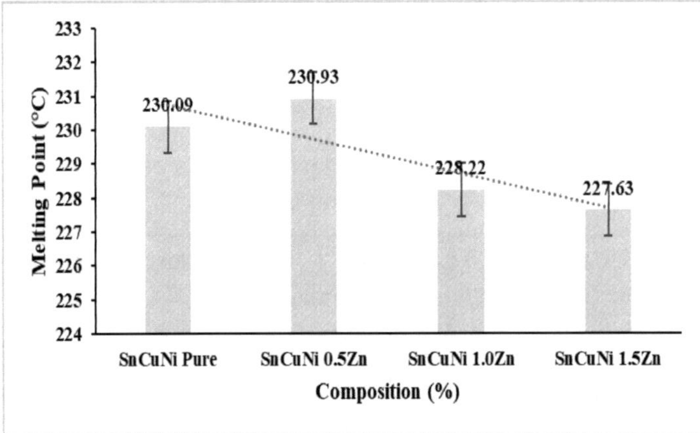

Fig. 4. Melting point of Sn-0.7Cu-0.05Ni solder alloy with addition of 0.5, 1.0 and 1.5 wt.% Zn.

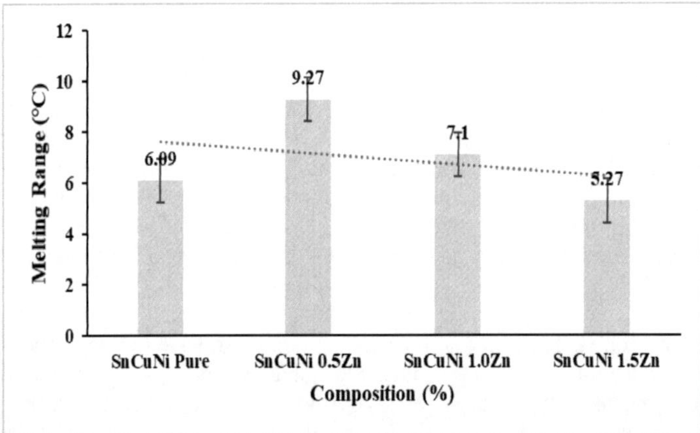

Fig. 5. Melting range of Sn-0.7Cu-0.05Ni solder alloy with addition of 0.5, 1.0 and 1.5 wt.% Zn.

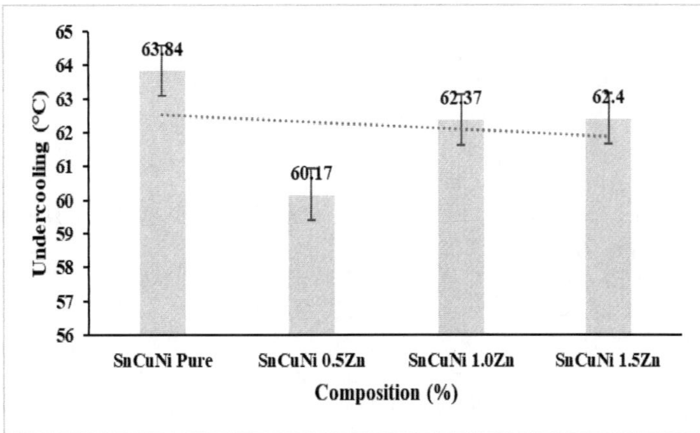

Fig. 6. Undercooling of Sn-0.7Cu-0.05Ni solder alloy with addition of 0.5, 1.0 and 1.5 wt.% Zn.

4 Conclusion

This study investigated the effect of Zn addition on intermetallic compound, wettability and morphology of Sn-0.7Cu lead free solder alloy. The conclusion was summarized as follows:

i. The addition of with 0.5 wt.% Zn in Sn-0.7Cu-0.05Ni could improve the wettability.
ii. The thermal properties including melting temperature, undercooling and melting range has been investigated. The addition of 0.5 wt.% has slightly increase the melting point with 230.93°C while the addition of 1.0 wt.% Zn can decrease the melting temperature with 227.63°C.
iii. The addition of 0.5 wt.% of Zn has a short time to transfer from liquid to solid thus reduce the undercooling.

The authors gratefully acknowledge the fundamental research grant scheme (FRGS) (FRGS) (FRGS/1/2017/TK05/UNIMAP/02/7) (9003-00635) from the Ministry of Higher Education, Malaysia and Universiti Malaysia Perlis (UniMAP)-Nihon Superior Co. Ltd. collaboration research project for the financial support throughout the research project.

References

1. M.A.A.M. Salleh, M.H. Hazizi, M.A.B.M. Abdullah, N. Noriman, R. Mayapan, and Z.A. Ahmad, Advanced Materials Research. **626** (2012)
2. M.I.I. Ramli, M.A.A.M. Salleh, F.A.M. Sobri, P. Narayanan, K. Sweatman, and K. Nogita, Journal of Materials Science: Materials in Electronics. **30**(4) (2019)
3. R.M. Said, M.A.A.M. Salleh, M.I.I. Ramli, and N. Saud, Solid State Phenomena. **280** (2018)
4. M.I.I. Ramli, M.S.S. Yusof, M.A.A.M. Salleh, R.M. Said, and K. Nogita, Solid State Phenomena. **273** (2018)

5. M.A.A.M. Salleh, S.D. McDonald, C.M. Gourlay, H. Yasuda, and K. Nogita, Materials & Design. **108** (2016)
6. M.I.I. Ramli, M.A.A.M. Salleh, M.N. Derman, R.M. Said, N.M. Nasir, and N. Saud, Key Engineering Materials. **700** (2016)
7. N.M. Nasir, N. Saud, M.N. Deraman, A.A.M. Salleh, M.I.I. Ramli, and R.M. Said, Materials Science Forum. **803** (2015)
8. K. Nogita, M.A.A.M. Salleh, E. Tanaka, G. Zeng, S.D. McDonald, and S. Matsumura, JOM. **68**(11) (2016)
9. M.A.A.M. Salleh, A.M.M.A. Bakri, F. Somidin, and H. Kamarudin, International Review of Mechanical Engineering. (2013)
10. M.A.A.M. Salleh, M.M.A.B. Abdullah, F. Somidin, A.V. Sandu, N. Saud, K. Hussin, S.D. McDonald, and K. Nogita, Revista de Chimie **64**(7) (2013)
11. L. Sun and L. Zhang, Advances in Materials Science and Engineering. (2015)
12. N.B. Duong, T. Ariga, L.B. Hussain, and A.B. Ismail, Materials Transactions. **49**(6) (2008)
13. K. Nogita, M.A.A.M. Salleh, S. Smith, Y.Q. Wu, S. D.McDonald, A.G.A. Razak, T. Akaiwa, and T. Nishimura. *Effects of Bi in Sn-Cu based lead-free solder alloys and interconnects*. in *Electronics Packaging (ICEP), 2017 International Conference on*. 2017. IEEE.
14. J. Shen, Y. Pu, D. Wu, Q. Tang, and M. Zhao, Journal of Materials Science: Materials in Electronics. **26**(3) (2015)
15. L. Zang, Z. Yuan, H. Zhao, and X. Zhang, Materials Letters. **63**(23) (2009)
16. M.I.I. Ramli, M.A.A.M. Salleh, N. Ibrahim, and R.M. Said, Solid State Phenomena. **280** (2018)
17. H. Wang, J. Fang, Z. Xu, and X. Zhang, J Mater Sci: Mater Electron. **26** (2015)
18. B.I. Noh, J.H. Choi, J.W. Yoon, and S.B. Jung, Journal of Alloys and Compounds. **499** (2010)
19. T.T. Bao, Y. Kim, J. Lee, and J.-G. Lee, Materials Transactions. (2010)
20. Y. Liu, F.L. Sun, T.L. Yan, and W.G. Hu, Electronic Packaging Technology & High Density Packaging, 2008. ICEPT-HDP 2008. International Conference on. (2008)

IOP Conf. Series: Materials Science and Engineering **701** (2019) 012025 doi:10.1088/1757-899X/701/1/012025

Effect of crumb rubber type on macroballoons towards physical and mechanical properties in syntactic foams

Zunaida Zakaria[1,*] and *Tiyagu* Sathivelu[1]

[1]School of Materials Engineering, Universiti Malaysia Perlis, Kompleks Pusat Pengajian Jejawi 2, Taman Muhibbah, 02600 Arau, Perlis, Malaysia

Abstract. The effect of crumb rubber types on physical and compression properties of macroballoons in syntactic foams was investigated. Expanded polystyrene (EPS) beads were used as initiation material for starting up the macroballoons. The epoxy and hardener system were prepared to be coated with EPS beads and these coated EPS beads were later cured and post-cured to shrink the EPS beads thus producing epoxy macroballoons (EM). The prepared epoxy macroballoons then were coated with different crumb rubber type which is rejected NBR glove crumb (CG-EM) and waste tire rubber crumb (CT-EM). The EM, CG-EM, and CT-EM were then filled in the epoxy mixture to produce syntactic foam. The fabricated syntactic foam was characterized through physical and compression properties, respectively. Results indicated that the syntactic foam embedded with CT-EM shows high density, reduced radius ratio, and high compressive properties. The syntactic foam embedded with CT-EM shows the highest compressive strength and compressive modulus as compared to syntactic foam embedded with CG-EM and EM. The inclusion of the crumb rubber as the coating layer results in high energy absorption and stress transfer in syntactic foam.

1 Introduction

The microballoons or hollow spheres are normally can be made up of glass, carbon, epoxy [1] and ceramic and widely used in thermal insulation, energy absorption and structural uses due to their unique properties, such light weight, high specific strength and thermal insulation. The hollow sphere in syntactic foam which is currently available is brittle in nature, expensive and involving complicating processing procedure. Waste tires discarded every year are rapidly reducing the available sites for disposal, which is turning out to be a serious environmental issue. Reuse of the waste tire by grounding into crumb rubber helps to reduce the unwanted waste. A few examinations have been completed trying to utilize crumb rubber as filler in development [2-3]. Gloves are commonly recycled such as natural rubber gloves and nitrile gloves. However, nitrile gloves are more preferable compared to the natural rubber gloves due to its excellent properties. The nitrile gloves are having a good solvent resistance and protection over a wide range of chemicals. Besides, it could acts as reinforcing

* Corresponding author: zunaida@unimap.edu.my

filler in the rigid composites by modifying the compressive properties. These investigations show that, despite the fact that the compressive and flexural quality of solid material is brought down by expansion crumb rubber because of the absence of holding, by addition of crumb rubber particles due to the lack of bonding, the crumb rubber particles are found to increase the toughness, crack resistance, shock wave absorption, noise level reduction and also the flexibility of the material. Therefore, in this research by incorporating a layer of rubber waste on macroballoons can produce excellent properties, reduce disposal issues, as well as reduce the production cost of macroballoons.

Syntactic foams are composite materials in which hollow microspheres are embedded in a matrix. Syntactic foams are also known as foam composites since the hollow microspheres can be viewed as reinforcements in a [4-5]. Matrix utilized in syntactic foam incorporates polymers, metals or earthenware production. Polymer matrices are generally classified into thermoplastics and thermosets. Syntactic foams are principally arranged by utilizing thermosetting frameworks on account of the good handling conditions avoiding breakage by gently mixing the hollow microspheres with the thermoset precursor of low density. Thermosetting syntactic foams can be handled at much lower temperatures compared with thermoplastic syntactic foam, accordingly lessening the material and costs for preparing. The blend of diglycidyl ether of bisphenol A (DGEBA), the most broadly utilized epoxy resin used as the matrix in the syntactic foam. Crumb rubber used as the coating layer on macroballoons to expand the strength and stiffness; the epoxy macroballoon serves to decrease the weight; and the rubber coating serves to absorb impact energy and capture microcracks.

2 Materials and methods

2.1 Materials

D.E.R. 331 or diglycidyl ether of bisphenol A (DGEBA), a clear liquid epoxy resin manufactured by Euro Chemo- Pharma Sdn Bhd was selected for the study. Clear epoxy hardener 8161 or isophorone diamine (IPD) was chosen as the curing agent for this study and this chemical was supplied by Euro Chemo-Pharma Sdn Bhd (Malaysia). When used with D.E.R 331 resin, the curing agent would provide a reasonably good curing process within 2 to 7 days at ambient temperature. The Potassium hydroxide (KOH) used for this study was obtained from Sigma – Aldrich Sdn.Bhd. KOH aqueous used as diluent with a concentration 3% (w/w) was added to reduce the viscosity of the resin mixture. Calcium carbonate ($CaCO_3$) was supplied from Sun Minerals Sdn.Bhd is used to ensure that the stickiness problem of the uncured epoxy-coated beads was addressed thus preventing the beads from clumping to each other. The EPS beads were supplied from San Yong Enterprise Sdn. Bhd and in the range of sizes between 3- 6 mm. The waste tire rubber crumb (0.4 mm) were supplied by Gcycle Tyre Recycling Sdn.Bhd and rejected NBR gloves (1.4 mm) were supplied by Top Glove Sdn. Bhd.

2.2 Methods

2.2.1 Preparation of EM

A resin blend containing clear epoxy resin (D.E.R 331) and polyamine hardener (clear epoxy hardener 8161) with 2:1 proportion was formulated and mixed by chopstick for 3 minutes to create the epoxy system. The EPS bead were later included into the epoxy system and filled the container at the half and were completely covered by the epoxy system. The epoxy covered EPS beads were then moved into a polypropylene tray which was readily filled with calcium carbonate powder. The epoxy covered EPS beads were then moved over the calcium carbonate by using wooden chopstick until the point when they completely covered with calcium carbonate to ensure that the stickiness of the uncured epoxy covered globules was tended to accordingly keep the beads from clustering to one another. These coated EPS beads were then cured in an oven for 15 minutes at 80 °C and post-cured at 120 °C for 1 hour 30 minutes to shrink all the EPS beads inside the epoxy-coated spheres to deliver empty structures inside the epoxy-coated spheres. The resultant cured EM was then splashed with compressed air to evacuate abundance $CaCO_3$ powder on their surface.

2.2.2 Preparation of crumb rubber coated EM

A clear epoxy resin (D.E.R 331) and polyamine hardener (clear epoxy hardener 8161) were mixed with 2:1 ratio were formulated and mixed by using chopstick about 3 minutes. The EM was later added into the prepared epoxy system and filled the polypropylene cup at the half. Then the EM was stirred gently in the cup of the mixed epoxy system and ensures they were fully coated with epoxy. Then EM coated with epoxy resin was transferred into polypropylene tray filled with sufficient waste NBR glove crumb. The EM was then rolled on the crumb rubber with a wooden chopstick and ensure they were fully coated to produce NBR glove crumb coated EM (CG-EM). The EM coated with waste NBR glove crumb was then cured in the oven for 15 minutes at 80°C. Similar techniques were used in the preparation of tire rubber crumb coated EM (CT-EM).

2.2.3 Fabrication of epoxy syntactic foam

The pre-determined amount of EM needed in the mould was done by completely filling a polypropylene mould. The EM was then removed from the mould and properly put aside. Next, the resin mixture was prepared by mixing the epoxy resin and the hardener continuously (2:1) together with the 3% KOH solution for about 15 minutes. The amount of KOH solution was fixed at 15% by weight concerning for to the resin mixture. The prepared cured EM was then added at regular intervals into the mixture. Using such a procedure, the uncured matrix compound consisting of EM dispersed in the epoxy matrix was achieved. The mixture was then poured evenly into the mould and a constant load with standard weight (2 kg) was placed on top of the mould lid to maintain the EM in their well-dispersed state. The mixture was left at room temperature to complete the curing process for 24 hours [1]. The cured syntactic foam was then demoulded and cut according to standard dimensions for respective testing after seven days. Cutting process was done after seven days of the composites were produced due to the typical cure schedule (given by supplier) for epoxy resin with polyamine curing agent is around 2 to 7 days at room temperature. Similar techniques were repeated to prepare syntactic foam embedded with CG-EM and CT-EM.

2.3 Testing and Characterization

The morphology of these macroballoons was captured by using Dino-Lite Digital Microscope. The outer and inner radii of the macroballoons were then analysed using the

Image J software to measure the wall thickness of macroballoons. The parameter of radius ratio, η was then calculated using equation below;

$$Radius\ ratio,\ \eta = Inner\ radius,\ r_i / Outer\ radius,\ r_o \quad (1)$$

The density of the syntactic foams was obtained by the mass divided to the volume of syntactic foam in accordance to the ASTM D3574. The mass of the syntactic foams was measured within ±1%. The density of the syntactic foams is calculated by using the following equation;

$$Density = Mass,\ (g)/Volume\ (mm^3) \quad (2)$$

The compression load deflection of the specimens was performed by using Shimadzu Universal Tensile Machine fitted with a compression jig and in accordance to the ASTM D3575 [1]. The specimens with dimension of 40 × 40 × 25 mm were tested with crosshead speed of 2.5 mm/min. The compressive properties of the samples were recorded and analyzed.

3 Results and Discussion

3.1 Wall thickness

The wall thickness of the epoxy macroballoons (EM) are varying with different layers of NBR glove crumb coated EM (CG-EM) and tire rubber crumb coated EM (CT-EM) are shown in Fig. 1. Based on the graph, the mean wall thickness of EM, CG-EM, and CT-EM are 0.67 mm, 1.35 mm and 2.07 mm. There are significant changes in the range of the wall thickness of the EM for a different layer of coating. The sample coated with rejected NBR glove crumb (CG-EM) is 0.68 mm thicker than the sample of control EM. While the wall thickness coated with waste tire rubber crumb (CT-EM) is 2.07 mm which is 1.4 mm thicker than control EM.

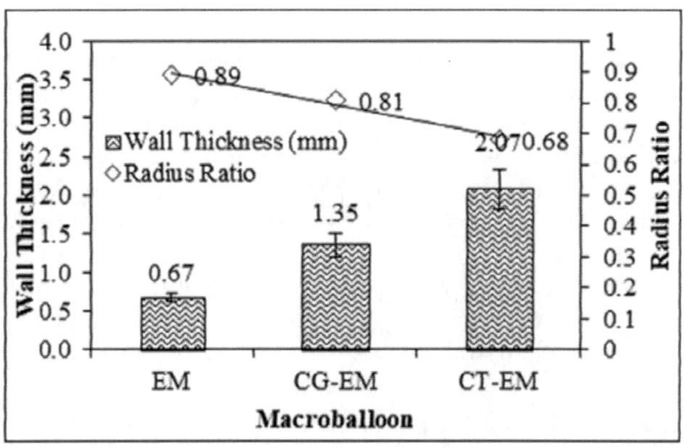

Fig. 1. Wall thickness of macroballoons against different type of coating layer.

The wall thickness of CT-EM is thicker compared to CG-EM and EM. This is because of the size of the coating particle playing a major role in the formation of a layer on the EM surface. The average size of the waste tire rubber crumb is 0.4 mm (40 mesh), whereby the average size of the rejected NBR glove crumb is 1.4 mm (14 mesh) therefore the macroballoons coated with tire crumb rubber forms thicker layer as compared to rejected NBR glove crumb. For the radius, the average radius ratio of control EM, CG-EM and CT-EM are 0.89, 0.81 and 0.68 respectively. It has shown the radius ratio of EM shows the highest and the CT-EM shows the lowest. This is because the wall thickness of macroballoons contributes to the radius ratio and it is inversely proportional to wall thickness [6]. The decrement of the radius ratio as the thickness increases is due to the increment of the outer radius of the macroballoon which can be explained by the formula of the radius ratio in methodology. The microscopic images of control EM, CG-EM and CT-EM are as shown in Fig 2.

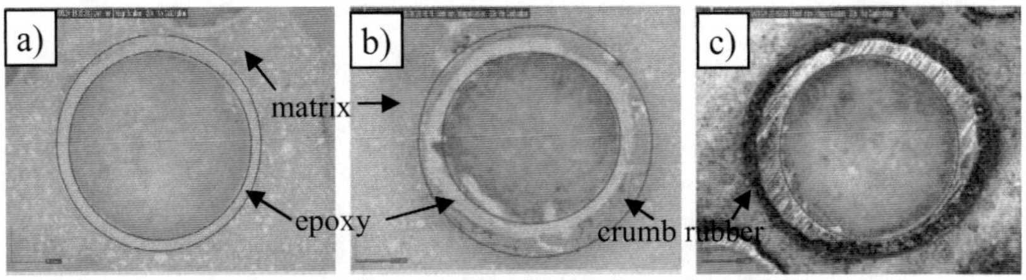

Fig. 2. Microscopic image of macroballoons (a) EM (b) CG-EM and (c) CT-EM at 20X magnification.

3.2 Density

Fig. 3 shows the effect of crumb rubber on the density of syntactic foams. From the Fig. 3, it has shown that the syntactic foam embedded with CT-EM has the highest density compared to the EM and CG-EM. The densities of the syntactic foam depend on the type of coating crumb rubber. As the specific gravity of the type of coating increase, the densities of the syntactic foam increase as well. The densities of the syntactic foam embedded CG-EM has increased by 17.24 % by adding a layer of crumb rubber coated on the EM while the densities of the syntactic foam embedded CT-EM have increased by 20.69 % by adding layers of crumb rubber on the EM. This determines by the specific gravity of the type of coating [7]. The specific gravity of the waste tire rubber crumb is 1.15 whereby the specific gravity of rejected NBR glove crumb is 0.92 therefore as the specific gravity of coating layers increases causes the density of syntactic foam increases.

Furthermore, the coating particle size on macroballoon also affects the density syntactic foam. This is because the finer the size of the coating particles, the higher the area of contact between the macroballoon and the coating layer. Therefore, more number of coating particle will attach to the macroballoon. The size of rejected NBR glove crumb is 1.4 mm (14 mesh) whereby the size of waste tire rubber crumb is 0.4 mm (40 mesh), which is much finer then rejected NBR glove crumb. Therefore more number of waste tire rubber crumb attached on the EM thus contributes to thicker of the wall thickness of the CT-EM and radius ratio decreases which results in higher in the density of syntactic foam [7-8]. Consequently, the thickness of the macroballoon has reduced the concentration of the macroballoons in the matrix of syntactic foams.

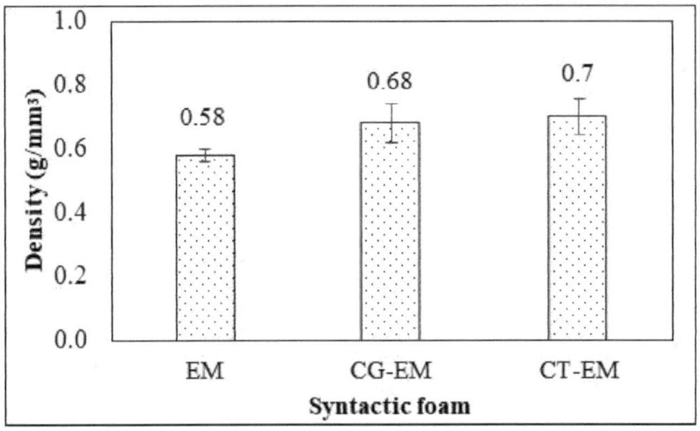

Fig. 3. Density of syntactic foam embedded with EM, CG-EM and CT-EM.

3.3 Compressive properties

The compressive properties of the syntactic foams are varied with a different type of crumb rubber layers on macroballoons are shown in Fig.4. From the Fig. 4, the syntactic foam embedded with CT-EM shows the highest compressive yield strength (14.98 MPa) followed by CG-EM (8.65 MPa) and EM (5.12 MPa). It can be claimed that the presence of crumb rubber layer could absorb deformation energy and delays the crack propagation in syntactic foam during compression process [6]. Therefore, syntactic foam embedded with CT-EM exhibited resistant to compression when a load was applied which resulted in enhanced compressive strength of the syntactic foam. Furthermore, the compressive yield strength of the syntactic foams is depending on the radius ratio, ŋ and size of the coating particles [9]. The decrement of ŋ results in the increment of wall thickness and consequently, it increases the density and thus increases the strength of the syntactic foams. The higher compressive yield strength in syntactic foams embedded with CG-EM and CT-EM have proved that they are stiffer than the syntactic foam embedded with EM. Besides, the epoxy itself is brittle in nature thus reduce the stiffness properties during compression.

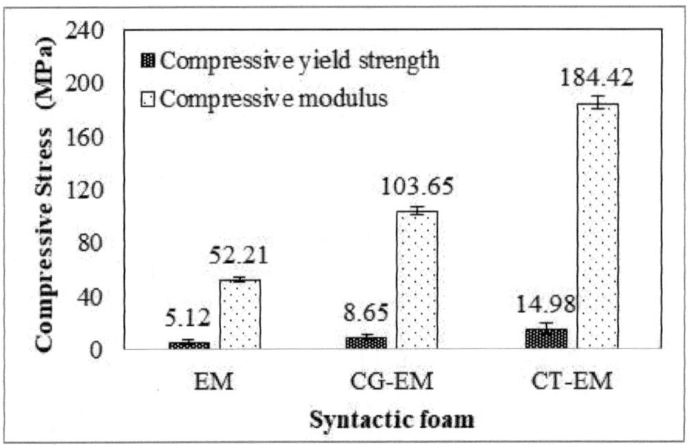

Fig. 4. The compressive yield strength and compressive modulus of syntactic foam embedded with EM, CG-EM and CT-EM.

The compressive yield strength of the syntactic foam is linearly dependent on the density and concentration of the macroballoon in the syntactic foams. The similar trend can be seen in compressive modulus where the compressive modulus increases in density. The syntactic foams embedded with crumb rubber are having significant higher modulus compared to syntactic foam embedded with EM. This is due to the brittle properties of epoxy, thinner wall thickness and containing high porosity in EM contributes to reducing the stiffness properties. The higher packing of macroballoons in the syntactic foam can result in the higher brittleness in syntactic foams [6,10]. Thus the non – permanent deformation of the control EM before the permanent deformation is only 5.12 MPa, while the CG-EM and CT-EM syntactic foam are 8.65 MPa and 14.89 MPa respectively.

Fig. 5 has shown the compressive properties of the syntactic foams up to 60 % deflection. From the Fig. 5, it can be seen that the syntactic foam embedded with CT-EM shows the highest compressive stress as compared to syntactic foam embedded with CG-EM and EM at all compressive strain. This indicated that the inclusive of crumb rubber as coating materials in macroballoons could contribute to the high compressive stress. The sample of the syntactic foam without the aid of crumb rubber (i.e, EM) has the longest and flattest plateau region in the stress-strain curve as compared to the CG-EM and CT-EM which result in slower strain densification. However, the area under the plateau region of the syntactic foam without crumb rubber is the lowest as compared to the coated with crumb rubber syntactic foam. This is due to the aid from crumb rubber tend to reduce the brittleness of the syntactic foam and initiation of stress. The higher energy absorption under the plateau region in the samples of CG-EM and CT-EM syntactic foam is due to ductility of the rubber particles and able to bridging the micro-cracks during compression and higher compression energy can be absorbed [11-12].

Furthermore, the high porosity in EM has extended the formation of the plateau regions of the stress -strain curve where there are more spaces in the syntactic foam to be consumed with constant stress [13]. From the Fig. 5, has shown the densification strain of the syntactic foam embedded with crumb rubber is seen while the densification strain of the syntactic foam embedded with EM is not that obviously occur.

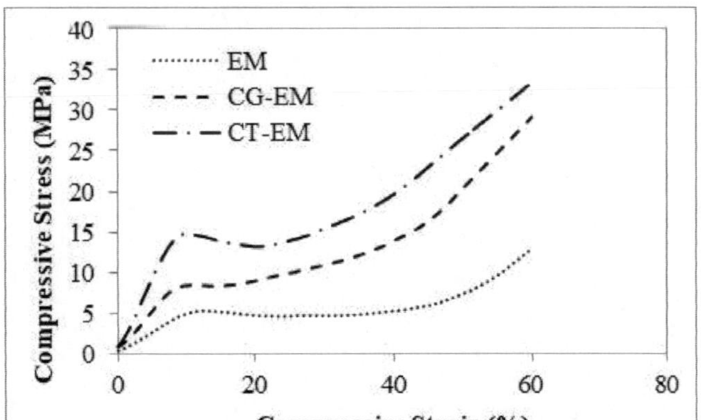

Fig. 5. The stress-strain curve of syntactic foam embedded with EM, CG-EM and CT-EM.

4 Conclusions

The prepared syntactic foam embedded with CT-EM showed great improvement in compressive properties as compared to syntactic foam embedded with CG-EM and EM. We found that the produced crumb rubber coated macroballoons had relatively contributes to high compressive modulus, yield strength, and can be used in various engineering applications. Moreover, increment of wall thickness and consequently, it increases the density in high compressive yield strength. The further studies regarding hybridizing coating between rejected NBR glove crumb and waste tire rubber crumb on the macroballoons could be considered in the future to enhance the utilization of these epoxy macroballoons in their desired applications.

References

1. S.S. Samsudin, Z.M. Ariff, Z. Zakaria, A.A. Bakar, Express Polym Lett, **5(7)**, (2011)
2. N. Sunthonpagasit, M. R. Duffey, Resour Concerv Recy, **40**,(2004)
3. D. Choqueuse, P. Davies, D. Perreux, L. Sohier, J.Y. Cognard, Appl Mech Mater, **24**, (2010)
4. C.S. Karthikeyan, S. Sankaran, J. Kumar, J Appl Polym Sci, **81(2)**, (2001)
5. N. Gupta, R. Nagorny, J Appl Polym Sci, **102(2)**, (2006)
6. Z. Zakaria, C.Y. Yao, Solid State Phenom, **280**, (2018)
7. R. Shanker, S. Sankaran , Mater Sci Eng A, **412(1-2)**, (2005)
8. X. Wu, B. Tang, J. Yu, X. Cao, C. Zhang, C. Y. , Fiber Polym, **19(1)**, (2018)
9. R.R. Maharsia, Development of high performance hybrid syntactic foams: structure and material property characterization (2005)
10. A.M. Zihlif, G. Ragosta, Polym Polym Compos, **9(5)**, (2001)
11. N. Gupta, W. Ricci, Mater Sci Eng A, **427**, (2006)
12. L. Anusha, Z. Zakaria, K.K.Chew, Mater Sci Eng, **548**, (2019)
13. C. Swetha, R. Kumar, Mater. Design, **32**, (2011)

The effect of Aluminium addition on the microstructure of lead-free solder alloys: A short review

Chi Ying Tan[1,*], and *Mohd Arif Anuar* Mohd Salleh[1]

[1]Center of Excellence Geopolymer and Green Technology (CEGeoGTech),
School of Materials Engineering, University Malaysia Perlis (UniMAP),
Kompleks Pusat Pengajian Jejawi 2, Taman Muhibbah, 02600 Jejawi, Arau, Perlis, Malaysia

Abstract. The ban on using lead in electronic packaging has been taken effect since the year 2006. The alternative to replacing lead (Pb) based solder has been given highly concerns by the electronic industry. Both Sn-Ag-Cu and Sn-Cu solder alloys were considered to be the most acceptable candidates to replace SnPb solder alloy. From the findings, it showed that β-Sn appears as island-liked shape and eutectic phases appear as a darker shade. The grain size is refined significantly after incorporated with a suitable amount of Al. When exorbitant amount of Al addition will cause Al-rich particles, which has high potential to acts as the heterogeneous nucleation site and result in a lower degree of undercooling. The most desired amount of Al addition is between 0.025 - 0.05 wt%. The primary intermetallic compounds found in Sn-based solder alloy are Cu_6Sn_5 and Ag_3Sn. New intermetallic compouds were found after addition of Al, which are Cu-Al and Ag-Al intermetallic. The formation of new intermetallic compounds causes the formation of Cu_6Sn_5 and Ag_3Sn to be suppressed.

Introduction

Soldering is a process of assembling or interconnecting two or more electrical components together with solder alloys. Pb based solder alloy has been used since the earlier Bronze Age [1, 2]. However, the ban of using hazardous materials such as leads, mercury, cadmium for electrical and electronic equipment, has been enforced by The Restriction of Hazardous Substances (RoHS) since the year 2006. Therefore, the development of lead-free solder is given extensive concerns. Amongst the lead-free solder, the near eutectic SnAgCu (SAC) solder alloy is found to be the most satisfied candidate to replace Sn-Pb solder, as Ag and Cu have excellent thermal and electrical properties, which fulfilled the primary requirements of the solder alloys. Currently, Sn-3.0Ag-0.5Cu (SAC305) and Sn-3.8Ag-0.7Cu (SAC387) are mainly used in electronic industries [3-7]. But Ag is costly; hence, some researchers are also working on SnCu solder alloy.

* Corresponding author: chiying95@outlook.com

Although SAC and SnCu solder alloys are compatible with Sn-Pb solder alloy, it has some drawbacks and still facing some challenges at the current stage. For instance, poor wettability, reliability and high melting point, are failed to meet the needs of current industry. Furthermore, the strength of the solder joint of current lead-free solder may degrade after undergoes a reflowing process or expose to external impacts or harsh enviroment condition [8-11]. This is because brittle fracture may happen as the rapid growth of the interfacial layer on the interfaces. Consequently, many researchers also have been added rare earth elements and metal alloy elements into SAC and SnCu solder alloys in order to achieve better performance and reliability of solder alloys [9, 12-24]. The performances of incorporated solder alloys can be enhanced by controlling the thickness of intermetallic compound (IMC), alter the morphology, prevent phases changes and increase the toughness.

The features that included in microstructure analysis are: (1) grains size, (2) grains boundaries, (3) interfaces, (4) phases precipitation and so on. These features can be altered with respond to surrounding conditions, such as external loads, impacts and temperature [25]. So, these make microstructure as one of the most accurate and direct reliability assessment of solder joint. Also, it enables the development of more reliable solder interconnection. This is why microstructure analysis is important in failure analysis of a material. The Cu_6Sn_5 and Ag_3Sn intermetallic (IMC) can easily be seen in the microstructure of SAC solder alloy. The formation of these IMC are due to the atomic diffusion between Cu substrate and Sn- based solder during the reflow process. Yet, these IMCs are brittle in nature, thus, undesire amount of IMCs can weaken the solder joint reliability. For that reason, researchers found that additional element can suppress the formation of Cu_6Sn_5 and Ag_3Sn by forming another new intermetallic. This paper is aimed to review the effect of Al addition on the microstructure of lead-free solder alloy.

2 Effect of Al Addition on the Microstructure

2.1 Effect on SAC solder alloys

Fig 1 shows the microstructure of Sn 3.0Ag 5.0Cu (SAC35) with different amount (0.025 – 0.2wt%) of Al addition. The presence of primary intermetallic dendrites and eutectic phase can be clearly seen in Fig 1. The primary intermetallic dendrites were known as Cu_6Sn_5, which appeared in the darker region, whereas the eutectic phase, the β-Sn appeared in the brighter region. The length of Cu_6Sn_5 was shortened and the growth of secondary dendrite was restricted after the addition of Al into SAC35 solder alloy. The shortest length of Cu_6Sn_5 was found at Al content of 0.025 wt%. When exorbitant (>0.05wt%) of Al added, the intermetallic became slightly longer. Furthermore, the exorbitant of Al addition may causes agglomeration, which has high potential to act as a heterogeneous nucleation site for Cu_6Sn_5 [26]. Al addition has boosted intergranular eutectic nucleation, which resulting in eutectic grain refinement. It is expected that the mechanical strength of a reasonable amount of Al added solder will be better than pure SAC solder, as the grain size is refined and it can impede dislocation motion [16].

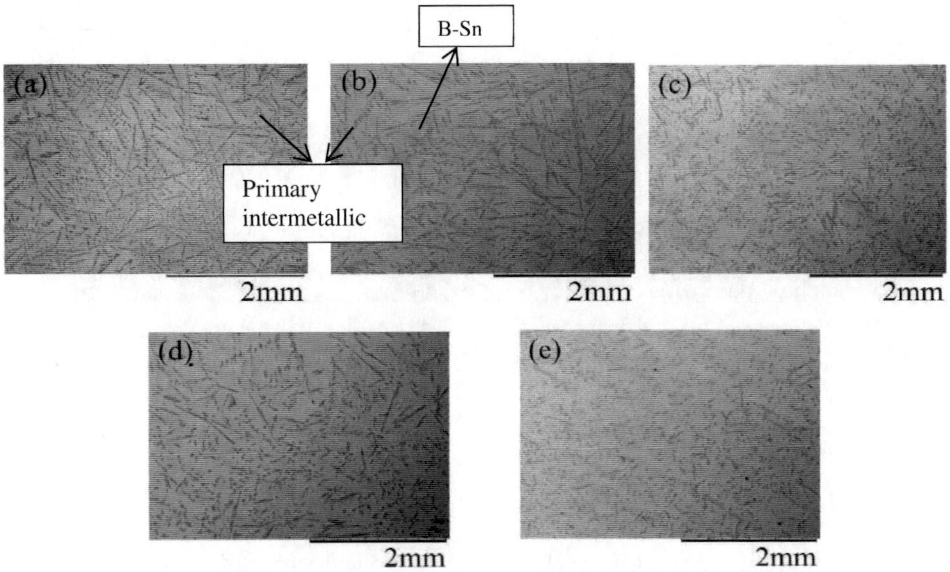

Fig 1: SEM micrographs of SAC35 solder alloys with addition of Al at magnification of 40x. (a) 0 wt% Al, (b) 0.025 wt% Al, (c) 0.05 wt% Al, (d) 0.1 wt% Al, and (e) 0.2 wt% Al [27].

In addition, the other intermetallic compounds such as CuAl, $CuAl_2$, AgAl, and Ag_3Al also can be found in Al added SAC solder alloys. These new IMCs can be seen in Fig 2 and they are more favorable to be formed compared to Cu_6Sn_5 and Ag_3Sn. This is because Cu-Al and Ag-Al IMC have bigger electronegativity [28], thus, they are more likely to be formed instead of Cu_6Sn_5 and Ag_3Sn. However, the $CuAl_2$ and AgAl are formed only at higher Al content sample. Moreover, Al-Sn-Cu IMC also can be appeared, this IMC is undesirable as it is expected to decrease the shear strength of the solder alloy.

Fig 2: SEM micrographs of SAC0305-2Al bulk solder at a magnification of 1000x [29].

Besides, the addition of Al into SAC solder alloys can suppress the formation of Cu_6Sn_5 also have been proven after reflowed on Cu substrate. As shown in Fig 3, the SAC0305 with 1 wt% Al were showed the thinnest IMC layer as compared with SAC0305 and SAC0305-2 wt% Al. The Cu_6Sn_5 layer appears in scalloped shaped, which is formed by diffusion process between solder alloy and Cu substrate [30].

Fig 3: Optical images at the interface of solder-substrate after relowing process, (a) SAC0305, (b) SAC0305-1Al, and (c) SAC0305-2Al [29].

2.2 Effect on SnCu solder alloys

Fig 4 reveals the optical microstructure of Sn-0.7Cu and Sn0.7Cu with different amount of Al addition (0.01 – 0.075wt%). Same as Al addition in SAC solder alloy, the β-Sn and primary intermetallic Cu_6Sn_5 phases can be found in the samples. From the findings, the microstructure of β-Sn and Cu_6Sn_5 are refined and the size becomes more uniform [31]. A similar result has been showed by N. Adli *et. al.* , the particle grains of Sn0.7Cu with 1 wt% Al are formed in almost same size and shape. Li Yang *et. al.* explained that microstructure refinement was caused by dissolution of Al atoms into β-Sn and eutectic [31]. The solid solution strengthening by Al modified the microstructure of the intermetallic to becomes finer and more uniform, which results in a better strength of solder joint [32, 33]. This is due to the fact that smaller grain size will give a larger surface of grain boundaries. Plevachuk *et. al.* [34] stated that, Cu-Al phase is considered softer than Cu_6Sn_5, which may increase the ductility characteristic to solder alloy. Nevertheless, some black particles are also can be seen in Fig 4 (b-e). These particles are isolated Al and agglomerated Cu.

Fig 4: The microstrucuture of Sn-0.7Cu-xAl (0.01 – 0.075 wt%) under optical microscope. (a) 0wt% Al, (b) 0.01wt% Al, (c) 0.025wt% Al, (d) 0.05wt% Al, and (e) 0.075wt% Al [31].

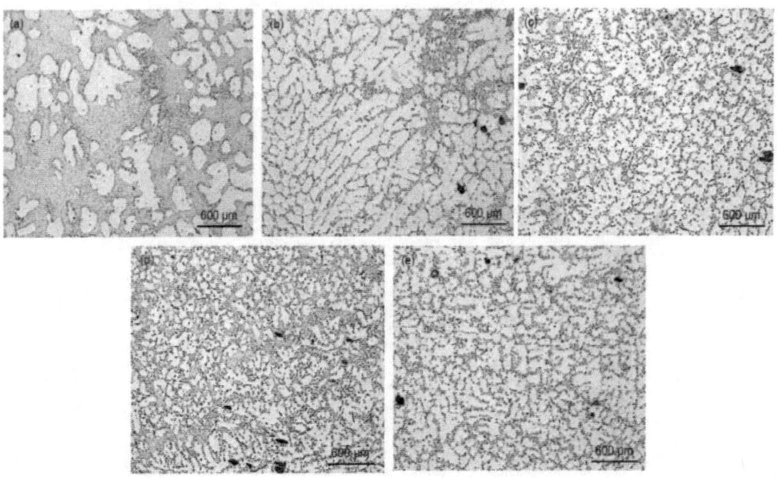

3 Summary

In summary, the microstructure of solder alloy with Al addition was systematically reviewed in this paper.

- It would seem that the microstructure of lead-free solder alloy was significantly refined with a proper amount of Al addition. The mechanical properties will be enhanced by the finer microstructure as it can impede dislocation motion of atoms.
- However, excessive Al addition will cause the formation of Al-rich phases. The Al-rich particles have high potential to act as a heterogeneous nucleation site, which will reduce the degree of undercooling.
- Lastly, the formations of Cu-Al and Ag-Al IMCs have suppressed the formation of Cu_6Sn_5 and Ag_3Sn IMCs. As a result, the interdendritic region is getting lesser.

This work was supported by an Institutional Links grant, ID 332397914, under the Newton-Ungku Omar Fund partnership. The grant is funded by the UK Department of Business, Energy and Industrial Strategy (BEIS) and Malaysia and delivered by the British Council.

Reference

1. R.M. Said, M.A.A.M. Salleh, M.N. Derman, M.I.I. Ramli, N.M. Nasir, and N. Saud, Key Engineering Materials. **700** (2016)
2. M.I.I. Ramli, N. Saud, M.A.A.M. Salleh, M.N. Derman, R.M. Said, and N. Nasir, Materials Science Forum. **803** (2015)
3. S. Cheng, C.-M. Huang, and M. Pecht, Microelectronics Reliability. **75** (2017)
4. H.R. Kotadia, P.D. Howes, and S.H. Mannan, Microelectronics Reliability. **54**(6-7) (2014)
5. M.A.A.M. Salleh, M.H. Hazizi, M.A.B.M. Abdullah, N. Noriman, R. Mayapan, and Z.A. Ahmad, Advanced Materials Research. **626** (2012)
6. M.A.A.M. Salleh, A.M.M.A. Bakri, F. Somidin, and H. Kamarudin, International Review of Mechanical Engineering. (2013)
7. M.A.A.M. Salleh, M.M.A.B. Abdullah, F. Somidin, A.V. Sandu, N. Saud, K. Hussin, S.D. McDonald, and K. Nogita, Revista de Chimie **64**(7) (2013)
8. S.G. Desmarest, Materials Science and Technology. **28**(3) (2012)
9. M.I.I. Ramli, M.S.S. Yusof, M.A.A.M. Salleh, R.M. Said, and K. Nogita, Solid State Phenomena. **273** (2018)
10. F. Somidin, H. Maeno, M.A.A.M. Salleh, X.Q. Tran, S.D. McDonald, S. Matsumura, and K. Nogita, Materials Characterization. **138** (2018)
11. K. Nogita, M.A.A.M. Salleh, S. Smith, Y. Q Wu, S.D. McDonald, A.G.A. Razak, T. Akaiwa, and T. Nishimura, 2017 International Conference on Electronics Packaging (ICEP). (2017)
12. K. Kanlayasiri and R. Kongchayasukawat, Transactions of Nonferrous Metals Society of China. **28**(6) (2018)
13. X. Tu, D. Yi, J. Wu, and B. Wang, Journal of Alloys and Compounds. **698** (2017)
14. H.-Y. Huang, C.-W. Yang, and Y.-C. Peng, Science and Engineering of Composite Materials. **23**(6) (2016)
15. Q. Tao, L. Benabou, L. Vivet, V. Le, and F. Ouezdou, Materials Science and Engineering: A. **669** (2016)

16. K. Maslinda, A.S. Anasyida, and M.S. Nurulakmal, Journal of Materials Science: Materials in Electronics. **27**(1) (2016)
17. Q.K. Zhang, W.M. Long, X.Q. Yu, Y. Pei, and P. Qiao, Journal of Alloys and Compounds. **622** (2015)
18. L. Li, Y. Cheng, G. Xu, E. Wang, Z. Zhang, and H. Wang, Materials & Design. **64** (2014)
19. A. Kantarcıoğlu and Y. Kalay, Materials Science and Engineering: A. **593** (2014)
20. A. El-Daly, A. Hammad, A. Fawzy, and D. Nasrallh, Materials & Design. **43** (2013)
21. W. Dong, Y. Shi, Y. Lei, Z. Xia, and F. Guo, Journal of Materials Science: Materials in Electronics. **20**(10) (2009)
22. M.G. Cho, S.K. Kang, D.-Y. Shih, and H.M. Lee, Journal of Electronic Materials. **36**(11) (2007)
23. K. Nogita, M.A.A.M. Salleh, E. Tanaka, G. Zeng, S.D. McDonald, and S. Matsumura, JOM. **68**(11) (2016)
24. M.I.I. Ramli, M.A.A.M. Salleh, I. Ibrahim, and R.M. Said, Solid State Phenomena. **280** (2018)
25. Tae-Kyu Lee, Thomas R. Bieler, Choong-Un Kim, and H. Ma, *Fundamentals of Lead-Free Solder Interconnect Technology*. 1 ed2015: Springer US.
26. M.F.M. Sabri, D.A. Shnawah, I.A. Badruddin, S.B.M. Said, F.X. Che, and T. Ariga, Materials Characterization. **78** (2013)
27. X.F. Tan, M. Salleh, M.A. Anuar, S.D. McDonald, and K. Nogita. Trans Tech Publ.
28. Y.M. Leong and A.S.M.A. Haseeb, Materials. **9**(7) (2016)
29. M.N. Nadhirah, K. Maslinda, M. Nurulakmal, and A. Anasyida, Procedia Chemistry. **19** (2016)
30. O.M. Abdelhadi and L. Ladani, Journal of Alloys and Compounds. **537** (2012)
31. L. Yang, Y. Zhang, J. Dai, Y. Jing, J. Ge, and N. Zhang, Materials & Design. **67** (2015)
32. M.F.M. Sabri, D.A. Shnawah, I.A. Badruddin, and S.B.M. Said, Journal of Alloys and Compounds. **582** (2014)
33. A. El-Daly, A. El-Taher, and S. Gouda, Materials & Design (1980-2015). **65** (2015)
34. Y. Plevachuk, V. Sklyarchuk, A. Yakymovych, S. Eckert, B. Willers, and K. Eigenfeld, Metallurgical and Materials Transactions A. **39**(12) (2008)

IOP Conf. Series: Materials Science and Engineering **701** (2019) 012027 doi:10.1088/1757-899X/701/1/012027

Preliminary study of zero charge corrosion protection on steel plate embedded in geopolymer paste

Siti Aisyah Razak[1*], *Farah Farhana* Zainal[1], and *Shaiful Rizam* Shamsudin[1]

[1]Center of Excellence Geopolymer Geopolymer and Green Technology (CEGeoGTech), School of Materials Engineering, University Malaysia Perlis (UniMAP) P.O. Box 77, D/A Pejabat Pos Besar, 01000 Kangar, Perlis, Malaysia.

Abstract. Steel reinforced concrete usually protected against corrosion by passivation of the steel arising from the high alkalinity of the pore solution within the geopolymer concrete. The ingress of chlorides to the steel or concrete interface or carbonation of the concrete reducing the alkalinity of the pore solution at the steel or concrete interface. This will cause loss of durability in reinforced concrete. The corrosion behavior of steel embedded in geopolymer paste was studied by open circuit potential, E_{ocp} and potential zero charge method E_{pzc} under 3.5% NaCI solution exposure. The E_{corr} (-660 mV vs SCE) more negative than E_{pzc} value (-630 mV vs SCE). It shows the steel plate have an excessive negative charges. It is noted that the zero charge corrosion protection (ZCCP) method successfully prevents the steel surface suffering from corrosion. Therefore, it is strongly believed that metal under E_{pzc} is not in contact with H+ and OH^- ionic charges due to the absence of electrical charge as well as electrostatic attraction on metal surface.

1 Introduction

The major durability issues faced by infrastructure components are corrosion of steel in reinforced concrete structures, and it also influences the lifespan of the structure. Nowadays, the prevention of corrosion prevention methods still in progress, such as one of the efforts being made is the development of components of materials such as geopolymers to replace, OPC.

The durability of concrete is important role in infrastructure, it depends on the interactions of the material with the environment. The application of geopolymers could enhance the corrosion protection by reducing the corrosion rate of steel in the concrete [1, 2]. The alkalinity environment of geopolymers which is pH of 12 to 13 will protect the steel reinforced on steel surface or passivation [3].

* Corresponding Author: ctaisyahrazak@gmail.com

According to ACI 222, the passive film does not actually stop corrosion, it reduces the corrosion rate to an insignificant level which is 0.1 µm per year and without the passive film, the steel would corrode at rates at least 1,000 times higher [4]. The destruction of the passivating layer occurs when the alkalinity of the concrete is reduced or when the chloride concentration in concrete is increased to a certain level [5].

Therefore, the steel reinforced exposed to the active environments such as sea bridge and costal area can corrode, it is because the sea water and it contains high chlorides which reduce the alkalinity of concrete and due to outside moisture in coastal area [6]. If the porous concrete is permeable the sea water in the form of moisture will enters in to the concrete form the porous surface.

Geopolymer binders have a nano-porous pore system, which are the reduction of ionic diffusion and low permeability. It is due to the steel rebar embedded Geopolymer concrete is low water absorption when less of porosity [7]. Previous researcher, have stated that the water absorption of fly ash geopolymer is less than that of OPC paste [8, 9]. Table 1 gives the results of water absorption tests for fly ash geopolymer and OPC paste.

Table 1. Water absorption of fly ash geopolymer and OPC paste.

Days	7	14	28	90
Water absorption%, geopolymer paste	16	12.4	9.3	7.2
Water absorption %, OPC paste	24	20	12	10

The result shows, a percentage of the water absorption geopolymer paste is smaller than OPC paste with long period which is at 90 days has different percentage of water absorption for both sample was 2.8%. Geopolymer paste has concluded the decrease of pore size or water absorption with long time [8].

The less of water absorption still affect the corrosion resistance, which are the low pH of sample, iron atoms lose electrons and move into the surrounding concrete as ferrous ions. Anodic reaction is represented as $2Fe \rightarrow 2Fe^{2+} + 4e$. [5].The metal surface charge is a positive charge. Meanwhile, at the high pH, a thin oxide layer forms on the steel and prevents metal atoms from dissolving. The reaction at the cathode is called a water reduction reaction $2H_2O + O_2 + 4e- \rightarrow 4OH$ [5]. The surface charge is negative charge and to maintain electrical neutrality. The ferrous ions migrate through the concrete pore water to these cathodic sites and combine to form iron hydroxides.

The corrosion protection of steel reinforced has been introduced by many researchers, but most of technique provide or donate electrons from anode to cathode such as cathodic protection and impressed current corrosion protection (ICCP). However, it is known that the metal will corrode and break down after a certain period of time.

Thus, this process makes corrosion protection very important in extending the lifespan of the metal from damage. The corrosion can be controlled and prevented by bringing the potential of the metal to a zero charge level. It is plays major role in eliminating of electric double- layer, ionics adsorption and neutralization of organic molecules on the electrode [10].

Syazwan et. al. has concluded that the pH variations with the presence of Cl- ions affect the E_{pzc} value. The high of pH value shows that the small value of E_{pzc} and steadily stable of current when a passive layer exist [11]. From the previous study, the value of open circuit potential and zero charge potential for mild steel, the Table 1 shows that, the value of E_{corr} is more than E_{pzc}, it is due to the acidic solution and low pH.

Table 2. The value of E_{corr} and E_{pzc} (mV vs SCE) for mild steel using EIS method.

Solution	E_{corr}	E_{pzc}	Rreferences
1.0MHCI + 0.5M H_2SO_4	-427	-483	Mallaiya et al. [12].

IN HCI	-505	-530	Kumar & Karthikeyan [13].
0.5M H_2SO_4	-503	-550	Prathibha et. al. [14].

2 Methodology

The geopolymer paste raw materials are fly ash class F, sodium hydroxide (NaOH) and sodium silicate (Na_2SiO_3) solution [12]. AISI 1016 steel plate used as reinforcement and embedded in geopolymer paste. The dirt and rust was removed and cleaned according to ASTM G1-03(2017) [16]. Sodium hydroxide (NaOH) solution with optimum molarity 12 M was prepared by dissolving of its flake into 1 litre of distilled water. A combination of sodium silicate (Na_2SiO_3) and NaOH solution (ratio was 2.5:1) as an alkaline activator for geopolymerisation prior to mixed with fly ash, where are solid/liquid ratio was 2:1. Then, the mixture was placed in a mold with dimension of 50 x 50 x 50mm. A retort stand was used to clamp the steel. It is because the paste was still in liquid form and steel cannot stand on its own as graphical illustration of sample preparation shown in Figure 1. The samples kept in the molds until its became hardened and after 24 hours, the samples were taken out from the molds and cured at 60^{0}C in the oven for 24 hours.

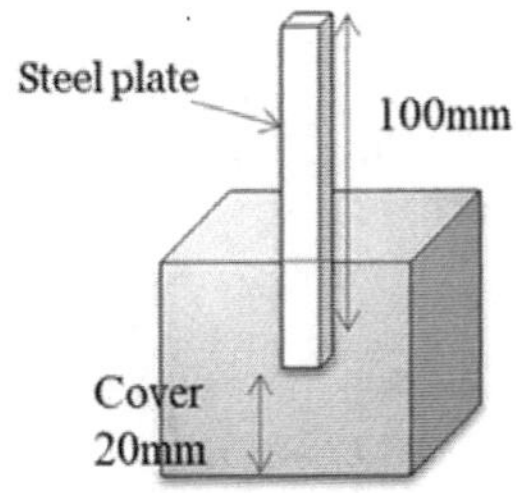

Fig. 1. Sample preparation.

The corrosion behaviour of steel embedded geopolymer was measured by open circuit potential, E_{ocp} using a high-impedance voltmeter based on saturated calomel electrode (SCE) [17]. Potential of zero charge, E_{pzc}, is measured by specific method in closed circuit using pulsed power supply at 100Hz of frequencies. The potential of the steel plate in geopolymer paste is measured by using the similar voltmeter and reference electrode as E_{corr}.

3 Results and Discussion

Fig. 2 shows that the first 6 days, the corrosion potential E_{ocp} of steel embedded geopolymer paste is drop sharply due to initial formation of passive layer in contact with 3.5 wt% of NaCl solution. After days 13, E_{corr} is decrease uniformly until days 28. It is because the growth of passive layer is slowly stable. After days 17, E_{corr} reach the steady state potential at -613 mV vs SCE. This potential is assumed as E_{corr}, of steel embedded geopolymer in NaCl solution.

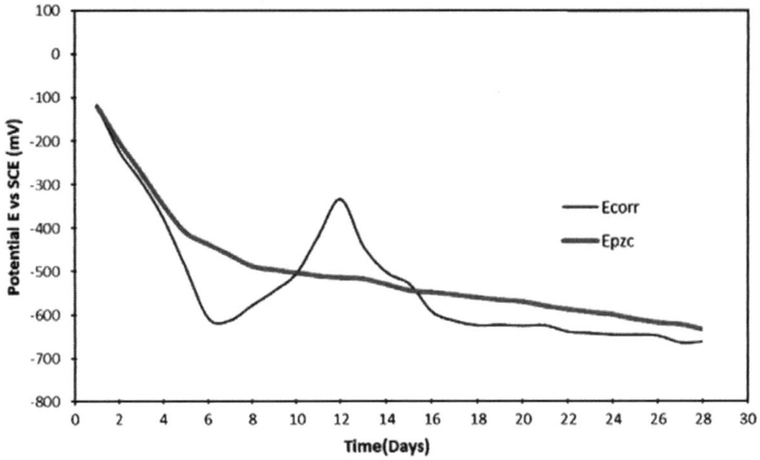

Fig. 2. Potential, E (volt vs SCE) against Time (Days).

On the other hand, E_{pzc} curves show a smoothly exponential reduction. The steel embedded geopolymer under zero charge corrosion protection (ZCCP) recorded -630 mV vs SCE. This -630mV vs SCE is stated as zero charge potential (E_{pzc}) is achieved just when the steady state curve achieved and the AC voltage recorded as V_{dc} = 0mV by voltmeter. This zero charges potentials, 0 mV, means that there is limited or no charges different on steel surface. In this condition, the electrode- electrolyte interface is no longer absorbed any negative or positive charges to the metal surface, hence prevents any charges transfer within the interface as reported by Shaiful [18]. The zero charge of the metal-solution interface is the elimination of electric double layer where it weakens the over-potential energy to react among negative and positive charges at the metal-solution interface. The absence of this reaction energy resulted in the slowdown of the charges reaction which would eventually eliminated the effects of corrosion.

IUPAC defines the potential difference with respect to the potential of zero charge, E_{pzc} [19]:

$$E_{pzc} = E_{corr} - E_{\sigma=0} \quad (1)$$

where, E_{pzc} is the electrode potential difference with respect to the point of zero charge, E_{corr} that the potential of the same electrode against a defined reference electrode and $E_{\sigma=0}$ is the potential of the same electrode when the surface charge is zero, where are in the absence of specific adsorption other than that of the solvent.

According to Figure 2, E_{pzc}, are found to be more positive as compared to steady state potential, E_{corr} (-660mV vs SCE). This situation is very different from previous findings in acid electrolytes (see Table 2). In acidic aqueous, it contains the positive charges (H+) that favor the metal surface to have excessive positive charges. However, in this experiments, the aqueous solution with a pH of ~12 have the negative charges (OH^-) that dominates the solution which are believed to be due to alkaline medium in a geopolymer paste. This condition causes the metal surface to be also surrounded by excessive negative charges. When the ZCCP method was applied to the metal embedded in the geopolymer paste, it was found that, $E_{pzc} > E_{corr}$. This means that the power supply supplies positive charge pulses to remove a negative charge on the metal surface. This process occurs until the metal surface loses all electrical charges. The removal of electrical charges at the electrode- electrolyte interface (as a capacitor discharging) leads the desorption of ionic charges which causes the metal surface no longer suffers a corrosion reaction due to the absence of a charge transfers. It is strongly believed that metal under E_{pzc} is not in contact with H+ and OH^{-} ionic charges due to the absence of electrical charge as well as electrostatic attraction on metal surface.

4 Conclusion

The corrosion can be protected by bringing the potential of the steel embedded geopolymer to a potential of zero charge. The corrosion potential E_{corr} (-660 mV vs SCE) is more negative than E_{pzc} (-630 mV vs SCE). It is shows the steel embedded in geopolymer paste in 3.5% NaCl solution have an excessive negative charge. In E_{pzc}, the steel surface does not suffer from corrosion reaction as it is not in contact to H+ and OH^{-} based molecules due to the absence of electrical charge as well as electrostatic attraction.

The authors would like to thank the staffs of Centre of Excellence Geopolymer & Green Technology (CEGeogGTech), School of Materials Engineering, Universiti Malaysia Perlis (UniMAP) for their involvement in the research. This work was supported and funded by the Fundamental Research Grant Scheme (FRGS) under a grant number of FGRS/1/2018/TK06/UNIMAP/02/2 from the Ministry of Education Malaysia.

References

1. Z. Farhana, H. Kamarudin, A. Rahmat, M. Al Bakri Abdullah, & S. Norainiza, Adv. Mater. Res., **795**, 509-512 (2013)
2. Y. Asmara, J. Siregar, C. Tezara, W. Nurlisa, & J. Jamiluddin, Inter. J. Cor., 2016, 1-5 (2016)
3. Z. F. Farhana, H Kamarudin, A. Azmi Rahmat, M. Mustafa Al Bakri, Aust. J. Basic Appl. Sci. **7**(5): 230-235 (2013)
4. ACI Committee 222, ACI 222R-01, American Concrete Institute, Farmington Hills, Michigan, (2001)
5. Types and Causes of Concrete Deterioration. Portland Cement Association PCA R&D Serial No. 2617 (2002)
6. Causes of Corrosion of Reinforcement Steel in Concrete. Retrieved 24 October 2019, from https://gharpedia.com/blog/causes-of-corrosion-of-reinforcement-steel/ (2019)
7. Z. Farhana, H. Kamarudin, A. Rahmat, & A. Al Bakri, *Materials Science Forum*, **803**, 166-172 (2014).
8. S. Alehyen, M. EL Achouri, M. Taibi , J Mater. Envi. Sci., **8**, (2018)
9. S. Luhar, U. Khandelwal, Inter. J Civ. Eng., **8**, (2015)
10. A.N. Frumkin, O.A Petrii., B.B. Damaskin, *Comprehensive Treatise of Electrochemistry* Springer, Boston, MA (1980)
11. M. Sanusi, S. Shamsudin, A. Rahmat, & R Wardan. , *Inter J Curr. Res. Sci. Eng. Tech.,* **1,** (2018)
12. Zainal, F.F., Hussin, K., Rahmat, A., Al Bakri Abdullah, M.M., Rizam, S., Selimin, M.T., Sandu, A.V, *Key Engineering Materials*, (2015)
13. K. Mallaiya, P.Thanapackiam, S. Rameshkumar, & S.S.Subramanian *Anti-Corrosion Methods And Materials*, **64(1)**, (2011)
14. S. Hari Kumar and S. Karthikeyan, J. Mater. Environ. Sci. **3(5)** (2012)
15. B.S. Prathiban, IOSR J Appl. Chem., **2(1)**, 45-53(2013)

16. ASTM G1-03 Standard Practice for Preparing, Cleaning and Evaluating Corrosion Test Specimens (2017)
17. ASTM WK66184, New Practice for Standard Practice for Revision of Standard that use the SCE (Calomel) Reference Electrodes for Determining Electrode potentials. (2018).
18. S. Shaiful Rizam, Thesis for: Doctor of Philosophy. University Malaysia Perlis (2016).
19. A.D. McNaught and A. Wilkinson. IUPAC (International Union of Pure and Applied Chemistry) *Compendium of Chemical Terminology*, 2nd ed. (the 'Gold Book'). *Blackwell scientific Publication* (2019).

IOP Conf. Series: Materials Science and Engineering **701** (2019) 012028 doi:10.1088/1757-899X/701/1/012028

Influence of NiO to SDC ratio on the properties of Ni-SDC cermet prepared via reduction process

Salmie Suhana Che Abdullah[1,2]* and *Siti Sarah* Ismail[1]

[1] School of Materials Engineering, Universiti Malaysia Perlis, Kompleks Pusat Pengajian Jejawi 2, 02600 Arau, Perlis, Malaysia
[2] Centre of Excellence Frontier Materials Research, Universiti Malaysia Perlis, Seriab, 01000 Kangar, Perlis, Malaysia

Abstract. This works presents the preparation of Ni-$Ce_{0.8}Sm_{0.2}O_{1.9}$ (Ni-SDC) ceramic-metal from reduction of NiO-$Ce_{0.8}Sm_{0.2}O_{1.9}$ (NiO-SDC) ceramics for application of solid oxide fuel cell. NiO-SDC powder with different weight percentage of NiO and SDC (100% NiO + 0% SDC), (80% NiO + 20% SDC), (70% NiO + 30% SDC) and (60% NiO + 40% SDC) was pelletized and sintered at 1400 °C for 4 hours. Then, pellet samples were undergone reduction process using mixed gas of hydrogen and nitrogen gas with concentration of 40% H_2 and 60% N_2 at 800 °C for 2 h. Phase analysis was done using X-ray Diffraction and it was found that for all samples, NiO was completely reduced to Ni, meanwhile SDC was remained in its phase. The density of the samples decreased after reduction, results in more porous samples were obtained. It was found that increasing NiO content lead to sample with higher porosity. The surface morphology of the samples was observed by using the Scanning Electron Microscopy.

1 Introduction

Solid oxide fuel cell (SOFC) is an electrochemical device that can generate electricity directly from chemical reaction. It has many advantages such as high efficient, low emission, and fuel flexibility. For that reason, it has been a subject for extensive research, both in development of materials, as well as for the systems [1].

One of the most important parts of SOFC is the anode. Anode materials must possesses good electrochemical activity to oxidize fuels, high electronic conductivity, and having thermal expansion compatibility with other components in SOFC. Furthermore, it is known that fine particle size, large surface area, adequate porosity and sufficient content of Ni in the anode are essential to achieve high performance [2-4]. Although common anode material for SOFC consist of NiO and yttria stabilized zirconia (YSZ), researchers found that anode material that consist of NiO and ceria (CeO_2) based oxide offer better performance than NiO-YSZ due to higher catalytic activity, has mixed electrical conductivity, less polarized and offer more stable operation [5-7].

Performance of anode dependence on its preparation method. Several techniques have been exploited to produce porous anode structure. Macrospores promote rapid gas transport

* Corresponding author: salmie@unimap.edu.my

through the porous electrode and nanopores provide a high surface area for gas adsorption or desorption. Available techniques that are frequently uses are ultrasonic spray pyrolysis [8, 9], plasma sprayed method [10-11], addition of pore-forming agent [12-14], using porous scaffolds [15, 16], and phase inversion [17]. An alternative approach to prepare anode is via reduction process. Reduction process is a simple, low cost and effective method to produce metal from ceramic metal oxide. Reduced ceramic metal oxide to metal during reduction process increases the porosity due to volume contraction between metal oxide and metal, resulted in high electrical conductivity and sufficient electrochemical activity of the anode due to high nickel activity for H_2 oxidation [18-20]. It has been reported that, NiO shows structural transformation when reduce under H_2, and the transformation of the metal oxide to metal is able to achieved in a very short time [21].

Properties of the reduced Ni powder are dependent on many parameters such as the temperature, holding time, and hydrogen gas flow [22]. Therefore, the aim of this study is to synthesis porous Ni-SDC cermet anode material from NiO-SDC by reduction process using H_2 gas. The influence of the ratio of NiO to the SDC on the phase transformation, porosity and microstructure was investigated.

2 Methodology

2.1 Sample preparation

2.1.1 Preparation of $Ce_{0.8}Sm_{0.2}O_{1.9}$ (SDC) *powder*

Stoichiometric $Ce_{0.8}Sm_{0.2}O_{1.9}$ (abbreviated as SDC) powder was prepared from Cerium Oxide, CeO_2 (Acros Organics, purity of 99.9%) and Samarium Oxide, Sm_2O_3 (purity of 99.95%). Raw materials were ground using agate mortar for 1 h with addition of ethanol to obtained homogeneous powder. Then, mixed powder was calcined at 800, 900 and 1000 °C for 3 h.

2.1.2 Preparation of NiO-SDC *pellet*

Table 1. Composition of NiO and SDC for samples.

Sample	Ratio (wt. %)	
	NiO	SDC
100NiO-0SDC	100	0
80NiO-20SDC	80	20
70NiO-30SDC	70	30
60NiO-40SDC	60	40

As prepared SDC powder was mixed with NiO (Acros Organis, purity of 97%) with various composition, as can be seen in Table 1. These powder were mixed using mortar for 1 h, and ethanol was used as mixing aid. Homogeneous powder were pelletized using uniaxial press under 3.44 MPa with holding time of 5 min, followed by cold isostatic press under 103

IOP Conf. Series: Materials Science and Engineering **701** (2019) 012028 doi:10.1088/1757-899X/701/1/012028

MPa for 5 min. The green pellets has dimension of 15 mm in diameter and 4-5 mm in thickness. Then, pellets were sintered at 1400 °C for 4 h.

2.2 Reduction process

Pellet samples were heated up to 800 °C with heating rate of 10 °C/min. Once furnace temperature reached 800 °C, H_2 and N_2 gases were supplied to the samples for 2 h. Concentration of gas mixture was fixed at 40% H_2 and 60% N_2 with total mass flow of 100 ml/min.

2.3 Characterization

2.3.1 Phase identification

X-ray Diffraction, XRD (Bruker D2 Phaser) was used to identify the phase of all pellet samples before and after reduction process. The scan range and scan rate is 20-80° and 1s/step, respectively.

2.3.2 Density

Apparent density of the samples before and after reduced under H_2 gas was measured using Archimedes method. Apparent density (ρ), theoretical density (ρ_{th}) and relative density (ρ_r) were obtained by using Equation (1), (2) and (3), respectively. Then percentage of density change after reduction process was calculated.

$$\rho = \rho_L \times (W / (W - (W_s - W_L))) \quad (1)$$

$$\rho_{th} = M Z / (N_A V) \quad (2)$$

$$\rho_r = \rho / \rho_{th} \quad (3)$$

where, ρ_L density of water, W is sample mass, W_S is sample mass in water, W_L is water mass, M is molecular weight, Z is number of atoms in unit cell, N_A is Avogadro's number and V is volume of unit cell.

2.3.3 Morphology observation

Scanning electron microscopy (JEOL JSM-6460LA) was used to observe the morphology of the samples. The samples were ground and coated with platinum by putter coating before tested. The morphology of the sample was viewed and recorded in 5000× magnification.

3 Results and discussion

3.1 Phase identification

3.1.1 $Ce_{0.8}Sm_{0.2}O_{1.9}$ (SDC)

XRD patterns of SDC powder calcined at different temperature which are 800, 900 and 1000 °C are shown in Fig. 1. Analyzed data show that four main peaks belongs to CeO_2 (ICCD file no: 01-080-5549). No significant peak shifting detected when calcination increased indicating that no changes in lattice parameter can be found. As all calcination temperatures could produce identical result, for the low energy consumption reason, lower calcination temperature, i.e 800 °C was choosen as calcination temperature for SDC powder in this work.

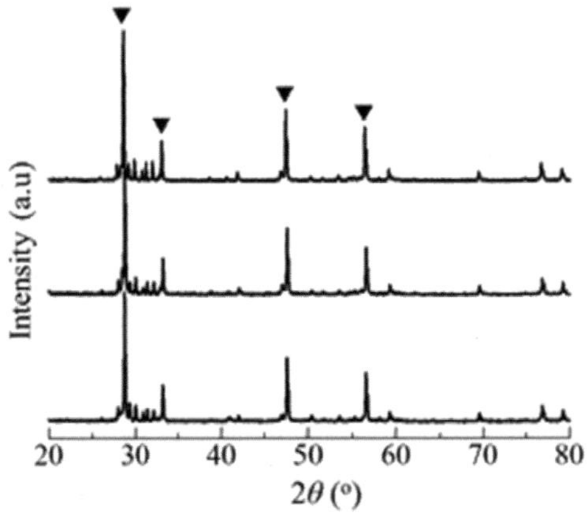

Fig. 1. XRD patterns of Ce0.8Sm0.2O1.9 (SDC) powder calcined at 800, 900 and 1000 oC for 3 h. ▼ indicates main peaks belong to ICDD 01-080-5549.

3.1.2 NiO-SDC before and after reduction

Phase confirmation was done for all samples before and after reduction. The results are shown in Fig. 2 (a)-(c). For all samples, before undergo reduction process, main peaks that corresponding to NiO (ICDD file no: 00-001-1239) at 37°, 43°, and 63° and CeO_2 (ICDD file no: 01-080-5548) at 28°, 33°, 47°, and 56° were detected. No additional peak for new compound is found. It shows that there is no reaction occur between NiO and CeO_2 to produce new compound and NiO and SDC remain in separate phase even though the samples were sintered at 1400 °C. After undergo reduction process using 40% H_2-60% N_2 gases at 800 °C for 2 h, phase identification was done to the same samples once again. For all samples, peaks that corresponding to CeO_2 remained. Meanwhile, peaks that corresponding to NiO was completely disappeared, and new main peaks that corresponding to Ni (ICDD file no: 01-088-2326) was identified at 45°and 52°.

IOP Conf. Series: Materials Science and Engineering **701** (2019) 012028 doi:10.1088/1757-899X/701/1/012028

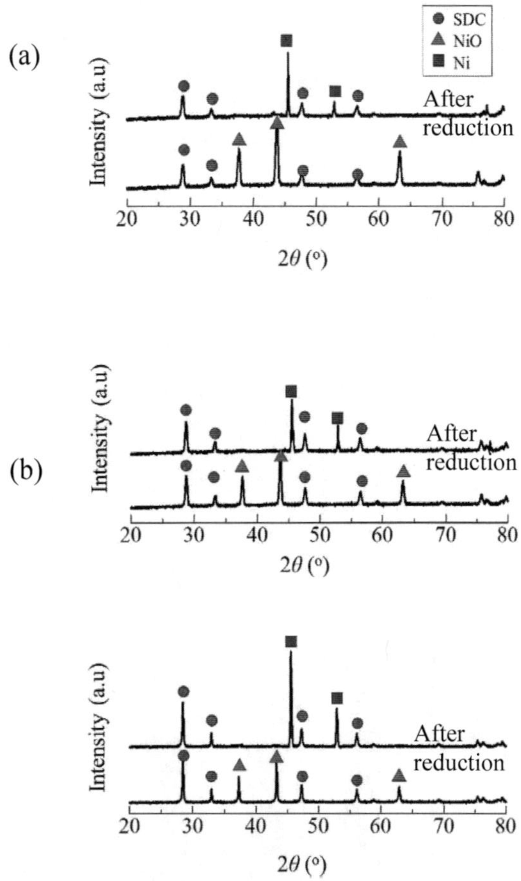

Fig. 2. XRD patterns of (a) 80NiO-20SDC, (b) 70NiO-30SDC, and (c) 60Ni-40SDC pellet before and after reduction at 800 °C for 2 h under 40% H_2 and 60% N_2 gas.

From these results, for transformation of NiO to Ni under H_2 gas, overall reaction can be written as follow;

$$NiO + H_2 \rightarrow Ni + H_2O \quad (4)$$

This reaction can be interpreted as, firstly, H_2 is dissociated, and then adsorption of hydrogen atom on nickel atom occurs, followed by diffusion of hydrogen atom. Desorption of H_2O take place after dissociation of nickel-oxygen bond, and finally, leave the nickel metal [19, 20, 23]. It has been reported that reaction between NiO and hydrogen gas adsorbed on NiO, and reaction at the boundary between the two solid phases NiO and Ni with hydrogen adsorbed on nickel occur simultaneously [24]. In addition, when increased NiO ratio, peak shifting to lower angle especially at 28 ° and 45° was observed, therefore it is expected that there was a small change in lattice parameter.

3.2 Density

Relative density of samples before and after undergo reduction process was calculated, and the result is tabulated in Table 2. Sintered pellets are sufficiently dense in the range of 92-96 % of theoretical density.

Table 2. Relative density and density change of samples before and after reduction process.

Sample	Relative density, %	
	Before reduction	After reduction
80NiO-20SDC	94.9	78.0
70NiO-30SDC	95.5	80.9
60NiO-40SDC	92.0	80.3

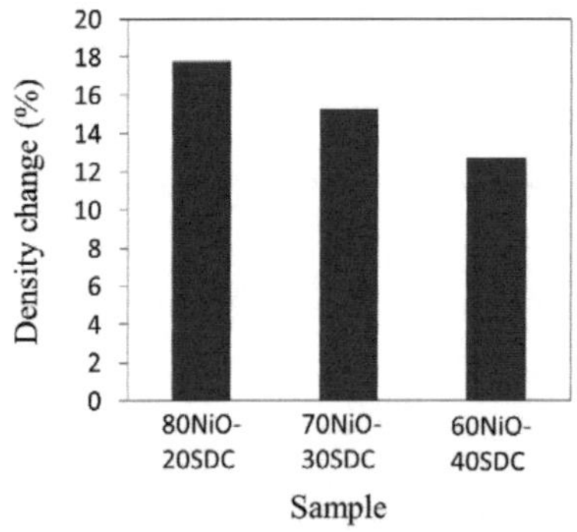

Fig. 3. Percentage of density change before and after reduction process.

The values of relative density after reduction are in the range of 78.0 – 80.9 %. This range meets the condition of porosity required for SOFC anode material especially for gas permeability [6, 25]. In addition, it can be said that parameters selected for reduction process in this work are suitable to produce porous Ni-SDC material.

Fig. 3 shows the percentage of density change before and after reduction for samples with different NiO and SDC content. From this Fig., it can be seen that the percentage of density change decreased gradually with decreasing the NiO content. This result indicates that gas permeability is directly proportional with the NiO content. Similar result was found for NiO-YSZ where, it is reported that for samples with low NiO content, reduction only occur at the surface of the sample [26]. The density change resulted from density change in NiO only, when NiO is reduced to Ni. Meanwhile, it is believed that the SDC compound does not involve in the density change of sample. This is proven by the XRD data where after reduction process, SDC remained in the SDC phase.

3.3 Morphology

Morphology observation were done with magnification ×5000, and the images are shown in Fig. 4. In general, for all samples, unreduced samples shows less pore compared to reduced samples showing that density of the unreduced samples is higher that the reduced samples. These results are in agreement with the density value shown in Table 2. Significant morphology change can be observed for sample that contain 100% of NiO, where unreduce sample possesses dense structure with only few pores. Meanwhile, for reduced sample, small pores were detected on the entire surface of the sample.

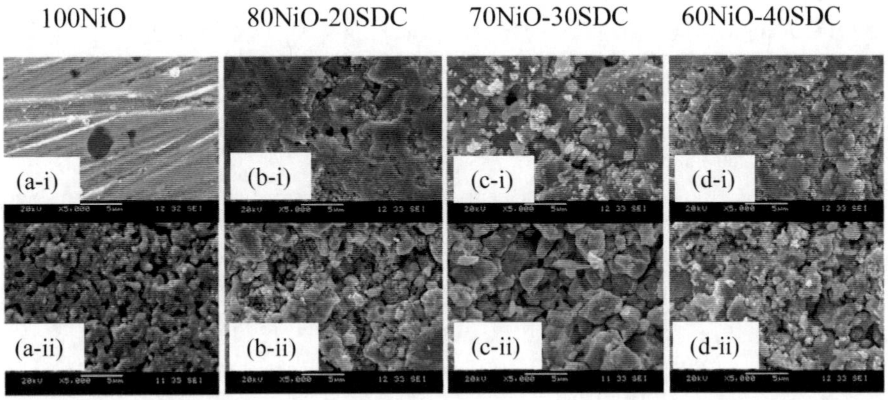

Fig. 4. SEM images of (a-i) unreduced 100NiO, (a-ii) reduced NiO, (b-i) unreduced 80NiO-20SDC, (b-ii) reduced 80NiO-20SDC, (c-i) unreduced 70NiO-30SDC, (c-ii) reduced 70NiO-30SDC, (d-i) unreduced 60NiO-40SDC, (d-ii) reduced 60NiO-40SDC.

4 Conclusions

$Ce_{0.8}Sm_{0.2}O_{1.9}$ (SDC) was successfully synthesized via conventional method and calcination process at as low as 800 °C. NiO-SDC with different NiO to SDC ratio was sintered at 1400 °C. All pellet NiO-SDC samples show identical phase which is mixture of NiO and SDC phase. Reduction of NiO-SDC at 800 °C for 2 h using 40% H_2- 60% N_2 completely reduced NiO to Ni. Larger NiO ratio to SDC resulted in higher percentage of density change. The porosity of the samples after reduction process is in the range that is required by SOFC anode material.

This work was partially supported by Short Term Grant: Anugerah Kecemerlangan Universiti Malaysia Perlis (Grant Number: 9009-00055).

References

1. B. C. H. Steele, J. Mater. Sci. **36**, 5 (2001)
2. A. J. Jacobson, Chem. Mater. **22**, 3 (2010)
3. S.P.S. Shaikh, A. Muchtar, M.R. Somalu, Renew. Sust. Energ Rev. **51** (2015)
4. Y. Li, Z.Y. Luo, C.J. Yu, D. Luo, Z.A. Xu, K.F. Cen, J Zhejiang Univ Sci B. **6**, 11 (2005)
5. A. Ideris, E. Croiset, M. Pritzker, A. Amin, Int. J. Hydrog. Energy. **42**, 36 (2017)
6. M. Chen, B. H. Kim, Q. Xu, O. J. Nam, J. H. Ko, J. Eur Ceram Soc. **28** (2008)

7. A.J.M Araujo, J.P.F. Grilo, F.J.A. Loureiro, L.F.A. Campos, C.A. Paskocimas, R.M. Nascimento, D.A. Macedo, Ceram. Int. **44** (2018)
8. C. H. Lim, K. T. Lee, Ceram. Int. **42**, 12 (2016)
9. M. Kawano, H. Yoshida, K. Hashino, H. Ijichi, S. Suda, K. Kawahara,T. Inagaki, J. Power Sources, **42**, 36 (2017)
10. Y. C. Yang, T. H. Chang, Y. C. Wu, S. F. Wang, Int. J. Hydrog. Energy, **37**, 18 (2012)
11. R. Nishida, K. Kakinuma, H. Nishino, T. Kamino, H. Yamashita, M. Watanabe, H. Uchida, Solid State Ion. **180**, 14-16 (2009)
12. N. B. Pavzderin, A. V. Nikonov, S. N. Paranin, K. A. Kuterbekov, K. Zh. Bekmyrza, Russ. J. Electrochem. **54**, 6 (2018)
13. J. J. Haslam, A. Q. Pham, B. W. Chung, J. F. DiCarlo, R.S. Glass, J. Am. Ceram. Soc. **88**, 3 (2005)
14. R. M.C.Clemmer, S. F. Corbin, Solid State Ion. **180**, 9-10 (2009)
15. E. Ruiz,-Trejo, A. Atkinson, N. P. Brandon, J. Power Sources, **280** (2015)
16. J. Kang, K. Lee, J. Y. Yoo, J. Bae, J. Hydrog. Energy. **43**, 12 (2018)
17. C. Ren, T. Liu, Y. Mao, P. Maturavongsadit, J. A. Luckanagul, Q. W.ang, F. Chen, Electrochim. Acta, **149** (2014)
18. R Chatterjee, S. Banerjee, D. Ghosh, Trans. Indian Inst. Met., **65**, 3, (2012)
19. K.D. Kim, C.S. Oh, S.C. Kil, J.P. Wang, Y.H. Kim, Rev.Adv.Mater. Sci. **28** (2011)
20. Q. Jeangros, T. W. Hansen, J. B. Wagner, C. D. Damsgaard, R. E. Dunin-Borkowski, C. Hebert, J. V. Herle, A. Hessler-Wyser. J. Mater Sci. **48** (2012)
21. K. V. Manukyan, A. G. Avetisyan, C. E. Shuck, H. A. Chatilyan, S. Rouvimov, S. L. Kharatyan, A.S. Mukasyan, J. Phys. Chem. **119**, 28 (2015)
22. M. Abdullohi, M. Sameezadeh, M. Vaseghi, J. Mater. Manuf. Pros., **33**, 8 (2018)
23. A. Faes, A. Hessler-Wyser, A. Zryd, and J.V. Herle. **2**,(2012)
24. J. Bandrowski, C. R. Bickling, K. H. Yang, O. A. Hougen, Chem. Eng. Sci. **17**, 5 (1962)
25. P.S.N. Baity, B. Budiana and S. Suasmoro, IOP Conf. Ser. Mater. Sci. Eng. **214** (2017.)
26. F. Iguchi, S. Akrasevee, Y. Miyoshi, Mater. Trans. **59**, 1 (2018)

IOP Conf. Series: Materials Science and Engineering **701** (2019) 012029 doi:10.1088/1757-899X/701/1/012029

Effect of surface finish on the wettability and electrical resistivity of Sn-3.0Ag-0.5Cu solder

Siti Farahnabilah Muhd Amli[1*], *Mohd Arif Anuar* Mohd Salleh[1], *Rita.* Mohd Said[1], *Nurul Razliana* Abdul Razak[2], *Juyana A.* Wahab[1] and *Mohd Izrul Izwan* Ramli[1]

[1]Center of Excellence Geopolymer & Green Technology (CeGeoTech), School of Materials Engineering, Universiti Malaysia Perlis (UniMAP), Taman Muhibbah, 02600 Jejawi, Perlis, Malaysia
[2]Department of Mechanical Technology, Faculty of Engineering, Kampus UniCITI Alam Sungai Chuchuh, Padang Besar, 02100 Perlis, Malaysia

Abstract. The effect of different surface finish with Sn-3.0Ag-0.5Cu (SAC305) solder was successfully investigated. The SAC305 solder was fabricated by using casting method and solder was placed on copper substrate that coated with different surface finish. The soldering process was carried out by using F4N reflow oven followed up with the mounted and metallographic steps. Wettability of SAC305 solder was observed through contact angle formed between solder and four different surface finish located on the copper substrate. Subsequently, the electrical resistivity of solder was studied by conducted the four-point probes. The results of wettability test was found to be in the accepted range which is below 45° for all different surface finish. In terms of electrical resistivity, the results showed that the ImAg surface finish had enhanced the electrical conductivity of SAC305 lead-free solder.

1 Introduction

In the past, tin-lead (Sn-Pb) solder have been widely used in electronic packaging due to numerous merits including their low melting point (183 °C), excellent wettability and also good mechanical properties. However, starting on July 1st 2006, the usage of electronic products with Pb content had been banned due to increasing awareness of Pb that can be harmful to human health and environment [1-5]. Consequently, the development to find the suitable replacement for Sn-Pb solder become an important in the electronic field [6-10]. Among the numerous of lead-free solder proposed, the Sn-Ag-Cu (SAC) solder has been the most promising solder for replace Sn-Pb solder owing to its modest melting temperature and [11]. However, there are still problems needs to be resolved for these SAC solder, such as the poor wettability and electrical resistivity [12].

* Corresponding author : sitifarahnabilah@outlook.com

Generally, there are numerous methods to enhance the solderability and electrical conductivity of the solder and one of a method is by using a surface finish [13]. In surface mount technology (SMT), surface finish that covered the printed circuit board (PCB) plays a significant role. There are several functions of surface finish which are to preserve the solderability layer [14], to act as a barrier to prevent the excessive dissolution of the element from substrate towards the solder [15] and also protect form the oxidation. Example of surface finish that had been used by previous researchers is Hot Air Solder Levelling (HASL) due to the a lower cost and low thermal shock resistance [16]. Furthermore, Arra et al. study the effect of nickel/ gold (Ni/Au), ImAg, OSP and ImSn finishes was effect the wetting property [17]. Thus, it is believed that the surface finish has the potential to improve the properties of solder in terms of wettability, conductivity and so forth.

Since the excellent of a solder joint required good wettability and excellent conductivity, this has been motivation for this research to investigated the effects of SAC 305 solder with various surface finish including Immersion Tin (ImSn), Immersion Silver (ImAg), Electroless Nickel Immersion Gold (ENIG) and copper Organic Solderbility Preservative (Cu-OSP). In addition, there is no single study reported that comparing the effects of four different surface finish on SAC305 solder.

2 Experimental Procedure

In this study, the material used was 96.5%Sn-3.0%Ag-0.5%Cu (SAC305) solder alloy and four different types of surface finish coated on the copper substrate including copper Organic Solderability Preservative (Cu-OSP), Electroless Nickel Immersion Gold (ENIG), Immersion Tin (ImSn) and Immersion Silver (ImAg). The fabrication of solder alloy was carried via casting method by melting the SAC305 solder in a clean graphite crucible and heated up to 350 °C using a furnace. Subsequently, the ingots performed cold roll to produce thin sheet of solder. Both wetting and electrical resistivity test, the samples were prepared by reflowed the SAC305 solder on different substrate with the of small amount rosin mildly activated (RMA) flux in a F4N reflow oven. The reflowed samples were mounted, grinded and polished to form a flat surface.

The wettability of solder was evaluated through wetting angle of solder on different substrate as shown in Fig.1. A 10 cross sectional samples of each substrate were prepared to measure the contact angle and measured by using Optical Microscope (ECLIPSE L300N) and Image-J software. Moreover, the electrical properties of SAC305 in different surface finish were evaluated by measuring the contact resistance of a sample by using multi-height four-point probe/2400 Source Meter, Keithley four-point probes. This electrical resistivity technique uses separate pairs of probes, the outer probes were used to supply current (I) and other pairs were used to measure a voltage (V) across the samples as illustrate in Fig.2. The resistivity of the solder sample was expressed as equation (1).

$$\rho = k(\frac{V}{I}) \quad (1)$$

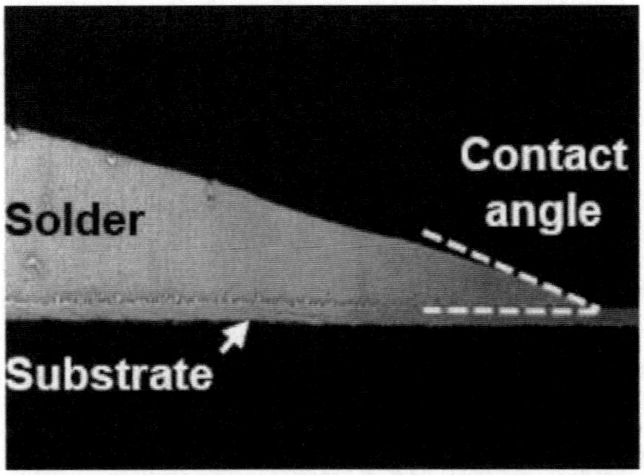

Fig. 1: Measurement of contact angle for SAC305 solder on different surface finish.

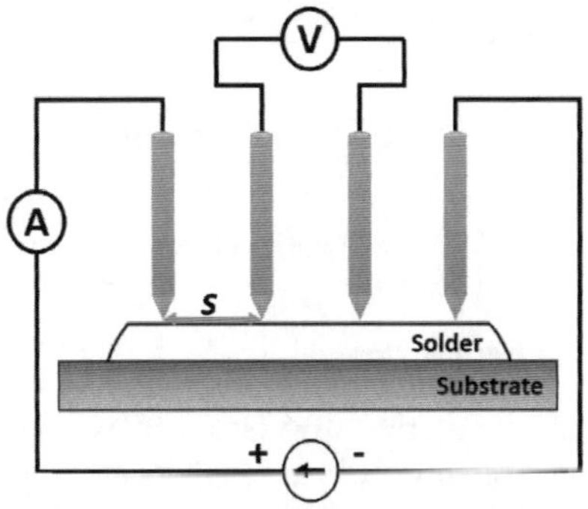

Fig. 2: Schematic diagram of four-point probe configuration.

3 Results and Discussion

3.1 Wettability Analysis

The wettability of solder was measured by the contact angle of solder form on substrate [18]. Wetting behaviour is crucial for solder alloys in ensuring an excellent metallic bonding in electronic packaging [11]. In soldering, the smaller the contact angle of solder, the better the wettability of the interconnections [19]. As in Fig. 3, the result showed different surface finish had different value of contact angle. The contact angle of SAC305 solder on Cu-OSP obtained as the largest with 28.64° followed by ENIG and ImAg with values 20.88° and 14.06° respectively. Similar result was found in studies by Siewiorek et al., that investigated the contact angle of the Sn-3.1Ag-0.8Cu solder in contact with Cu-OSP and ENIG surface finish [16].

Meanwhile, it was observed that the optimal contact angle achieved when used the ImSn as a surface finish with the smallest value 11.54°. According to the result attained, the decreasing of contact angle could be attributed with the surface tension of substrate. Studied by M. Arra et al. reported that the ImSn has better wetting property compare on the ImAg and Cu-OSP surface finish. Therefore, it can conclude that the different surface finish will influence the wetting angle. Overall, smaller contact angle represents excellent wettability.

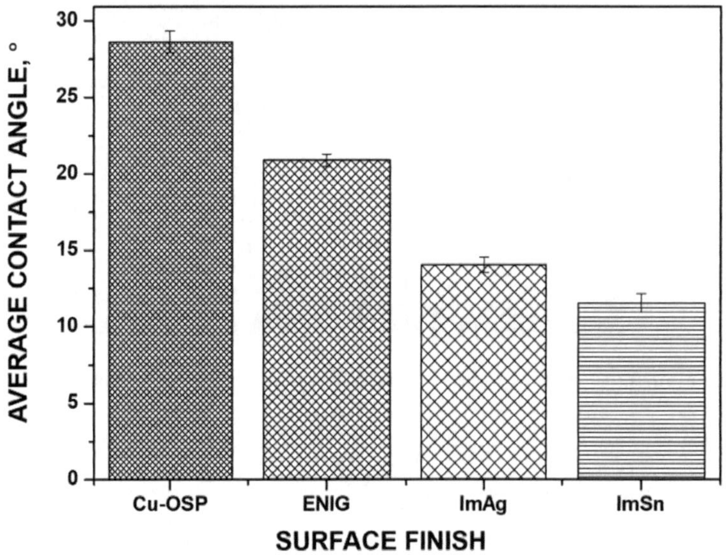

Fig. 3: Average contact angle of SAC305 solder alloys with different surface finish.

3.2 Electrical Resistivity Properties

Resistivity is the ability of a material to resist the flow of current. When the value of resistivity is lower, it means the specimens has higher conductivity and can easily allow an electric to flow through it. The electrical resistivity measurements of SAC305 on different surface finish were shown in Fig. 4. The result reveals that with the ImAg surface finish, electrical resistivity was reduced with 11.82 μΩ.cm due to the excess formation of Ag_3Sn intermetallic compound increase after the soldering process. Previous researcher reported that the presence of silver element will reduce the electrical resistivity of solder [20-21]. Moreover, resistivity value for Cu-OSP, ENIG and ImSn was 28.9, 19.5 and 13.92 μΩ.cm. Hence, as the electrical resistivity decreased when used ImAg as surface finish, it will enhance and formed a good electrical conductivity of solder.

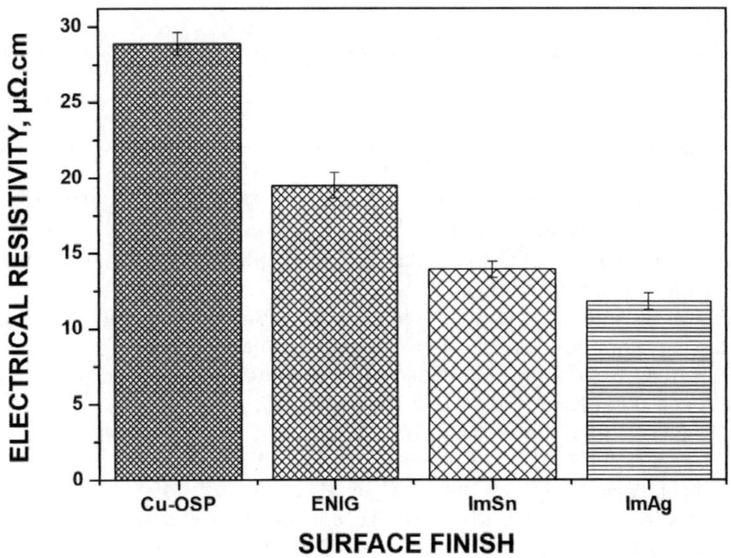

Fig. 4: Electrical resistivity of SAC305 solder alloy with different surface finish.

4 Conclusion

The study on difference surface finish underlying on copper substrate is found to influence on the wettability and electrical resistivity of SAC305 solder. The following conclusion can be made:

i. The solderability of SAC305 solder was successfully improved with the presence of ImSn surface finish.
ii. Immersion silver layer on copper substrate decreased the electrical resistivity of SAC305 due to the numerous of Ag_3Sn particles formed.

The authors gratefully acknowledge the Universiti Malaysia Perlis (UniMAP) and Nihon Superior Co. Ltd. collaboration research project and the fundamental research grant scheme (FRGS) FRGS/1/2017/TK05/UNIMAP/03/3 (9003-00621) from the Ministry of Higher Education, Malaysia for the financial support throughout the research project.

References

1. R.M. Said, M.A.A.M. Salleh, M.N. Derman, M.I.I. Ramli, N.M. Nasir, and N. Saud, Key Engineering Materials. **700** (2016)
2. M.I.I. Ramli, N. Saud, M.A.A.M. Salleh, M.N. Derman, R.M. Said, and N. Nasir, Materials Science Forum. **803** (2015)
3. M.I.I. Ramli, M.S.S. Yusof, M.A.A.M. Salleh, R.M. Said, and K. Nogita, Solid State Phenomena. **273** (2018)
4. F. Somidin, H. Maeno, M.M. Salleh, X.Q. Tran, S.D. McDonald, S. Matsumura, and K. Nogita, Materials Characterization. **138** (2018)

5. D. H. Jung, A. Sharma, D. U. Lim, and J. H. Yun., Metal Mater Trans A, **48**: p. 4372 (2017)
6. M.A.A.M. Salleh, M.H. Hazizi, M.A.B.M. Abdullah, N. Noriman, R. Mayapan, and Z.A. Ahmad, Advanced Materials Research. **626** (2012)
7. K. Nogita, M.A.A.M. Salleh, S. Smith, Y. Q Wu, S.D. McDonald, A.G.A. Razak, T. Akaiwa, and T. Nishimura, 2017 International Conference on Electronics Packaging (ICEP).
8. K. Nogita, M.A.A.M. Salleh, E. Tanaka, G. Zeng, S.D. McDonald, and S. Matsumura, JOM. **68**(11) (2016)
9. M.A.A.M. Salleh, A.M.M.A. Bakri, F. Somidin, and H. Kamarudin, International Review of Mechanical Engineering.
10. M.A.A.M. Salleh, M.M.A.B. Abdullah, F. Somidin, A.V. Sandu, N. Saud, K. Hussin, S.D. McDonald, and K. Nogita, Revista de Chimie **64**
11. D. H. Jung and J. Jung., Solid State **44**: p. 324 (2018)
12. L. Sun and L. Zhang., Adv Mater Sci Eng, (2015)
13. Milad. G., Circuit World, **34**(4): p. 4 (2008)
14. R. Berni, et al., IEEE T Reliab, **65**(1): p. 274 (2016)
15. R. A. M. Anuar and S.A. Osman., Materialwiss Werkst, **48**: p. 235-240 (2017)
16. A. Siewiorek, A. Kudyba, N. Sobczak, M. Homa, Z. Huber, Z. Adamek, and J. W. Budka., J Mater Eng Perform, **22** (8): p. 2247 (2013)
17. M. Arra, D. Shangguan, D. Xie, J. Sundelin, T. Lepisto, and E. Ristolainen., J Electron Mater. **33**(9): p. 990 (2004)
18. N. Rodrigues, A. C. Ferreira, and S.F. Teixeira., J Mater Sci-Mater El, **27**: p. 8942 (2016)
19. E. E. Mhd Noor, N. F. Mhd Nasir, and S. R. A. Idris., Solder Surf MT Tech, **28**: p. 127 (2016)
20. M. A. A. Mohd Salleh, M. H. Zan@Hazizi, Z. A. Ahmad, K. Hussin and K. R. Ahamd., Adv. Mater. Res, **277**: p. 109 (2011)
21. N. A. A. Mohd Amin, D. A. Shnawah, S. Mohd Said, M. F. Mohd Sabri and H. Arof., J. Alloys Compd, **599**: p. 116 (2014)

IOP Conf. Series: Materials Science and Engineering **701** (2019) 012030 doi:10.1088/1757-899X/701/1/012030

Physical properties of Sn-3.0Ag-0.5Cu lead-free solder with the additional of SiC particles

Zawawi Mahim[1,*], *Mohd Arif Anuar* Mohd Salleh[1], *Norainiza* Saud[1], *Nurul Razliana* Abdul Razak[2], *Khor* Chu Yee[2], *Romisuhani* Ahmad[2], and *Muhammad Mahyiddin* Ramli[3]

[1]School of Materials Engineering, Kompleks Pusat Pengajian Jejawi 2, Universiti Malaysia Perlis (UniMAP) Taman Muhibbah, 02600 Arau, Perlis, Malaysia.
[2]Faculty of Engineering Technology, Universiti Malaysia Perlis (UniMAP), Sungai Chuchuh Padang Besar, Perlis, Malaysia
[3]School of Microelectronic Engineering, Universiti Malaysia Perlis (UniMAP), Pauh Putra Campus, 02600 Arau, Perlis, Malaysia

Abstract. Composite solder has gained researcher's attention due to its promising improvement in physical and mechanical properties for lead-free solder. To improve the properties of Sn-3.0Ag-0.5Cu (SAC) with the promising lead-free candidate, addition of silicon carbide (SiC) as a reinforcement was used in this study. This study was carried out to identify the effect of SiC particle on microstructure evolution and physical properties of SAC based solder alloys. SAC-SiC composite solder was synthesized by powder metallurgy method (PM), which consists of several processes such as mechanical blending, compaction and sintering. Three different weight percentages of SiC particles; 0.00, 0.50, and 1.00 were mechanically blended with SAC lead-free solder. The results show that the additional of particle SiC was able to refine the microstructure and reduced the size of β-Sn.

1 Introduction

It has long been well-known that tin-based solders, including SnAg, SnAgCu, and Sn-.07Cu, were the most significant solders for electronic applications in industries [1, 2]. Many researchers still study the selection for Pb-free alternatives. SAC have attracted considerable attention as leading alloy Pb-free solder because of its greater mechanical properties [3].Besides, SAC solder alloy was chosen due to their melting point which is relatively low and closer to Sn-Pb. However, formation of large Ag_3Sn IMC within the bulk solder matrix duet tto the high amount of Ag is known to easily initiate cracks. Ag_3Sn IMC which is brittle in nature can cause degradation in term of hardness performance and commonly will causes failure in portable electronic devices that used the solder [4] . In this research, SiC particles were chosen as reinforcement for SAC solders due to its light weight and ease of handling. The additional of SiC particles is believed be able to improve the microstructural, physical, electrical, mechanical properties, solder joint and IMC formation

* Corresponding author: mahimzawawi@gmail.com

issues. There is still insufficient data on the effect of the nano-size or micron-size SiC particle as reinforcement for SAC in the microstructure, thermal behaviour and mechanical properties when applied in soldering application [5]. Due to SiC have no solubility in β-Sn matrix, it is predicted that SiC will have significant effect on the SAC alloys. Thus, the aim of this study is to investigate the influence of micron-size SiC particles addition on microstructure, and microhardness properties of SAC[6].

2 Materials and Processing

SAC and SiC powder with particle size of < 45 μm were used in this research for the base matrix material and reinforcement respectively. The various composition of SiC reinforcement, 0.00 wt% as reference, 0.50 wt% and 1.00 wt% were used to fabricate new SAC-SiC composite solder via powder metallurgy technique (PM) [7,8]. The matrix of SAC solder was mixed with different composition of SiC in an airtight container using a planetary mill rotated at a speed 200 of RPM for an hour. After that, 12 mm diameter mould was used for compaction at a load of 4.5 tons (44.21 × 106 MPa) by using manual hydraulic press [9]. Next, the samples were sintered using a microwave at 185 °C for 2 minutes [10]. The sample of pure SAC which is without SiC also was prepared as reference sample [11].

The comparison between both pure and composite lead-free solders was studied on Cu-substrate (PCB FR-4 type). All the samples were cut into smaller pieces approximately 1 gram. Next, the pieces of bulk samples were placed on Cu-substrate with a no-clean (NC) paste flux type AFM037A that was provided by ASAHI Flux Medium. Finally, the samples were heated to a maximum peak temperature 250 °C in desk lead-free reflow oven [12].

3 Microstructure Analysis and Hardness Test

All samples were mounted for microstructure observation under the optical microscope. To observe the microstructure, mounted samples were ground by using sand paper from grit 600 to 2000. The samples that already ground need were soaked in ultrasonic cleaning to remove the impurities that remains on the surface of the samples. To reveal the microstructure, all samples were etched. The solution needs to be prepared for etching is the mixture of 2 % of HCl, 3% of HNO_3 and 95% of methanol. Each of the samples were etched for 5 second to reveal the microstructure. This testing was refer based on ASTM B933-09 standard [13]. The compacted 12-mm diameter samples were tested using the Vickers Microhardness Test Model FV-700e. The testing was measured on the flat surface sample with 1 Kgf indenting load and 10 second dwell time [14].

4 Results and Discussion

4.1 Microstructure

Fig. 1 shows the optical images of pure SAC and SAC-*x*SiC (x = 0.5 and 1.0 wt. %) composite solder. Based on result, β-Sn region for composite solder was decreased by increasing the amount of SiC particles. These findings proved that the addition of SiC particles able to retard the grain growth and gave rise to a finer β-Sn phase. Reinforcement concentration between grains will tend to prevent and limit grain boundary sliding and retard the grain growth[15]. Gain et al., [16] studied the addition of ZrO_2 nano-particles into

SAC found that the grain size was decreased due to precipitation of ZrO_2 at the grain boundary and change relative relationship of the growth velocity between the crystalline direction of the IMC particles, which reduces the IMC particles size [17]. Moreover, the SiC particles also functioned as the diffusive barrier to suppress metal atom diffusion on the surface of the grains. This process also contributed to delaying the growth of the grains, leading to finer microstructures [18]. The refined microstructure also contributed to the enhancement of microhardness value in the composite solder. Besides, Chen et al., [19] also found the addition of fullerene nanoparticles (FNSs) into the SAC were able to refine microstructure of the composite solder. The microstructure of SAC composite lead-free solders was observed by using SEM backscattered image.

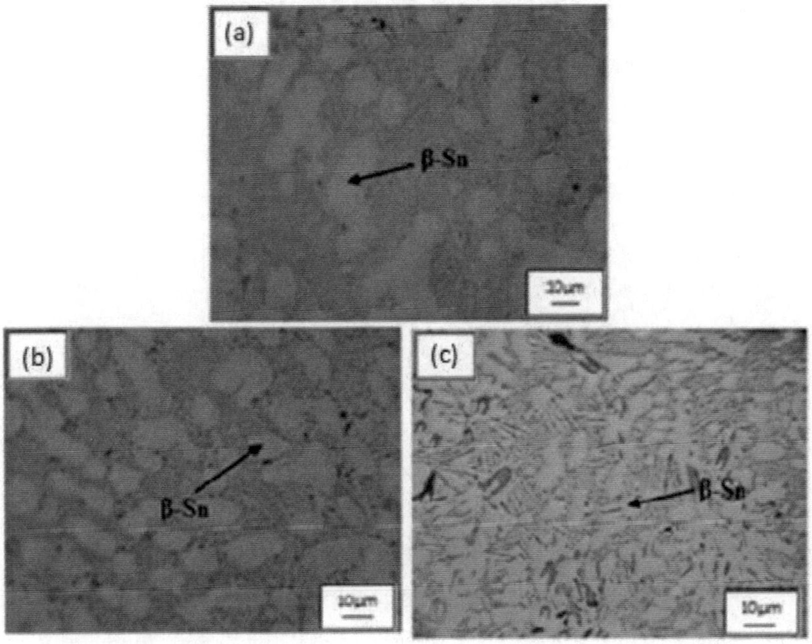

Fig. 1. Optical images of SiC as a reinforcement in; (a) pure SAC, (b) SAC-0.5SiC, and (c) SAC-1.0SiC

Fig. 1 (a), (b) and (c) shows the grain boundary and β-Sn matrix on the surface of pure SAC, SAC-0.50SiC and SAC-1.0 Fig. 2 (b) shows the observation area of SiC particle that were located between grain boundaries for SAC-SiC composite solders. Fig. 3 illustrates the location of SiC particles in SAC-SiC composite solder. The dark element is SiC particles. The SEM-EDX analysis result also support the existence of SiC particles within solidified SAC solder. The existence of Cu_6Sn_5 and β-Sn were also detected in SAC-SiC composite lead-free solder.

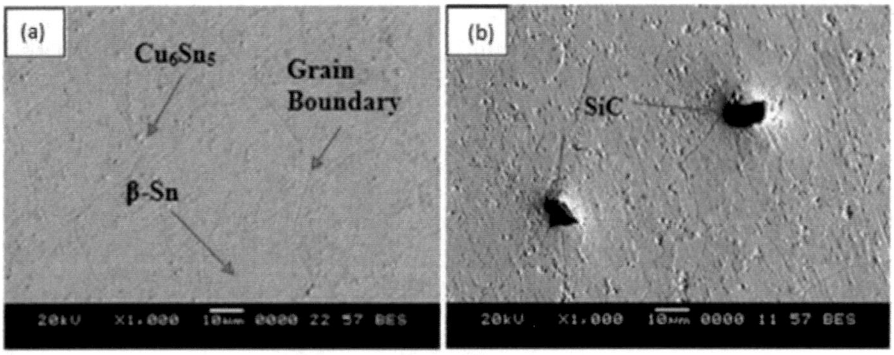

Fig. 2. SEM microstructures of; (a) pure SAC solder, and (b) SAC-1.0SiC composite solder.

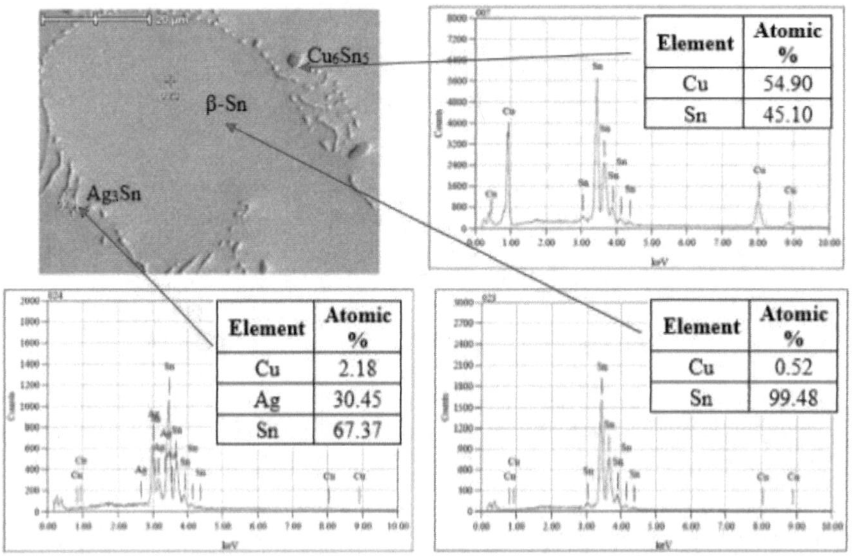

Fig. 3. Point analysis of SAC-SiC.

4.2 Microhardness Test

Fig. 4 showed the comparison of microhardness's data between the various compositions of SiC as a reinforcement for SAC lead-free solder. According to the Fig. 4, the microhardness of the composite solder rises with the increased of SiC particles compare to the pure SAC. The properties of SiC, a ceramic type material with high strength and high hardness, is the cause for enhancement of the properties of the solder. It believed the decreasing size of β-Sn region which has increased the microhardness of composite solder due to the classical dispersion strengthening theory[20]. From that theory, the foreign particles that located in the solder alloy matrix and grain boundaries are likely to change the solder alloy's deformation characteristics by retarding dislocation movement as well as impeding grain-boundary sliding in solder matrix, resulting in enhancement in microhardness [19].

The additional of SiC particles influence the growth and size of the β-Sn region in SAC-SiC composite lead-free solder. Apart from that, the presence of reinforcement contributes to decreasing of solder matrix grain size. The microhardness of composite lead-free solder was increased because of dislocation path are disturb by reinforcing particle.

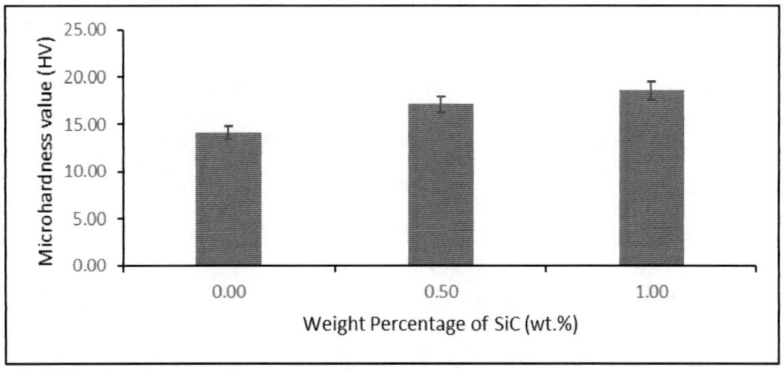

Fig. 4. Microhardness value comparison

The enhancement of microhardness value was attributed by the role of SiC particles as the pinning grain boundaries. The reinforcement that located in the solder matrix and grain boundaries were able to change the solder alloy's deformation characteristics by retarding dislocation movement, as well as impeding grain boundary sliding in solder matrix, resulting improved in the microhardness [21]. That mechanism become an obstacle to the movement of dislocations and increased dislocation densities. The microhardness value of composite lead-free solders displayed consistently higher value than both pure SAC lead-free solders due to homogeneous distribution of SiC particles and refinement of the IMC particles[22]. In conjunction with the dispersion strengthening mechanism, the fine IMC particles and ceramic reinforcement can improve the microstructure and mechanical properties of a lead-free solder [16].

5 Conclusion

This study has proven with the addition of SiC particle in SAC lead-free solder are able to retard the grain growth and gave rise to the number of finer β-Sn phase. With that, the soft area of β-Sn matrix was decreased that able to enhance the microhardness properties of composite lead-free solder.

The author gratefully acknowledges the financial support from Ministry of Higher Education (MoHE) Malaysia under the Fundamental Research Grant Scheme (9003-00612) (FRGS/1/2017/TK10/UNIMAP/03/2) and Center of Excellence Geopolymer & Green Technology (CEGeoGTech), School of Materials Engineering, Universiti Malaysia Perlis (UniMAP) for supporting the research effort through materials and facilities.

References

1. T. Laurila, V. Vuorinen, and J. K. Kivilahti, *Mater. Sci. Eng. R Reports*,. **49,** (2005)
2. M. I. I. Ramli, N. Saud, M. A. A. M. Salleh, M. N. Derman, and R. M. Said, *Microelectron. Reliab.*, **65**, (2016)
3. S. W. R. Lee and S. Fubin, *2006 Int. Microsystems, Packag. Conf. Taiwan, IMPACT - Proc. Tech. Pap.*, no. c, (2006)
4. D. A. A. Shnawah, S. B. M. Said, M. F. M. Sabri, I. A. Badruddin, and F. X. Che, *Mater. Sci. Eng. A*, **551**, (2012)
5. P. Liu, P. Yao, and J. Liu, *J. Electron. Mater.*, **37**,(2008)
6. R. M. Said, M. A. A. M. Salleh, M. I. I. Ramli, N. Saud, M. M. A. B. Abdullah, and A. V Sandu, *AIP Conf. Proc.*, **1835**, Apr. (2017)
7. M. A. A. Mohd Salleh, M. H. Hazizi, M. M. Al Bakri Abdullah, N. Z. Noriman, R. Mayapan, and Z. A. Ahmad, *Adv. Mater. Res.*, **626**, (2013)
8. M. A. A. Mohd Salleh, E. Tasnaka, G. Zheng, S. D. McDonald, and S. Matsumura, *Rev. Chim. -Bucharest- Orig. Ed.*, **64**, pp. 725–728, Jul. (2013)
9. F. Somidin, M. A. A. Mohd Salleh, and R. A. Khairel, *Adv. Mater. Res.* **620**, (2013)
10. R. Mohd Said, N. Saud, M. A. A. Mohd Salleh, M. N. Derman, M. I. Izwan Ramli, and N. M. Nasir, *Appl. Mech. Mater.*, **754–755**, (2015)
11. M. A. A. Mohd Salleh, M. M. A. B. Abdullah, F. Somidin, and H. Kamarudin, *Int. Rev. Mech. Eng.*, **7**, pp. 53–59, (2013)
12. M. A. A. Mohd Salleh, M. M. A. B. Abdullah, *et al.*, no. **1**, (2013)
13. N. R. Abdul Razak, N. Adli, N. Saud, and S. A. Musa, *Appl. Mech. Mater.*, **754–755**, (2015)
14. M. A. A. Mohd Salleh, M. H. Hazizi, Z. A. Ahmad, K. Hussin, and K. R. Ahmad, *Adv. Mater. Res.*, **277**, 2011.
15. M. A. A. M. Salleh *et al.*, *Sci. Rep.*, vol. **7**, Jan. (2017)
16. A. K. Gain, Y. C. Chan, and W. K. C. Yung, *Microelectron. Reliab.*, **51**, (2011)
17. K. Nogita *et al.*, in *2017 International Conference on Electronics Packaging (ICEP)*,(2017)
18. Y. Yao, J. Zhou, F. Xue, and X. Chen, *J. Alloys Compd.*,. **682**, (2016)
19. G. Chen, F. Wu, C. Liu, and Y. C. Chan, in *2015 IEEE 65th Electronic Components and Technology Conference (ECTC)*, (2015)
20. F. Somidin *et al.*, *Mater. Charact.*, (2018)
21. G. Chen, H. Peng, V. V Silberschmidt, Y. C. Chan, C. Liu, and F. Wu, *J. Alloys Compd.*, **685**, (2016)
22. K. Nogita, M. A. A. Mohd Salleh, E. Tanaka, G. Zeng, S. D. McDonald, and S. Matsumura, *JOM*, **68**, (2016)

IOP Conf. Series: Materials Science and Engineering **701** (2019) 012031 doi:10.1088/1757-899X/701/1/012031

Effect of rare-element (Ga) addition on the microstructure and mechanical properties of Sn-0.7Cu and Sn-0.7Cu-0.05Ni lead-free solder alloys

Sufian Firdaus Nazri[1] and *Mohd Arif Anuar* Mohd Salleh[1,*]

[1]Centre of Excellence Geopolymer and Green Technology (CEGeoGTech), School of Materials Engineering, Universiti Malaysia Perlis (UniMAP), Taman Muhibbah, 02600 Jejawi, Arau, Perlis, Malaysia.

Abstract. The effects of rare element Ga on wettability, microstructure, and mechanical properties of Sn-0.7Cu (SC) and Sn-0.7Cu-0.05Ni (SCN) lead-free solder alloys were investigated. The experimental results show that Ga plays a positive role in improving the wettability and the microstructure of the solder. When the content of Ga is at 0.05 wt%, the grain size of the solder is smaller and the hardness is enhanced greatly. It is also found that the thickness of the IMCs at the solder/Cu interface is reduced with proper addition of Ga. The increase of mechanical properties may be related to the refining of IMCs of the solder due to Ga addition.

1 Research Background

Conventional Sn-Pb solders in microelectronic devices have been generally used since decades ago. The lead-free solder has been fully realized into electronic products due to environmental and toxicological issues [1-6]. Sn-0.7Cu-0.05Ni solder alloys free of lead are regarded to be one of Sn-Pb's most successful substitute candidates. [7, 8] In developing the electronics industry there were greater demands for lead-free electronic assembly. Because of reliability and property issues, the current Sn-0.7Cu solder does not fulfill the recent trends in electronic packaging. Therefore, the Sn-0.7Cu-0.05Ni solder idea for the next generation solder alloy was suggested [9-15], its prominent feature is the reduction of β-tin, which directly improves the mechanical reliability of the solder alloy. In the meantime, good soldering performance and mechanical properties of the solder can be achieved. Ga has a low melting point of 29.78 °C and a 3.4% increase in volume during solidification [16]. Moreover, it can be a very appealing property for microelectronic systems to weigh the majority of metals and oxides with no flux [17]. Previously, Ga was chosen by many researchers as a lead-free solder modifier. [18-20].

*Corresponding author: arifanuar@unimap.edu.my

Ga played a key role in enhancing the solders' microstructure, melting performance and other characteristics. The objectives of this work are to investigate the effect of the Ga addition on the microstructure of Sn-0.7Cu and Sn-0.7Cu-0.05Ni solder alloys and to compare it directly with the based solder alloy. Furthermore, to explore the wettability and wetting reactions between the solder and the Cu substrate, and to prepare a new solder alloy with a fine microstructure and decent stability of properties.

2 Experimental

2.1 Solders design and preparation

The SC-xGa and SCN-xGa alloys were prepared from Sn-0.7Cu and Sn-0.7Cu-0.05Ni solder alloys supplied by Nihon Superior (M) Sdn. Bhd. (NSM) and 99.9 % Ga. The solders have been introduced to the proper amount of pure Ga (0 and 0.05wt%). The solder was kept for approximately 5 minutes, while every 1 minute the mechanical stirring was done carefully. For Ga addition samples, Ga was added to the melted alloys after the oxide layer formed on the top was removed and thoroughly mixed to ensure homogenous mixture occurred. The reflowed samples were cross-sectioned and mounted in resin. The polished bulk morphology of the solder joint has been observed using an optical microscope. The cast metals were then further processed for a microstructural study. The casting was carried out to allow the solder mixture liquid to attained the desired shape, and then let it solidify. The list of Sn-based welding alloy samples is shown in Table 1.

Table 1. List of Sn-based Solder Alloys Samples

Base materials	Solidus Temperature (°C)	Weight of base alloy (g)	wt.% of Ga	Materials Designation
Sn-0.7Cu	227	150g	0 Ga	SC
			0.05 Ga	SC-0.05Ga
Sn-0.7Cu-0.05Ni	227	150g	0 Ga	SCN
			0.05 Ga	SCN-0.05Ga

2.2 Wettability test

The wetting balance method was adopted to test the solderability of the solder with a must 3 Wettability Tester Machine (GEN3 System) according to Japan's Industry Standard JIS Z3198-4 for lead-free solders. The Cu strips (30 mm x 10 mm) were ultrasonic cleaned in acetone for 3 min, then cleaned in ethanol and dried before testing. Wetting force and wetting time were tested at 250 °C temperature to assess the wetting characteristics of the solder. Besides, the water-soluble flux was used to enhance wettability during the experiment and the temperature, velocity and moment of immersion were 2 mm, 4 mm / s and 10 s respectively. The tested result here was a mean value of five tests.

IOP Conf. Series: Materials Science and Engineering **701** (2019) 012031 doi:10.1088/1757-899X/701/1/012031

2.3 Microstructure analysis

The samples of solders were prepared by polishing with 0.3 μm of Al_2O_3 dust according to the United States Department of Defence MILSTD-883 [21]. Optical camera and imaging microscopy (SEM) have been used for observing the microstructure. The samples were perpendicular to the solder / Cu interaction on the solder joints in the first interaction study. SEM was used to test the morphologies of the surfaces, and power dispersive X-ray (EDX) was used to determine the chemical characteristics.

2.4 Mechanical property test

Due to geometric effect as well as the formation and evolution of the brittle IMC at the interface, the solder/substrate interface is the weakest part of a solder joint. The nature of the force of the solder joints is usually the interfacial shear stress caused by the expansion failure. To assess the reliability of the solder joints, the Vickers hardness test was carried out. The machine operated according to the ASTM B933-09 standard test method. Before been test, the sample will be grinded using Si-C sandpapers (200, 320, 400, 600, 800 and 1200) grit size and indention load of 1kg for a 10s dwell time. The precision of the median outcomes is calculated using 10 indention marks for each sample. Without checking the hardness values of Vicker's output resistance and the final tensile strength can be measured. Using formulas, yield strength (YS) and ultimate tensile strength (UTS) can be calculated.

3 Results and Discussion

3.1 Wettability of solder alloys

To get a magnificent solder joint, decent wetting properties are a prerequisite. The Wettability and wetting reactions between the solder and Cu substrate have been investigated in this research work. Figure 1 shows the wetting time and wetting forces of SC-xGa and SCN-xGa solders at 250 °C temperature. At this temperature, it can be seen that the wetting time significantly increase at Ga addition of 0.05 %, and then gradually increases. The most significant increase is when added to the SCN solder alloy the wetting force exhibits the same trend. Klein et al. [22] pointed out that the short wetting time and high wetting force are considered to be excellent. Consequently, the wetting properties of SC and SCN solder alloys can be enhanced momentously with the addition of Ga and the best wetting properties can be attained with a Ga addition of 0.05wt%. The liquid solder takes capillary caulking during the soldering process and it diffuses mutually with the base metal at the same time, liquid solder must penetrate the surface of the base metal to fill the solder joints. The droplet placed on a solid surface, if the change in the droplet and the solid interface lowers the liquid-solid system free energy, the droplet will flow automatically along the solid surface. This is based on the basic principles of physical chemistry.

Any element that can improve the interfacial pressure of solid gases or decrease the interfacial pressure of liquid gases or liquid gases can enhance solder's wetting characteristics. Besides, Ga can easily accumulate at the solder interface in the molten state because of its surface-active element characteristic. Therefore, the surface tension of the liquid solder is reduced, enhances flexibility, which results in better solderability [23]. Moreover, Ga segregated seems to react with Sn, decrease the function of Sn and suppressing the excessive reaction of the interface. In the meantime, the wettability is enhanced caused by the diffusion between the solder and Cu substrate becomes effortless.

Additionally, the diffusion movement between the liquid solder atoms and the surface atoms at high-temperature increases, i.e. the velocity of diffusion is increasing and the solderability improves.

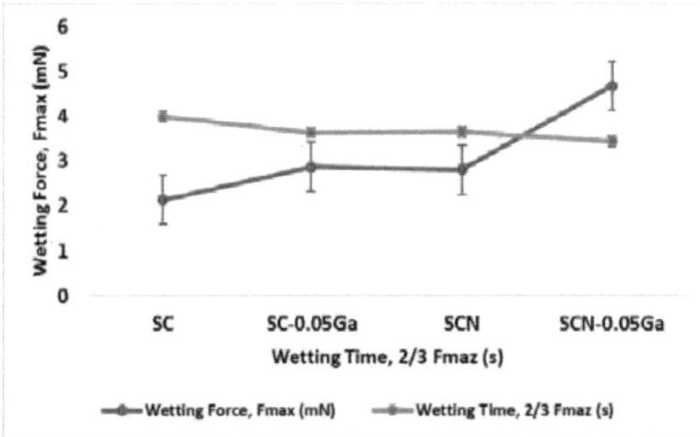

Fig. 1. Wetting properties of Sn-0.7Cu and Sn-0.7Cu-0.05Ni solder alloys with the addition of Ga at 0.05wt%

Table 2. List of the SC and SCN solder alloys that involved in the wetting balance test

Solder Alloy Designation	**Wetting Force, Fmax (mN)**	**Wetting Time, 2/3 Fmax (s)**
SC	2.14	4.00
SC-0.05Ga	2.87	3.64
SCN	2.81	3.66
SCN-0.05Ga	4.69	3.44

3.2 Microstructure analysis of the solder alloys

The microstructure morphology from the optical microscope of SC-xGa and SCN-xGa is shown in **Fig. 2.** In **Fig. 2.**, the microstructure of SC-xGa and SCN-xGa base solder consists of β-Sn matrix phase, bulk intermetallic and eutectic structures [7]. Compared with the microstructures of based Sn-0.7Cu and Sn-0.7Cu-0.05Ga solder alloy, the solder alloys with the Ga addition shows the smaller dendritic formation of crystal β-Sn phase. This is mainly because the nearly eutectic composition is at Sn-0.7Cu-0.05Ni [24], of which the content of Ga is in quite a great difference from Sn-0.7Cu. Large IMCs formation may well affect the mechanical reliability of the solder because of their high hardness [25]. Not only that, the matrix structure is refined-the coarse IMCs become smaller and distribute more uniformly with the addition of Ga. Although, this paper only shows 0.05wt% addition of Ga, a range of a proper amount of Ga addition will be a later study to get the optimum addition percentage. But, for now at least the Ga show a positive effect to especially to the microstructural formation. When the content of Ga reaches 0.05wt%, the most uniform structure is obtained and the grain achieves the highest degree of refinement. There are two main primary explanations, according to the adsorption theory. First and foremost, the adsorbed Ga reduces the surface energy difference of the crystal surfaces. Secondly, due to the pinning effect, the growth rate of the crystal surface adsorbing Ga is reduced, thus the formation of the grains get smaller and distribute homogeneously. The IMC particles are small and uniformly dispersed instead of large and coarse, and therefore the mechanical properties of the solder can achieve the best in this component as well. However, the optimum value of the Ga addition still will be carried out. As depicted in **Fig. 3.**, the result of the SEM-EDX analysis for the Sn-0.7Cu-0.05Ga solder alloys is shown. The EDX analysis shows that the black phases consist primarily of Sn, Cu, and Ga, the existence of which may also affect the mechanical properties of SC and SCN solder joints. Based on the result, Ga mostly distributed at the primary IMC at 0.30 atomic percentage, wt% while at the bulk solder interface the reading recorded 0.02 atomic percentage wt%. **Fig. 4.** shows the elemental mapping of SC and SCN solder alloys.

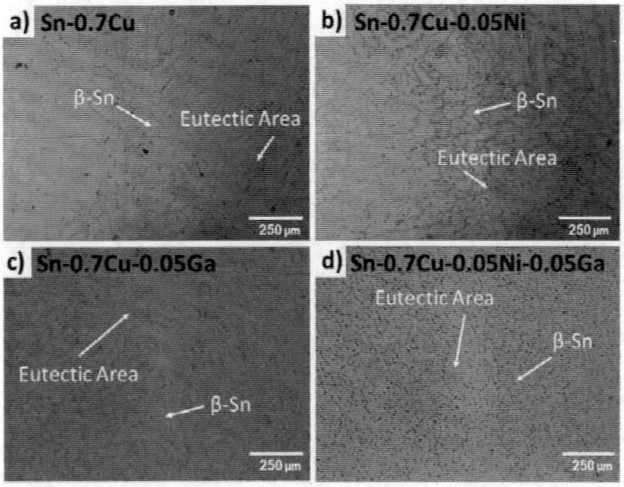

Fig. 2. Optical micrographs show the microstructure of designated lead-free solder alloys: a) SC, b) SCN, c) SC-0.05Ga, and d) SCN-0.05Ga solder alloys

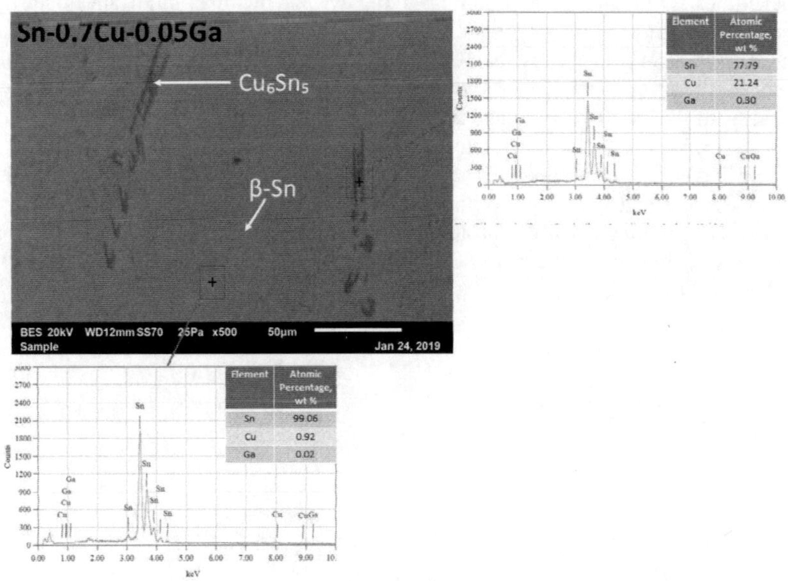

Fig. 3. SEM-EDX analysis of the Sn-0.7Cu-0.05Ga solder alloys

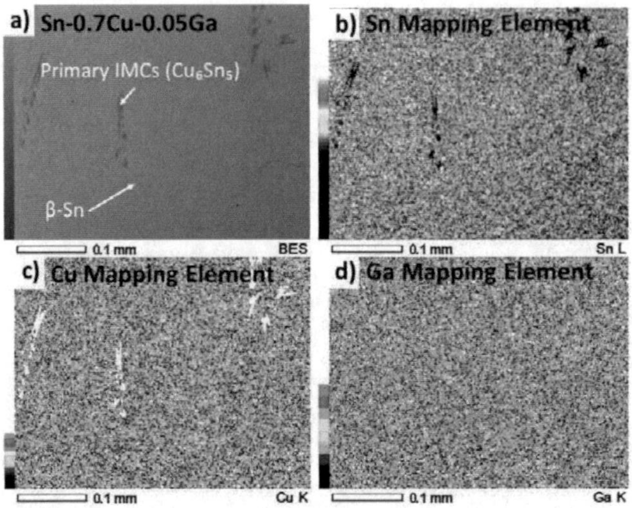

Fig. 4. SEM-EDX elemental mapping of SC-0.05Ga solder alloy on Cu OSP for different elements: (a) SEM image of SC-0.05Ga, (b) Sn, (c) Cu and (d) Ga.

3.3 Mechanical properties of the solder alloys

In this research, as shown in **Fig. 5.**, the maximum hardness impact of the solder alloys evaluated. The hardness can be improved significantly with the addition of Ga. When the addition of Ga was 0.05 %, the shear strength increased linearly. The reason for this is primarily the refinement of the bulk solder microstructure and soldering interface by Ga. In the meantime, when the Ga addition is at 0.05wt %, the maximum shear strength is obtained. Although this is only one range of the parameter, the best addition of Ga incorporated at SCN solder alloy. Not only that, the modified solder matrix is good for the mechanical properties of the solder alloy based on the fine-grain strengthening theory [26]. On the basis of the Ga-Sn phase diagram, Sn exhibits a maximum solubility of 6.4 % of Ga, so Ga may form a solid solution strengthening with Sn in this study. The dissolution of Ga in Sn contributed by the mechanical strengthening effect. Though, the produced Ga-rich phases of Sn, Ga and Cu may have an adverse effect on mechanical properties as the addition of Ga continue to raise. For now, it cant be concluded how much the appropriate amount of Ga content in SC and SCN solder alloys. But further research work and parameters will tackle this issue soon.

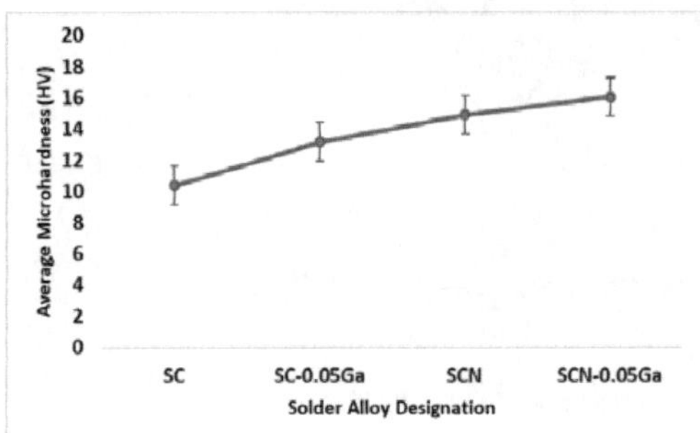

Fig. 5. Average microhardness for of Sn-0.7Cu and Sn-0.7Cu-0.05Ni solder alloys with addition of 0.05 Ga %

4 Conclusion

From this preliminary study, this research work investigated the effect of Ga addition on the microstructure morphology and mechanical properties of Sn-0.7Cu and Sn-0.7Cu-0.05Ni lead-free solder alloys. Below are the early subsequent conclusions can be obtained:

1. The solderability of Sn-0.7Cu and Sn-0.7Cu-0.05Ni lead-free solder alloys was can be significantly improved by adding Ga and the solder achieves the peak values at the addition of 0.05 %. Besides, when applying the suitable temperature, the wetting performance can be enhanced greatly as well.

2. For this early result, the correct amount of Ga addition still in research. But the addition of 0.05 Ga can effectively improve the mechanical properties of the solder joints as a result of the smaller dendritic formation of crystal β-Sn.

With all due respect, this research project was mainly under the collaboration between Universiti Malaysia Perlis (UniMAP) and Nihon Superior (M) Sdn. Bhd. under CREST's grant. The authors wish to deliver their gratitude to Higher Education Malaysia, Universiti Malaysia to Higher Education Malaysia, Universiti Malaysia Perlis (UniMAP), Nihon Superior (M) Sdn. Bhd. and CREST Grant Scheme no. P14C1-17/001 for financial supports, equipment-wise, and experiment setup.

References

1. M. Abtew, G. Selvaduray, Mater. Sci. Eng. R 27, 5–95 (2000)
2. M.A. Dudek, R.S. Sidhu, N. Chawla, M. Renavikar, J. Electron. Mater. 35, 12–2088 (2006)
3. R.M. Said, M.A.A.M. Salleh, M.N. Derman, M.I.I. Ramli, N.M. Nasir, and N. Saud, Key Engineering Materials. **700** (2016)
4. M.I.I. Ramli, M.S.S. Yusof, M.A.A.M. Salleh, R.M. Said, and K. Nogita, Solid State Phenomena. **273** (2018)
5. M.A.A.M. Salleh, C.M. Gourlay, J.W. Xian, S.A. Belyakov, H. Yasuda, S.D. McDonald, and K. Nogita, Scientific Reports. **7** (2017)
6. F. Somidin, H. Maeno, M.M. Salleh, X.Q. Tran, S.D. McDonald, S. Matsumura, and K. Nogita, Materials Characterization. **138** (2018)
7. L. Zhang, S.B. Xue, L.L. Gao, Y. Chen, S.L. Yu, Z. Sheng, G. Zeng, Solder. Surf. Mt. Tech. 22, 2–30 (2010)
8. K. Nogita, Intermetallics 145–149 (2010).
9. M. A. A. M. Salleh, S. D. Mcdonald, C. M. Gourlay, S. A. Belyakov, H. Yasuda and K. Nogita, J. Electron
10. M. A. A. Mohd Salleh, M. M. A. A.Bakri, M. H. Zan@Hazizi, N. F. M. Alui, and Z. A. Ahmad, Mater. Sci.
11. M.A.A.M. Salleh, M.H. Hazizi, M.A.B.M. Abdullah, N. Noriman, R. Mayapan, and Z.A. Ahmad, Advanced Materials Research. **626** (2012)

12. K. Nogita, M.A.A.M. Salleh, S. Smith, Y. Q Wu, S.D. McDonald, A.G.A. Razak, T. Akaiwa, and T. Nishimura, 2017 International Conference on Electronics Packaging (ICEP). (2017)
13. K. Nogita, M.A.A.M. Salleh, E. Tanaka, G. Zeng, S.D. McDonald, and S. Matsumura, JOM. **68**(11) (2016)
14. M.A.A.M. Salleh, A.M.M.A. Bakri, F. Somidin, and H. Kamarudin, International Review of Mechanical Engineering. (2013)
15. M.A.A.M. Salleh, M.M.A.B. Abdullah, F. Somidin, A.V. Sandu, N. Saud, K. Hussin, S.D. McDonald, and K. Nogita, Revista de Chimie **64**(7) (2013)
16. Eng. A 633-637 (2012).F.J. Cheng, F. Gao, J.Y. Zhang, W.S. Jin, X. Xiao, J. Mater. Sci. 46, 10–3424 (2011)
17. H.J. Caul, D.L. Smith, J. Am. Dent. Assoc. 53, 3–315 (1956)
18. S. Bhattacharya, D. Baldwin, Adv. Pack. 9, 8–61 (2000)
19. T.C. Hsuan, K.L. Lin, J. Alloys Compd. 409, 350 (2009)
20. K.I. Chen, K.L. Lin, J. Electron. Mater. 32, 10–1111 (2003)
21. K.I. Chen, S.C. Cheng, C.H. Cheng, S. Wu, Y.L. Jiang, T.C.
22. Cheng, Adv. Mater. Sci. Eng. 2014, 1 (2014)
23. R.J. Klein Wassink, (Isle of Man, British Isles, 1989), pp. 149–150
24. N.S. Liu, K.L. Lin, Scr. Mater. 54(2), 219 (2006)
25. Y.W. Shi, J. Tian, H. Hao, Z.D. Xia, Y.P. Lei, F. Gao, J. Alloys Compd. 453, 1–180 (2008)
26. L.L. Gao, S.B. Xue, L. Zhang, Z. Sheng, G. Zeng, F. Ji, J. Mater. Sci. Mater. Electron. 21, 7–643 (2010)

IOP Conf. Series: Materials Science and Engineering **701** (2019) 012032 doi:10.1088/1757-899X/701/1/012032

A review of geopolymer ceramic as a potential reinforcement material in solder alloys.

Nur Nadiah 'Izzati Zulkiflii[1,*], *Mohd Mustafa Al-Bakri* Abdullah , and *Mohd Arif Anuar* Mohd Salleh[1]

[1]School of Materials Egineering, Kompleks Pusat Pengajian Jejawi 2, Universiti Malaysia Perlis (UniMAP) Taman Muhibbah, 02600 Arau, Perlis, Malaysia.

Abstract. Nowadays, lead-free solder has been currently used in electronic packaging technology as part of soldering material. Since SnPb was detected to produce toxicity and might harm the consumers, the usage of Pb solder has been banned by WEE and RoHS. Therefore, various studies have been developed as alternatives to replace the usage of SnPb. Since lead-free solder might not perform as well as their traditional SnPb, researchers suggested to add some elemental reinforcement particles in matrix alloy. Previously, addition of ceramic reinforcement has been widely known in enhancing the properties of solder-substrate. This paper reviews about the presence studies of ceramic as solder reinforcement, the characteristics of geopolymer ceramic as potential solder reinforcement, and their properties in providing a superior solder joint. In this review, the characterization is divided into two stages; 1) characterization of geopolymers in terms of microstructural and crystallographic; 2) characterization of solder properties in terms of intermetallic layer growth (IMC), wettability, and its mechanical properties

1 Introduction

Soldering has been widely known in electronic packaging industries for electronic assembly where the solder materials serves as a conductive materials that bond together with the substrate through reflow soldering process at controlled temperature forming one or more electrical interconnection within the circuits [1], [2]. Few years ago, lead (Pb) solder was typically known as the most superior solder material in electronic packaging industries as it provide good soldering properties, reliability, and manufacturability with an affordable price [3]. In early 2000, Needleman found out that the usage of Pb solder might lead to health and environmental issues for a long period of time. The toxicity of Pb older may cause harm if in direct contact with skin, mouth, nose, and eyes if it is not properly controlled [4]. Therefore, European Union's Waste in Electrical and Electronic Equipment (WEE) and Restriction of Hazardous Substances (RoHS) has banned and restricted the used of Pb solder in electronic industries [1], [3]. Since the usage of Pb has been restricted, National Electronics Manufacturing Initiatives (NEMI) has recommend to find other alternatives in replacing lead solder by inventing new lead-free solder material [5]. As a result, the elimination of Pb solder

* Corresponding author: nadiahizzati.zulkifli@gmail.com

has lead to a great response by electronic consumers industries in EU markets and later from around the world such as China, Japan, South Korea, Turkey, and United States [3], [4].

According to Xu et. al. lead-free solder must be designed to form electrical and mechanical connection to maintain integrity through manufacturing process and service conditions [1]. Various types of lead-free solder has been fabricated to replace SnPb solder such as tin silver (SnAg), tin bismuth (SnBi), tin copper (SnCu), tin zinc (SnZn), tin rhenium (SnRe), tin zinc aluminium (SnZnAl), tin silver copper indium (SnAgCuIn), tin silver copper (SAC) and many others but it still beneath the wide range availibility of SnPb [2, 6-10]. In manufaturing the new lead-free solder, its future properties must be comparable with it traditional SnPb solder material. Since Pb is widely known as the least expensive elements on earth, the overriding concerns is one of its replacement material will have an increasing intrinsic cost and whether the new lead-free solder can have equally or higher performance level to their traditional lead solder [11].

2 Solder reinforcement

Numerous researchs have been studied for the replacement of harmful SnPb that might have better potential than the previous one [12-15]. Researchers discovered that the perfomance of solder material becomes higher when some small amounts of particles element such as transition metal or rare earth elements were added in matrix alloy as reinforcement compared to plain matrix alloy [16-17].

Previously, most conducted research illuminated that addition of ceramic reinforcement produced high performance in terms of mechanical and electrical properties compare to plain matrix lead-free alloy [16-17]. Kotadia et. al. stated that the new elemental addition towards solder alloys should improves the wettability by reducing the surface tension of pure Sn, improves IMC thickness by suppressing intermetallic layer growth, and boost their mechanical and electrical properties such as hardness, shear strength, fatigue, creep, and thermal ageing [11]. Ceramic materials demonstrate enhanced properties in terms of microstructure, thermal properties, tensile strength, liquidous temperature and others under demanding conditions became one of the factors in choosing ceramic as reinforcement as it is reported to decompose β-Sn microstructure [18].

Ceramics are divided into several types which are nitride, oxide, silicate, and oxide ceramic were widely used in most studies. Addition of amorphous nanosilica ($Si0_2$) in Sn-0.7Cu could effectively provide optimum mechanical properties in terms of yield and tensile strength, and elastic modulus as the weight percentage of $Si0_2$ increased up to 1.5 wt.%. The following performances can be due to ceramic properties that usually is harder than the matrix alloy [19]. The wettability and the thickness of IMC layer has been reported to improve when 0.75 wt.% SiC was added in monolithic Sn-0.7Cu [20]. The wettability shows reduction in contact angle while the thickness of IMC layer changed from scallop shape to combination of planar and scallop shape. Chellvarajoo and Abdullah studied that the addition of NiO nanoparticles increased the melting point of SAC305 solder paste and gradually decrease the IMC formation after several addition [21].

The presence of diamond nanoparticles in crystalline phase in SAC305 was observed by Chellvarajoo et. al. as Fe_2Ni0_4 was used as composite solder [22]. Before addition, the IMC shape was thick and formed as elongated scallop-shaped. After Fe_2Ni0_4 nanoparticles were added, it reacts as catalyst to reduce Sn activities and therefore the nucleation and growths of both Cu_6Sn_5 and Cu_3Sn phase on Cu substrate were suppressed.

2.1 Geopolymer ceramic as reinforcement

Formerly, there were not much research conducted on geopolymer ceramic as reinforcement in monolithic solder alloy. Geopolymer was discovered by Professor Joseph Davidots in early 1980s as one of concrete materials that gives higher mechanical and durability properties. Other studies found that geopolymer materials gives excellent resistance to chemical attack [23]. The word geopolymer came from geopolymerization process where geopolymerization occurs when chemical reaction between aluminosilicate sources and alkaline silicate solution was mixed together and cures or sinter at several temperature depends on its properties [23], [24].Therefore, geopolymerisation happened when the raw materials rich in alumina and silica or pozzolanic materials dissolved in an alkaline activator solution [25-27]. Basically

In geopolymer industries, there are various mix design for each type of geopolymer. The alkaline activator solution that mostly used in geopolymer industries were the combination of Na_2SiO_3, and NaOH or KOH but NaOH was mostly used due to their high properties. The rate of geopolymerization are depends on molarity of alkaline activator solution, ratio of Na_2SiO_3/NaOH, ratio of solid/liquid, and the time for curing or sintering per temperature [28-29]. Different types of raw materials geopolymer gives different optimum mix design to ensure their performances. In this paper, the structure of fly ash (FA), kaolin, and slag as raw material for geopolymer were reviewed.

Fly ash-based geopolymer was obtained by coal burning fire plants and have been widely used due to its good quality coal available and the low cost of producing this source [30]. Based on Mustafa et. al. the microstructure of fly ash appeared to be in the shape of spherical, hollow, and glassy which also known as cenospheres [31]. On the other hand, kaolin is reported to have plate-like structure with some needle-like shape [32]. The morphological of slag when observed under SEM as slag tend to have coarse and edgy shape compare to FA and kaolin just like Fig 1 below [33]. Slag can be classified into two group which is from ferrous and non-ferrous. Ferrous slag are such as blast furnace slag, iron steel slag, and ferroalloy slag while non-ferrous which is lead, nickel, cadmium, tin, copper, and zinc slag [34-35] With different microstructure of geopolymer ceramic reinforced monolithic matrix alloy, it might change the performance of solder joint.

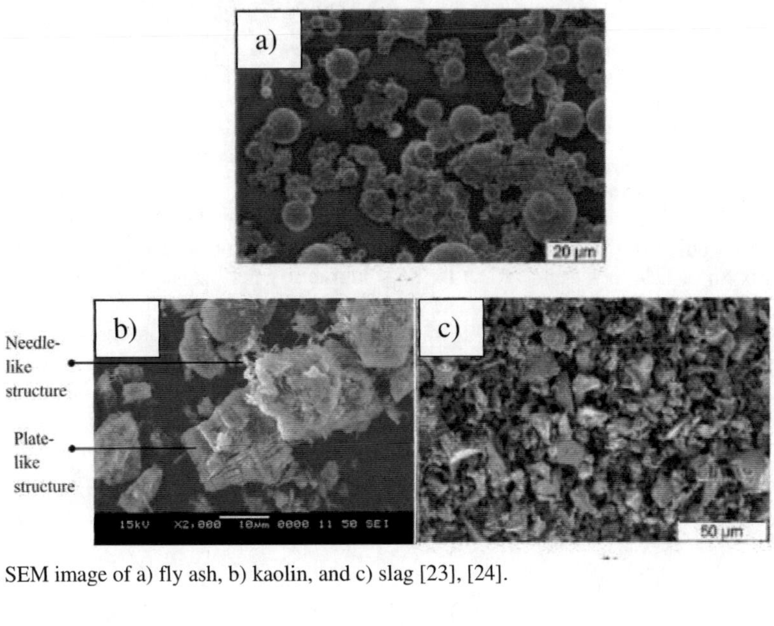

Fig. 1. SEM image of a) fly ash, b) kaolin, and c) slag [23], [24].

IOP Conf. Series: Materials Science and Engineering **701** (2019) 012032 doi:10.1088/1757-899X/701/1/012032

Geopolymer ceramic is produced by sintering the raw geopolymer at 600°C temperature as the crystalline phase already detected from the XRD result [36]. Based on Liew et al., the changed of geopolymer colour occurs for each sintering temperature. This is due to dehydration of moisture and phase transformation during sintering. The geopolymer ceramic reached their optimum strength when sintered it at 1200°C compare to 900°C, 1000°C, 1100°C and 1300°C. This is because the amorphous phase of geopolymer ceramic which tend to form brittle effect transformed into nepheline phases which is more stable [32]. In most research, the size of particles reinforcement needs to be smaller or the same size as the matrix alloy to ensure both materials pinned together during mixing process. The bigger size of particle reinforcement can lead to unhomogenous distribution mixing and causing agglomeration of specific material [37].

3 Properties of solder

3.1 Interfacial reactions between solder substrates

During soldering, intermetallic layer phase will grow due to interaction between solder alloy and solder substrate forming solder joint interface. However, rapid Cu dissolution between solder and Cu substrate caused high intermetallic layer thickness. Usually, high IMC thickness gives disadvantages to the solder joint as it is typically brittle and therefore contribute to failures joint [11]. In another words, the smaller the thickness of IMC layer growth, the higher performance provided since the initiation and propagation of cracks become less [38]. Figure 2 show the area of IMC layer produced during solder reflowed.

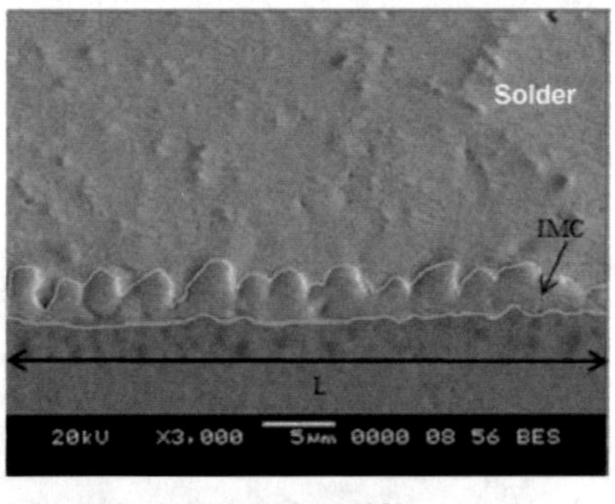

Fig. 2. Area of IMC phase at the solder joint interface [38].

A few studies clarified that longer period of time for reflow solder and dwell time gave rapid growth of IMC layer and reflect to the solderability performance of solder joint [21]. Mohd Salleh stated that the formation of intermetallic layer may leads to the formation of Kirkendall voids around the Cu_6Sn_5/Cu substrate interface and therefore may deteriorate the reliability of the solder joint [39]. The presences of other reinforcing elements appear to limit the growth of neighbouring grains making the IMC layer more stable [6]. During reflow soldering, the mobility of reinforcing particles embed in matrix solder grain structure and substrate. Thus, the movement of Cu atoms will slowly reduced the formation of IMCs layer as it can act as an obstacle in the diffusion of Sn and Cu [16].

3.2 Wettability ability

In soldering, wettability refers to the ability of the molten solder to spread on the Cu substrate to create a solder joint layer when the solder is subjected to high temperature [18]. Wettability can be determined by the contact angle (θ°) between molten solder alloy and the Cu substrate such as Figure 4 below. The lower the contact angle, the great wettability achieve by the solder on its application [40]. M. Ramli clarified that the contact angle of Sn-0.7Cu-0.05Ni decrease from 35.6° to a minimum 22.38° after added 1.0 wt.% of TiO_2 powder which indirectly improves the wettability of solder [38].

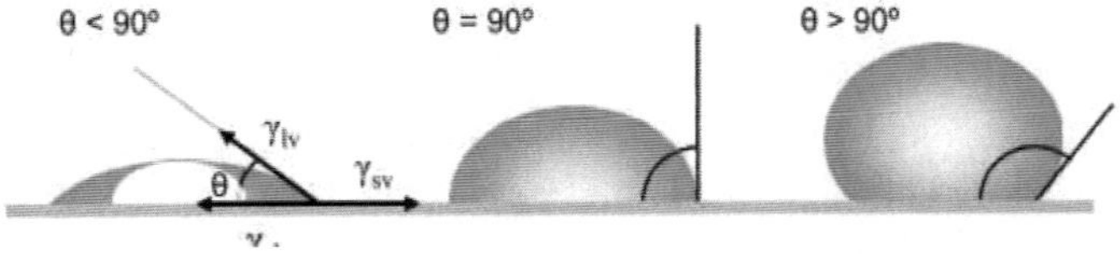

Fig. 3. Illustration of contact angle formed by molten solder alloy on a smooth solder-substrate [32].

Rate of wettability depends on the solder alloy proportion and surface properties of the substrates [41]. Dusek et. al. elucidate that there are difference in wettability between lead solder and lead-free solder where lead-free solder has smaller wettability angle but are prone to obtain Tombstoning effect compare to lead solder. Tombstoning effect are frequently take place in lead-free solder material when one side of soldered component are not same or risen during remelting process [41].

3.3 Mechanical properties

Hardness and shear strength can be categorized as mechanical properties of solder alloys. Based on Salleh et. al., higher values of hardness (Hv) reduced deformation and crack to the solder part. The microhardness values of Sn-Cu-Ni with TiO_2 became higher with average of 16.64 Hv when sintered in microwave oven compare to plain Sn-Cu-Ni [42]. This result shows that addition of particles reinforcement might alter the hardness of solder sample. Besides, the particles size of the reinforcement also plays an important roles since fine distributed size of particle reinforcement of solder influenced the hardness value due to high pining effect [38, 42].

In terms of shear strength, high intermetallic layer growth contributes to low shear strength since the presence of high intermetallic layer produce brittle effect to the solder joint. Several researchers proved that addition of In as reinforcement in Sn-Ag-Cu could modified their hardness, shear stress as well as their tensile strength [43].

4 Conclusion

Geopolymer ceramic reinforcement show significant potential to be reinforced in matrix alloy for the future. The different shape of geopolymer particles and chemical composition also might lead to different solder-joint performance. On the other hand, the high thermal might improved the solder joining. Therefore, further research is needed to investigate the suitable criteria for reinforcing particles in solder alloy.

This work was supported by Universiti Malaysia Perlis, Nihon Superior Co. Ltd and Institutional Links grant, ID 332397914, under the Newton-Ungku Omar Fund partnership. The grant is funded by the UK Department of Business, Energy and Industrial Strategy (BEIS) and Malaysia and delivered by the British Council.

References

1. S. Xu, A. H. Habib, A. D. Pickel, M. E. McHenry, Prog. Mater. Sci., **67** (2015)
2. M. Aamir, R. Muhammad, N. Ahmed, M. Sadiq, M. Waqas, **36** (2017)
3. S. Cheng, C. M. Huang, M. Pecht, Microelectron. Reliab., **75** (2017)
4. O. A. Ogunseitan, Jom, **59** (2007)
5. J. W. Yoon, J. Alloys Compd., **381** (2004)
6. N. Nadiah 'Izzati Binti Zulkifli, M. Mustafa Al Bakri Abdullah, M. A. A. Mohd Salleh, IOP Conf. Ser. Mater. Sci. Eng., 551 (2019)
7. M.I.I. Ramli, N. Saud, M.A.A.M. Salleh, M.N. Derman, R.M. Said, and N. Nasir, Materials Science Forum. **803** (2015)
8. R.M. Said, M.A.A.M. Salleh, M.I.I. Ramli, and N. Saud, Solid State Phenomena. **280** (2018)
9. M.A.A.M. Salleh, M.H. Hazizi, M.A.B.M. Abdullah, N. Noriman, R. Mayapan, and Z.A. Ahmad, Advanced Materials Research. **626**
10. F. Somidin, H. Maeno, M.M. Salleh, X.Q. Tran, S.D. McDonald, S. Matsumura, and K. Nogita, Materials Characterization. **138** (2018)
11 H. R. Kotadia, P. D. Howes, S. H. Mannan, Microelectron. Reliab., **54** (6–7) (2014)
12. K. Nogita, M.A.A.M. Salleh, S. Smith, Y. Q Wu, S.D. McDonald, A.G.A. Razak, T. Akaiwa, and T. Nishimura, 2017 International Conference on Electronics Packaging (ICEP).
13. K. Nogita, M.A.A.M. Salleh, E. Tanaka, G. Zeng, S.D. McDonald, and S. Matsumura, JOM. **68**(11) (2016)
14. M.A.A.M. Salleh, A.M.M.A. Bakri, F. Somidin, and H. Kamarudin, International Review of Mechanical Engineering
15. M.A.A.M. Salleh, M.M.A.B. Abdullah, F. Somidin, A.V. Sandu, N. Saud, K. Hussin, S.D. McDonald, and K. Nogita, Revista de Chimie **64**(7) (2013)
16. M. A. A. Mohd Salleh, A. M. Mustafa Al Bakri, H. Kamarudin, M. Bnhussain, M. H. Zan Hazizi, F. Somidin, Phys. Procedia, **22** (2011)
17. R. M. Said, M. A. A. Mohd Salleh, M. I. I. Ramli, N. Saud, M. M. A. Al Bakri Abdullah, A. V. Sandu, AIP Conf. Proc., **1835** (2017)
18. E. A. Eid, A. N. Fouda, E. S. M. Duraia, Mater. Sci. Eng. A, **657** (2016)
19. Z. Fathian, A. Maleki, B. Niroumand, Ceram. Int., 43(6) (2017)
20. N. S. M. Zaimi, M. A. A. M. Salleh, M. M. A. Al Bakri, R. M. Said, N. Saud, Solid State Phenom., **280** (2018)
21. S. Chellvarajoo M. Z. Abdullah, Jmade, **90** (2016)
22. Srivalli Chellvarajoo, M.Z. Abdullah, Z. Samsudin, Mater. Des., 67 (2015)
23. O. A. Abdulkareem, A. M. Mustafa Al Bakri, H. Kamarudin, I. Khairul Nizar, A. A. Saif, Constr. Build. Mater., **50** (2014)
24. M. M. A. B. Abdullah, L. Y. Ming, H. C. Yong, M. F. M. Tahir, Cem. Based Mater.,

(2018)

25. I. H. Aziz, M. M. A. B. Abdullah, C.Y. Heah, and Y.M. Liew, D.Panias, K.Sakkas. IOP Conference Series : Materials Science and Engineering, 209,(2017).
26. Z. Yahya, M. M. A. B. Abdullah, K. Hussin, K. N. Ismail, R. A. Razak, A. V. Sandu, Materials (Basel)., **8** (5) (2015)
27. D. D. Burduhos Nergis, M. M. A. B. Abdullah, P. Vizureanu, M. F. Mohd Tahir, Mater. Sci. Eng., **374** (2018)
28. C. Y. Heah et al., Phys. Procedia, **22** (2011)
29. G. F. Huseien, J. Mirza, M. Ismail, M. W. Hussin, M. A. M. Arrifin, A. A. Hussein, Indian J. Sci. Technol., **9** (2016)
30. I. H. Aziz et al., Proc. Inst. Mech. Eng. Part L J. Mater. Des. Appl., **233**(4) (2019)
31. A. M. Mustafa et al., J. Eng. Technol. Res., **3** (2011)
32. Y. M. Liew et al., Constr. Build. Mater., **156** (2017)
33. M. A. Yazdi, M. Liebscher, S. Hempel, J. Yang, V. Mechtcherine, Constr. Build. Mater., **191** (2018)
34. I. H. Aziz et al., OALib, **4**(11)(2017)
35. I. H. Aziz, M. M. A. B. Abdullah, C.-Y. Heah, Y.-M. Liew, Adv. Cem. Res., (2019)
36. K. Zulkifly, H. C. Yong, M. M. A. B. Abdullah, L. Y. Ming, D. Panias, K. Sakkas, IOP Conf. Ser. Mater. Sci. Eng., **209**(1) 2017
37. P. Sharma, S. Sharma, D. Khanduja, Mater. Manuf. Process., 30(11) (2015)
38. M. I. I. Ramli, N. Saud, M. A. A. M. Salleh, M. N. Derman, R. M. Said, Microelectron. Reliab., **65** (2016)
39. M. A. A. Mohd Salleh, S. D. McDonald, K. Nogita, J. Mater. Process. Technol., **242** (2017)
40. Y. Yuan and T. R. Lee, Surface Science Techniques - Google Livros. 2013.
41. M. Dušek, I. Szendiuch, J. Bulva, M. Zelinka, “Wettability – SnPb and Lead-free,” Http://Www.Umel.Feec.Vutbr.Cz/, pp. 2–6, 2011.
42. M. A. A. M. Salleh, S. D. Mcdonald, Y. Terada, H. Yasuda, and K. Nogita, Mater. Des., **82** (2015)
43. K. Kanlayasiri, M. Mongkolwongrojn, T. Ariga, J. Alloys Compd., **485**(1–2) (2009)

IOP Conf. Series: Materials Science and Engineering **701** (2019) 012033 doi:10.1088/1757-899X/701/1/012033

Influence of kaolin geopolymer ceramic additions to the wettability and electrical properties of Sn-3.0Ag-0.5Cu (SAC305) lead free solder

Nur Syahirah Mohamad Zaimi [1] , *Mohd Arif Anuar* Mohd Salleh [1,*], *Mohd Mustafa Al Bakri* Abdullah[1], *Marliza* Mostapha[1] and *Romisuhani* Ahmad[2]

[1]Centre of Excellence Geopolymer and Green Technology, School of Materials Engineering, Universiti Malaysia Perlis (UniMAP), Taman Muhibbah 02600, Jejawi, Arau, Perlis, Malaysia
[2]Faculty of Engineering Technology, Uniciti Alam Campus, Universiti Malaysia Perlis (UniMAP), Perlis, Malaysia

Abstract. The effect of kaolin geopolymer ceramic addition to the wettability and electrical resistivity of Sn-3.0Ag-0.5Cu (SAC305) lead free solder was successfully explored. Powder metallurgy with microwave sintering method was used to fabricate SAC305 composite solder. Five different weight percentage (wt.%) of kaolin geopolymer ceramic (0, 0.5, 1.0, 1.5 and 2.0 wt.%) were used in this study. The wettability of composite solder revealed the optimum contact angle was achieved at 1.0 wt.% with 20.8 °. Then, the electrical resistivity of composite solder showed significant change in the resistivity value.

1 Introduction

The world nowadays is facing on a fast-track in technology transformation. The fundamental factors for this was the ever developing and advancing in electronic packaging industry. The emergence of advance electronic devices that are getting smaller and thinner [1] with more complicated functions demand a very reliable and high-performance solder interconnections. In solder interconnections, solder which employed as interconnections materials play a vital role in producing the ever-reliable solder joints that determine the overall functionality [2].

Tin-Lead (Sn-Pb) solder alloys are widely used in the electronic packaging industry as an interconnection material since more than 50 years ago due to its outstanding reliability and solderability. However, as the health and environment matters had been taken seriously, the usage of lead in any electric and electrical equipment had been prohibited. Therefore, a new lead free solder system with excellent properties need to be introduced [3-10]. There are numerous of studies had been done by previous researchers in order to improve the performance and properties of lead-free solders such as microalloying, composite and transient liquid phase (TLP) bonding [11]. Among all those methods, the

* Corresponding author: arifanuar@unimap.edu.my

most attractive, promising and viable way to enhance the performance and properties of solder is by introduce a second phase or reinforcement to an existing lead-free solder, forming a lead-free composite solder. In general, a composite solder is form when the solder alloy was intentionally corporate or reinforced with metals or non-metals materials [12]. Ceramics is one type of reinforcement which can be incorporated in lead free solder forming composite solders. An extensive studies on the ceramic particles incorporating with solder matrix had been carried out by many researchers [13-18]. According to Wang et.al., [19], the addition of Silicon Oxide (SiO_2) into SAC solder give positive effects to the performance of solder. The addition of SiO_2 results in refining the microstructures, improve wettability, shear force and suppress the growth of IMC [19]. Besides that Tsao [20] discussed possibility of adding Titanium Oxide (TiO_2) into SAC solder systems in terms of the performance of solder during isothermal ageing.

Therefore, with the improvements shown in the lead-free solder as it was reinforced with various reinforcements, it is very interesting to study on how the other type of ceramic reinforcement could affect the performance of solder in terms of the microstructure, solderability, thermal properties and electrical properties. In this research, an attempt was made by incorporating Sn-3.0Ag-0.5Cu (SAC305) lead free solder with kaolin geopolymer ceramics. There are no research reporting on the effects of kaolin geopolymer ceramics to SAC305 lead free solder to date.

2 Methodology

In this research, the base material, Sn-3.0Ag-0.5Cu (SAC305) lead free solder powders with an average size of 25-45 μm was supplied by Sigma Aldrich. Kaolin geopolymer ceramic (KGC) with particle size less than 38 μm was used as the reinforcement material. Powder metallurgy with microwave sintering method was performed to fabricate composite solders with the addition of KGC particles. Five different wt. %. of KGC (0, 0.5, 1.0, 1.5 and 2.0 wt.% KGC) were homogeneously mixed with SAC305 powder by using planetary mill machine at 200 rpm for one hour. Subsequently, the solder mixtures for each composition were uniaxially compacted at 4.5 tonnes by using a Specac 15-ton Manual Hydraulic Press. Then, each of compacted pellets were sintered by a microwave sintering method at a temperature of 185 °C under ambient conditions in an 800 W, 50 Hz Panasonic microwave oven.

Then, the samples were reflowed in a F4N reflow oven with the aid of rosin mildly activated flux (RMA). The samples were mounted with epoxy resin, grinded and polished. For the solder, about 10 samples for each of compositions were measured in order to ensure the accuracy of the results. The wettability of solders can be determined respected to the contact angle to the Cu substrates. The contact angle was measured by using J- Image software.

The electrical resistivity of composite solders was measured by using 4-point probe as shown in Fig. 1. The samples used for these testing was compacted into a cylindrical shape with approximately 12 mm diameter and 5 mm thick. A set of 5 samples for each monolithic and composite solders system were prepared. A probe spacing (s) of 3 mm was used in this measurement. The current applied in this measurement was fixed to 1 A and the corresponding voltage (V) drop across the samples were measured. As the sample thickness (t) was greater than the probe spacing, the following formula equation (1) was used to measure the electrical resistivity (ρ) of SAC 305 composite solder system.

$$\rho = 2\pi s \left(\frac{V}{I}\right) \quad (1)$$

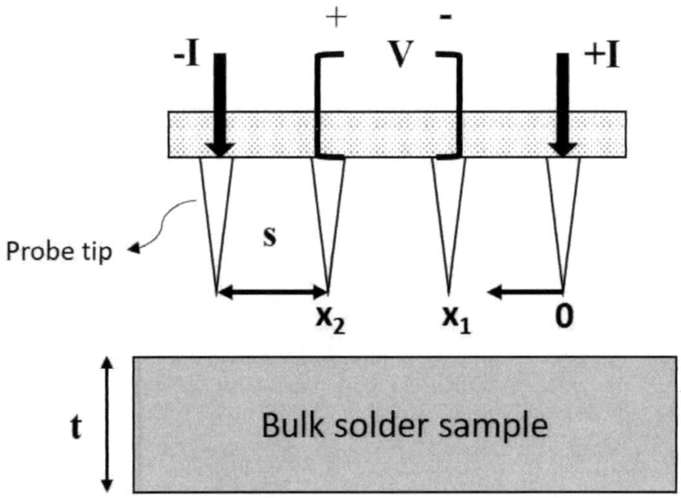

Fig. 1. Four-point probe schematic diagram.

3 Results and Discussions

3.1 Wettability

Wettability is one of crucial parts in the soldering which determine the solderability classification. Forming a reliable solder bonding with substrate was very important in order to produce electrical and electronic device with an excellent performance. The wettability of the composite solder can be measured with respect to the contact angle (ϴ°) of the solder materials and Cu substrate [21]. To create a reliable solder bonding, the contact angle (ϴ°) formation must be smaller and it was acceptable if the contact angle measured was less than 45° [17].

The influence of various weight percentage (wt. %) kaolin geopolymer ceramic (KGC) reinforcement on the wettability of SAC305 lead free solder was shown in Fig. 2. The error bars on the graph represent the standard deviations. Results shows that the average contact angles of solder is firstly decreased and then starts to increase with the increasing content of KGC reinforcement. Specifically, the average contact angle firstly decreased from 24.0 µm for SAC305 without the addition of KGC to 22.3 µm and 20.8 µm at 0.5 wt.% and 1.0 wt.% KGC addition, respectively. Then, the increasing trend of contact angle was shown as the value of contact angle increased to 22.0 µm and 23.1 µm for SAC305/1.5 wt.% and SAC305/2.0 wt.% KGC, respectively. Although the contact angle of SAC305 solder with the addition of 1.5 wt.% and 2.0 wt.% KGC particles shown an increasing trend, the value was still lower than the contact angle of SAC305 without the addition of KGC. These results indicate that the appropriate amount of KGC reinforcement added on the SAC305 lead free solder influenced the wettability performance.

The improvement in the wettability of SAC305 composite solder as compared to SAC305 without the addition of KGC may be attributed to the facts that the KGC particles added accumulated at the interface between the molten solder and the flux during soldering process. The accumulation of KGC particles causes the surface tension between molten solder and flux to be lowered whereas the surface tension between substrate and solder will be high. Therefore, the contact angle formed for SAC305 composite solder at 0.5 wt.% and 1.0 wt.% KGC were smaller than SAC305 without addition of KGC. However, an excess addition of KGC particles beyond than 1.0 wt.%, increased the contact angle of composite solder. The possible reason for this was that the higher fraction of KGC particles may inhibited the flow of molten solder by increase the viscosity of the solder and 'pins' the leading edge of the molten solder from further spreading [22].

This result was in consistent with the previous finding by Wang et.al., [19] whereby the wettability of SAC 305 was improved with the addition of SiO_2 up to 0.1 wt.%. However, as 0.5 wt.% and 1.0 wt.% SiO_2 were added into SAC305 solder, the wettability declined with the contact angle value lower than the pure solder.

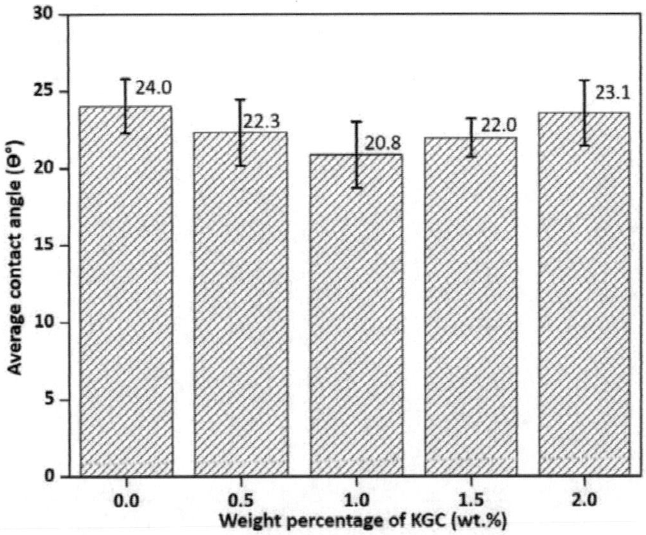

Fig. 2. Bar graph showing the influence of geopolymer ceramics addition to the wetting angle of SAC 305 lead free composite solder.

3.2 Electrical properties

Solder plays an important role in the electronic industry such that it acts as an electrical interconnection in the electronic devices. Thus, it was necessary for the composite solders to provide good electrical properties. In this research, an electrical resistivity of the composite solders was measured. Fig. 3 depicts the average electrical resistivity of SAC305 lead free solder with addition of various weight percentage (wt.%) of kaolin geopolymer ceramic (KGC) particles. The result on the electrical resistivity of SAC305 lead free solder at various wt.% of KGC particles shows an showing an insignificant change in electrical resistivity values. The electrical resistivity of SAC305 lead free solder at 0 wt.% KGC was 4.190 μΩ.cm. With the addition of KGC particles, the electrical resistivity value shows slightly increased. The electrical resistivity of 0.5 wt.%, 1.0 wt.%, 1.5 wt.% and 2.0 wt.% KGC were 4.219, 4.220, 4.222 and 4.229 μΩ.cm, respectively.

According to Chen et.al., [23], the electrical resistivity value was determined by type, size, shape and volume fraction of reinforcements added. Based on Marthiessen's rule, the total electrical resistivity of materials consists of three parts which are impurity resistivity, thermal resistivity and deformation resistivity. The existence of these three types of resistivity will disturb the normal electron motions in which it affects the impurity scattering and lattice scattering. In the case of composite solder, the electrical resistivity value was strongly depending on the impurity resistivity. The reinforcement particles added serves as electron scattering centres. Therefore, high amount of reinforcement particles can greatly affect the electrical resistivity of composite solder system. However, in this research, the changes in the electrical resistivity was not significant between monolithic SAC 305 solder and composite SAC 305 solder which corresponds to the small addition of KGC particles. Therefore, the addition of KGC particles would not disturb the ability of the solder to perform as electrical interconnect with lower electrical resistivity value. The results obtained was in good agreement with the research done by Babaghorbani et.al., [24] where the addition of ZrO_2 insignificantly influence the electrical resistivity of Sn-3.5Ag and Sn-3.5Ag-0.7Cu.

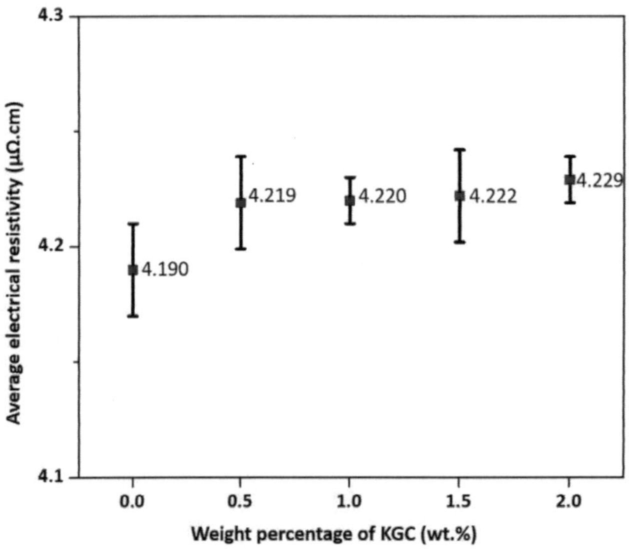

Fig. 3. Average electrical resistivity of SAC305 lead free solder at various weight percentage (wt.%) kaolin geopolymer ceramic.

4 Summary

A new lead-free composite solder system with the addition of various wt.% KGC to the wettability and electrical properties of SAC305 composite solder were investigated. The addition of KGC improved the wettability of the solder. The optimum value of contact angle was achieved at 1.0 wt.% with contact angle value of 20.8 °. Then, in terms of electrical resistivity, the addition of various wt.% KGC do not significantly change the resistivity value as compared to SAC305 without the addition of KGC. Based on the results obtained, it is interestingly to explore and investigate more on the influence of the KGC to the microstructure, thermal properties and mechanical properties.

The authors acknowledge the financial support through the Ministry of Higher Education, Malaysia, under the Fundamental Research Grant Scheme (FRGS/1/2017/TK10/UNIMAP/03/2) (9003-00612).

This work was also supported by an Institutional Links grant, ID 332397914, under the Newton-Ungku Omar Fund partnership. The grant is funded by the UK Department of Business, Energy and Industrial Strategy (BEIS) and Malaysia and delivered by the British Council.,. The authors also acknowledge Center of Excellence Geopolymer & Green Technology (CEGeoGTech) and School of Materials Engineering, Universiti Malaysia Perlis.

References

1. M.I.I. Ramli, N. Saud, M.A.A.M. Salleh, M.N. Derman, R.M. Said, and N. Nasir, Materials Science Forum. **803** (2015)
2. Che, F.X. and J.H.L. Pang,. J Alloy Compd, **541**, 6-13 (2012)
3. R.M. Said, M.A.A.M. Salleh, M.I.I. Ramli, and N. Saud, Solid State Phenomena. **280** (2018
4. M.A.A.M. Salleh, S.D. McDonald, and K. Nogita, Journal of Materials Processing Technology. **242** (2017)
5. F. Somidin, H. Maeno, M.M. Salleh, X.Q. Tran, S.D. McDonald, S. Matsumura, and K. Nogita, Materials Characterization. **138**
6. M.A.A.M. Salleh, M.H. Hazizi, M.A.B.M. Abdullah, N. Noriman, R. Mayapan, and Z.A. Ahmad, Advanced Materials Research. **626**
7. K. Nogita, M.A.A.M. Salleh, S. Smith, Y. Q Wu, S.D. McDonald, A.G.A. Razak, T. Akaiwa, and T. Nishimura, 2017 International Conference on Electronics Packaging (ICEP).
8. K. Nogita, M.A.A.M. Salleh, E. Tanaka, G. Zeng, S.D. McDonald, and S. Matsumura, JOM. **68**(11) (2016)
9. M.A.A.M. Salleh, A.M.M.A. Bakri, F. Somidin, and H. Kamarudin, International Review of Mechanical Engineering. (2013)
10. M.A.A.M. Salleh, M.M.A.B. Abdullah, F. Somidin, A.V. Sandu, N. Saud, K. Hussin, S.D. McDonald, and K. Nogita, Revista de Chimie **64**(7) (2013)
11. Guo, Q., et al., Microelectron Reliab, **80**, 144-148 (2018)
12. Mohd Salleh, M.A.A., et al., Mater. Des, **82**, 136-147 (2015)
13. Mohamad Zaimi, N.S., et al., Solid State Phenomena, **280**, 169-174 (2018)
14. Wattanakornphaiboon, A., R. Canyook, and K. Fakpan, Mater Today-Proc, **5**, 9213-9219 (2018)
15. Gain, A.K. and L. Zhang,. Materialia, **3**, 64-73 (2018)
16. Mohd Salleh, M.A.A., S.D. McDonald, and K. Nogita,. J. Mater. Process. Technol, **242**, 235-245 (2017)
17. Ramli, M.I.I., et al., Microelectron Reliab., **65**, 255-264 (2016)
18. Chellvarajoo, S., M.Z. Abdullah, and Z. Samsudin,. Mater. Des,**67**, 197-208 (2015)
19. Wang, Y., et al., J. Mater. Sci.: Mater. Electron., **26**(12), 9387-9395 (2015)
20. Tsao, L.C., J Alloy Compd, **509**(33),8441-8448 (2011)
21. Zaimi, N., et al.,. IOP Conf Ser Mater Sci Eng, **551**, 012081 (2019)
22. Zhang, L. and K.N. Tu, Mat Sci Eng R **82**,1-32 (2014)
23. Chen, G., et al.,.Mater. Sci. Eng. A, **636**(Supplement C), 484-492 (2015)
24. Babaghorbani, P., S.M.L. Nai, and M. Gupta, J Alloy Compd, **478**(1), 458-461 (2009)

Synthesis of alumina nanoparticles by sol-gel method and their applications in the removal of copper ions (Cu^{2+}) from the solution

Siti Nur Syakirah Mohamad[1], *Norsuria* Mahmed[1,2,*], *Dewi Suriyani* Che Halin[1,2], *Kamrosni* Abdul Razak[2], *Mohd Natashah* Norizan[2,3] and *Ili Salwani* Mohamad[2,3]

[1]School of Materials Engineering, Universiti Malaysia Perlis, Kompleks Pusat Pengajian Jejawi 2, Taman Muhibbah, 02600, Jejawi, Perlis
[2]Center of Excellence, Geopolymer and Green Technology, School of Materials Engineering, Universiti Malaysia Perlis, Kompleks Pusat Pengajian Jejawi 2, Taman Muhibbah, 02600, Jejawi, Perlis
[3]School of Microelectronic Engineering, Universiti Malaysia Perlis, Pauh Putra Campus, 02600 Arau, Malaysia

Abstract. Two different phases of alumina (Al_2O_3) nanoparticles (γ-alumina and α-alumina) have successfully been synthesized by using a sol-gel method. During the process a mixture of aluminum nitrate and citric acid (C/N=0.5) was heated at 60°C followed by 80°C until a gel was formed. The amorphous gel structure then was dried and sintered from 600°C to 1200°C. From the X-ray diffraction (XRD) analysis, crystalline structure of γ-alumina started to form at 800°C with average crystallite size of 11.5 nm, followed by the formation of the mixture phase of γ-alumina and α-alumina at 1000°C. The transformation from γ- to α-alumina occured at 1100°C of sintering temperature and above with the average crystallite size of 49 nm. The efficiency of the synthesized alumina nanoparticles as an adsorbents was tested by immersing the powder into the copper ions solution. The percentage of the copper removal was measured by using atomic absorption spectroscopy (AAS). It was found that, the efficiency of the alumina nanoparticles as an adsorbent was not depending on their phases, but might due to the increased of the particle size at higher sintering temperature. The highest percentage of removal 82.1% was obtained when using the alumina sintered at 1200°C.

1 Introduction

Alumina (Al_2O_3) has been used in different applications because of its various importance and beneficial properties. The chemical compound of alumina composed of aluminum and oxygen and most widely used ceramic materials among others ceramic material such as aluminum nitride, zirconia, silicon carbide, etc. Various applications that used alumina are as a biomedical implants, catalyst support and absorbents, fire retardants, polymer matrix composite, insulator and in clinical field, electronic fields, etc. [1-6]. The reason of alumina most widely attractive used material in many application is due to its greater properties such

* Corresponding author: norsuria@unimap.edu.my

as chemical and thermal stability, relatively good strength, good wear resistance, high hardness, high melting points, and good electrical and chemical resistance [7].

Alumina can be synthesized into different phases including alpha, beta, gamma, and delta. All of these phases can be attained at different temperature during synthesizing of alumina. Each phase have its own features which are different for application used. However, among these phases the alpha alumina is most stable structure and popular among researchers due to its superior properties such as high hardness, high stability, high insulation and transparency [8]. The basic unit cell structure of alpha-alumina is hexagonal. While internal crystal structure of alpha-alumina for oxygen is in close packed hexagonal with aluminum in 2/3 of octahedral sites.

The synthesis of alumina nanoparticle can be used by using variety of methods such as, vapor phase reaction, mechanical milling, sol-gel, hydrothermal, combustion and, precipitation method. The synthesis of alumina by using mechanical milling are easily introduces contaminations. The preparation of alumina via vapor phase reaction demand high temperature above 1200°C. The precipitation method need long washing and aging time which are contribute to long time consuming. Hydrothermal process needs high pressure and temperature that can cause the increasing in cost. While, alumina powders that yield via combustion method contains nano sized primary particles. Sol gel is a method ordinarily use precursor that good control over stoichiometry and morphology of nanoparticles. Thus, among these methods sol gel are the most favorable and mostly investigated because it produces high purity and high specific surface area of solid particles [9].

Waste water treatment are quite challenging due to the development of the industrial activities. For example, heavy metal ion is the one material content in the waste water include copper (Cu^{2+}), cadmium (Cd^{2+}), zinc (Zn^{2+}). These hazardous materials are come from the waste product industry activity such as electroplating, galvanizing and mine drainage. Therefore, several of treatments have been done in order to overcome this problem. This include by using the metal oxide as an adsorbent metal through adsorption process instead of filtering the metal ion. Alumina is one of the effective metal oxide which it can be as adsorbent material [10].

2 Experimental

The reagents that were applied during the experiments were aluminum nitrate ($Al(NO_3)_3 9H_2O$, Merck) citric acid ($C_6H_8O_7 H_2O$, Sigma-Aldrich) and deionized water. All the materials were used as-received without further purification.

2.1 Synthesis of alumina nanoparticles

Alumina nanoparticles were synthesized by using a sol-gel method. About 18.76 g of aluminum nitrate was dissolved with citric acid in the deionized water. The molar ratio of the citrate/nitrate (C/N) was 0.5. The solution was stirred continuously for several hours at 60°C until it turned into a yellowish sol. Then, the solution was heated up to 80°C under constant stirring until the transparent gel was formed. The gel was dried at 90°C in the oven for 12 hours. The dried gel was ground and sintered at 600°C, labeled as Sample A.

The synthesis of alumina nanoparticles was repeated by using the same process as described above and were sintered at temperature of 700°C, 800°C, 900°C, 1000°C, 1100°C, and 1200°C, labelled as Sample B, Sample C, Sample D, Sample E, Sample F, and Sample G, respectively.

2.2 Characterization of the materials

The phases of the synthesized sample were determined by using X-ray diffraction (XRD) model Shimadzu XRD-6000 at room temperature. The X-ray copper (Cu) tube was used with radiation wavelength of $K\alpha_1$= 1.54Å in the range of 2θ = 10°-80° at scanning speed of 30 kV and 10mA. The average crystallite size of alumina nanoparticles was calculated according to the Scherrer equation as shown in Eq. (1) by using the FWHM data of each phase after correcting the instrumental broadening.

$$t = \frac{0.9\lambda}{B\cos\theta} \quad (1)$$

where,
t = estimate of the crystallite size
λ = X-ray wavelength of the source material
Θ = Bragg angle.
B = spectral breadth due to the crystallite size effect.

The morphology and elemental analysis of the alumina nanoparticles were identified by using a scanning electron microscopy (SEM) JOEL-JSM 6460 LA with an energy dispersive analyzer (EDX).

2.3 Application of the synthesized alumina particles in adsorption of copper ions (Cu^{2+})

For the adsorption of Cu^{2+}, about 1 g of alumina nanoparticles (Sample A) was added into the 100 ml of 150 mg/L copper sulphate ($CuSO_4$) solution. The mixture was mixed for 60 minutes by using an orbital shaker at the speed of 200 rpm. Then, the solution was filtered by using a Whatmann No.1 filter paper, followed by the Cu analysis of the filtrate by using an atomic absorption spectrometry (AAS). For the analysis, the filtrate was dissolved in 25 ml of hydrochloric acid and few drops of nitric acid, followed by the dilution with 1L of deionized water in the volumetric flask. The presence of residual alumina in filtrate was also identified by using AAS to ensure the minimum leaks of alumina from the samples into the filtrate. The adsorption process was repeated by using sample B-G.

3 Results and discussion

Alumina nanoparticles is expected to be physically white after the sintering process. However, from Fig.1 only samples E, F and G turned into white powders, which were sintered at 1000°C, 1100°C and 1200°C respectively. Sintering at temperature below than 1000°C resulted in the black powders. The black appearance of the powder is expected due to the existence of the residual carbon inherited from the sol that was not fully burnout during sintering below 1000°C [11].

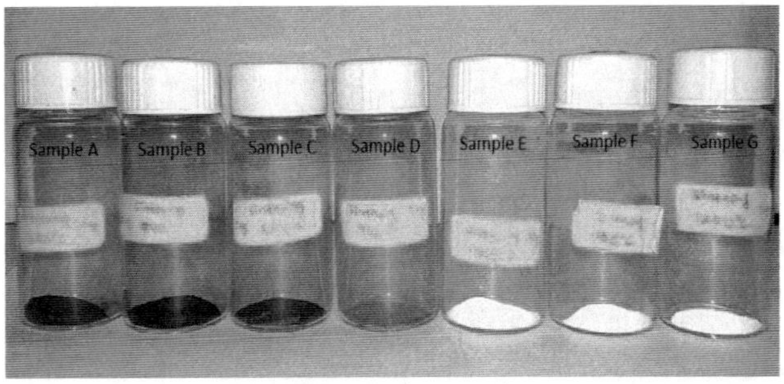

Fig.1. The sintered samples at different temperature (A) 600°C (B) 700°C (C) 800°C (D) 900°C (E) 1000°C (F) 1100°C and (G) 1200°C.

Fig. 2 shows XRD pattern of the phases formed after the sintering process. The amorphous structures are observed for the powder sintered at 600°C and 700°C. When the sintering temperature was increased to 800°C, there are peaks appeared at 2θ values, 61.3°(440) and 46.7° (400), which reflects to the γ-alumina peaks that indicates the transition of amorphous structure into γ-alumina phase. The formation of γ-alumina becomes stronger at 900°C where additional of γ-alumina peak at 38.5°(222) appeared. At 1000°C there are peaks of α-alumina started to grow at 44.3°(113), 35.1°(104) and 26.4°(012). Transformation of γ-alumina into α-alumina completely occured at 1100°C. At sintering temperature of 1200°C, the α-alumina phase is remain unchanged.

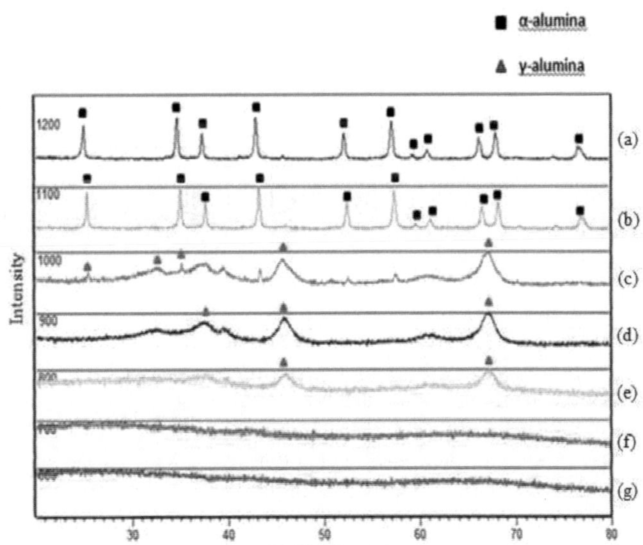

Fig.2. XRD pattern of the synthesized sample and heat treated at (a) 1200°C, (b) 1100°C, (c) 1000°C, (d) 900°C, (e) 800°C, (f) 700°C, (g) 600°C.

The crystallite size for α- and γ-alumina nanoparticles were calculated from few most intense XRD peaks by using a Scherrer equation. The average of the crystallite size is presented in Table 1. At sintering temperature 600°C and 700°C, since the structure was amorphous, no crystallite size can be determined. For Sample C, which is sintered at 800°C, the average crystallite size is 11.5 nm. When the temperature was increaed to 900°C, the crystallite growth is up to 15.5 nm in average and this value is not far difference with Sample E which consist of mixture phases of alumina. The obvious crystallite growth can bee seen when the sintering temperature was raised to 1100°C and 1200°C where the average crystallite size are 46.6 nm and 49.1 nm, respectively. The rapid growth of the crystallite due to particles coarsening of α-alumina at higher temperature [12].

Table 1. The crystallite size and phase structure of the synthesized Al_2O_3 sintered at different temperatures.

Sample	Sintering temperature (°C)	Crystallite size (nm)	Phase structure
A	600°C	-	Amorphous
B	700°C	-	Amorphous
C	800°C	11.5	γ-alumina
D	900°C	15.5	γ-alumina
E	1000°C	19.2	α- + γ-alumina
F	1100°C	46.6	α-alumina
G	1200°C	49.1	α-alumina

Fig. 3 shows the morphology of the alumina nanoparticles before and after sintering. The images were captured under 2000x of magnification. It can clearly be observed the big agglomeration of alumina nanoparticles, thus the particle size cannot be calculated. This might due to no surfactant added during the sol-gel synthesis [13]. The EDX analysis also was conducted for sample A, D and G. From Fig. 4, the percentage of carbon decreased as the sintering temperature icreased. The weight percent of carbon at 600°C was 15.4 and then decreased to 13.40 at 900°C and the carbon peak disappeared at 1200°C. The present of residual carbon in the alumina nanoparticles is due to the citric acid used during the experimental work. At 1200°C the carbon was fully burnout resulted in white color powder. Additionally, since the sample was coated by platinum before the SEM analysis in order to make the sample conductive, the small amount of element Pt also has been detected. Nevertheless, percent of element Al still has the highest amount of percentage among the samples.

IOP Conf. Series: Materials Science and Engineering **701** (2019) 012034 doi:10.1088/1757-899X/701/1/012034

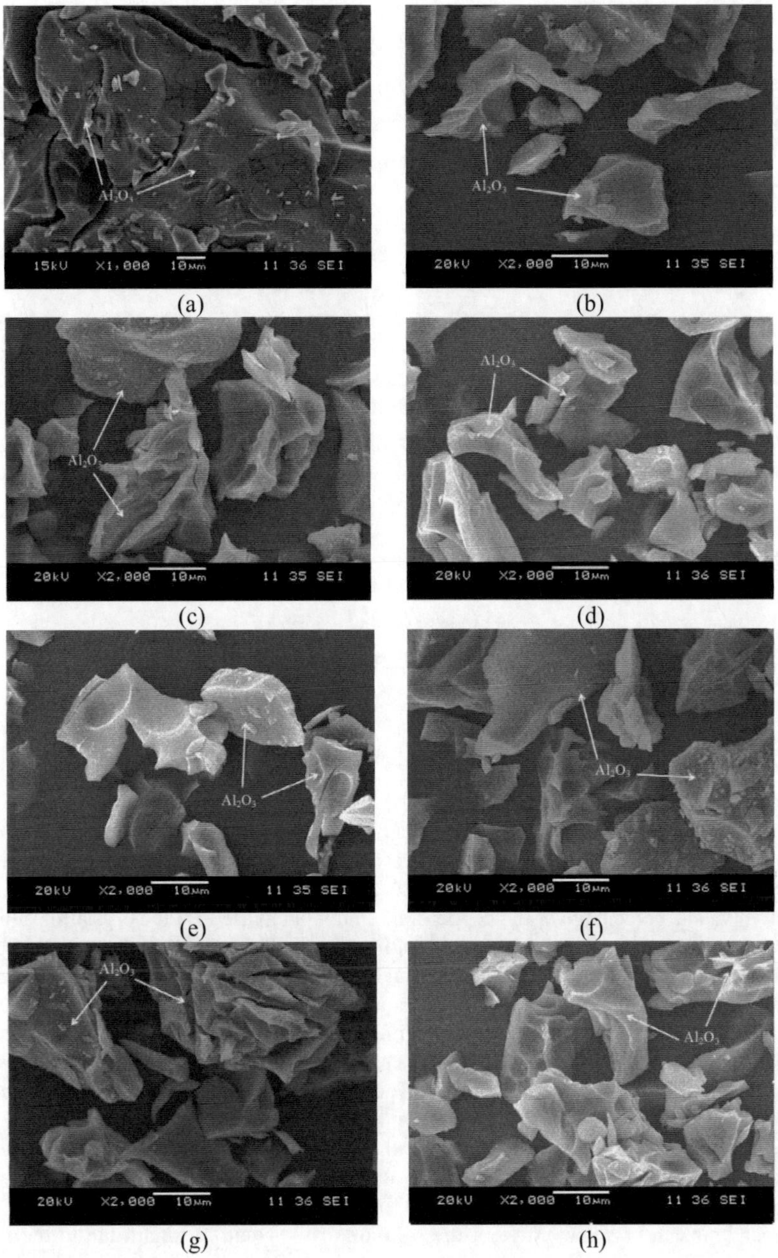

Fig.3. SEM micrographs of alumina nanoparticles sintered at different temperature (a) 0°C, (b) 600°C (c) 700°C (d) 800°C (e) 900°C (f) 1000°C (g) 1100°C and (h) 1200°C.

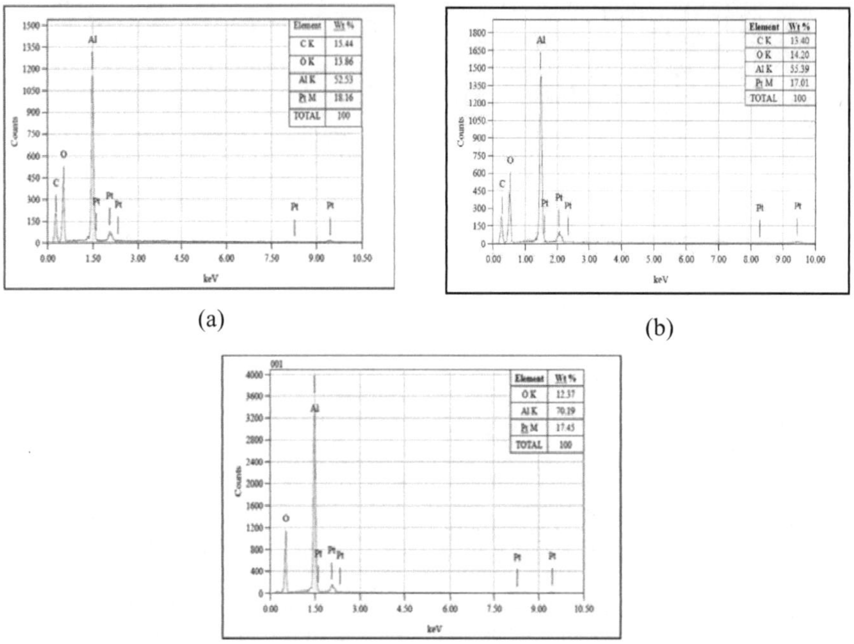

Fig. 4. The EDX analysis of samples sintered at (a) 600°C (b) 900°C and (c) 1200°C.

The efficiency of the synthesized alumina nanoparticles as an adsobents for the removal of copper ions (Cu^{2+}) from the solution was tested. From the results, the amorphous samples (sintered at 600°C and 700°C), about 70.4% and 71.6% of copper ions can be removed, respectively. The percentage of the copper ion removal increased for the crystalline alumina. For the γ-alumina nanoparticles which was formed at 800°C and 900°C, about 76.7% and 78.4% of copper ions can be removed from the solution. The percentage of the copper ions removal was increased as the sintering temperature increased with maximum of the removal about 82.1%. From Table 1, the sintering temperature caused the crystallite size to grow. A particle may made up from several crystallites except for the single crystal [14]. For this case, the synthesized alumina is not a single crystal and it is expected that their particle size consist of several crystallites. Thus, it is expected that as the crystallite size growth, the particle size also becomes bigger. Based on this results, the increased of the percentage of the copper removal might not depending on the phase of alumina, but might due to the particle size since the bigger the size, the higher the surface area for the absorption to occur. However, not 100% of copper ions can be removed from the solution. This situation might due to the several factors that affected the adsorption process including the heavy metal ion concentration, adsorbent dosage, pH of the solution, temperature, etc. [15].

4 Conclusion

Alumina nanoparticles was successfully synthesized via the sol-gel method by using a citric precursor derived from aluminum nitrate and citric acid (C/N=0.5) mixed solution. Sintering at 800°C change the amorphous structure into crystalline γ-alumina with the average crystallite size of 11.5 nm. The γ-alumina continue to growth at higher sintering temperature until the white powder of α-alumina formed at 1100°C and 1200°C with rapid growth of the crytstallite size about 48 nm in average. For the application as an adsorbent, the efficiency of the copper ions removal from the solution is not depending on the alumina phases, but might due to the increased of particle size at higher sintering temperature. The highest percentage of removal 82.1% was obtained when using the alumina sintered at 1200°C.

This work was financially supported by the Fundamental Research Grant Scheme FRGS/1/2014/TK04/UNIMAP/02/2 from Ministry of Education Malaysia (MOE). The supports from School of Materials Engineering, Universiti Malaysia Perlis are greatly acknowledged.

References

1. G. Schierano, F. Mussano, M. G. Faga, G. Menicucci, C. Manzella, C. Sabione, Biomed. Res. Int. (2015)
2. R. J. Bhargavi, U. Maheshwari, and S. Gupta, Int. J. Ind. Chem. **6** (2014)
3. Laachachi, M. Ferriol, M. Cochez, J. M. Lopez Cuesta, and D. Ruch, Polym. Degrad Stabil. **94** (2009)
4. Omrani, L. C. Simon, and A. A. Rostami, Mater. Chem. Phys. **114** (2009).
5. M. Touzin, D. Goeuriot, C. Guerret-Piécourt, D. Juvé, and H. J. Fitting, J Eur Ceram Soc. **30** (2010)
6. T. Joshi, J. Prakash, A. Kumar, J. Gangwar, A. K. Srivastava, S. Singh, J Phys. D **44** (2011)
7. Sedaghat, E. Taheri-Nassaj, G. Soraru, R. Ceccatob, and T. Ebadzadeh, J Adv. Mater Process **2** (2014)
8. V. Piriyawong, V. Thongpool, P. Asanithi, and P. Limsuwan, J Nanomater. **2012** (2012)
9. M. Shojaie-Bahaabad and E. Taheri-Nassaj, Mater. Lett **62** (2008)
10. N. H. Yarkandi, Int. J. Curr. Microbiol. App. Sci **3** (2014)
11. M. Rahaman and M. N. Rahaman, *Ceramic processing* (CRC press, 2006)
12. J. Li, Y. Pan, C. Xiang, Q. Ge, and J. Guo, Ceram Int. **32** (2006)
13. F. Mirjalili, L. C. Abdullah, H. Mohamad, A. Fakhru'l-Razi, A. Dayang Radiah, and R. Aghababazadeh, ISRN Nanotech. **2011** (2011)
14. B. Akbari, M. Pirhadi Tavandashti and M. Zandrahimi, Iran. J. Mater. **2011** (8)
15. S. S. Thavamani and R. Rajkumar, Res J Chem Sci **2231**(2013)

IOP Conf. Series: Materials Science and Engineering **701** (2019) 012035 doi:10.1088/1757-899X/701/1/012035

Properties of eco-friendly composites: palm kernel shell treated with sodium bicarbonate filled recycled high-density polyethylene.

Nadiatul Husna[1], *Bee Ying* Lim[1,*], *Chun Hong* Voon[3] *and Pei Ling* Teh[1]

[1]School of Materials Engineering, Universiti Malaysia Perlis, Taman Muhibbah, 02600 Jejawi, Perlis, Malaysia.
[2]Center for Frontier Materials Research, School of Materials Engineering, Universiti Malaysia Perlis, 02600 Arau, Perlis, Malaysia.
[3]Institute of Nano Electronic Engineering (INEE), Universiti Malaysia Perlis, 01000, Kangar, Perlis, Malaysia.

Abstract. Palm kernel shells (PKS) treated by commercial sodium bicarbonate (SB) filled recycled high-density polyethylene (rHDPE) were produced using melt compounding. These environmentally friendly composites were prepared using twin-screw extruder by varying untreated and treated filler loading (0, 10 and 30 php). In this study, the tensile properties of untreated rHDPE/PKS and treated rHDPE/PKS-SB were investigated. The composites were subjected to tensile test and the fracture surfaces were observed under scanning electron microscope (SEM). The results showed that SB treatment improved the filler-matrix adhesion and interaction of composites. Meanwhile, the SEM micrographs on the fracture surfaces confirmed that SB treatment which may influence the tensile properties of these composites.

1 Introduction

The increasing polymer waste in modern days has led to plenty environmental issues. With the effort to reduce the usage of non-degradable petroleum-based polymer, addition of natural filler is one of the solutions. Thermoplastic composites filled with natural fillers offer attractive properties such as lightweight, high specific mechanical properties, biodegradability, recyclability and low cost [1]. There are researches report on the use of natural fillers to produce composites and their respective applications [2]. The common natural filler are chitosan [3], hemp [4, 5], sisal [6], rice husk [7] and etc.

The palm kernel shell (PKS) is the waste residue generated from palm fruit processing in Malaysia. The palm kernel nuts that contain kernel oils are cracked down to kernel and kernel shell. The kernel is served for oil extraction; while the kernel shell is considered as waste. Therefore, the huge amount of PKS generated annually has led to disposal issue. In order to resolve this issue, PKS can be processed and utilized as an alternative to synthetic filler. The

* Corresponding author: bylim@unimap.edu.my

PKS has high potential to be used as filler in polymeric composites as it contain 50.70 % lignin, 20.80 % cellulose and 22.70% hemicellulose [8].

Though the natural fillers demonstrate attractive properties, their hydrophilic properties are not compatible with the hydrophobic properties of polymers. This cause the poor adhesion between filler and matrix. Surface modification offers solution to improve the interaction between filler and matrix [9]. Alkali treatment is one of the widely used surface modification that can modify the filler surface and increase the interfacial bond strength between filler and matrix [10]. However, the inappropriate disposal of chemical waste after treatment may endanger the environment. Thus, Fiore and co-workers [11] had proposed an environmentally friendly treatment that involved the use of sodium bicarbonate solution instead of strong alkali. Their study discovered the treated fibre composites showed notable improvement of mechanical properties.

The main objectives of this research were to study the effect of PKS loading on the tensile and morphological properties of rHDPE/PKS composites. Surface modification using commercial sodium bicarbonate (SB) was used with the aim to improve the properties of composites.

2 Experimental

2.1 Materials

The recycled high-density polyethylene (rHDPE) flakes were supplied by Titan Petchem (M) Sdn. Bhd., Pasir Gudang, Malaysia. The palm kernel shell (PKS) was provided by Malpom Industries Sdn. Bhd., Nibong Tebal, Malaysia. The fibres on raw PKS was first removed and cleaned thoroughly. Then, it was dried in the oven at 80 °C for 24 hours to remove the moisture. A grinding machine was used to grind the PKS into fine particles. The PKS powder was then subjected to sieving process and only the PKS with size smaller than 75 μm was selected.

2.2 Filler treatment

Commercial sodium bicarbonate (SB) as known as baking soda, $NaHCO_3$ was chosen as surface treatment. It was acquired from a local retail shop. Sodium bicarbonate surface coating was applied using a solution method. 20 g of SB was dissolved in 500 mL distilled water. The solution was manually stirred at regular interval until all the SB was dissolved. Thereafter, 200 g of PKS were added into solution. The mixture was stirred manually for about one minute and left to stand for 120 hours under room temperature. After that, the mixture was rinsed with distilled water and dried in oven for 24 hours at 50 °C.

2.3 Preparation of eco-friendly composites

The rHDPE/PKS composites were prepared by using twin-screw extruder. The filler and matrix were pre-mixed before compounding. Then, they were poured into hopper of extruder, with a fixed feeding rate that continuously feeding the filler and matrix from hopper into extruder barrel. The processing temperature was set at 180 °C with screw speeds of 50 rpm. The extrudates were cooled in water bath and cut into pellet by the means of a pelletizer machine. By using compression moulding, the rHDPE/PKS pellets pressed into sheets of 1 mm. The pellets were preheated for 12 minutes, compressed for 4 minutes and cooled for 4 minutes. Then the composites were cut into dumbbell shape. The formulation of untreated rHDPE/PKS and treated rHDPE/PKS-SB composites at different filler loading was shown in Table 1.

Table 1. Formulations of Untreated rHDPE/PKS composites and Treated rHDPE/PKS eco-friendly composites.

Materials	Untreated rHDPE/PKS Composites	Treated rHDPE/PKS-SB Composites
rHDPE (php)	100	100
PKS (php)	0, 10, 30	10, 30

*php = part per hundred of polymer

2.4 Tensile Test

Tensile tests were conducted by using an Instron 5569 and according to ASTM D638 [12]. A gauge length of 50 mm was used, with cross head speed of 15 mm/min. The test was performed under room temperature. For each formulation, 10 samples were tested, and an average was obtained. Tensile strength, elongation at break and Young's modulus for the composites were recorded.

2.5 Scanning Electron Microscope

The fracture surfaces of the composites were observed under a scanning electron microscope (SEM), model JSM 6260 LE JOEL. The fractured ends of specimens were mounted on aluminium stubs and sputter-coated with a thin layer of platinum to avoid electrostatic charging during examination.

3 Results and discussion

3.1 Tensile Properties

Fig. 3.1(a) illustrates the average values of tensile strength of untreated rHDPE/PKS and treated rHDPE/PKS-SB composites with sodium bicarbonate. The increasing of PKS content in rHDPE matrix had decreased the tensile strength of untreated rHDPE/PKS composites and treated rHDPE/PKS-SB composites. The weak interaction between PKS and rHDPE resulted in poor stress transfer thus reduced the tensile strength of the composites. However, it was discovered that the tensile strength of treated rHDPE/PKS-SB composites was higher than untreated rHDP/PKS composites. The mildly alkaline sodium bicarbonate solution was capable to remove the lignin on the surface of PKS [10]. The removal of lignin on the surface of filler after alkali treatment contributes to change the tensile strength of filler [13].

Meanwhile, Fig. 3.1(b) shows the elongation at break of rHDPE/PKS-SB composites decreasing with the rise of filler loading. The presence of rigid PKS had decreased the ductility of rHDPE/PKS composites. Besides, the elongation at break of treated rHDPE/PKS-SB composites were higher compared to untreated rHDPE/PKS composites. This can be attributed to the enhanced interaction of matrix molecular chain onto the surface of treated PKS filler as a consequence of improved compatibility on the interface [9].

On the other hand, the Young's modulus (Fig. 3.1(c)) of treated rHDPE/PKS-SB composites were higher than that of untreated rHDPE/PKS composites and continued to increase with increase of PKS loading. The addition of PKS increased the stiffness of rHDPE/PKS composites, thus led to higher Young's Modulus. It was noticed that the SB treated PKS improved interaction between PKS and rHDPE and showed stiffening effect.

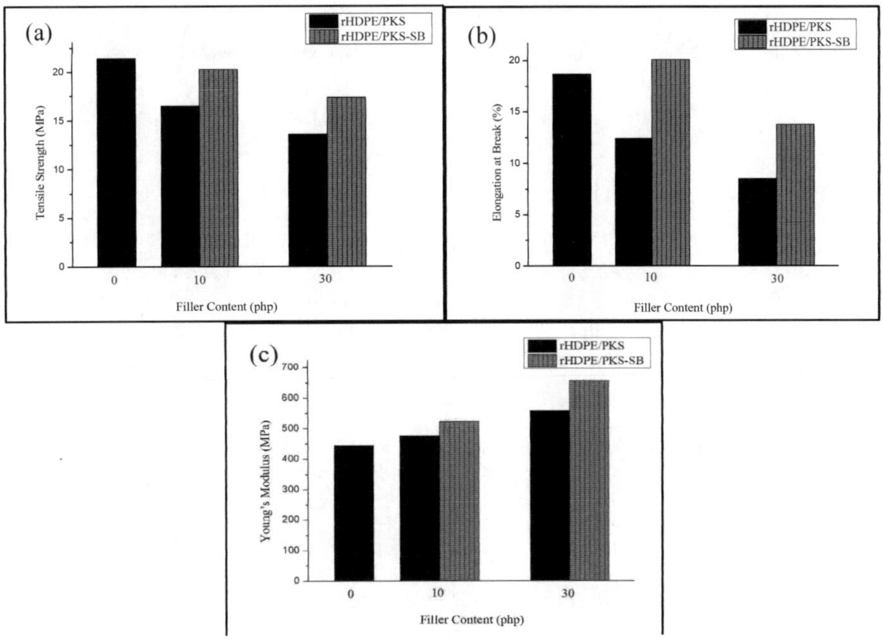

Fig. 3.1 Effect of filler content on the (a) tensile strength, (b) elongation at break, (c) Young's Modulus of untreated rHDPE/PKS composites and treated rHDPE/PKS-SB composites.

3.2 Morphological Study

Fig. 3.2 (a) shows the SEM micrographs of tensile fracture surface of treated rHDPE/PKS-SB composites at 10 php of PKS loading. It can be seen in the micrograph of treated composites that PKS fillers were embedded in the matrix. Under the tensile stress, some of the PKS filler were detached from the matrix. This stress concentration weakened the tensile properties of rHDPE/PKS composites. The SB treatment had successfully improved the interfacial bonding between PKS and rHDPE. As a result, it enhanced the tensile strength and modulus of elasticity of composites. Fig. 3.2 (b) presented the fracture surface of rHDPE/PKS-SB composites at 30 php PKS loading. It showed the rough fracture surface and the detachment of PKS filler from the rHDPE matrix. This is indicated that imperfect bonding occurred between the rHDPE matrix with increasing of PKS filler loading. The interaction and adhesion were also poor, which led to detachment of PKS. This result showed the lower tensile strength of the composites at higher filler loading.

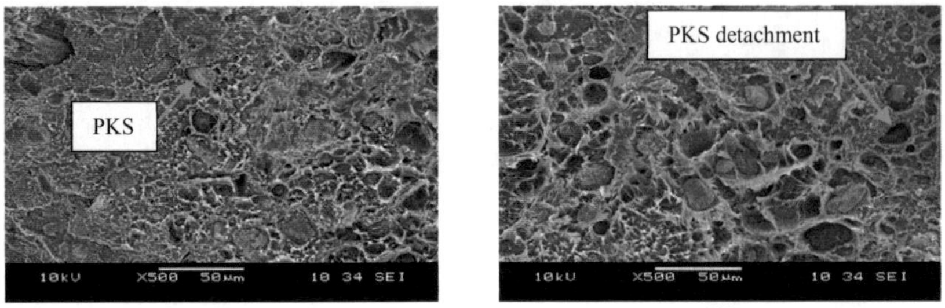

(a) (b)

Fig. 3.2 SEM micrograph of (a) 10 php and (b) 30 php treated rHDPE/PKS-SB composites.

4 Conclusion

In this study, the rHDPE/PKS were successfully prepared at different PKS loading. It was discovered that presence of PKS had deteriorated the tensile strength and elongation at break of the composites but enhanced the Young's modulus of composites. The environmentally friendly alkali treatment, sodium bicarbonate was capable to enhance the properties of rHDPE/PKS composites. The improved interfacial bonding of rHDPE and PKS were evidenced in SEM micrographs.

The authors gratefully acknowledge the funding support for this work from the Fundamental Research Grant Scheme (FRGS/1/2014/TK04/UNIMAP/02/6) under Ministry of Education Malaysia (MOE).

References

1. N. F. Zaaba and H. Ismail, Journal of Physical Science. **30**.(2019)
2. N. F. S. Arshad, S. Husseinsyah, and L. B. Ying. in *MATEC Web of Conferences*, (2017)
3. B. Lim, C. Voon, H. Salmah, and H. Nordin. in *AIP Conference Proceedings*, (2016)
4. R. Sepe, F. Bollino, L. Boccarusso, and F. Caputo, Composites Part B: Engineering, **133**., (2018)
5. A. Q. Dayo, A. Zegaoui, A. A. Nizamani, S. Kiran, J. Wang, M. Derradji, W.-a. Cai, and W.-b. Liu, Mater. Chem. Phys., **217**, (2018)
6. P. Sahu and M. Gupta, Journal of Reinforced Plastics and Composites. **36**(24), (2017)
7. R. Arjmandi, A. Hassan, K. Majeed, and Z. Zakaria, Inter. J Polym. Sci., (2015)
8. S. Abdullah, S. Yusup, M. M. Ahmad, A. Ramli, and L. Ismail, (2010)
9. M. A. D. Hasnan, S. Husseinsyah, L. B. Ying, and M. F. A. Rahman, BioResources. **11**(3), (2016)
10. J. C. d. Santos, L. Á. d. Oliveira, L. M. Gomes Vieira, V. Mano, R. T. S. Freire, and T. H. Panzera, Construction and Building Materials. **211**, (2019)
11. V. Fiore, T. Scalici, F. Nicoletti, G. Vitale, M. Prestipino, and A. Valenza, Compos., (2016)
12. A. International, *ASTM D638-14, Standard Test Method for Tensile Properties of Plastics*.: ASTM International (2015)
13. B. Bakri, A. E. E. E. Putra, A. A. Mochtar, I. Renreng, and H. Arsyad, Inter. J Auto. Mech. Eng. **15**(3), (2018)

IOP Conf. Series: Materials Science and Engineering **701** (2019) 012036 doi:10.1088/1757-899X/701/1/012036

Synthesis of SiC nanowhiskers by microwave heating: effect of size of graphite

Suhaimi Mat Kahar [1], *Chun Hong* Voon [1,*], *Bee Ying* Lim[2], *Mohd Khairuddin* Md Arshad [1], *Kai Loong* Foo [1], *Nor Azizah* Parmin [1], *Subash C. B.* Gopinath [1], *and A. Rahim* Ruslinda [1]

[1]Institute of Nano Electronic Engineering, Universiti Malaysia Perlis, 01000, Kangar, Perlis, Malaysia
[2]School of Materials Engineering, University Malaysia Perlis, Jejawi, 02600 Arau, Perlis, Malaysia.

Abstract. Silicon carbide is a promising material for various applications, especially for applications in high frequency, high power and high temperature and harsh conditions such as highly oxidative and corrosive environment. Currently, silicon carbide is produced by conventional heating process. For silicon carbide nanomaterials, various methods have been attempted. In this study, SiC nanowhiskers were synthesized by microwave heating of blends of graphite and silica in the ratio of 3: 1. Graphite of two different size, namely micro graphite and nano graphite were used. The blends firstly prepared by ultrasonic mixing graphite and silica in ethanol followed by drying. The blends were then heated by using laboratory microwaves furnace to 1450 °C at heating rate of 20 °C/min for 20 minutes. Finally, the SiC nanowhiskers was characterized by using scanning electron microscope.

1 Introduction

Silicon carbide (SiC) is one of the promising and attractive ceramic due to its fascinating properties such as high melting point, excellent oxidation resistance, excellent thermal stability, high mechanical strength, chemical inertness, wide energy band gap, and unique optical properties. Acheson process is the most widely used method for the production of SiC. This process involves carbothermal reduction of SiO_2 by coke at 2200-2500 °C for about one week [1]. However, this method required high temperature and long heating duration, which render the process time consuming and costly. For this reason, numerous other synthesis route have been studied and developed for production of SiC including mechanical milling [2], sol-gel processes [3], chemical vapor deposition (CVD) [4], rapid carbothermal synthesis [5] and thermal plasma [6]. However, these methods have similar limitations and drawbacks such as presence of impurities in the product, large variation of grain size and time-consuming process. In recent years, various nanomaterials were extensively synthesized and studied for their unique and novel properties. SiC nanomaterial is one of the most important nanomaterials for current and future research and applications for their unique properties.

* Corresponding author: chvoon@unimap.edu.my

Comparing to conventional heating which heats from outside to the inside of the materials, microwave heating can volumetrically heat materials with favourable dielectric properties. Therefore, microwave heating is more uniform and efficient comparing to conventional heating which causes temperature gradient across the materials. Furthermore, microwave heating is also more energy efficient than conventional heating due to higher heating rate and lower heat loss to the surrounding. Previous studies have shown that graphite is an excellent microwave absorber [7] and 1:3 for silicon dioxide to carbon source (carbon nanotube) was the suitable ratio for the preparation of high purity SiC nanomaterials [8]. Thus, the aim of this study is to produce silicon carbide nanowhiskers (SiCNWs) by microwave heating the blends of silicon dioxide and graphite in the ratio of 1:3 at 1400 °C with heating rate of 20 °C/min for 20 minutes. Besides, the effect of size of graphite, namely micro graphite and nano graphite was also studied.

2 Experimental

2.1 Sample preparation

Blend of SiO_2 powder (particle size 4 μm, purity: >99.9%) and micro graphite powder (particle size 4 μm, purity: >99.9%) were used as starting material. Both SiO_2 powder and micro graphite powder were obtained from Sigma Aldrich while nano graphite powder was obtained from Guangzhou Hongwu Materials Technology Co., Ltd. The ratio of the amounts of silica to graphite was fixed at 3:1 with total of 1 gram of blend was acquired. Ethanol was used as liquid medium for the mixing of the raw material. Ultrasonic mixing bath was conducted for 1 hour generate vibration in the mixture of silica and graphite for the homogeneous mixing of the raw material. Mixture of silica and graphite was dried using the hot plate to vaporize ethanol. The blend of SiO_2 and micro graphite were compressed to become a pellet to prevent the escape of SiO gas during the heating process. The same procedure was repeated for nano graphite powder (particle size 100 nm, purity: >99.9%).

2.2 Synthesis of SiCNWs by microwave irradiation

Microwave heating was performed in microwave furnace with multimode cavity in which 2.45 GHz microwave irradiation was brought out through a waveguide. The set up for the microwave heating process was reported previously [9]. Blend of micro graphite and SiO_2 was placed in silica crucible and the crucible was placed in microwave cavity. Silica sand was used as a heat insulator to prevent heat lost. SiC susceptors was used as microwave absorber to absorb and convert electromagnetic energy to heat because SiC susceptor is a good microwave absorber material. This allowed rapid heating of the blend of micro graphite and SiO_2. The blend of SiO_2 and micro graphite was heated to 1450°C with heating rate of 20 °C/min and soaked for 20 minutes. The blend was then allowed to cool down to room temperature. The same procedure was repeated for blend of nano graphite and SiO_2.

2.3 Characterization of SiC nanowhiskers

After the SiCNWs were synthesized, characterization was conducted using scanning electron microscopy (SEM). The morphology of the SiCNWs was observed by scanning electron microscopy (SEM JEOL JSM6010LV) at magnification of 30000X and accelerating voltage of 20 kV.

3 Result and discussion

3.1 Characterization of graphite using SEM

SEM images of micro graphite and nano graphite were shown in Fig. 1. As shown in Fig. 1 (a), the particle size of micro graphite ranges from 1 to 4 μm. The shape of the micro graphite is irregular polygonal, as shown in Fig. 1 (b). Fig. 1 (c) shows the SEM image of nano graphite in which nano graphite are agglomerated to form larger particles. Fig. 1 (d) shows the SEM image of nano graphite at higher magnification and it is observed that the particle size of nano graphite ranged from 100 – 200 nm.

Fig. 1. SEM images of (a) and (b) micro graphite and (c) and (d) nano graphite.

3.2 Characterization of SiCNWs using SEM

SEM images of SiCNWs synthesized from micro graphite and nano graphite were shown in Fig. 2. As shown in Fig. 2 (a), SiCNWs were successfully synthesized when micro graphite was used. SiCNWs grew randomly on the surface of micro graphite particles. SEM images of higher magnification in Fig. 2 (b) shows the SiCNWs are straight and have uniform diameter along the length. The diameter of the SiCNWs ranges from 100-300 nm. Although SiCNWs were successfully synthesized, the reaction is incomplete with unreacted micro graphite. Fig. 2 (c) shows that very few SiCNWs were synthesized when nano graphite was used. Higher magnification in Fig. 2 (d) shows that the diameter of SiCNW ranges from 50 - 100 nm. Comparing to the SiCNWs synthesized from micro graphite, the amount of SiCNWs synthesized by using nano graphite as carbon source is very small. Further study is required

IOP Conf. Series: Materials Science and Engineering **701** (2019) 012036 doi:10.1088/1757-899X/701/1/012036

to understand the mechanism of formation of SiCNWs from both micro graphite and nano graphite.

Fig. 2. SEM images of SiCNWs synthesized from (a) and (b) micro graphite and (c) and (d) nano graphite.

4 Conclusions

In this study, synthesis of SiCNWs was achieved by using microwave heating of blends of SiO_2 and graphite at 1450°C for 20 minutes. It is found that SiCNWs were formed when both micro graphite and nano graphite were used as carbon source to react with silica. However, the amount of synthesized SiCNWs were significantly larger when micro graphite was used, compared to when nano graphite was used.

The authors are grateful to the Department of Higher Education, Ministry of Higher Education, Malaysia for funding this research through the Fundamental Research Grant Scheme (FRGS) with the grant number 9003- 00441. The author also would like to acknowledge all the team members in Institute of Nano Electronic Engineering (INEE), Universiti Malaysia Perlis (UniMAP) for their guidance and help.

References

1. S.M. Kahar, C.H. Voon, C.C. Lee, U. Hashim, M.K. Md Arshad, B.Y. Lim, S.C.B. Gopinath, W. Rahman, Mat. Sci-Poland **34**, 4 (2016)
2. D. Chaira, B. K. Mishra, S. Sangal, Mat. Sci. Eng. A **460-461**, (2007)

3. L. Xin, Q. Shi, J. Chen, W. Tang, N. Wang, Y. Liu, Y. Lin, Mater. Charact. **65**, (2012)
4. L. Cao, H. Jiang, H. Song, Z. Li, G. Miao, J. Alloys Compd **489**, 2 (2010)
5. S. Pan, J. Zhang, Y. Yang, G. Song, Ceram. Int. **34**, 2 (2008)
6. Z.Károly, I.Mohai, Sz.Klébert, A.Keszler, I.E.Sajó, J.Szépvölgyi, Powder Technol **214**, 3 (2011)
7. S.M. Kahar, C.H. Voon, C.C. Lee, S.C.B. Gopinath, M.K.M. Arshad, B.Y. Lim, K.L. Foo, U. Hashim, Mater. Res. Express **4**, 1 (2017)
8. V.C.S. Tony, C.H. Voon, C.C. Lee, B.Y. Lim, S.C.B. Gopinath, K.L. Foo, M.K.M. Arshad, A.R. Ruslinda, U. Hashim, M.N. Nashaain, Y. Al-Douri, Mater. Res. **20**, 6 (2017)
9. V.C.S. Tony, C.H. Voon, C.C. Lee, S.C.B. Gopinath, Ying, B.Y. Lim,, M.K.M. Arshad, K.L. Foo, U. Hashim, A.R. Ruslinda, M.N. Nashaain, N.A. Parmin, Sains Malays. **46**, 7 (2017)

IOP Conf. Series: Materials Science and Engineering **701** (2019) 012037 doi:10.1088/1757-899X/701/1/012037

Mitigation techniques for crosstalk in ICs

Rifhan Amira Mohd Razif[1], *Siti Marwangi* Mohamad Maharum[1*], *Azri Husni* Hasani[2] and *Zuhanis* Mansor[1]

[1]Advanced Telecommunication Technology, Universiti Kuala Lumpur British Malaysian Institute, 53100 Gombak, Selangor, Malaysia
[2]Universiti Tenaga Nasional 43000 Kajang Selangor, Malaysia

Abstract. Crosstalk is a major concern among integrated circuit (IC) designers as it may contribute significantly to the amount of noise presented in a device. Hence, there is a need to mitigate the impact of crosstalk in order to preserve the signal integrity between interconnections. This paper will review several methods to mitigate crosstalk in high speed ICs. The methods include the usage of repeater, Schmitt trigger, shielding, skewing, space adjustment between signal wires and setting the coupling length of the signal wires. Although these methods are effective for crosstalk mitigation, the features might only be applicable due to certain constraints. Consequently, this paper will provide a sufficient review on each method as a guideline for crosstalk mitigation method selection in which the crosstalk is largely regulated in long interconnections.

1 Introduction

Nowadays, integrated circuits are designed in high densities and operated at high frequencies in the gigahertz region. The high densities feature corresponds to Moore's Law that states the number of transistors in an IC is doubled for every two years [1]. Technically, an IC is a set of electronic circuits which consist of thousands to billions of transistors on one small flat piece of semiconductor material that is normally silicon [2]. Figure 1 illustrates the evolution of total number of transistors in high speed ICs from 1971 until 2007. Based on the timeline, it can be seen that with the shrinkage of the process technology, more transistors can be integrated over the same die.

Continuous invention of advanced smart devices increases the need for high data rate transmission and support for large network capacity [3], [4-5]. The evolution of transceiver technology has allowed multiple wideband transmitter and receiver to be integrated on a single chip. Such integration of billions of transistors in single chip shall result reduction in terms of power consumption, chip size and weight without affecting the overall performance.

The next phase of mobile telecommunications is the fifth-generation (5G) technology. 5G is a system designed to fulfil the prerequisites of International Mobile Telecommunications (IMT-2020) set by the International Telecommunication Union Radio communication sector (ITU-R) Specification M.2083 [1-2].

* Corresponding author: sitimarwangi@unikl.edu.my

Research on the communication system 5G millimetre wave (mmWave) has begun in terms of system development in order to provide market-ready systems by 2020 [2]. In 2020, the communication system of 5G mmWave has a clear vision of getting user devices in the Internet of Things (IoT) and Big–Data setting [6-7]. Moreover, communications system 5G mmWave is capable in providing low latency high-speed data connections with gigabit speed for billions of consumers.

5G technology has to be established, developed, and deployed by a range of industry players. The upcoming service and implementation environment will involve associated applications such as smart city, multiple personal devices, wearable computing, education, and lifeline systems [6], [8-9]. Figure 2 shows the timeline of the third generation partnership project (3GPP) standardization towards 5G implementations in the industry.

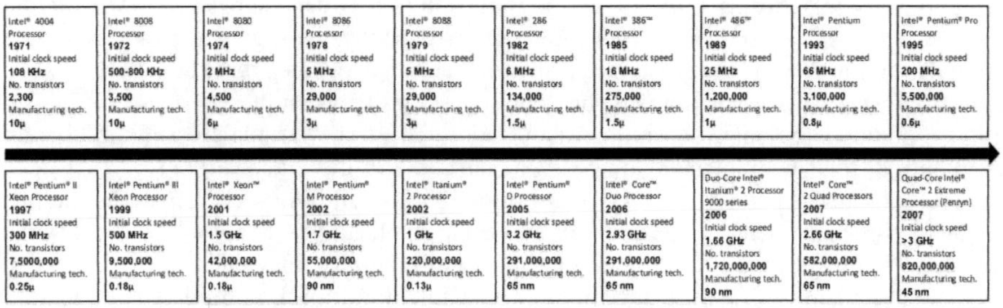

Fig. 1. 60 years of the transmissions: 1971-2007 [1].

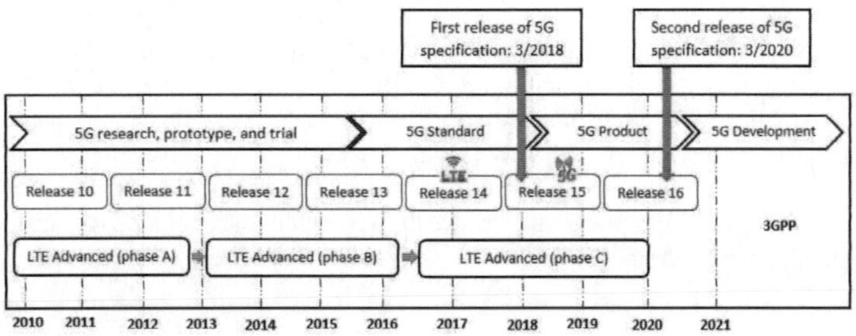

Fig. 2. The timeline of third generation partnership project (3GPP) standardization towards 5G implementations in the industry [10].

It is expected 5G technology will fundamentally change the way we see the role of telecommunications technology, which will have an impact through the pervasive digitalization of a hyper-connected society, as well as further economic growth. MmWave technology is just one of the techniques for the future 5G networks to be implemented. In mmWave technology, the spectrum band ranges from 30GHz-300GHz, in which the wavelength is between 10mm to 1mm [11]. Figure 3 shows the spectrum for both super high frequency (SHF) with a range of 3-30GHz and spectrum for extremely high frequency (EHF) is 30-300GHz.

IOP Conf. Series: Materials Science and Engineering **701** (2019) 012037 doi:10.1088/1757-899X/701/1/012037

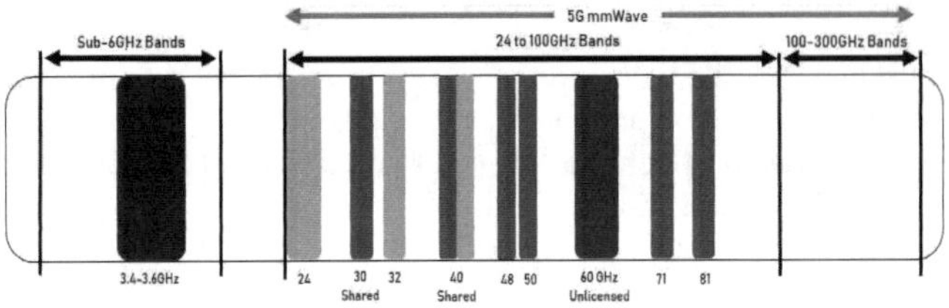

Fig. 3. The 5G mmWave spectrum.

Today, one of the most challenging issues faced by circuit designers is to design a system that is resilient to crosstalk. Generally, crosstalk is recognized as a hypercritical issue in high-speed IC design. Crosstalk refers to the interaction between signals on two parallel interconnects. An "aggressor" is the interconnect carrying the signal which creates the crosstalk, and a "victim" is the adjacent interconnect that suffers from the crosstalk [12]. Crosstalk appears when electromagnetic coupling occurs between the interconnects, which include mutual capacitance and mutual inductance. There are various crosstalk mitigation techniques that can be implemented in accordance to the IC design constraints.

The subsequent parts of this paper are organized as follows. Section 2 highlights the challenges in high-speed IC design for mmWave applications. Section 3 discusses the numerous techniques used to mitigate the impact of crosstalk. In section 4, the impact of crosstalk in IC will be discussed and finally, the concluding remarks terminate this paper.

2 Crosstalk issues in high speed integrated circuit design

ICs in nanoscale CMOS technology need to be designed in high densities and they operate at high frequencies, mainly in the gigahertz region. Based on these factors, it is essential to have circuits with smaller wire pitch that are capable to process data faster. Such requirements have caused ICs operating at mmWave frequencies to suffer from an electromagnetic phenomenon known as crosstalk. Crosstalk is a signal integrity (SI) issue where adjacent interconnect lines are susceptible to crosstalk when they are placed close enough for electromagnetic coupling to occur.

3 Crosstalk mitigation technique

There are numerous design techniques which can be implemented to reduce the effect of crosstalk in high-speed ICs. This section provides a brief review on the techniques which were agreed by researchers to reduce the impact of crosstalk.

3.1 Repeater insertion

As the size of complementary metal-oxide-semiconductor (CMOS) IC continues to shrink, interconnections became progressively vital. The interconnect delay will increase quadratically due to a linear increase in each interconnect resistance and capacitance with a linear increase in length [13-14]. Repeaters are traditionally used to linearize interconnect delay increases and improve signal transition times. Repeaters makes time delay linear with length by dividing the interconnection into smaller subsection [15].

To reduce interconnect delay, paper [16] suggest to implement a repeater. They used a linearized form of the Shichman-Hodges equations method [17]. At a particular operating point to determine the suitable repeater insertion locations. Meanwhile, paper [18] introduce a parallel regeneration concept which added precharge circuitry to the repeaters to reduce the evaluation time. Though this method needs fewer repeaters, extra area is critical and a precharge signal is needed to operate correctly. They also consider optimal method for driving resistive interconnect in [19].

Although the repeater insertion function is to reduce the delay of the interconnects in a circuit, the increasing number of repeaters will increase the slew rate and the operating frequency but in the same time it will decrease the design performance in terms of area and power. Paper [20] proposed a method of the repeaters and improve the slew rate.

3.2 Schmitt trigger

With the help of buffer insertion, very large-scale integration (VLSI) circuit interconnect delay has always been addressed. Various buffer insertion methods have been suggested for single-line interconnections and tree structures. In [21], the impacts of inductance on propagation delay and repeater insertion in VLSI circuits were observed and shown. Paper [22], stated that interconnections are not always linear in shape and handling the delays in tree structure is also a problem. This is attributed to the effects of crosstalk due to the neighbouring branches.

Paper [22] proposed a novel method for buffer insertion which is based on Schmitt trigger. The Schmitt trigger can be intended to have a threshold voltage lower than $V_{dd}/2$ and therefore can be made to respond faster. For their analysis, a four transistors Schmitt trigger is used. The findings indicate that the proposed methods surpass current methods in terms reduce of delay, power and crosstalk noise. They also proved that the proposed method works even in nanometer designs.

Besides, in [12] stated that Schamitt trigger is intended to function at a greater frequency as a buffer. It is used to decrease energy and delay in interconnections. Compared to standard CMOS inverters as a buffer, the Schmitt trigger buffer switches faster, thus leading to decrease in delay. Coupling noise is decreased by concurrent sizing of the buffer and wire in order to propagate the signal with decreased noise and delay.

Furthermore, the author in [23] designed a Schmitt trigger as a buffer which operates at a frequency of 20GHz to minimize crosstalk noise and interconnection delay. They also state that buffer helps to reduce power and delay. They have noted that crosstalk noise is effectively reduce by 60% and the delay is reduced by 56%.

3.3 Shielding

Crosstalk generally happens due to magnetic or inductive coupling between an aggressor and victim trace. The best way to mitigate it is to increase the spacing between the high-speed traces; but that will end up eating premium board real estate. Another popular implementation is to add a guard trace in between adjacent traces which reduces the crosstalk [24].

The guard trace, also known as shield is used to decrease the shared inductance owing to which the present return path is formed by the shield between the two interconnecting lines. Shielding is one of the most efficient and popular ways to decrease uncertainty about crosstalk and signal delay [12]. The author stated that shielding is nothing but a wire that is straight attached to V_{dd} or ground. One of the efficient shielding methods is to place ground or power lines between two cables to reduce noise. Different parameters that can be considered when shielding are the shield length, shield width, and the separation between shield and signal.

Some research has stated that two entire layers of metal are dedicated to shielding even in a 600MHz Alpha microprocessor [25-26]. Some research has also been conducted on integrated shielding and net ordering in paper [27-28]. Besides that, impact of different structures of shielding under separate standardized energy pitches as well as under restricted power grid disturbances were investigated in [28]. Furthermore, integrated shielding and signal net ordering algorithms without addressing any power grid design issues were studied in [27].

On the other hand, the authors in [24] proposed "Elevated guard trace technique" that is capable to reduce more crosstalk and electromagnetic interference in order to increase overall system performance. They vary the guard trace thickness to enhance the protection between the traces to further decrease the crosstalk compared to standard guard trace application.

3.4 Skewing

Repeater insertion methods are commonly used to decrease the capacity and strength of lengthy interconnections in order to reduce the wire delay and to decrease crosstalk [29]. Nevertheless, since simultaneous transition at opposite directions still occur; crosstalk increases the worst-case bus delay. The authors in [30] proposed a delay reduction technique to prevent simultaneous opposite transition by skewing signal transition timing of adjacent wires.

Skewing is a time-consuming technique that can be implemented to reduce crosstalk [12]. Time or timing skew has been previously used in buses where the delay of a coupled bus was decreased by skewing the timing of adjacent wires or interconnections. Besides, skewing was used to decrease the energy dissipation, delay of a coupled bus, and decrease the peak power.

However, skewing is nothing but a static delay in signal propagation [26]. Thus, driver skewing triggered the time decrease in driver switching, thereby decrease the crosstalk. Also, skewing is known as the spatial variation in the arrival time of the signal edge [31]. With the increasing skew of the victim line with regard to aggressor lines, the window for the overlapping part of the signals decreases. The overlapping window minimizes with the increased skew, resulting in weaker inductive coupling into the victim wire and also resulting in lower voltage overshooting [32]. The authors in [32] also stated that skewing the victim line (when the aggressors switch in the opposite direction) significantly improves voltage shooting and propagation delay to some degree.

Skewing was implemented to decrease a coupled bus energy dissipation where it was used to decrease bus peak power [33-34]. It should be noted that deliberate skewing has previously been used to reduce crosstalk effect such as crosstalk delay and crosstalk energy dissipation in buses [30], [33]. Paper [35] proposed reducing crosstalk voltage by induced on quite victim wires.

3.5 Space between signal wires and coupling length of the signal wires

Crosstalk refers to the unintended coupling of adjacent transmission wires. It is an electromagnetic phenomenon caused by the inductive and capacitive coupling between the adjacent wires [36]. Therefore, suppressing the crosstalk to an acceptable margin is crucial. To attain that goal, designers increasing increase the separating distance between the aggressor and victim wires. In addition, increasing the physical distance between the wires of the aggressor and the victim can reduce the mutual inductance and mutual capacitance between two adjacent lines [12][36].

It is stated that to minimize the crosstalk coupling between the two wires, the distance between the separation wires must be maximized [37]. [37]. As this increases the wire pitch, it is required to balance out between crosstalk and the wire pitch. Researchers also agreed that the distance separation between traces must be three times the width of a single trace measured from centreline to centreline, or the distance separation between two traces, edge-to-edge, must be greater than two times the trace width [37-38].

The authors in [39] investigated the effect of crosstalk on channel performance based on the IC package and the motherboard breakout routing by varying the differential pair spacing and coupling length. From these models, the differential pair spacing affects the crosstalk reading from the frequency and time domain.

4 Discussion

With the advancement of technology, IC designers were facing a challenge to design ICs with smaller form factor that are capable to operate at high operating frequencies. This is caused by one of the SI issues known as crosstalk. This section will discuss on the impact of crosstalk in IC.

Based on the articles reviewed, there are an abundance of researches on methods to mitigate crosstalk and the impact of crosstalk in IC. Crosstalk is a major issue because it can contribute significantly to the amount of jitter present in a device. Briefly, jitter is an edge deviation from its expected location. A large amount of jitter in a coupling wires can trigger bit errors in the victim line. It can cause a failure in the timing budget. These impacts have decreased the interactions between signals and reduce the noise immunity of digital CMOS circuits.

Furthermore, crosstalk may trigger undesirable impacts including excessive overshooting, extra signal delay, even a decrease in signal delay and undershooting. Crosstalk will cause noise on other lines, which may further degrade the integrity of the signal and increase noise margins.

Crosstalk can cause a signal to assume inaccurate value. This is particularly critical when the signal is about to be latched, as an inaccurate value can be loaded into a storage component and crosstalk may delay signal settling to the accurate value. This is often called "noise-on-delay".

However, there are inadequate investigations done on crosstalk in IC for mmWave applications. There is a need to carefully investigate the behaviour and characterize the crosstalk to prevent further iterations caused by failing to address the crosstalk issue during the IC design stage. Other than that, the impact of crosstalk from the IC for mmWave applications is also inadequate.

5 Conclusion

In conclusion, there are several techniques that can be used to minimize the impacts of crosstalk. However, crosstalk cannot be minimized totally. Further research on crosstalk mitigation techniques for designers to overcome the limitations on millimetre wave applications is inevitable. In the future, the finding of this study is expected to enable aggressive high-speed signal density for premium packages and client platforms.

This work is funded by Fundamental Research Grant Scheme with a Grant No. of FRGS/1/2018/TK04/UNIKL/02/9.

References

1. I. Corp., 60 Years of the Transistor, (2007)
2. S. K. Agrawal, K. Sharma, 5G millimeter wave (mmWave) communications, *Proc. 10th INDIACom; 2016 3rd Int. Conf. Comput. Sustain. Glob. Dev. INDIACom,*

(2016)

3. S. M. M. Maharum, N. Fisal, A. Lo, A. S. . Ghafar, F. A. Saparudin, N. Katiran, A demand-based spectrum orthogonalisation scheme for interference avoidance in LTE-Advanced heterogeneous networks, (2017)
4. F. A. Saparudin, N. Fisal, R. A. Rashid, A. S. A. Ghafar, S. M. M. Maharum, Cooperative communication and Cognitive Radio (CR) technology in LTE-advanced, *Commun. Comput. Inf. Sci.*, **253** CCIS, PART 3, (2011)
5. K. Yang, Interference management in LTE wireless networks, *IEEE Wirel. Commun.*, **19**, (2012)
6. S. Borkar, H. Pande, Application of 5G next generation network to Internet of Things, *2016 Int. Conf. Internet Things Appl. IOTA 2016*, (2016)
7. V. Gazis, Short Paper : IoT : Challenges , Projects , Architectures, (2015)
8. W. Roh , Millimeter-Wave Beamforming as an Enabling Technology for 5G Cellular, (2014)
9. F. C. Commission, Evaluating Compliance with FCC Guidelines for Human Exposure to Radiofrequency Electromagnetic Fields OET Bulletin 65, *Eng. Technol.*, **65**, (2001)
10. E. AlMousa, F. AlShahwan, Performance Enhancement in 5G Mobile Network Processing, *Lect. Notes Inf. Theory*, **3**, (2015)
11. GSMA, Road to 5G : Introduction and Migration, **54**, (2018)
12. P. Patil, S.Pable, Crosstalk Delay Analysis in Very Deep Sub Micron Vlsi Circuits, (2016)
13. V. Adler, E. G. Friedman, Repeater design to reduce delay and power in resistive interconnect, *IEEE Trans. Circuits Syst. II Analog Digit. Signal Process.*, **45**, (1998)
14. S. Bothra, B. Rogers, M. Kellam, C. M. Osburn, Analysis of the Effects of Scaling on Interconnect Delay in ULSI Circuits, *IEEE Trans. Electron Devices*, **40**, (1993)
15. H. B. Bakoglu, J. D. Meindl, Optimal Interconnection Circuits for VLSI, *IEEE Trans. Electron Devices*, **32**, (1985)
16. C.H. Wu, M.C. Shaiu, Accurate speed improvement techniques for RC line and tree interconnections in CMOS VLSI, (2002)
17. H. Shichman, D. A. Hodges, Modeling and Simulation of Insulated-Gate Field-Effect Transistor Switching Circuits, *IEEE J. Solid-State Circuits*, **3**, (1968)
18. M. Nekili, Y. Savaria, Parallel Regeneration of Interconnections in VLSI & ULSI Circuits, (1993)
19. M. Nekili, Y. Savaria, Optimal Methods of Driving Interconnections in VLSI Circuits, *Electr. Eng.*, (1992)
20. N. Ahmed, M. H. Tehranipour, D. Zhou, M. Nourani, Frequency driven repeater insertion for deep submicron, *2004 IEEE Int. Symp. Circuits Syst. (IEEE Cat. No.04CH37512)*, **5**, (2004)
21. Y. I. Ismail, E. . Friedman, Effects of inductance on the propagation delay and repeater insertion in vlsi circuits, *IEEE Circuits Syst. Mag.*, **2**, (2005)
22. S. Saini, A. M. Kumar, S. Veeramachaneni, M. B. Srinivas, Schmitt trigger as an alternative to buffer insertion for delay and power reduction in VLSI interconnects, *IEEE Reg. 10 Annu. Int. Conf. Proceedings/TENCON*, (2009)
23. S. Singh, V. S. Verna, Crosstalk Noise and Delay Reduction in VLSI Interconnects, **2**, (2011)

24. G. T. Technique, R. Balakrishnan, Crosstalk and EMI Reduction using enhanced, *2018 IEEE Electr. Des. Adv. Packag. Syst. Symp.*, (2018)
25. T. Zhang, S. S. Sapatnekar, Simultaneous shield and buffer insertion for crosstalk noise reduction in global routing, *IEEE Trans. Very Large Scale Integr. Syst.*, **15**, (2007)
26. D. Kaur, V. Sulochana, Crosstalk Minimization in VLSI Interconnects, *Int. J. Reconfigurable Embed. Syst.*, **2**, (2013)
27. K. M. Lepak, M. Xu, J. Chen, L. He, Simultaneous shield insertion and net ordering for capacitive and inductive coupling minimization, *ACM Trans. Des. Autom. Electron. Syst.*, **9**, (2004)
28. J. D. Z. Ma, L. He, Formulae and applications of interconnect estimation considering shield insertion and net ordering, *IEEE/ACM Int. Conf. Comput. Des. Dig. Tech. Pap.*, (2001)
29. D. Li, A. Pua, P. Srivastava, U. Ko, A repeater optimization methodology for deep sub-micron, high-performance processors, (2002)
30. K. Hirose, H. Yasuura, A bus delay reduction technique considering crosstalk, *Proc. -Design, Autom. Test Eur. DATE*, (2000)
31. G. Khanna, R. Chandel, A. K. Chandel, Impact of skew and jitter on the performance of VLSI interconnects, *IEEE Asia-Pacific Conf. Circuits Syst. Proceedings, APCCAS*, (2010)
32. A. Roy, J. Xu, M. H. Chowdhury, Analysis of the Impacts of Signal Slew and Skew on the Behavior of Coupled RLC Interconnects for Different Switching Patterns., **18**, (2010)
33. M. Ghoneima, Y. I. Ismail, Utilizing the effect of relative delay on energy dissipation in low-power on-chip buses, *IEEE Trans. Very Large Scale Integr. Syst.*, **12**, (2004)
34. Y. M. Lee, K. H. Park, Mesochronous bus for reducing peak I/O power dissipation, *Electron. Lett.*, **37**, (2002)
35. P. Liljeberg, J. Tuominen, S. Tuuna, J. Plosila, J. Isoaho, Self-timed approach for noise reduction in noc reduction in NoC, *Interconnect-Centric Des. Adv. SoC NoC*, (2005)
36. J. Xiaosong, X. Runjing, Crosstalk Analysis and Simulation in High-Speed PCB Design, (2007)
37. F. D. Mbairi, W. P. Siebert, H. Hesselbom, High-frequency transmission lines crosstalk reduction using spacing rules, *IEEE Trans. Components Packag. Technol.*, **31**, (2008)
38. M. I. Montrose, *Printed Circuit Board Design Techniques for EMC Compliance*, (2010)
39. A. H. Hasani, A. M. Shahar, A. J. Yusof, J. Kong, Investigation of crosstalk impact on channel performance from IC package and motherboard breakout routing, *2012 10th IEEE Int. Conf. Semicond. Electron. ICSE 2012 - Proc.*, (2012)

IOP Conf. Series: Materials Science and Engineering **701** (2019) 012038 doi:10.1088/1757-899X/701/1/012038

Effect of zero charge corrosion protection (ZCCP) on the carbon steel surface in 3.5% NaCl solution

Shaiful Rizam Shamsudin*, *Zin Wen* Sim, *Rajaselan* Wardan, *Nadia Nadhira* Mohd Jamalullalel and *Azmi* Rahmat

Center of Excellence Geopolymer & Green Technology, School of Materials Engineering, Universiti Malaysia Perlis, Jejawi 02600 Arau, Perlis

Abstract. The application of the zero charges concept is the latest and promising corrosion protection technique as compared to the cathodic protection (CP) technique. The study was conducted using carbon steel as metal to be protected. The effectiveness study was carried out on carbon steel in 3.5% NaCl solution with and without corrosion protection system. The results showed that carbon steel surface under zero charge corrosion protection (ZCCP) had no corrosion effect and consumed a very low electrical current. The protection effectiveness of ZCCP technique is found to be comparable to impressed current corrosion protection (ICCP), while unprotected carbon steel in the same corrosive solution undergoes severe corrosion attacks.

1 Introduction

The losses caused by corrosion are about 3.4% of the country's GDP excluded an indirect cost that finally will affect the quality of life [1]. Existing corrosion protection techniques are coating technology, corrosion inhibitor, and cathode protection (CP). All of these techniques are complementing each other, and of course, have their own weaknesses. The basic principle of CP just manipulation of polarization of the two-electrode system that requires a large electrical current and high voltage to overcome the electrolyte resistance in order to provide ionic flow paths. In some places and circumstances, CP may not be used for a number of specific reasons as discussed at the actual stage of implementation [2].

The zero charge corrosion protection (ZCCP) is a new and promising technology to control metal structures damage caused by oxidation reaction due to climate changes, environmental factors, and operational harsh conditions. Potential of zero charge (E_{pzc}) is a characteristic value of the electrode potential for any metal at which a clean surface of the metal will not acquire any electrical charge when it comes into contact with an electrolyte [3]. At E_{pzc}, a metal surface is not in contact with an aqueous solution due to the absence of electrical charge and no electrostatic attraction [4].

The effectiveness of ZCCP is studied by current/ voltage pulses on the exposed carbon steel in the corrosive environment. The power supply is customized to eliminate the excessive charges on the metal surface to be zero and prevent metal-solution corrosion reaction [5].

The objective of this study is to prove that ZCCP technique can act as corrosion protection of carbon steel in the corrosive electrolyte of 3.5% sodium chloride (NaCl) solution.

2 Methodology

The AISI 1016 grade carbon steel plate was ground and polished to remove the oxide layers, rust and corrosion deposits to get a smooth and shiny surface. The steel samples were then painted by leaving 1 to 2 cm^2 exposed area on the surface. The test to determine the effectiveness of ZCCP was performed by comparing it with the steel sample that soaked in 3.5% NaCl consisting of the test system; (i) without corrosion protection (Cond-A), (ii) impressed current cathodic protection, ICCP, (Cond-B) and, (iii) zero charge corrosion protection, ZCCP, (Cond-C). Cond-A and Cond-B tests were conducted for 3 and 5 days respectively, whereby Cond-C was carried out for 7 and 14 days. In Cond-B, ICCP was performed by applying an electric current to the exposed area of steel sample. The protected potential (On and Off potential) was controlled at −780 to -1030 mV vs SCE. In Cond-C, ZCCP was applied by using a customized power supply that capable of supplying current/ voltage pulses to steel samples. The ammeter with USB connection was connected in series in the circuit to measure electrical current consumption for both techniques at 1 reading/ hour interval.

All corrosion protection tests were performed at room temperature with 1-atmosphere pressure. After immersion testing, the sample was rinsed with distilled water while the paint was removed from the steel surface. The surface condition was visually inspected and documented by digital DSLR camera (D3100, Nikon) to record the original condition. The topography of the steel sample surface was cleaned by stain remover solution and analysed by stereomicroscope (SZX16, Olympus).

3 Results and discussion

A visual examination of the surface conditions of steel samples that immersed in 3.5% NaCl solution with and without corrosion protection are shown in Fig. 1. Fig. 1(a) shows a fresh steel sample. It looks smooth and shiny after being ground and polished. The immersed steel sample without corrosion protection for 3-days (Cond-A) is covered by a non-homogeneous stain or deposits of reddish-brown on its surface (Fig. 1b). This rust layer is quite thick. The white residue on the green paint-coated steel surface shows the deposition of NaCl salts from the solution. At Cond-B, ICCP-prevented sample is observed to have no corrosion effect (Fig.1c). Corrosion protection through ICCP technique is quite perfect with the experimental setup condition (Fig. 1c). On the other hand, the surface condition of the steel sample immersed under ZCCP (Cond-C), shows that it is protected from the corrosion process (Fig. 1d). However, upon closer inspection, the surface of the ZCCP protected steel is somewhat faded due to a slight oxidation reaction at the early stages of immersion at which, it was the time when the steel potential has not yet reached the E_{pzc}. This observation is initially acknowledged to be a bit disappointing as its protective effects are less clear effective. Further observations are therefore made more closely using stereomicroscopes.

The topographical of steel surface as the immersion and corrosion protection effect are shown by the stereomicroscope images in Fig. 2. To demonstrate the effectiveness of corrosion protection through ZCCP, the stereomicroscope images focused on the boundary of the removed paint-coated and the exposed areas of steel (marked as red X). From the above point of view, it is noted that the steel soaked in 3.5% NaCl solution for 3-days without corrosion protection showed rough and irregular surfaces with the residual stubborn oxide deposits left on the steel surface (Fig. 2a). The steel sample in this Cond-A experimental setup suffered from severe corrosion. As for steel immersed under the ICCP and ZCCP, there was no significant effect of corrosion reaction (Figure 2b, c, d).

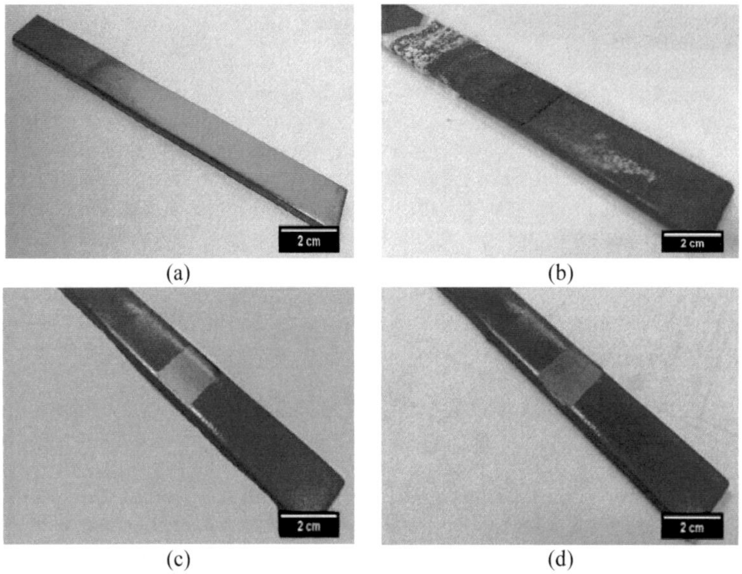

Fig. 1. Camera photographs of (a) fresh, and immersed samples; (b) without & with (c) ICCP and (d) ZCCP corrosion protection.

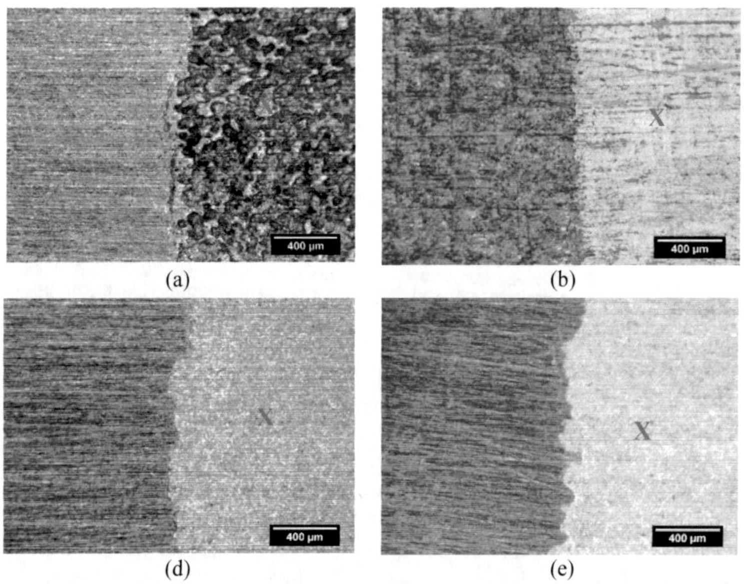

Fig. 2. Stereomicroscope images of immersed samples (a) without, and with (b) ICCP, and (c) 7 days ZCCP) and 14 days ZCCP corrosion protection.

The ICCP technique has proven to protect against corrosion, nevertheless, ZCCP is also comparable to the established ICCP. The efficacy of the ICCP technique can be seen in the topography of ZCCP samples immersed at 7 and 14 days that appear almost identical. Fig. 2 (c-d), is the answer where the surface of the steel is faded and somewhat disturbed. At the initial stage of immersion, it takes about 12 to 48 hours before the ZCCP goes into effect. At this period, there is a spontaneous reaction between the metal and the solution by which the charge transfer process takes place in the electrode-electrolyte interface. Once the steel potential reaches at E_{pzc}, the corrosion process is continuous to be prevented from invading the surface and penetrating the next steel atoms.

ZCCP is different from a conventional cathodic protection method. This protection technique applies an alternating current (ac) mode as compared to ICCP which applies a direct current system. Referring to Table 1, the ICCP records a current consumption of 90 uA/ cm^2. It is a high reading compared to 20 to 30 mA/ cm^2 in stagnant seawater as stated by Baboian & Treseder [6]. This current reading is acceptable without doubted because the current consumption at the initial stage of ICCP is known to be very high. Oppositely, ZCCP was found to record significantly lower current consumption readings. The decreases of current consumption as protection extend from 7 to 14 days reflected the significant economic savings by the ZCCP.

Table 1. Current consumption of ICCP & ZCCP system.

Corrosion Protection	Average Current Density (μA/cm^2)
ICCP	90
ZCCP (7 days)	0 (2-6)
ZCCP (14 days)	0 (0-2)

* ( ) AC current mode

4 Conclusion

Zero charge corrosion protection (ZCCP) is proven can hinder corrosion processes on carbon steel in the corrosive electrolytes such as 3.5% NaCl solution. It consumed a multiple smaller amount of electric current than impresses current cathode protection (ICCP) techniques.

The author would like to acknowledge the support from the Fundamental Research Grant Scheme (FRGS) under a grant number of FRGS/1/2015/TK10/UNIMAP/02/5 from the Ministry of Education Malaysia. Thanks to the School of Materials Engineering & Universiti Malaysia Perlis for providing a special financial support for enabling the publication of this manuscript.

References

1. G. Koch, J. Varney, N. Thompson, O. Moghissi, M. Gould, J. Payer,. *NACE Intl*, (2016)
2. A. Groysman, *Koroze a ochrana materialu*, **61**(3), (2017).
3. J. Kaur, R. Kant, *J. Phys. Chem. C,* **121**(4), (2017)
4. S. R. Shamsudin. *PhD Thesis. UniMAP*, (2016)
5. A. Rahmat, S.R. Shamsudin, M.N. Derman, *M'sia Patent*, PI 2015703887, (2015)
6. R. Baboian, R. Treseder, *NACE International*, (2002)

IOP Conf. Series: Materials Science and Engineering **701** (2019) 012039 doi:10.1088/1757-899X/701/1/012039

Current advancement in electrically conductive polymer composites for electronic interconnect applications: A short review

Farah Badrul[1,2], *Khairul Anwar* Abdul Halim[1,2*], *Mohd Arif Anuar* Mohd Salleh [1,2], *Mohd Firdaus* Omar [1,2], *Azlin Fazlina* Osman[1,2] *and Mohd Salihin* Zakaria[1,2]

[1]School of Material Engineering, Universiti Malaysia Perlis (UniMAP), Kompleks Taman Muhibbah, Jejawi 2, Perlis, Malaysia
[2]Centre of Excellence Geopolymer and Green Technology (CEGeoGTech), Universiti Malaysia Perlis (UniMAP), Kompleks Taman Muhibbah, Jejawi 2, Perlis, Malaysia

Abstract. The development of conductive polymer composites along with their advantages are rapidly growing to meet current demands in electronic applications. There are many types of matrix and filler that had been extensively researched in order to find the most suitable materials that can be utilized for electronic interconnect applications. Previous works carried out by researchers within the field revealed that by using melt blending techniques such as twin screw compounding and compression moulding can be used to develop conductive composite polymer such as from polypropylene (PP) incorporated with graphite as conductive filler. The conductivity of the composite can be measured using the 4-point probe technique. This short review aims to provide the latest insight in the area of electrically conductive polymer composites focused on the types of matrix and filler, processing and utilisation in electronic interconnect and other potential applications.

1 Introduction

Polymers such as low density polyethylene (LDPE), High-density polyethylene (HDPE), polypropylene (PP), polyvinyl chloride (PVC), and polystyrene (PS) are known with their ability as an insulator rather than conductor material. Polymers are electrically insulating materials with conductivity values as low as 10^{-7}-10^{-14} S cm^{-1}[1]. According to the electronic industries association (EIA) standard 541, a plastic material would be classified as conductive one if it has got the ability to protect against electrostatic discharge (ESD; surface resistivity between 10^5 and 10^{12} ohms sq^{-1}) or electromagnetic interference/radio

* Corresponding author: kanwar@unimap.edu.my

frequency interference (EMI/RFI; surface resistivity of <10^5 ohm sq^{-1}) [2]. This problem can be solved by incorporation of conductive filler with the matrix or known as conductive polymer composites (CPC's). The method currently used to increase the electrical conductivity of polymers is to fill them with specific conductive additives, such as metallic powder, metallic fibers, carbon black ionic conductive polymers and intrinsically conductive polymeric powders such as polypyrrole.

1.1 Interconnect applications

Recently, the electronic technology is growing rapidly. Non-degradable polymer-metal composites have been studied for interconnect materials in electronic packaging. Basically, conductive composite materials provide many benefits compared to traditional interconnect technology such as low-temperature processing conditions that reinforce the use of heat-sensitive and low cost of components and substrates, less hazardous environmental waste, lower cost processing, reduced stress on the substrates, and the ability to pattern fine-pitch interconnects, facilitating the miniaturization of electronics system [3].

1.2 Conductive polymer composites

In the past decades, conductive polymer has attracted great attention among researcher due to application such as electromagnetic interference shielding where it used in cell phone cases, antistatic protectors [4,5] which it used for electrostatic mats, packaging of sensitive electronics), sensors in the microelectronic field [4] and conducting adhesives in electronic devices [5]. This had been discovered by Hideki Shirakawa, Alan McDiarmid, and Alan Heeger in 1977, where the polymer material was polyacetylene. The investigation for more potential material has been found such as polypyrrole (PPy), polyaniline (Pani), polythiophene (PT), poly(3,4-ethylenedioxythiophene) (PEDOT), trans –polyacetylene, and poly (*p*-phenylene vinylene) (PPV) [6]. Polymer composites made up of two components which is a polymer matrix and filler. For conductive polymer composites, it consists of polymer matrix with conductive filler such as organic filler, inorganic filler and intrinsically conducting polymer. The conductive fillers contribute for the electrical properties and the polymer matrix contribute to the physical and mechanical properties. The conductivity of composites depends critically on the volume content of the filler [7]. Fig 1 shows the schematic diagram of possible conductivity pathways in composite polymer. Conductivity pathways is formed when conductive particle form a path along the other with particle and it allows electricity to flow.

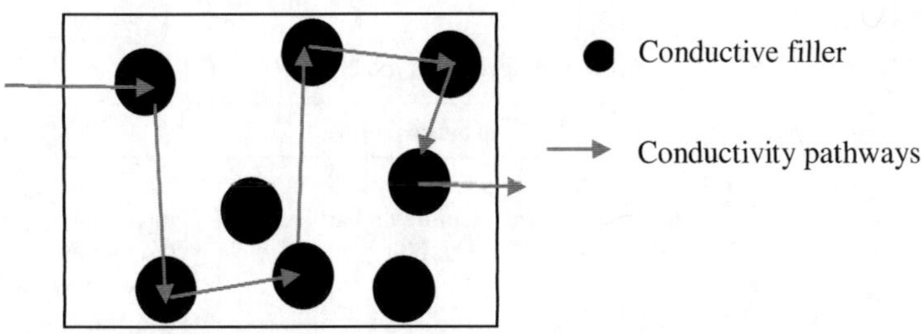

Fig. 1. Schematic diagram of possible conductivity pathways in composite polymer

Conductive polymer composites can be divided into two which is intrinsically conductive polymer and an extrinsically conductive polymer. The intrinsically conductive polymer is polymeric materials that conduct electric current without any additional conductive substance and it represent a special class of material with electronic and ionic conductivity [8] while for the extrinsically conductive polymer composite.

1.3 Flexible/stretchable conductive polymer composites

Flexible is defined as the capability of being flexed or bend without any braking or known as brittle while stretchable is a capability of being stretched. Research on flexible or stretchable polymer conductive become interesting due to the need of development of electronics technology. It has been utilized to have its own interesting properties, and diverse applications such as electrode material in electrostatic actuators, impregnation material for making conductive fabric, and circuit components in microelectronics or making electronic polymer devices had been found [9]. Jeroen van den Brand et al. found the basic principle of stretchable technology where the electronic functionality is distributed onto islands that are interconnected to each other by meander-shaped interconnects and this system was embedded into a stretchable rubber in order to keep the rubber together and to protect the device [10]. The advantages of this technology is that it uses flex foil manufacturing technologies [11]. This proved that this technology has gained its own interest in the electronic industry.

1.4 Fillers for conductive polymer composites

Fillers have been widely used and studied widely for their potential use in many applications. The conductive filler is one of the filler that has been used in preparing conductive polymer composites where it can make composite become conductive. There are many types of conductive filler that used to increase the conductivity of polymers composites for examples carbon based such as graphite, carbon nanotube (CNT), and carbon black, and metal materials such as copper Among of the materials, CNT, carbon nano fiber (CNF), and graphene has received greater attention of a research community because of their superior multifunctional properties compared to conventional fillers [12-15].

1.4.1 Graphite

Generally, graphite is used as an electro conductive filler because the cost is moderate and it also has good conductivity [16]. It is crystal form of pure carbon that much rembling of mica sheets of strongly linked atoms with very weak bond between the sheets. Graphite can be divided into 3 types which are natural graphite, synthetic graphite, and modification of graphite.

Table 1 Different of type of graphite [17]

Natural graphite	Synthetic graphite	Modification graphite
Can be divided into 3 categories based on its shape which is amorphous, flake and highly crystalline	It has two categories which are electro-graphite and artificial graphite	There three classes of modified graphite: graphite intercalated compounds (GICs), graphene oxide (GO) and expanded graphite (EG)
Originates from crude oil deposits that through time, temperature and pressure	Produced by treating carbonaceous precursors such as coal, petroleum and synthetic or natural organic	

	chemical in inert atmosphere	
Highly degree of crystallinity, Good electrical and thermal conductivity	Low density, High electrical resistance and porosity	

According to Rzeczkowski et al. [18], the highest conductivity of 36.4 S/cm was obtained for PP-based composites that consist of 55 wt% of graphite and 25 wt% carbon black prepared by melt mixing and 23.2 S/cm for PP-based composites that prepared by using solution mixing. The effect of different type of graphites grades and particles also play important role in order to get good conductivity. Graphite with flake-like shape has highest conductivity where the conductivity is 20.6 S/cm [19].

1.4.2 Carbon nanotube (CNT)

Carbon nanotubes are great interest as a filler for polymer composites due to their unique properties such as high electrical and thermal conductivity, ultrahigh mechanical strength, high ratio length/diameter (~1000) at nano-sized value of diameter. It has low density, high specific surface area and also have high aspect ratio. It can divide into two type which are single well or multiple wells. Nanotubes with single well are known as single-wall carbon nanotubes (SWCNTs) while the ones with more than one well are multiwall carbon nanotubes (MWCNTs) [25]. CNTs is one of the good material that we can observed due to the resistivity that relatively low (10^{-2}-10^{-3} Ω cm) [20]. Although, CNT is a good conductive filler, CNT have very high aspect ratios and strong Van der Waals attractions between each other, and this make the CNT have the tendency to agglomerate

1.4.3 Copper

One of the metallic material which is copper is the most common conductive filler had been used to produce conductive polymer composites due to the cost is low than others metal material. It is a metallurgy powder that rank at third after iron and steel. important uses for copper materials include friction materials, electrical parts, filters, additives to iron, aluminium, powder alloying, catalysts, paints, and pigments The four cardinal techniques to producing copper powders are atomization, oxide reduction, electrolysis, and hydrometallurgy. Of the above methods, atomization and oxide reduction are presently applied on a large global scale [21].

Table 2 Advantages and disadvantages of conductive filler [22-24]

Fillers	Advantages	Disadvantages
Graphite	Positive influence on the mechanical properties, thermal and dimensional stability	Poor filler matrix dispersion
Carbon Nanotube	Extreme mechanical properties , excellent thermal conductivity, outstanding electrical conductivity	Homogeneity of the material that contains wide distribution of the nanotubes diameters
Copper	Low resistivity, and improved electro-migration performance	Easy to oxidizes at low temperature and has self-protective layer to prevent further oxidation.

1.5 Processing of conductive polymer composite

Variety methods and processing can be used to produce conductive polymer composites. However, each of methods give different result at the end properties of composites. To obtain low percolation threshold and boost the composite's conductivity, conductive fillers should disperse well within the non-conductive polymer matrix without degrading the aspect ratio.

1.5.1 Twin Screw Extruder

Twin screw extruder is one machine that use for melt blending process that attractive to be used use to the advantage of being free of solvents. It made up of a machine that has a screw where material will be feed inside the hopper, and the material will be melt inside the chamber before it was push out from the die by the screw and undergoes pelletizing process and form pellet. Mainly this machine is use to premix the polymer matrix and mix homogenously with conductive filler before it undergoes manufacturing process. Researchers have frequently reported that the conventional process, as well as injection and compression molding, is capable to boost the mechanical properties of the end product [25]. Nevertheless, the mechanical properties increase, the electrical conductivity be liable to decrease.

1.5.2 Compression Moulding

One of popular technique to produce polymer composite products is compression moulding. In early 1950s, this process growth especially in automotive and appliance applications. This technique was first applied to application of glass fiber reinforced polymer in automotive industry which are they produce front panel of GM Corvette in 1953[26]. They are four basic step for compression moulding process which are precharge preparation and placement, mould closure, curing and part release. Compression moulding enable the moulding of complex shapes in the case of very high viscosity of the polymeric mixture, due to loading. In the past, compression moulding was used to assess the influence on the electrical resistivity and morphology of multi walled carbon nanotube/Polycarbonate (MWNT/PC) composites. It one of the process that have low cost process and able to high production rate process.

1.6 Potential application of conductive polymer composites

In market nowadays, conductive polymer composites have become one of the important product that can be applied in various applications. This include in medical, electronic, and automotive application. Several applications that attracts industry has been choose below.

1.6.1 High capacity batteries

The development of the technology has rapidly developed by the time goes. The impact of this rapid development had seen revolution in the manufacturing of devices such as hand phone, laptop and television. Lithium-ion batteries store or release energy by passing lithium ions back and forth through a microporous separator between the positive and negative electrodes (cathode and anode) that intercalate and deintercalate these ions respectively.

1.6.2 Artificial muscle

The needs of surgical robot interacting with medical team and patient or team of robot picking fruit from tree are different from those robots that use to assembles cell phone or electronic devices in a clean room. Nowadays, the development of artificial muscles become an important thing in medical fields. Artificial muscle is an actuator that generally defined by analogy with the skeletal muscle that characterized by its ability to contact in response to a chemical or physical stimulus [27]

1.6.3 Biosensors

The first actual biosensor was developed by Leland C. Clark, Jr in 1956 for oxygen detection [28]. The characteristics of a biosensor are selectivity, reproducibility, stability, sensitivity, and linearity. It has wide range of application such as to detect biomolecules that either indicator of disease or target of drug, prosthetic devices, water quality management, food quality monitoring, soil quality monitoring, and environmental monitoring [29]. Despite of the wide range of application, itself has challenges where difficulties in translating academic research into commercial viable prototypes by industry.

2 Conclusion

Although, the development of conductive polymer composite has been rapidly developing, there is still many parameters such as different type of filler, effect of filler loading and effect of different type of matrix that can be study in order to have a better electrical conductivity with good mechanical properties. From the previous study, there is many type of filler that has been used as conductive filler for example, graphite, carbon nanotube, copper and silver. Future research on others conductive filler materials need to done in order to find the best conductive filler that can give higher conductivity. This is to ensure that the conductivity can support the needs in interconnect application. Therefore, more effort need be taken to fulfil the new methods, materials and technology in interconnect application.

This work was supported by an Institutional Links grant, ID 332397914, under the Newton-Ungku Omar Fund partnership. The grant is funded by the UK Department of Business, Energy and Industrial Strategy (BEIS) and Malaysia and delivered by the British Council.

References

1. S. K. H. Gulrez, M.E. Ali Mohsin, H. Shaikh, A. Anis, A. M. Pulose, M. K. Yadav, E. H.P. Qua, S.M. Al-Zahrani, Polym.Comp., E **1** (2013)
2. J. Margolis, Conductive plastics for electrical and electronic applications, Reinforced Plastics, E **49**, 121 (2000)
3. Y. Li and C. P. Wong, Mater. Sci. Eng., R, Rep., E **51**, 1–3 (2006)
4. H. G. Elias, *An Introduction to Polymer Science* (VCH) (1997)
5. T.K. Das, S. Prusty, Polym. Plast. Technol, Eng. E **51**, 1487 (2012)
6. S. A. Ghani, S. H. M. Din, J. A. Jalil, J. of Vinyl & Add Tech (2018)
7. N. Hall, Chem Commun. E **7**, 1-7 (2003)
8. R. Strumpler, J. Glatz-Reichenbach, J. of Electroceramics. E **3:4**, 330 (1999)
9. S. Shang, W. Zeng, X. Tao, Adv. Materials Research. E **123-125** (2010)

10. J. V. D. Brand, M.D. Kok, M. Koetse, M. Cauwe, R. Werplancke, F. Bossuyt, M. Jablonski, J. Vanfleteren, Solid state Electronics. E **113** (2015)
11. F. Dogan, H. Hadavinia, S. J. Barton, P. J. Mason, Mech.Eng. E **3**, 174 (2010)
12. I.R. Wheeler, *Metallic Pigments in Polymers* (Rapra Technology Limited) (1999)
13. O. Breauer, U. Sundararaj, Polym. Compos. E **25**, 630 (2004)
14. M. Mu, A.M. Walker, J.M. Torkelson, K.I. Winey, Polymer. E **49**, 1332 (2008)
15. M. Moniruzzaman, K.I. Winey, Macromelecules, E **39**, 5194 (2006)
16. Beaucage G., Rane S., Schafer DW, Long G., Fischer D., J. Polym Sci Part B, Polym. Phys. E **37**, 1105 (1999)
17. P. Rzeczkowski, B. Krause, P. Potschke, Polymers. E **11** (2019)
18. T Derieth, G.Bandlamudi, P. Beckhaus, C. Kreuz, F. Mahlendorf, A. Heinzel, J. New Mater. Electrochem. Syst. E **11** (2008)
19. M. Wen, X. Sun, L. Su, J. Shen, J. Li, S. Guo, Polymer. E **53** (2012).
20. A. Eatemadi, H. Darae, H. Karimkhanloo, M. Kouhi, N. Zarghami, A. Akbarzadeh, M. Abasi, Y. Hanifehpour, S. W. Joo, Nanoscale Res Lett. E **9**, 1 (2014)
21. O. D. Neikov, S. S. Naboychenko, I. B. Murashova, *Handbooks of Non-Ferrous Metal Powders* (Elsevier Science Ltd) (2019)
22. I. Novak, I Krupa, Eur. Polym. J. E **40** (2004)
23. Y. Nakaramontri, C. Nakason, C. Kummerlowe, N. Vennemain, Rubber Chem. and Tech. E **89**, 2 (2016)
24. L.N. Ho, H. Nishikawa, T. Takemoto, Effect of different copper fillers on the electrical resistivity of conducting adhesives, *J. Mater. Sci,* **22,** (2011)
25. W.S.W. Harun, M.S.I.N. Kamariah, N. Muhamad, S.A.C. Ghani, F. Ahmad, Z. Mohamed, Powder Tech. E 327 (2017)
26. B. A. Davis, P. Gramann, A. C. Rios, T. A. Osswald, *Compression Molding, Applied Plastics Engineering Handbook: Processing and Materials (*Elsevier Inc.) (2003)
27. B. Tondu, Actuators. E **4** (2015)
28. W.R. Heineman, W.B. Jensen, C. Leland, Clark Jr. Biosens. Bioelectron. E **21** (2006)
29. N. Bhalla, P. Jolly, N. Formisano, P. Estrela, Essay in Biochem. E **60** (2016)

IOP Conf. Series: Materials Science and Engineering **701** (2019) 012040 doi:10.1088/1757-899X/701/1/012040

Effect of Laser Loop on Surface Morphology of Copper Substrate and Wettability of Solder Joint

S. Faqihah Roduan[1], *J. A.* Wahab[1*], *M. H.* Aiman[2], *A.Q.* Zaifuddin[2], *M A A* Mohd Salleh[1], and *M.* Ishak[2]

[1]Center of Excellence Geopolymer & Green Technology (CeGeoGTech), School of Materials Engineering, Kompleks Pusat Pengajian Jejawi 2, Universiti Malaysia Perlis (UniMAP), Taman Muhibbah, 02600 Arau, Perlis, Malaysia.
[2]Faculty of Mechanical and Automotive Engineering Technology, Universiti Malaysia Pahang, 26600 Pekan, Pahang Darul Makmur, Malaysia.

Abstract. The effect of different laser loop parameter on the geometry of micro-groove pattern on copper substrate and its effect on the wettability was investigated. The micro-grooves pattern was fabricated on the copper surface through laser surface texturing process. 3D measuring laser microscope and contact angle measurement test was conducted to measure the geometry of the micro-grooves pattern and wettability of the solder joint respectively. The results showed that the improvement in laser loop parameter increased the depth of the micro-grooves due to the more exposed time which allows more material ablation. It also showed that the contact angle of textured substrate is smaller than the untextured substrate which results in better wettability.

1 Introduction

Surface texturing is one of the current growing concerns in the materials industry, which involves modifying the surface characteristics resulting in an enhancement in the material's performance. The arrays of patterns can be either micro-scale, nano-scale, or its combination. Textured surfaces provide improved tribological, biological, optical or wetting characteristics after tailoring of the texture for specific purposes. The textured surface has gained considerable interest due to its excellent performance and its use as an anti-reflection, anti-fouling, anti-corrosion, low surface energy surface and self-cleaning material [1–4].

As the interest in tribological effects of surface texturing increases, various methods has been developed over the decades to fabricate surface textures. At a very early age, etching technique was once used to produce a physical textured sample for the purpose of analysing the most efficient parameters discovered from theoretical analysis [5]. Vibro-rolling method, where a hard indenter was used to produce shallow grooves on metallic parts, was later developed by Schneider [6]. Due to technological advancement in the past decades, the techniques of surface texturing was improved. Other surface texturing techniques include

*Corresponding author: juyana@unimap.edu.my

IOP Conf. Series: Materials Science and Engineering **701** (2019) 012040 doi:10.1088/1757-899X/701/1/012040

laser surface texturing, reactive ion etching, lithography and anisotropic etching and micro-casting [7–11].

Throughout the years, laser surface texturing (LST) has emerged as a viable means in research study as it is one of the most advanced surface texturing techniques in producing micro-textured patterns and enhanced the tribological performance and biomedical applications. The laser used is extremely fast which allows short processing times. It also provides high precision control of the size and shape of the geometrical parameters to generate holes and complex shapes on a wide range of materials including metals, ceramics and polymers [12–13].

Laser surface texturing (LST) is a process to generate periodic micro or nano or micro/nano structures which involves very thin layer of material surface. LST is usually used to enhance the tribological performance of a mechanical component or create functional surfaces for certain applications. This process is applied in the tribology [14–16], biomedical [17] and optical fields [18] as it offers great surface wettability and high adhesion strength for the material parts [17–20].

Wettability is one of the criteria that contribute to the reliability. It is defined as the bond between the solder and the base metal is formed only when the solder wets the base metal properly. Solder wettability is determined from contact angle (°) to the copper (Cu) substrate. The smaller of contact angle, the better wettability of the solder interconnection [21]. To create reliable joint, it is necessary to reach good wettability on contact pads as well as on component. Smaller wetting angle will cause the excellent wetting. Measurement of the wetting angle is the easiest way to characterize a wetting process. Wetting angle depends on several important factor such as solder alloy, flux, and surface roughness [22].

This present study focuses to evaluate the effect of laser loop on surface morphology of copper substrate and wettability of solder joint. The effect of laser loop on the microstructure of the micro-groove on the surface of copper substrate is analysed using 3D measuring laser microscope to evaluate the surface profiles. The wettability of the solder joint on the non-textured and textured copper substrate were investigated using optical microscope and J-image software.

2 Experimental

The micro patterned surface of copper substrate was produced through laser ablation to verify the structure geometry and contact angle. The effects of laser loop on the micro-texture dimension of samples will also be studied.

2.1 Material details and pattern design

A flat square-shaped copper sheet specimens are used in this research. Each specimen has a dimension of 15 mm × 15mm × 1 mm. The specimens were ground using 1000-grit sandpaper and then polished with 1 μm alumina diamond polishing paste.

2.2 Laser surface texturing

A micro-groove texture was developed on the polished copper surface by using a laser surface modification technique. The Ytterbium Fiber Laser Marking Machine was used to produce the closed-pore type texture, with a fixed groove distance of 100 μm. A laser beam with a size of approximately 50 μm was introduced directly to the copper substrate surface using different parameter of laser loop. The loop (L) varies from Loop 1 to Loop 4 for the specimens. But for each specimen, the main parameter was fixed and shown in Table 1.

Table 1: Laser surface texturing parameters.

Parameter	Value
Name	Ytterbium Fiber Laser Marking Machine
Model	ML-MF-A01
Wavelength (nm)	1064
Scanning speed (mm/s)	300
Laser power (W)	18
Repetition rate (KHz)	20

2.3 Reflow soldering

The solder alloy, Sn-0.7Cu ingots which was supplied from Nihon Superior Co Ltd., Japan, was used as solder material. The solder alloy was prepared by melting the ingots at a temperature of 350°C in a solder pot. The ingot was then cold roll to produce thin solder sheets. The thin solder sheet was cut and then divided into approximately 0.5 g using a weighing balance. The 0.5 g of Sn-0.7Cu was then reflowed onto micro-textured Cu-substrates in a F4N reflow oven in making the solder joints with the aid of small amount of resin mildly activated (RMA) flux.

2.4 Characterisation and testing of samples

The microstructural characteristics of the textured copper substrates was observed using 3D measuring laser microscope model Olympus OLS5000. The wettability of the solder alloy on micro-textured substrate was then observed by measuring the contact angle of the solder joint using optical microscope and Image-J software.

3 Results and discussion

3.1 Analysis of surface morphology

Laser surface texturing is a localized thermal process where the material will absorb the laser beam energy and results in material ablation. This process will then change the surface morphology and topography of the material involved. The dimensions of the micro-groove on the copper substrates differ with the laser loop parameter. The 2D images of the surface morphology and profiles of the micro-groove copper substrates for laser loop 1 is shown in Fig. ***1***(a) and Fig respectively. When the laser beam in contact with the copper substrate, the material will absorb the laser energy which will heated the material up and then evaporated. This process is called as laser ablation which means that the part that are in contact with the laser beam are going to be removed. In this case, the micro-groove are formed via the laser surface texturing process where the depth of the micro-groove depends on the laser loop parameter. Based on the figure, the average depth of the micro-grooves of laser loop 1 is around the range of 2-4 μm. The depth of the micro-grooves of laser loop 1 of the copper substrate is small due to the short exposed time of the material to the laser beam.

As we can see in Fig. ***1***(b) and Fig, the average depth of the micro-groove formed by using laser loop 2 is in the range of 6-8 μm. The depth of the micro-groove of laser loop 2 is increased compared to the laser loop 1. This is due to the copper substrate of laser loop 2 absorb more laser energy which gave the material to melt and evaporates more compared to loop 1.

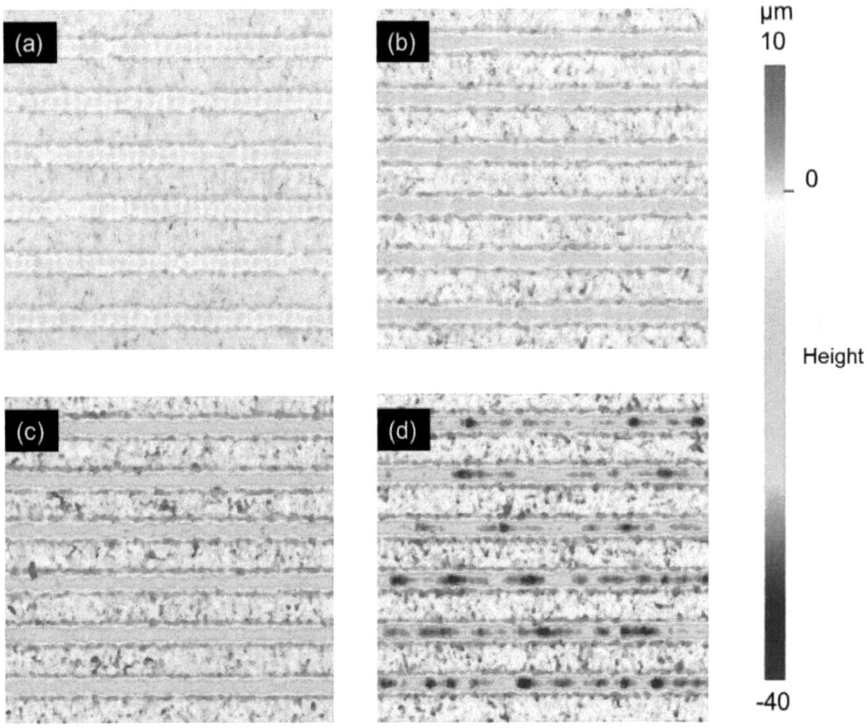

Fig. 1. 2D images of copper substrate of ; a) Loop 1, b) Loop 2, c) Loop 3 and d) Loop 4.

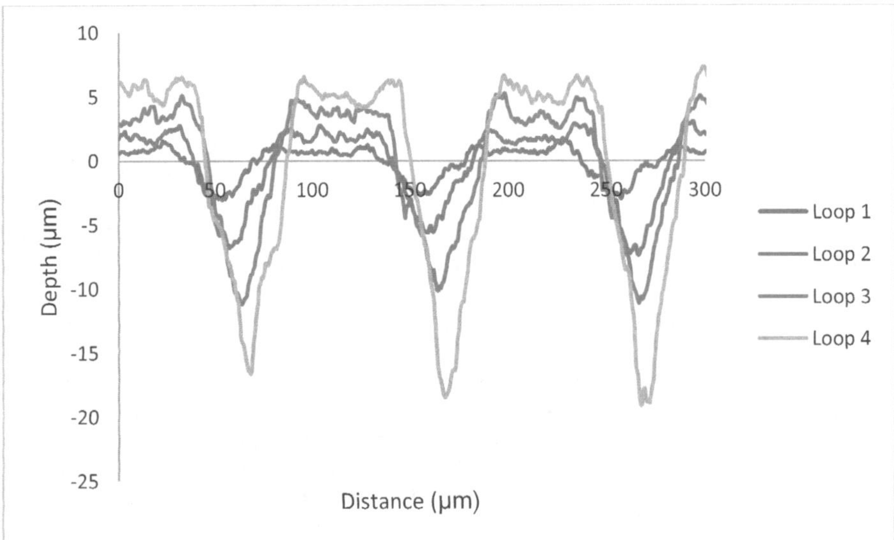

Fig. 2. Surface profiles under different laser loop of textured copper substrate.

Based on Fig. *1*(c) and the surface profiles, it is shown that the average depth for laser of the micro-groove is higher compared to loop 1 and 2 where it is in the range of 10-12 μm. This value begins to show that the depth of the micro-groove keep increasing when the laser loop increased.

Fig. *1*(d) and Fig display the 2D images and surface profile of the laser loop 4 micro-groove on the copper substrate respectively. Based on the graph of the surface profile, the average depth of the micro-groove is 17-19 μm which makes the laser loop 4 has the deepest depth among all the laser loop that has been discussed before. This is due to the copper substrate of laser loop 4 has the longest time to absorb the laser energy which means that more material can be ablated. Thus, it can be concluded that the depth of the micro-groove of the copper substrate depends on the laser loop parameter.

3.2 Analysis of wettability of solder joint

Wettability is an important property in a soldering system for electronics as the reliable joint of the solder depends on the wetting angle. In this study, the wettability is quantitatively measured via the contact angle measurement. Solder wettability is determined from contact angle (°) to the copper (Cu) substrate where the better wettability of the solder interconnection comes from a smaller contact angle [21]. Based on Fig. *2*, the highest measured contact angle is 17.84° for the non-textured copper substrate while the lowest contact angle is 10.4° for the micro-groove laser loop 4. The figure also revealed that the increasing laser loop parameter which increased the depth of micro-grooves are able to reduce the degree of contact angle. When the solder alloy (Sn-0.7Cu) is reflowed on the textured substrate, the solder alloy flows into the valleys of the micro-grooves by capillary action. The micro-groove on the copper substrate increases the surface area which indirectly increase the surface tension between the solder alloy and the substrate itself. The higher surface tension will allow the solder alloy to flow more into the valley and decrease the contact angle of solder joint. It can be concluded that the good wettability of the solder joint can be achieved by using textured substrate.

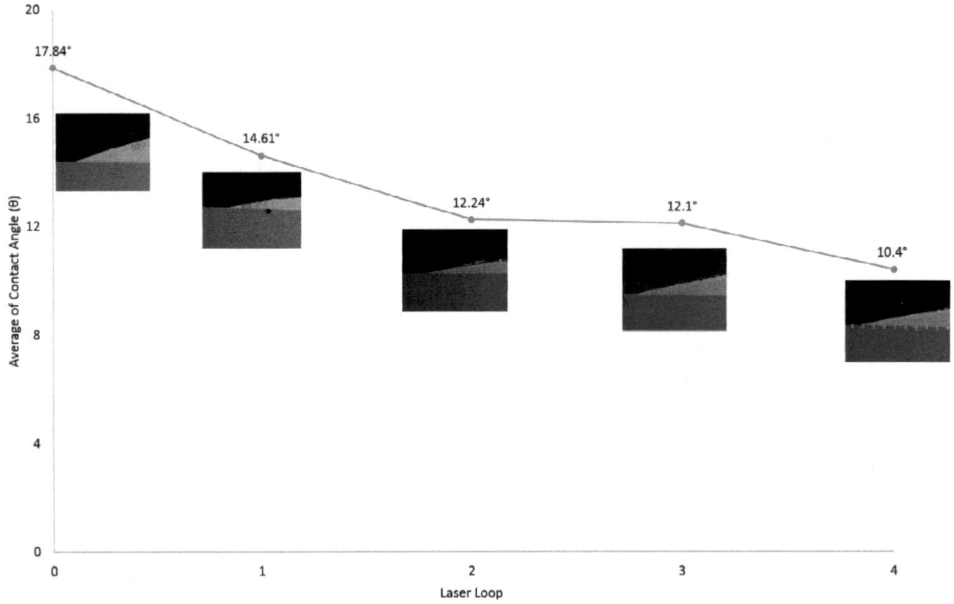

Fig. 3. Influence of micro-groove textured substrate on the contact angle in Sn-0.7Cu lead-free solder.

4 Conclusion

The effect of laser loop on surface morphology of copper substrate and wettability of solder joint have been investigated. The conclusion that can be made from this study are the microstructural observations from the surface profiles revealed that the increment in laser loop parameter affect the depth of the micro-groove formed on the copper substrates. The wetting properties decrease with the increasing depth of the micro-groove pattern and lowest contact angle of 10.4° was observed with laser loop 4 of micro-groove textured substrate.

The authors gratefully acknowledge the financial support through the Ministry of Education (MoE), Malaysia, under the Fundamental Research Grant Scheme (FRGS/1/2017/TK05/UNIMAP/03/3), and the School of Material Engineering, University of Malaysia Perlis (UniMAP) for supporting this research effort through materials and facilities. The authors would also acknowledge Nihon Superior Co. Ltd. and Universiti Malaysia Pahang (UMP) for the support given throughout the research project.

References

1. G. S. Watson, D. W. Green, L. Schwarzkopf, X. Li, B. W. Cribb, S. Myhra, J. A. Watson, Acta Biomater. E **21** (2015)
2. Z. Liu, W. Yin, D. Tao, Y. Tian, Biotribo. E **1–2** (2015)
3. A. Malshe, K. Rajurkar, A. Samant, H. N. Hansen, S. Bapat, W. Jiang, CIRP Ann. - Manuf. Technol. E **62**, 2 (2013)
4. L. Yao and J. He, Prog. Mater. Sci. E **61** (2014)
5. D. B. Hamilton, J. A. Walowit, C. M. Allen, J. Basic Eng. E **88**, 1 (2011)
6. Yu.G.Schneider, Precis. Eng. E **6** (1984)
7. U. Pettersson, S. Jacobson, Tribol. Int. E **36**, 11 (2003)
8. A. H. Cannon, W. P. King, J. Micromechanics Microeng. E **19**, 9 (2009)

IOP Conf. Series: Materials Science and Engineering **701** (2019) 012040 doi:10.1088/1757-899X/701/1/012040

9. H. W. Deckman, J. Vac. Sci. Technol. B Microelectron. Nanom. Struct. E **1**, 4 (2002)
10. E. Circuit, O. Lithography, and I. Fundamentals, (2018)
11. J. A. Wahab, M. J. Ghazali, Wear. E **432–433** (2019)
12. T. Ibatan, M. S. Uddin, and M. A. K. Chowdhury, Surf. Coatings Technol. E **272** (2015)
13. T. L. See (2015)
14. Y. Wan , D. S. Xiong, J. Mater. Process. Technol. E **197**, 1-3 (2008)
15. J. A. Wahab, M. J. Ghazali, W. M. W. Yusoff, Z. Sajuri, Trans. Inst. Met. Finish. E **94**, 4 (2016)
16. J. A. Wahab, M. J. Ghazali, M. N. Derman, Y. Otsuka, Solid State Phenom. E **280** (2018)
17. A. G. Demir, P. Maressa, B. Previtali, Phys. Procedia. E **41** (2013)
18. F. Müller, C. Kunz, S. Gräf, F. A. Müller, C. Kunz, S. Gräf, Materials (Basel). E **9**, 6 (2016)
19. D. Xiong, Y. Qin, J. Li, Y. Wan, R. Tyagi, Tribol. Int. E **82** (2015)
20. B. Kim, Y. H. Chae, H. S. Choi, Tribol. Int. E **70** (2014)
21. M. Abtew, G. Selvaduray, Mater. Sci. Eng. R Reports. E **27**, 5-6 (2000)
22. S. V. Sattiraju, B. Dang, R. W. Johnson, Y. Li, J. S. Smith, M. J. Bozack, IEEE Trans. Electron. Packag. Manuf. E **25**, 3 (2002)

IOP Conf. Series: Materials Science and Engineering **701** (2019) 012041 doi:10.1088/1757-899X/701/1/012041

Suitable materials selection that act as a retardant agent against microwave radiation and electronics packaging to human life

Ahmed I Mohammed[1] , and *Mohabattul* Z Bukhari[1*]

[1]School of Materials Engineering, Universiti Malaysia Perlis, 02600 Jejawi, Arau, Perlis, Malaysia

Abstract. Recently there is a high attempts to investigate the possibility of protecting human body from microwave radiation and electronics packaging. In this study an electromagnetic radiation shielding film by reflecting and/or absorbing the incident wave were prepared as a promising solution. Basically PVC and iron oxide nanoparticles were synthesized using solution casting method to make the protection film. Analysis via VNA has been conducted to understand the shielding behaviour during the exposure to the radiation. The prepared films exhibit ability of absorbing microwave radiation in the range of 8 GHz – 11 GHz, with a maximum value of -10 dB at 30 wt. % of iron oxide nanoparticle. Furthermore microwave absorption performance can be improved by putting two layers or more in order to increase the thickness of the shield. It is suggested that the results are mainly attributable to good effective properties of iron oxide and their homogeneous dispersion in the PVC matrix, also from the analysis it will propose the optimum solutions for this problem.

1 Introduction

Since microwave is an electromagnetic radiation, it is a combination of two types of wave energies electrical energy and magnetic energy, move together in the spaces. It stays between the range of Radio and IR spectrums with about 300 MHz to 300 GHz as its frequency range and 10 mm to 100 μm as rang of wavelength [2]. It can be generated from many different sources such as antennas and vacuum tubes, also there is natural source sun generate microwave radiation within its electromagnetic (EM) spectrum.

This radiation may have bad side effects on human life [3, 4] so; they can be protected against this radiation whether individually or in group. Supporting the buildings with materials that have abilities of absorbing or reflecting microwaves radiation during the construction is advised especially for those who leave near to high microwaves radiation resources such as communication antennas. Where radiation proof suits are a good example for individual protection cases. Even though there is no any proved conformation that the microwave radiation has direct harm to human body, we have to find such a way that may reduce human interruption with this radiation as possible as can.

*Corresponding author: mzaman19704@gmail.com

If possible to provide way which is easy to apply and efficient economically it will be an important step to assure human safety. Since many of this era technologies are derived to be used by electromagnetic (EM) radiation.

Such an application must be funded by selecting a suitable material considering many factors. Cost must be affordable, simple processability, and no side effects. At the same time it can adsorb or reflect electromagnetic radiation of microwaves. Composites that have high electrical conductivity and dielectric constant have a good electromagnetic (EM) blocking property, especially nanocomposites have better compatibility and lower overall mass where at the same time show excellent EM shielding performance with larger frequency bandwidth. Many researches on EM shielding effectiveness of nanocomposites have been conducted during the past decade. Joo and his colleagues added nano-sized silver into epoxy matrix and obtained the EM shielding value of 46 dB in the range of 10 MHz – 1 GHz. They also carried out the study of the spectra attributes of multilayer absorber composed of materials with different electrical conductivity. [5]

Gairola et al. reported a 2.5 mm polyaniline based nanocomposites with nano ferrite $Mn_{0.2}Ni_{0.4}Zn_{0.4}Fe_2O_4$ which shows high shielding effectiveness up to 50 dB in the frequency band of 8–12 GHz (X-band).
[6] In recent years, the unique electrical mechanical and magnetic attributes of carbon nanotubes (CNTs) enable it to play an important role in nano EM shielding materials preparation. [7] Currently, EM shielding composites prepared by CNTs compounding with organic polymers have become significant field in new generation electromagnetic interaction (EMI) shielding material research. [10–14]

Thermal, mechanical, electrical and microwave radiation absorbing or reflection properties of conductive composites based on dodecylbenzenesulfonate doped polyaniline / organoclay nanocomposites and propylene – ethylidene–norbornene rubber have been investigated with special interest on the effect of nanocomposite concentration. Composites were prepared by melt blending using an internal mixer. [15]

2 Materials and methods

2.1 Materials

Materials that used in this study are commercially rated without any further characterization or purification. Those materials are polyvinyl chloride (PVC, pr- 1069), tetrahydrofuran (THF, AR), dioctyl phthalate plasticizer (DOP), and iron oxide nanoparticles (Fe_2O_3) with purity higher than 95%.

2.2 Preparation of the composites

An iron oxide nanoparticles and PVC polymer are prepared by dissolving and homogenizing them in a tetrahydrofuran solvent (THF) solvent separately at 25 °C for 90 min, DOP plasticizer is added to PVC solution. Then the two mixtures are stirred together in one flask until a viscous solution was obtained to form the composites. The solution was spread with a film maker (Casting knife) to form a thin film ~50 μm thick and then dried at 60 °C for one day. Finally, the obtained films were annealed (near the glass transition temperature) at 82°C for 3 h to eliminate the residual THF and cut the samples film according to the desired requirements for characterization techniques.

2.3 Characterization

Samples of synthesized material made in sheet forms to fit the dimensions of DUT; sample holder in used Vertical Network Analyzer. The blend of composites applied on a mold with 1 mm in thickness. Shielding efficiency measurement will be taken using vector network analyzer to determine the S parameters reading, using the equations 3.0 where the shielding properties of the selected materials at the applied microwave frequency range will get higher shielding efficiency with increasing of the reinforcement.

3 Results and Discussion

The measurements are done using S- Parameters by comparing the incident wave directed to the sample with the responses that can be gain as transmission or reflection waves. From these two resulted waves samples characteristics can be calculated. In this study we are using complex impedance phase measurement types. Complex impedance data is information such as resistance, reactance, phase, and magnitude that can be determined from an S_{11} or S_{22} measurements. Complex impedance data can be viewed using either the Smith Chart format or the Polar format.

In order to investigate the frequency properties of obtained composites an electromagnetic spectrum transmission and reflection coefficients are measured. The values were measured using the rectangular cavity working in excitation mode at 8 GHz. The reflection coefficient of the cavity was recorded as a function of applied field. The maximum value of the reflection coefficient is related with the maximum value of energy absorbed by sample at the ferromagnetic resonance.

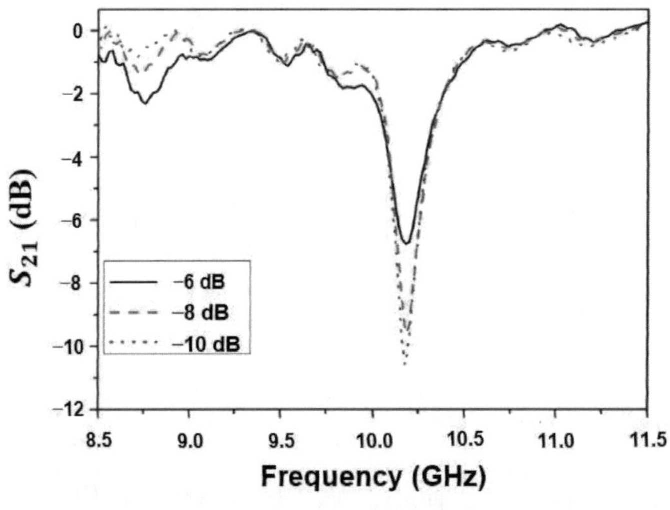

Fig. 1. VNA results presentation for the three samples with their different compensations, sample A (with -10dB), sample B (-8dB), and sample C (with -6dB).

Reflectivity coefficient of the samples is presented in decibels, dB as a function of frequency, in GHz and absorber thickness of the samples as shown in figure 1. In this study permittivity data of the samples are not available so, there will not be any comparison with theoretical reflection curves. High radiation attenuation values (−10 dB) for sample A, in the frequency range of 8 – 11 GHz, were obtained when the composites contained 30 wt.% of the Fe_2O_3 nanoparticles. This value corresponds to a dissipation of higher than 70% of the incident radiation. Sample B of the composition that contain of 20 wt. % of Fe_2O_3

nanoparticles shows a value of radiation attenuation at about -8 dB in the range of 8 GHz to 11 GHz of the frequency. For this case the sample have a broadband behavior of microwave radiation absorption of about 70 % as shown in figure 1. Hence, the reflectivity properties of PVC/Fe_2O_3 conducting composite depends on the composition of the composite and microstructure attained after processing. In this case the Fe_2O_3 nanoparticles will agglomerate between the elastomeric phases act on the wave matter interaction.

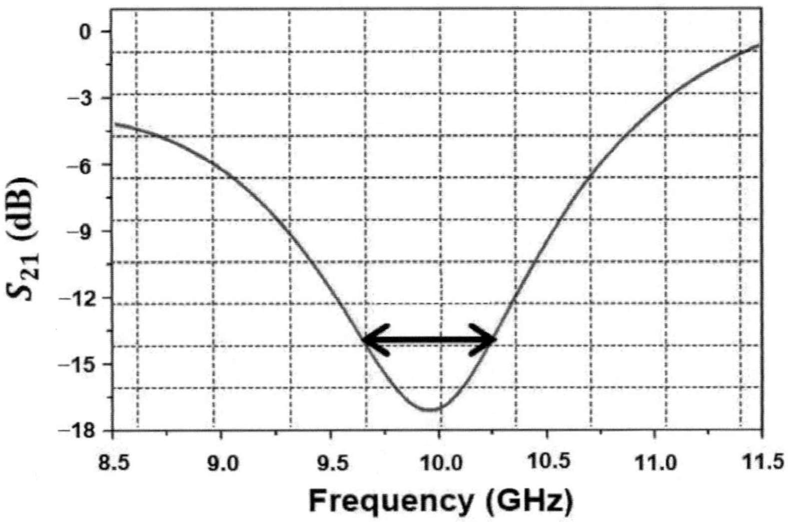

Figure 2: Result of two 20 wt. % layers at the range of 8 GHz – 11 GHz.

The affection of the sample thickness on the absorption, transmission and reflection was investigated, by obtaining variation of samples thicknesses whiter by making a thicker samples or attaching layers together and place them in the waveguide. As shown in figure 2 for two layers of the composite film. This shows that the attenuation peak frequency of composites can be manipulated by changing the thickness of materials, like in the two layers of 20 wt. % got a value of higher than -15 db. The minimum value of transmission coefficient ($S12$) was obtained for the sample with the highest concentration of iron oxide nanoparticles (sample A) which is about -10 dB at less than 11 GHz wave frequency. The greatest value of this coefficient was measured for sample B was approximately -8 dB at the same range of the frequency. The absorption of high frequency radiation is due mainly to metallic and ferromagnetic losses in the nanoparticles samples. At the increase of the nanoparticles amount in PVC based nanocomposites high frequency absorption ability increases too. The transmission coefficient decreases at the increases of the nanoparticle concentration suggesting a more shielding effect but this is associated with increasing of the reflection coefficient as well.

4 Conclusion

PVC/Fe_2O_3 nanocomposite films were produced by solution casting. They were observed only as thin film samples about 0.5 μm with 30 wt. % as a higher concentration and 10 wt. % as the lower one. Absorption was found to reach a maximum at an optimum Fe_2O_3 loading concentration. Obviously can be observed that the sample C exhibits the highest radiation transmission value of the others, due to the lowest amount of conductivity in the films composition, where the sample B got shifted to higher value, and the sample A have the highest value of radiation absorption comparing to sample C and sample B with the value of -10 dB at the tested wave range. Overall, the thickness variations also affect the shielding ability. All the samples prepared in this work can present microwave radiation absorbing and tuning properties with the ability of being applied in many ways during the use time, better thermal properties and flexibility. Thus a good microwave absorbing materials with lower content of conducting polymer was prepared.

References

[1] Limits of Human Exposure to Radiofrequency Electromagnetic Fields in the Frequency Range from 3 kHz to 300 GHz. Published by authority of the Minister of Health. Cat. H46-2/99-237E, ISBN0-662- 28032-6

[2] Kumar, Sanjay; Shukla, Saurabh. Concepts and Applications of Microwave Engineering (2014)

[3] V. PA, van D. TE, R. MH. Workgroup report: base stations and wireless networks radiofrequency (RF) exposures and health consequences. Environ. Health Prospect. 2007.

[4] Cecie Starr Biology: Concepts and Applications. Thomson Brooks/Cole. ISBN 0-534-46226-X. (2005).

[5] Electromagnetic Spectrum facts, information, pictures Encyclopedia.com articles about Electromagnetic Spectrum. www.encyclopedia.com

[6] Elert, Glenn. "The Electromagnetic Spectrum, The Physics Hypertextbook". Hypertextbook.com. Retrieved 2017-10-26

[7] Kumar, Sanjay; Shukla, Saurabh (2014). Concepts and Applications of Microwave Engineering

[10] "hertz". American Heritage Dictionary of the English Language (3rd ed.),Boston: Houghton Mifflin (1992)

[11] Kumar, Sanjay; Shukla, Saurabh Concepts and Applications of Microwave Engineering (2014)

[12] John C. D. Brand (1995). Lines of light: the sources of dispersive spectroscopy, 1800–1930. CRC Press. pp. 30–32. ISBN 978-2-88449-163-1

[13] Analytical Chemistry: An Introduction (Saunders Golden Sunburst Series) 7th Ed., by Douglas A. Skoog, Donald M. West, F. James Holler. 1999

[14] Electromagnetic noise inhibits radiofrequency radiation-induced DNA damage and reactive oxygen species increase in human lens epithelial cells. K.Yao, W.Wu, K.Wang, S.Ni, P.Ye, Y.Yu, J.Ye, and L.Sun. Mol Vis. 2008; 14: 964–969. Published online 2008 May 19

[15] Faez, R., Reis, A., Soto-Oviedo, M., Rezende, M., & De Paoli, M. (2005). Microwave Absorbing Coatings Based on a Blend of Nitrile Rubber, EPDM

Rubber and Polyaniline. Polymer Bulletin, 55(4), 299- 307. doi: 10.1007/s00289-005-0433-y

Graphitization damage on seamless steel tube of pressurized closed-loop of steam boiler

Nur Farhana Hayazi[1], *Shaiful Rizam* Shamsudin[1,2,*], *Rajaselan* Wardan[1,2], *Mohd Syazwan* Muhamad Sanusi[1,2] and *Farah Farhana* Zainal[1,2]

[1]School of Materials Engineering, Universiti Malaysia Perlis, Kompleks Pengajian Jejawi 2, Jejawi 02600 Arau Perlis.
[2]Center of Excellence Geopolymer & Green Technology (CEGEOGTECH), School of Materials Engineering, Universiti Malaysia Perlis, Kompleks Pengajian Jejawi 2, Jejawi 02600 Arau Perlis.

Abstract. The pressurized closed-loop of steam boiler has been in operation for five years has experienced leakage problems at the bottom of the reducer and at the center of the tube on weldment. The metallographic analysis found that the leakage was due to the degradation of microstructure as a result of long-term overheating. The degraded microstructure of the tube boiler material dominated by graphitization. The graphitization normally applicable at 450°C, but with high-pressure water flow (90 bar) in the combination of internal stress on the tube material due to the welding process as well as the formation of the reducer, making the boiler operating temperature is sufficient to produce graphitization. The leakage occurred when the chain and accumulated graphitization happen because the cavities that covered the graphite facilitates the cracking propagation due to the presence of stresses.

1 Introduction

The pressurized closed-loop of steam boiler has been in operation for five years. The boiler uses a reverse osmosis (RO) water without any chemical additives. Pressure and operating temperatures are 90 bar and 290 - 320°C respectively. The boiler tube was found leaked in April 2018 at the bottom reducer. Subsequently, the second leak was also found at the bottom reducer in opposite direction to the first leak some times in May 2018. Followed by the third leak on July 2018 on the bottom plug due to the air trap. Finally, the fourth leak was found on October 2018 at the center tube on weldment. The failed boiler has been dismantled from its location (Figure 1).

Pressure and temperature play an important role in degradation mechanism for alloys such as plain carbon, medium carbon and high alloy steels which has been used in different process of plant application. The microstructure of steel stipulates the properties of metallurgical, physical, mechanical and corrosion. The degradation will occurred, when microstructure is exposed to high pressure-high temperature [1].

*Corresponding author: farhanahayazi@unimap.edu.my

The principle of degradation mechanisms that could be detected by metallography analysis is the microstructural degradations for example graphitization [1], spheroidization of pearlite [2], creep, thermal fatigue, hydrogen attack, carburization, grain boundary oxidization and embrittlement of microstructure etc [3]. Typically, the condition of a boiler is measured by in-situ metallography technique (known as replication testing). In the current situation, as the failed tubes are quite severe, therefore the sampling was carried on the adjacent tube (Figure 2a). The objective of this analysis are to investigate the failure evidence presented by the boiler tube, to understand the graphitization of the graphite nodules through the microstructural changes.

Fig. 1. Camera photograph of the failed boiler component that been dismantled from its location.

2 Methodology

As-received boiler tube was cross-sectioned into 9 workpieces in according to Figure 2b. The samples for chemical composition analysis was marked as OES whereas metallography analysis and hardness measurement were marked as A, B, C, D, E, F, G and H respectively. The chemical composition was analyzed by using optical emission spectrometer, OES (Q8 Magellan, Bruker) at five different spots on the metal surface. The hardness and metallography samples were ground and polished. Hardness measurement was carried out by Rockwell hardness tester (Brevetti Affri, 206 RTD) in HRB scale for each section vertically (Figure 3). Then the polished samples were etched using 2% nital solution to reveal the microstructural pattern on the samples. The analysis was carried out by metallurgical microscope (BX41, Olympus). The morphological analysis was carried out by scanning electron microscope (JSM-6490LV, Jeol).

IOP Conf. Series: Materials Science and Engineering **701** (2019) 012042 doi:10.1088/1757-899X/701/1/012042

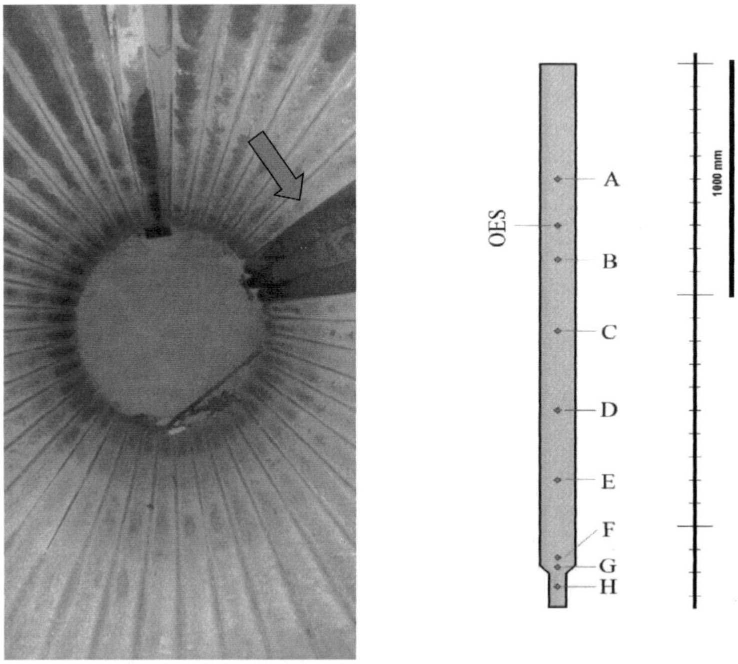

(a) (b)

Fig. 2. (a) Sampling was carried adjacent to the failed tube as shown by red arrow marks and (b) schematic diagram of a sampling section of the as-received boiler tube.

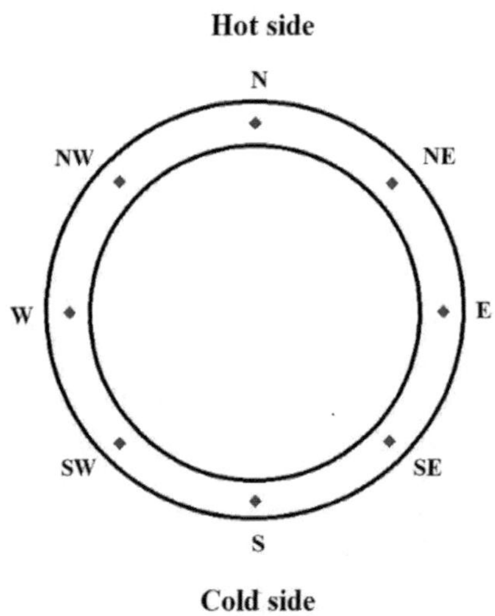

Fig. 3. The location of hardness measurement points.

3 Results and Discussion

The chemical composition of boiler tube material is in compliance with both DIN 17175 St 35.8 and ASTM A192 of the seamless steel tubes for boiler industry where the C has a maximum value of 0.17%. The hardness profile was presented in the Figure 4 below. The results were then compared with standard specification of ASTM A192 (HRB 77). In overall, the hardness level of each sections and directions are significantly inconsistent. The highest hardness is recorded on G sample (HRB 82.4 ± 1.6) which is at the reducer section. It is above the standard specification of ASTM A192. A significant uneven hardness also observed along the circumference of the tube cross-section. At certain direction, the hardness of vertical cross-section is slightly higher than the standard specification of ASTM A192 especially on sample D and H. This proved that there are significant microstructural changes in each vertical sections and the directions of the boiler tube wall due to the non-uniform of the heat distributions along the boiler tube from top to the bottom.

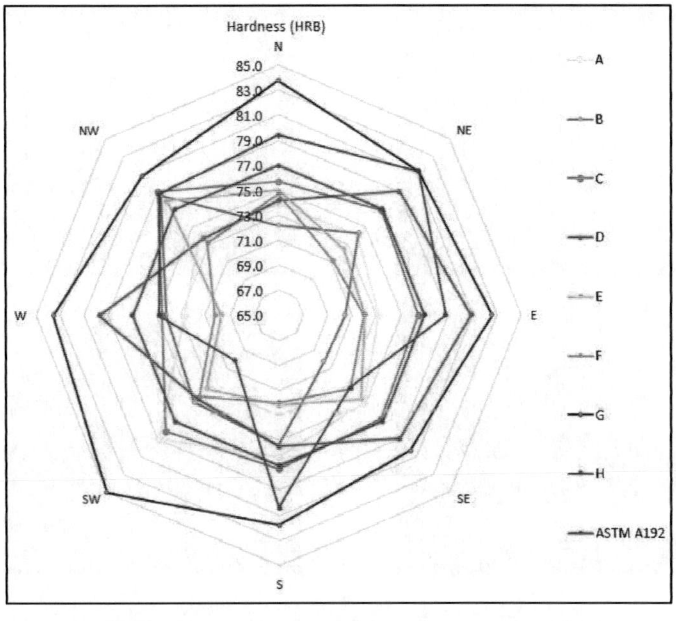

Fig. 4. Hardness profile (HRB) on as-received boiler tube for every vertical section

Note: *The circumference of the tube cross-section on 8 points namely north (N), northeast (NE), east (E), south-east (SE), south (S), south-west (SW), west (W) and north-west (W). It should be noted that the front tube surface facing the fire flame (N direction) is considered as a hotside whereas the back onto tube surface (S direction) is marked as the cold side.*

The microstructure of boiler tube is found to be degraded in term of graphitization and spheroidization of pearlite. Figure 5 shows the degradation begins when iron carbide particles in the ferrite matrix decompose into relatively small graphite nodules (A), growth to be globular graphites (B) followed by coarsening of carbides (C). The significant graphitization is depicted by the planar distribution of graphite nodules occurred at (D) section that located near the weldment. The chain graphitization is detected on (A) section.

IOP Conf. Series: Materials Science and Engineering **701** (2019) 012042 doi:10.1088/1757-899X/701/1/012042

Figure 6 shows the enlarging of graphitization from nodular (E) to globular graphite (F). The spheroidization of pearlite also detected in the (F) metal matrix as depicted by the sphere black dot. It occur when the shape of pearlite was migrated from an irregular to a sphere left behind the ferrite ductile matrix. The most extreme graphitization occur on reducer section (G) followed by small radius tube (H). The graphitization occurs at strange microstructure and a higher level of stress region at (G) and (H) sections. These microstructure observation were found to be consistent with the hardness profile in Figure 4. The hardness results can be summarized to be in the range of HRB 73 and HRB 83, whereby the average hardness from the ASTM A192 is HRB 77.

Graphitization is sort of cavity that contains nodules graphite in it (Figure 7a). This cavity is the root cause to the point of weakness of the steel tube matrix. The resulting spheroidite structure is a microstructure that contains sphere-like cementite particles. Spheroidite structure is easy to be ground and polished because of its ductile properties as compared to graphite nodules (Figure 7b).

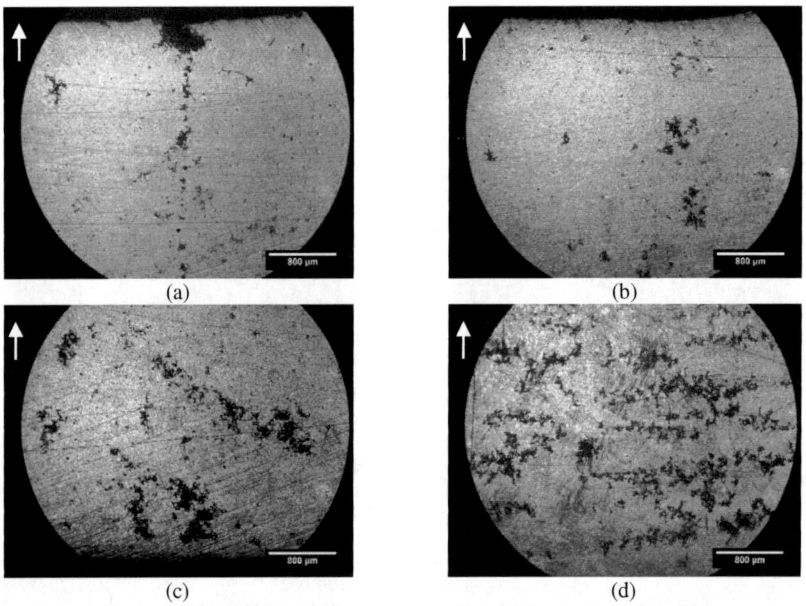

(a) (b) (c) (d)

Fig. 5. Optical metallographs on sample sections at hot side N direction (↑=shell side): (a) A, (b) B, (c) C & (d) D.

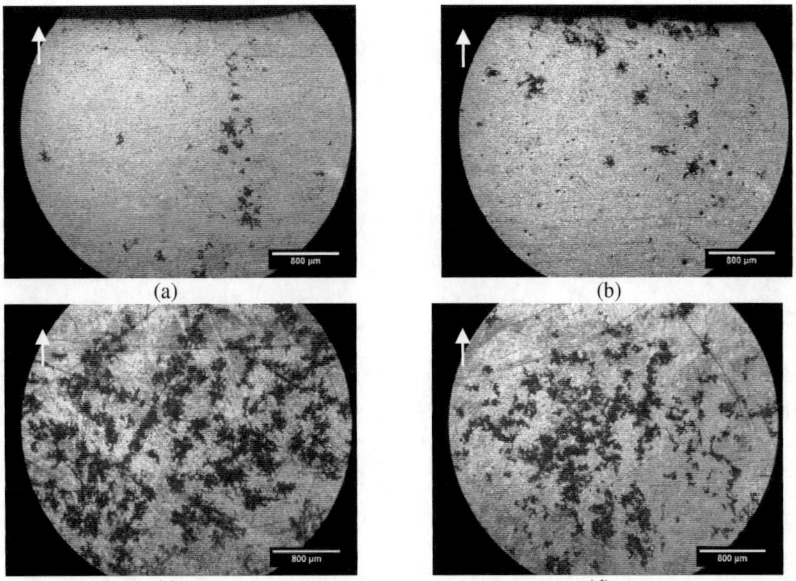

Fig. 6. Optical metallographs on sample sections at hot side N direction (↑=shell side): (a) E, (b) F, (c) G & (d) H.

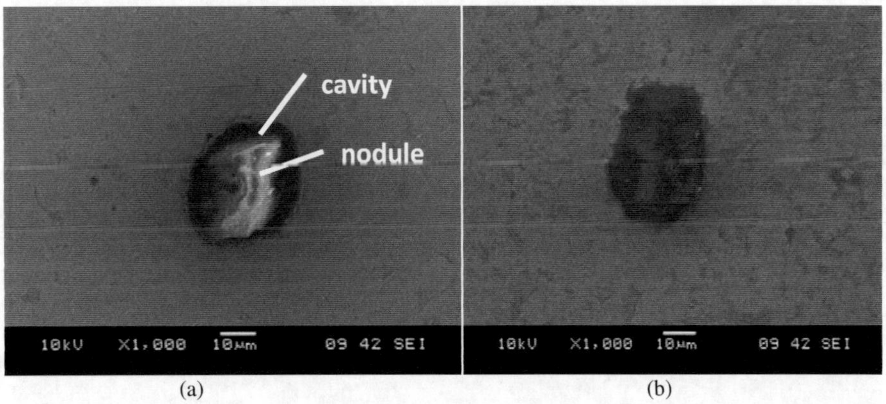

Figure 7: SEM micrographs on (a) graphite nodule and (b) spheroidite of pearlite structure.

4 Conclusion

The boiler tube is in compliance with either DIN 17175 St 35.8) or its equivalent (ASTM A192). The microstructural degradation of tube boiler materials is dominated by graphitization rather than spheroidation of pearlite. Generally, graphitization is applicable at 450°C, but with high-pressure of water flow (90 bar) and internal stress on the tube material due to the welding process and the formation of the reducer, the boiler operating temperature is sufficient to produce the graphitization itself. Graphitization forming cavities containing the structure of the carbide iron graphite is uniformly dispersed in the ferrite matrix rarely causes failure. Nonetheless, the most dangerous is when chain and accumulated graphitization occur because the cavities that covered the graphite facilitates the cracking propagation due to the presence of stresses as described above. Different hardness in each vertical section and tube direction is due to the different microstructure changes lead by graphitization.

The author would like to acknowledge the support from the Prescott Engineering & Services Sdn Bhd, Pasir Gudang, Johor, Malaysia. Thanks to the School of Materials Engineering & Universiti Malaysia Perlis for providing a special financial support for enabling the publication of this manuscript.

References

1. I. U. Perez, T. L. Silveira, T. F. Silveira, H. C. Furtado, J. Fail. Anal. and Preven. E **11** (2011)
2. N. Nutal. C. J. Gommes, S. Blacher, P. Pouteau, J. P. Pirard, F. Boschini, K. Traina, R. Cloots, Image Anal Stereol. E **29** (2010)
3. G. R. Lobley, W. L. Al-Otaibi, Adv. Mats. Res. E **41-42** (2008)

IOP Conf. Series: Materials Science and Engineering **701** (2019) 012043 doi:10.1088/1757-899X/701/1/012043

Influence of filler content and mixing time on the properties and morphology of kenaf/polyurethane foam composites

Nur Suhaili Mohd Soberi[1], A. R Rozyanty[1], Firuz Zainuddin[2] and A.F Osman[1]

[1]Center of Excellence Geopolymer and Green Technology (CEGeoGTech), School of Materials Engineering, Universiti Malaysia Perlis, 01000 Kangar, Perlis, Malaysia.
[2]School of Materials Engineering, Kompleks Pusat Pengajian Jejawi 2, Taman Muhibbah, Universiti Malaysia Perlis, 02600, Jejawi, Arau, Perlis, Malaysia.

Abstract. Kenaf/polyurethane foam composites were prepared at percentage of 2, 4, 6 and 8 wt% of kenaf fiber. Free rising polyurethane foam was prepared at constant (NCO/OH) ratio 1.5:2, while water was used as blowing agent. The mixture was stirred at 400 rpm using mixing times of 20 and 25 seconds. Rise time and hardness of the composites was determined. From the results obtained, rise time and hardness of polyurethane foam composites were reduced at 25s mixing time. However, as kenaf core fiber increase, the rise time decreased and the hardness of the foam composite increased. The Scanning Electron Microscopy (SEM) analysis evidenced the presence of regular cell shape and uniform cell size in the kenaf/polyurethane foam composites. The findings indicate the potential of using kenaf fiber to enhance the properties of the polyurethane foam.

1 Introduction

For over half a century, flexible polyurethane foams has been used in various applications such as footwear, construction [1], cushioning [2], furniture, packaging, bedding and automotive interior [3]. The polyurethane foams are famous with some of its favourable properties like low production cost [4], excellent abrasion resistance and hardness. The main advantage of foams rather than other materials are its light weight, which leads to the production of low density products [5]. These properties results from the complex generation of cellular structure that is made up of a solid and gas phase mixture [6]. Mostly, the basic ingredients to produce polyurethane foam are polyol and polyisocyanate via addition polymerization [7]. Polyurethane foam possesses cellular structure that was created by the expansion of either physical or chemical blowing agent. Usually chemical blowing agent such as water was used because it gives less effect to the environment rather than physical blowing agent such as chlorofluorocarbon (CFC) and hydrofluorocarbon (HCFC) [8]. The blowing agent, i.e. water works by generating carbon dioxide from its reaction with isocyanate and subsequently expand the polyurethane [9].

*Corresponding author: rozyanty@unimap.edu.my

Polyurethane foams can be tailored into different cell characteristics, known as closed and open cell structure [10]. Open-cell foams are usually flexible while closed-cell foams are generally more rigid [11]. Recently, researchers have investigated the potential of filler addition into polyurethane foam to improve its physical, mechanical and chemical properties [12][13] [14][15][16][17].

In the production of polymer composites, natural fibers were proposed to substitute synthetic fiber as filler material due to its advantages such as reduce cost, less abrasiveness to equipment, lower density and less irritation to skin and respiratory [18]. In addition, they are available in large amounts, renewable and biodegradable [19]. Locally available natural fiber such as kenaf (Hibiscus cannabinus L) can be used as the filler in the polyurethane foam. The kenaf plant consists of a single, straight and branchless stalk. While the stalk is made up of an inner woody core and outer fibrous bark [20], the inner woody core which is the kenaf core fiber comprises of 60% of the stalk's dry weight. Kenaf has been drawing world attention due to its application, such as paper products [21], composites, absorbent and building materials. Realizing on the possibility to exploit and derived products from kenaf, the National Kenaf Research and Development Program has been formed to develop kenaf as a new industrial crop in Malaysia [22]. In addition, kenaf plantations in Malaysia can be harvested three times in a year, which can be considered as the renewable source.

With regards to the development of well-optimized foam composite, filler content and its dispersion quality in polyurethane matrix are of great importance to improve the cellular structure of the resultant foam composites. Cellular structure of polyurethane foam composed of struts, nodes and cell wall [23]. However, fillers can cause various structural effects and lower the mechanical properties of the foam if the dimension of the filler is bigger than the cell's struts and nodes. In addition, the dispersion degree of the filler also gives significant effect on the properties of the polyurethane foam composites. Therefore, in this study, different mixing time was employed to optimize the fiber dispersion in polyurethane foam. In addition, the effects of filler loadings on the cellular structure and foam composite properties were also investigated. Herein, we report the influence of both mixing time and filler loading on rise time, density, hardness and morphology of the polyurethane foam composites.

2 Experimental

2.1 Materials

Polyethylene Glycol (PEG) and 4, 4-diphenylmethane diisocyanate (MDI) was supplied by Dr. Rahmatullah Holdings Sdn. Bhd. While kenaf (Hibiscus cannabinus L) core fiber was provided by Lembaga Kenaf Tembakau Negara (LKTN). Distilled water was used as blowing agent in this research. Kenaf core was dried in an oven at 80°C overnight to remove its moisture content. Then, the kenaf core was ground and sieved into 63 μm size. The percentage of kenaf fiber was varied from 2% to 8%.

2.2 Preparation of kenaf/polyurethane composite

Kenaf/polyurethane foam composites were prepared by using free rising method. Two different mixing times were employed, which were 20s and 25s. Firstly, PEG was mixed with three drops of water and stirred by using a mechanical stirrer at speed of 400 rpm Then, kenaf fiber was added and stirred until a homogenous mixture was obtained. Next, MDI was added into the mixture and stirred continuously for 20s and 25s. The mixture was cast into an open mould with dimension of 11.50cm× 6cm×6cm, while the rising time of

IOP Conf. Series: Materials Science and Engineering **701** (2019) 012043 doi:10.1088/1757-899X/701/1/012043

the polyurethane foam composites was recorded. The foam composites were left to cure at room temperature. Table.1 represents the formulation used to produce polyurethane foam composites.

Table 1. Formulation for polyurethane foam filled with kenaf fiber based on ratio 1.5:2.

Amount of kenaf fiber (%)	MDI (g)	PEG (g)	Water (Drops)	Mixing time (Sec)	
0	37.5	50	3	20	25
2	37.5	50	3	20	25
4	37.5	50	3	20	25
6	37.5	50	3	20	25
8	37.5	50	3	20	25

2.3 Hardness Test

Hardness test on soft foam was conducted based on ASTM D2240 by using SHORE A Durometer. This test measures the penetration of a specific indentor into polyurethane foam under specific conditions of force and time. The measurement was taken from five spots of the specimen.

2.4 Scanning electron microscopy study

The model used is JEOL JSM 6460 Scanning Electron Microscope (SEM). The samples were cut into 0.5cm X 0.5cm X 0.5cm. Then, the samples were mounted on an aluminium stub by using double-sided tape and then coated with a very thin layer of platinum by using Auto Fined Coater to attain observation of the samples. The samples were analyzed under 50 and 300 X magnifications.

3 Results and discussion

Fig.1 shows the effects of filler loading and mixing time on the rising time of polyurethane foam. Based on the result, the rising time of polyurethane foams becomes shorter as the amount of kenaf fibers increases. This may due to the higher amount of lignin content in kenaf fiber rather than other natural fiber. Lignin was responsible to provide free OH groups to react with the isocyanate and consequently increase the rising rate Furthermore, addition of kenaf fiber in the polyurethane foam resulted in faster cell nucleation and leading to more uniform cell sizes within the foam [3]. Besides that, rising time of the polyurethane foam was also affected by mixing time used in the preparation of the foam composites. The longer the mixing time, the shorter the rising time taken by the polyurethane foam because the filler dispersion was improved throughout the polyurethane foam matrix. In fact, at 25 seconds mixing time, the foam already started to rise in the mold.

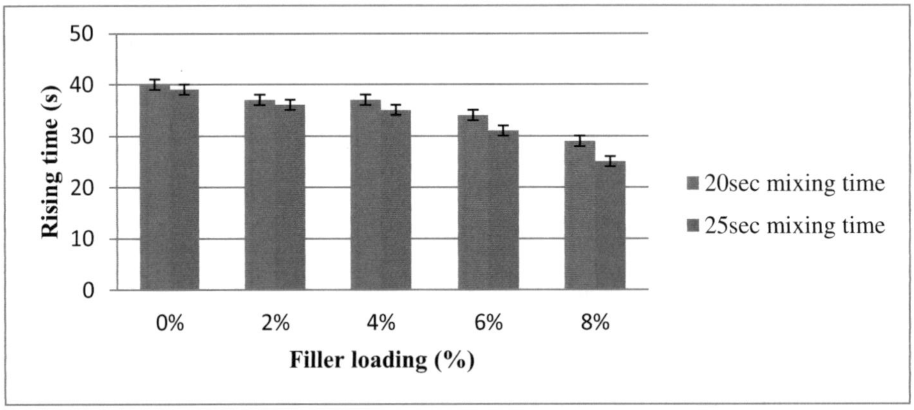

Fig.1: Effects of filler loading and mixing time on rising time of polyurethane foam.

Hardness is defined as the resistance to indention [24] Fig. 2 represents the effect of filler loading and mixing time on the hardness of polyurethane foam composites. Based on the results, hardness of polyurethane foam composites with addition of filler show 134% increments at 20s mixing time, while only 110.53% increments at 25s mixing time due to randomly cell size formation for higher gas releases in foaming process. Random cell size in foam composites will lower the hardness properties of the composites. From the result obtained, hardness of the polyurethane foam shows increased in trend as filler was added into it. As the amount of filler loading in the polyurethane foam increases, the foam cell size and cell structure become smaller which offer more resistance to compression. Furthermore, the addition of micro-sized fillers can increase the hardness of the foam composites due to the higher total surface area for a given particle loading. This can contribute to efficient stress transfer mechanism in the polyurethane foam structure [25].

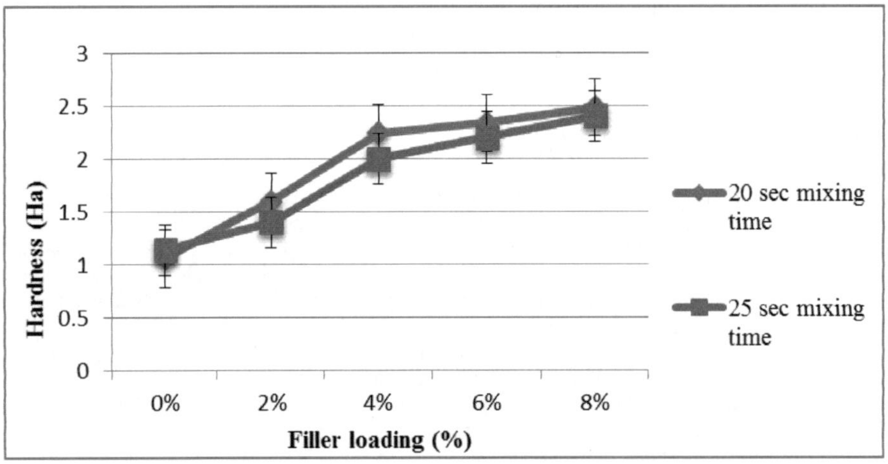

Fig.2: Effects of filler loading and mixing time on polyurethane foam composite hardness.

SEM analysis was carried out to investigate the morphology of pure polyurethane foam and polyurethane foam composites and simultaneously support the findings of hardness and density tests. Micrograph in Fig. 4 shows that thicker struts were formed on the cell walls of pure polyurethane foam. This is due to bigger cell size formed that take large surface areas of cell walls obtained [26]. It can be seen that the estimated cell size diameter of pure polyurethane foam is bigger than others which is about 1867μm. Based on the results obtained in Fig.4 (A) and (B), foam with bigger cell size shows lower hardness. However, the hardness can be improved with the addition of kenaf fiber which is able to act as nucleating agents in the composites. In addition, uniform cell size within the foam can be formed with the present of negligible nucleation energy and the presence of preexisting microvoids [27]. Fig.4 (B), (C) and (D) proved that the foam cell size become smaller and more uniform with the addition of filler in the polyurethane foam.

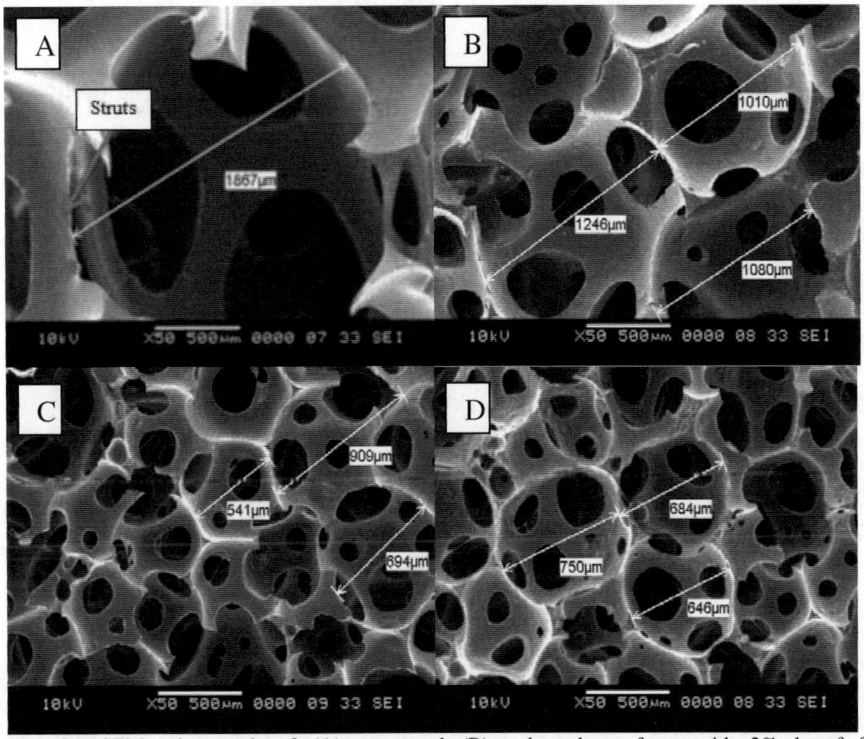

Fig. 3 : SEM micrograph of (A) pure and (B) polyurethane foam with 2% kenaf fiber (C) polyurethane foam with 4% kenaf fiber and (D) polyurethane foam with 6% kenaf fiber at (50X) mag at 20 sec mixing time.

4 Conclusion

In this work the effects of incorporation of kenaf fiber in polyurethane foam composites with different mixing time were investigated on rise time, hardness and morphological studies. From the studies, it show that polyurethane foam with shorter rising time have faster curing time. In addition and hardness of polyurethane foam increased as the amount of filler in the foam composites increased. Furthermore, the SEM micrographs showed polyurethane foam with regular and uniform cell size were obtained as the amount of kenaf fiber increases. As a conclusion, kenaf fiber plays a crucial role in determining the properties of the polyurethane foam

The authors would like to thank University Malaysia Perlis (UniMAP) for supporting this research and Lembaga Kenaf Tembakau Negara (LKTN) for raw materials.

References

[1] H.M Nakhoda , Dahman Y). Polym Bull **73(7)**:2039–2055. (2016)
[2] M. A Azmi , MFC Yusoff , H.Z Abdullah , M.I Idris. Int J of Integ Eng **4(1)**:11–15. (2012)
[3] GK Latinwo, DS Aribike , AA Susu , SA Karee . Nature and Sci **8(6)**:23-26. (2010)
[4] AB Frances, MVN Banon . Mater Sci and Eng **64:**1–6. (2014)
[5] ZM Ariff, NAA Rahim, LC Low. In Proceeding of polymer processing society 24th Annual Meeting. Italy. (2008)
[6] P Kekicheff, D Mouhamadou, G Dalongeville, C Gauthier, T Roland. Macromol Symp 369:56–66. (2016)
[7] P Li , Y Tao , QS Shi Bioresources . **9(2)**:2681–2688. (2014)
[8] M Modesti , V Adfuani , F Simioni . Polym Eng and Sci **40(9)**: 2046–2057. (2000)
[9] KH Cheo , DS Lee, WJ Seo, WN Kim. Polym J **36(5)**:368–373. (2004)
[10] C Webber , M Nar, N Dsouza . Polym Eng and Sci:1–13. (2014)
[11] E Aram, S Mehdipour-ataei. Inter J of Polym Mater **65 (7)**: 358-375. (2016)
[12] Y.M Daud, K Hussin, AF Osman, CMR Ghazali, MM Al Bakri Abdullah, AV Sandu, Materiale Plastice, **54 (3)**, pp. 543-545(2017)
[13] Y.M Daud, K Hussin, AF Osman, CMR Ghazali, MM Al Bakri Abdullah, Key Engineering Materials, **673**, pp. 55-63(2016).
[14] F.F Zainal, K Hussin, A, Rahmat, M.M.A.B Abdullah, S.R Shamsudin,. AIP Conference Proceedings, 1756, art. no. 040001(2016)
[15] N.S Azhar,.D.M., F.F Zainal,., M.M.A.B.Abdullah, AIP Conference Proceedings, 2129, (2019)
[16] M.F.A Hashim, M.M.A.B Abdullah, A.V Sandu, A Puskas, Y.M Daud, F.F Zainal, M.A Faris, Hasri, Hartati. IOP Conference Series: Materials Science and Engineering 572 (1) (2019).
[17] Y Mat Daud, A. Fazlina Osman., F Farhana Zainal, M Firdaus Abu Hashim, , H.M Si, H.S Si, N Soo Li,S Arzuria Adnan,. IOP Conference Series: Materials Science and Engineering , 551 (1) (2019)
[18] S Shibata, Y Cao, I Fukumoto. Polym testing 25:142-148. (2006)
[19] S Ochi. Mechanics of Mater: 40, 446–452. (2008)
[20] MR Ishak, Z Leman, SM Sapuan, AMM Edeerozey, IS Othman. Mater Sci and Eng: 1–6. (2010)
[21] R Shmulsky, PD Jones .. Wiley Blackwell, UK. (2011)
[22] HPSA Khalil, NL Suraya. Bioresources ***6* (2)**: 1122–1135. (2011)
[23] V Yakushin, L Belkova, I Sevastyanova . In ECCM15-15TH European Conference On Composite Materials (pp. 1–8). Venice, Italy. (2012)
[24] M Thirumal , D Khastgir, KNSingha, BS Manjunath. & Y.P. Naik,. Applied polym sci: 108, 1810-1817. (2007)
[25] J Njuguna, F Silva, S Sachs In: Lin T (ed) Production, properties and functional applications. UK, pp. 401-434. (2011)
[26] CW Shan, MI Idris, MI Ghazali . Applied Physics and Mathematics: 2(2). (2012)
[27] I Javni, K Song, J Lin, ZS Petrovi .J of Cellular Plastic: 358-372. (2011)

Electrochemical migration and corrosion behaviours of SAC305 reinforced by NiO, Fe_2O_3, TiO_2 nanoparticles in NaCl solution

Norinsan Kamil Othman[1*], *Fakhrul Rifdi Omar*[1,2] & *Fakhrozi Che Ani*[3]

[1] Centre for Frontier Sciences, Faculty of Science and Technology, Universiti Kebangsaan Malaysia, 43600 UKM Bangi, Selangor Darul Ehsan, Malaysia
[2] Maktab Rendah Sains MARA Kuching, Jalan Sultan Tengah, 93050 Kuching, Sarawak, Malaysia
[3] Western Digital Sdn Bhd, Penang, Malaysia

Abstract. Pb-free solders incorporated with nano-sized particles have been identified, as potential Pb-free nanocomposite solders that could provide higher microstructure stability and better mechanical properties than the conventional solders. Nowadays, the miniaturizations of electronic devices increase in their usage have increased the risk of failure due to ion migration and corrosion phenomena. This research investigates the effects the electrochemical migration (ECM) and corrosion behaviour of Sn-3.0Ag-0.5Cu (SAC305) reinforced by NiO, Fe_2O_3 and TiO_2 nanoparticles in 0.1M NaCl solution. The SAC305 solder alloys were doped with different percentages of NiO, Fe_2O_3 and TiO_2 at nominal percentages of 0.01, 0.05 and 0.15 wt% in producing nanocomposite solder paste. Then, the solder paste has flowed in the reflow oven at the temperature of 240°C. After that, the Water Drop Test (WDT) was performed using 0.1 M NaCl solution to investigate the ion migration behaviour and Potentiodyamic Tafel Polarization technique to determine the corrosion behaviour on the solder alloys. In general, the results showed the mean-time-to-failure (MTTF) of SAC305 solder alloy increased when added of the nanoparticles of NiO, Fe_2O_3 and TiO_2. Besides, SAC305-TiO_2 has the longest MTTF in 0.1 M NaCl solution followed by SAC305-Fe_2O_3 and SAC305-NiO nanoparticles. This results is believed due to the presence of nanoparticles as the alloying element in the solder that hinder the process of ECM to suppress the growth of the dendrites accelerating for the short circuit occurs and causes the failure of the IPC-24 test board. The potentiodynamic polarization curve results exhibit the SAC305 with nanoparticles decreased the corrosion rate compared with SAC305 solder alloy only. SAC305 with NiO and TiO_2 have slightly lower corrosion rates compared with Fe_2O_3 nanoparticles. Therefore, SAC305 reinforced by TiO_2, Fe_2O_3 and NiO nanoparticles exhibited excellent ECM and corrosion resistance behaviours in the 0.1 M NaCl solution.

* Corresponding author: insan@ukm.edu.my

IOP Conf. Series: Materials Science and Engineering **701** (2019) 012044 doi:10.1088/1757-899X/701/1/012044

1 Introduction

Soldering offers important technology on microelectronic packaging industries. It allows electrical current to flow from one point to another and become a supporter of the electrical components. Recently, the lead-free, Sn-Ag-Cu solder alloy has been commercially used as an alternative to Sn-Pb solder due to the vital issue on hazardous effect [1–4]. This Sn-Pb solder is due to poisonous qualities of lead which is harmful to human health and also to the environment [5].

Tin based solder alloy becomes the replacement for lead solder alloy [4]. Tin containing binary and ternary alloying elements has been most directed towards it [6]. Eutectic tin-silver-copper has a melting point around 216°C, which is 30°C little higher than that of eutectic lead-tin (T_m = 183°C), The addition of silver and copper reduces about 16°C of the melting point of tin (232°C). Tin-silver solders are already used with success in high-temperature applications [7].

Others researcher had a focus on some lead-free content solders, which may cause high risk in electronic devices due to electrochemical migration (ECM) failure phenomenon [8]. ECM commonly occurs in all electronic devices in the electronics industry. This phenomenon involves the dendritic growth [11, 12]. It can be led to a short circuit to occur in the electronic device issue is related to the susceptibility of SAC305 to corrosion [11].

The growing of miniaturization of the electronic system causes the use of Pb-free solder materials have led to increase many challenges in electronics installation. These challenges are because of the higher density packaging lead the components event smaller, closer spacing, thinner metallic parts and higher gradient of the electrical field [13]. Furthermore, the devices sometimes have to service in high humidity, salty and polluted environment. Due to the advances in technology, this led to electronic devices are becoming more and more susceptible to corrosion failure, even a small amount of weight loss [14]. Therefore, in this paper, the SAC305 reinforced by NiO, Fe_2O_3 and TiO_2 nanoparticles solder paste toward ECM and corrosion behaviours were investigated by using the Water Drop Test (WDT) and electrochemical polarization technique in 0.1M NaCl solution that simulating seawater environment as a medium.

2 Materials and Methodology

The lead-free SAC305 solder alloy containing 96.5%Sn, 3%Ag and 0.5%Cu were doped in different percentages of NiO, Fe_2O_3, TiO_2 nanoparticles at a nominal percentage at 0.01, 0.05 and 0.15wt% to produce nanocomposite solder paste. Then, all nanoparticles (Aldrich, ≥ 98 per cent trace metal basis, particle size: <25 nm) were mixed with the 96.5Sn–3.0Ag–0.5Cu solder paste (SAC305; Alpha OM-353) using a mechanical stirrer brand Fritsch Planetary Mill PULVERISETTE 5 for approximately 10 min (300 rpm) to achieve homogeneity before the assembly. The methodology of preparing nanocomposites can be referred at [15,16,17]. Then, the nanocomposite solder pastes were then well printed in a standard comb pattern with horizontal lines on the conductor strip is 0.4mm and the gap size about 0.2mm. After the printing with the solder paste, the test board is heated using *Desktop free-Lead Reflow Oven TYR108C*. The test board was designed according to IPC-B-24 test board, as shown in Fig. 1.

Before the WDT, the test board, voltmeter, resistor in series and power supplier have been connected, as shown in Fig. 2. During WDT, 15µml 0.1M sodium chloride (NaCl) solution was dropped by micropipette onto four comb patterns and 10 Volts DC was applied to each of the samples. During the WDT, short circuit formation is detected by a voltmeter connected with one resistor in series through average Time-To-Fail (MTTF). Par value failure is 0.1 VDC and resistor value (R) is 1 kOhm. After the WDT, the mean-time-to-failure (MTTF) of each sample should be computed using this formula [1]:

$$\text{MTTF} = \frac{y1 + y2 + y3}{n} \tag{1}$$

where y is time to failure and n the is number of time taken. After the WDT, the test boards were cut to dimension size about 1.0 x 1.0 cm^2 in order to investigate its surface morphology and corrosion product by using optical microscopy.

Fig. 1. The IPC-B-24 Standard comb pattern test board.

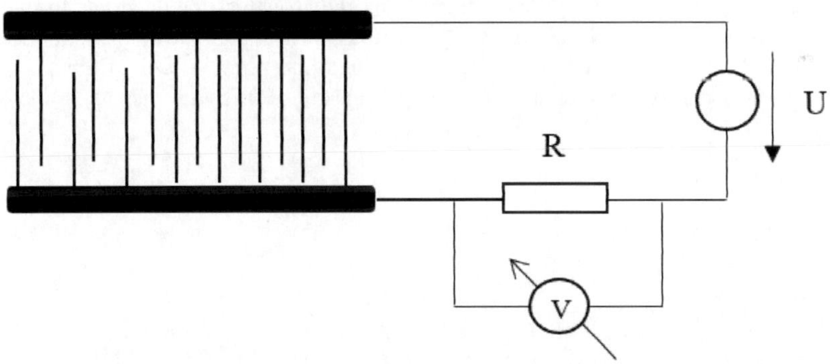

Fig 2. Schematic diagram of the test platform of Water Drop Test (WDT)

3 Results and Discussion

WDT was performed on SAC305 solder paste alloy and SAC305 reinforced nanoparticles solder alloys that have been printed on the test board by using 0.1M NaCl solution with the power voltage. The droplets volume and the resistance was set constant. The voltage value drops to 0 V and the time-to-failure was recorded. Mean-time-to-failure (MTTF) was the average value of the measured failure times. The test board will be experienced electromigration (ECM) led the short circuit occurs. Average MTTF can be obtained, as listed in Table 1.

Table 1. MTTF data of SAC 305 solder paste reinforced by TiO_2, Fe_2O_3 and NiO nanoparticles.

Types of Solder Alloy	wt% of nanoparticles	Mean-Time-To-Failure (s)
SAC305	0	857
SAC305-TiO_2	0.01	1168
	0.05	1389
	0.15	1548
SAC305-Fe_2O_3	0.01	1130
	0.05	1345
	0.15	1526
SAC305-NiO	0.01	1071
	0.05	1285
	0.15	1498

In general, The MTTF reveals that the addition of nanoparticles in SAC305 increased the time to failure due to short circuit compared to SAC305 solder alloy in NaCl solution. The MTTF results indicate that SAC305-TiO_2 has the most extended values, which is 1548s followed by SAC305-Fe_2O_3 (1526s) and SAC305-NiO (1498s) in the range of 0.15 wt% of nanoparticles. Meanwhile, SAC305 solder alloy shows faster MTTF valued at 857s. This phenomenon can be attributed due to the chemical reaction of solder paste alloy and NaCl solution with electrical potential as know as ionic migration or electromigration [. This effect shows the SAC305 solder is believed to have ECM which metal ions migrate from anode to cathode cause the voltage drops to zero and continue to fail the overall devices. However, by addition of nanoparticles TiO_2, Fe_2O_3 and NiO into SAC305 solder alloy hinder the process of ECM from suppressing the growth of the dendrites for the short circuit occurs and causes the failure of IPC-24 test board. Fig. 3 shows that summarize of MTTF of SAC305 reinforced by nanoparticles in different wt% also exhibits a similar pattern where increasing percentages of nanoparticles until 0.15wt% increasing the MTTF values.

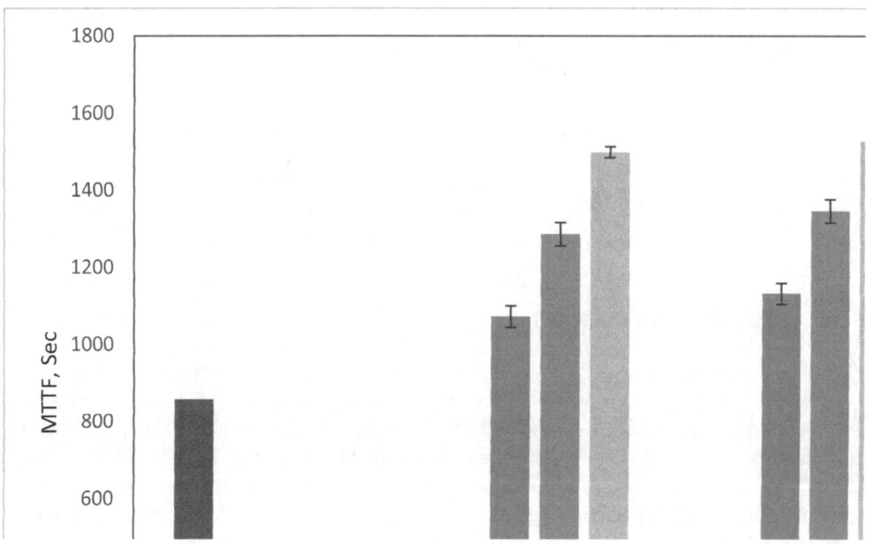

Fig 3. Mean Time for Failure of SAC305 reinforced of NiO, Fe_2O_3 and TiO_2 nanoparticles

In this study, the morphology and microstructure of SAC305 solder paste with addition of nanoparticles were investigated. Fig. 4 shows the optical micrograph of SAC305 solder paste and SAC305 reinforced nanoparticles after WDT experienced the formation of dendrites growth. As can be seen in Fig. 4a, severe dendrite formation found in SAC305 solder compared to SAC305 with doped nanoparticles which have less severe of dendrites growth (Fig.4b,c,d). It was found than nanoparticles doping increased the MTTF values. This outcome is believed that the dendrites formation have been suppressed by precipitated of nanoparticles in the solder paste.

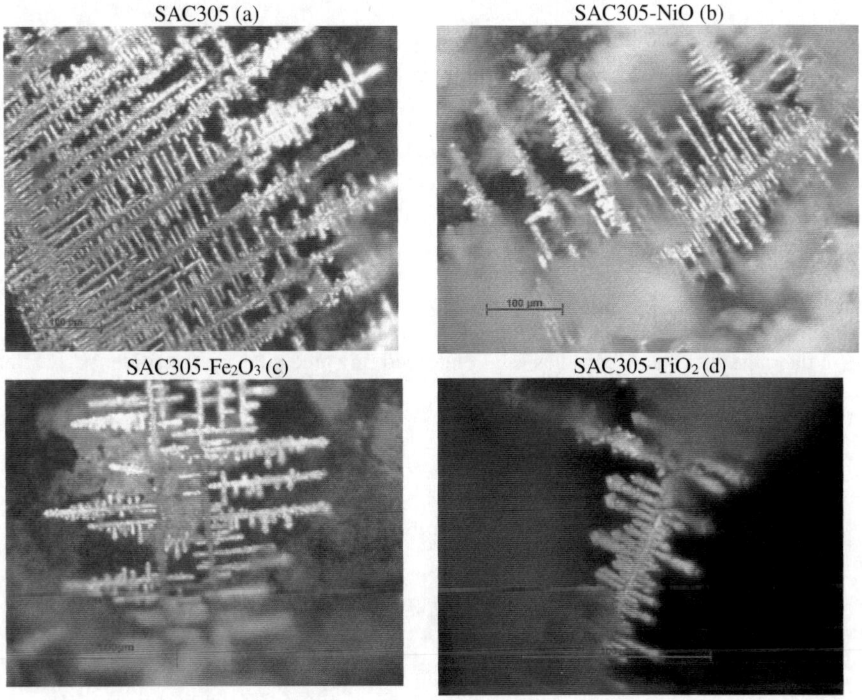

Fig. 4 Optical micrographs of SAC305 and SAC305 reinforced of nanoparticles solder paste alloy after WDT in 0.1M NaCl solution.

Potentiodynamic polarization curves for SAC305 and SAC305 reinforced nanoparticles for 0.05%wt in a solution containing 0.1M NaCl extrapolates in Fig. 5. The corresponding polarization parameters such as corrosion current (I_{corr}), corrosion potential (E_{corr}), passivation current (i_{corr}) values obtained from the polarization curves listed in Table 2. The Tafel fit using Gamry Echem Analyst Version 7.07 were used to determine the corrosion parameter. At cathode region corresponds to reduction reaction of dissolved O_2 to produce hydroxyl ions (OH^-). The reaction can be illustrated as follow [13,18]

$$O_2 + 2H_2O + 4e^- \rightarrow 4OH^- \quad (2)$$
$$2H_2O + 2e^- \rightarrow H_2\,(g) + 2OH^- \quad (3)$$

The H_2(g) will be released at cathode from the reduction reaction. Nevertheless, the oxidation at the anode is rather complicated. The reaction at the anode can be described as follows [13,18,19]:

$$Sn + 2OH- \rightarrow Sn(OH)_2 + 2e^- \quad (4)$$
$$Sn + 4H_2O \rightarrow Sn(OH)_2 + 4e^- + 4H^- \quad (5)$$

Later, $Sn(OH)_2$ and $Sn(OH)_4$ will be suffered from dehydration and each of them form SnO and SnO_2 according to the following reactions [13,18, 20]:

$$Sn(OH)_2 + 2OH^- + 2e^- \rightarrow Sn(OH)_4 \quad (6)$$
$$Sn(OH)_2 + 2OH^- \rightarrow SnO + 4H_2O \quad (7)$$
$$Sn(OH)_4 + SnO_2 + 2H_2O \quad (8)$$

Due to the dissolution $Sn(OH)_2$, Sn^{2+} ions released and move to cathode region. Then, these ions arrived at the cathode, ECM deposition occurs. Deposition of this cathode will form a dendrite growth towards the anode led to short circuit after dendritic growth reached to anode [21].

The E_{corr} values after addition of 0.05%wt nanoparticles shifted towards more negative in from -178 mV to more than -278 mV. As it can be noticed, both cathodic and anodic reactions of SAC305 reinforced nanoparticles of NiO, Fe_2O_3 and TiO have reduced the corrosion rates about 0.189, 0.131 and 0.126 mpy, respectively from 1.431 mpy on SAC305 only. This result indicates that the addition of nanoparticles into SAC305 solder somehow is reduced E_{corr}, I_{corr} and corrosion rate values. SAC305-TiO_2 exhibits a lower corrosion rate followed by SAC305-Fe_2O_3 and SAC305-NiO. This influence is most likely due to the presence of nanoparticles as a reinforced tiny precipitate in the solder paste retard the corrosion attack from aggresive ion Cl^- [22].

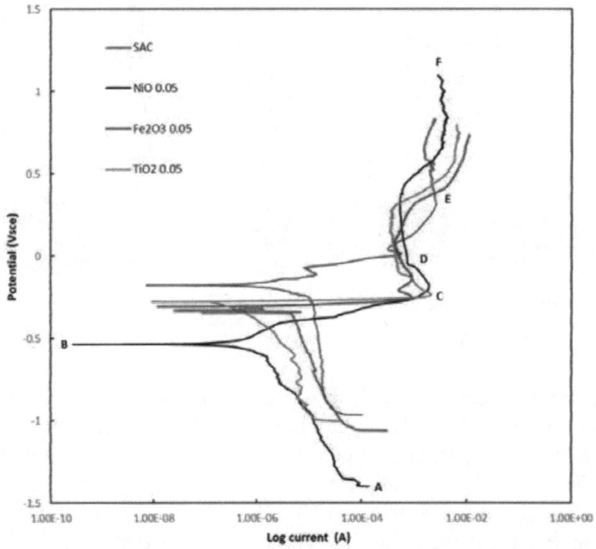

Fig 5. Potentiodynamic polarization curves of SAC305, SAC305-0.05NiO and SAC305-0.05Fe_2O_3 and SAC305-TiO_2 in 0.1M NaCl solution.

Table 2. Potentiodynamic polarization parameters for SAC alloys solder in 0.05wt% nanoparticles in 0.1M NaCl solution

Solder	E_{corr} (mV)	I_{corr} (mA/cm2)	Beta a (V/decade)	Beta c (V/decade)	Corrosion rate (mpy)
SAC305	-178	12.640 x 10^{-6}	297 x 10^{-3}	3.317	1.432
SAC305-NiO	-535	1.604 x 10^{-6}	356 x 10^{-3}	563 x 10^{-3}	0.189
SAC305-Fe_2O_3	-309	1.161 x 10^{-6}	243 x 10^{-3}	433 x 10^{-3}	0.131
SAC305-TiO_2	-278	1.113 x 10^{-6}	222 x 10^{-3}	700 x 10^{-3}	0.126

4 Conclusion

The SAC305 reinforced composite nanoparticles by NiO, Fe_2O_3 and TiO_2 were investigated to study the behaviour of electromigration dan corrosion in 0.1M NaCl solution.

1. The results show that the MTTF of SAC305-TiO_2 has slowest the mean-time-to-failure, followed by SAC305-0.15Fe_2O_3 and SAC305-NiO compared to SAC305 attributed to ECM after WDT.
2. The addition of nanoparticles in solder paste suppressed the formation of dendrites to growth from cathode to anode, which can short the circuit, ultimately causing the failure of the test board.
3. Potentiodynamic polarization measurement showed a better corrosion resistance of SAC305-TiO_2 followed by SAC305-0.15Fe_2O_3 and SAC305-NiO compared to SAC305 solder paste. Addition of nanoparticles in solder paste enhanced corrosion resistance due to the nanoparticles act as precipitate to hinder the ion migration.

This work was supported by the Ministry of Higher Education, Malaysia under the Fundamental Research Grant Scheme DIP-2018-032 and the authors thanks Western Digital Sdn Bhd for their collaboration in this research.

References

[1] E. Havia, H. Montonen, E. Bernhardt, M. Alatalo, *NEXT Symposium*, Finland (2005).

[2] P. Biocca and C. Rivas, *Proc. Int. Symp. Exhib. Adv. Packag. Mater. Process. Prop. Interfaces* (2008).

[3] W. Loss, “A Study of Lead-Free Wave Soldering,” Am., pp. 7–8.

[4] E. Efzan Mhd Noor, A. Singh, Solder. Surf. Mt. Technol. E **26**, 3, (2014)

[5] J. Y. Jung, S. B. Lee, H. Y. Lee, Y. C. Joo, and Y. B. Park, J. Electron. Mater. E **37**, 8, (2008)

[6] I. Ahmad, A. Jalar, B. Y. Majlis, and R. Wagiran, E **4**, 1 (2007).

[7] S. L. Lewis, *Microstructural Evolution in Tin- Eutectic Alloys*. May 2003 (2003).

[8] B. Medgyes, B. Illes, and G. Harsanyi, Period. Polytech. Electr. Eng. E **57**, 2, (2013)

[9] L. T. F. Mendes, V. F. Cardoso, and A. N. R. da Silva, J. Integr. Circuits Syst. E **6**, 2 (2011)

[10] G. A. Wayman D. Yang, D. D. Bose, A. Lesiak, V. Ledoux, D. Bruun, I. N. Pessah, P.J. Lein Environ. Health Perspect. E **120**, 7 (2012)

[11] C. W. See, M. Z. Yahaya, H. Haliman, and A. A. Mohamad, Procedia Chem. E **19** (2016)

[12] D. Li, P. P. Conway, C. Liu, Corros. Sci. E **50**, 4 (2008).

[13] X. Zhong, L. Chen, B. Medgyes, Z. Zhang, S. Gao, L. Jakab Royal Society of Chemistry. E **7**

(2017).

[14] D. Minzari, M.S.Jellesen, P. Moller, R.Ambat, Corrosion Science. E **53** (2011)

[15] F. Che Ani, A. Jalar, A. A. Saad, C. Y. Khor, R. Ismail, Z. Bachok, M. A. Abas, N. K. Othman, The Inter. J Adv. Manufacturing Tech. E **96**, 1-4 (2018).

[16] F. Che Ani, A. Jalar, A. Aziz Saad, C. Yee Khor, R. Ismail, Z. Bachok, M. Aizat Abas, N. K. Othman, Soldering & Surface Mount Technology. E **30**, 1 (2018)

[17] F. Che Ani, A. Jalar, A. Saad, C. Khor, C. Abas, Z. M. Bachok, N. Othman, Soldering & Surface Mount Technology. E **31**, 2 (2019)

[18] E. M. Salleh, N. K. Othman, A Jalar, Sains Malaysiana. E **47**, 8 (2018)

[19] N. K. Othman, K. Y. Teng, A. Jalar, F. Che Ani, and Z. Samsudin, Mater. Sci. Forum. E **846**, (2016)

[20] C. Dong Zou, Y. Lai Gao, B. Yang, X. Zhi Xia, Q. Jie Zhai, C. Andersson, J. Liu, J. Electron. Mater. E **38**, 2 (2009)

[21] M. T. T. Nguyen, *Reliability Assessment of Ion Contamination Residues on Printed Circuit Board,*" January, (2013).

[22] F. R. Omar, E. Marina Salleh, N. K. Othman, F. Che Ani, Z. Samsudin, *AIP Conference Proceedings* (2019)

Effects of ultrasonication process on crystallinity and tear strength of thermoplastic starch/cellulose biocomposites

Ismail Ibrahim[1,*], *Azlin Fazlina* Osman[1] , and *Tan* Leong Ping[1]

[1]Center of Excellence Geopolymer and Green Technology (CEGeoGTech), School of Materials Engineering, Universiti Malaysia Perlis, Arau, Perlis 02600, Malaysia.

Abstract. The thermoplastic starch (TPS)/cellulose biocomposites were manufactured as green biocomposite films. TPS was produced through plasticization process in which the mixture of starch granule and plasticizers (water and glycerine) went through the stirring and heating process. Cellulose was used as filler and prepared in suspension form through ultrasonication process involving isolation of cellulose in water medium. The ultrasonicated cellulose suspension was added into the TPS matrix to produce TPS/cellulose biocomposites by film casting. The effects of ultrasonication duration (1, 2, 3, 4 and 5 hours) on the crystallinity and tear strength of the TPS/cellulose biocomposite films were investigated. Based on the XRD results, crystallinity of the cellulose and biocomposite were altered by the cellulose ultrasonication process. The tear strength of the TPS biocomposites were greatly enhanced when 5-hours ultrasonicated cellulose was employed as filler. The biocomposites demonstrated 241% higher tear strength than the pure TPS film. This could be related to the transformation of the cellulose into more isolated fiber, with enhanced crystallinity for better reinforcing effect to the TPS matrix.

1 Introduction

Nowadays, synthetic plastics are regular daily basis used materials. It is undeniable, although petroleum derived synthetic plastics possess good mechanical properties, they are apparently non-biodegradable and non-renewable [1]. In fact, many analysts emphasized that plastic waste prompts to environmental issues. Therefore, the development of environmental friendly yet sustainable packaging plastic has become major research of today's scientists and researchers. [2, 3]. Thermoplastic starch (TPS) is a renewable, biodegradable bioplastic, used to replace the synthetic plastics, especially in film products like packaging. Nonetheless, the fundamental inconveniences of TPS plastics include their low tensile and tear properties, which can restrict their long term application, especially when exposed to water/humidity and weathering. The naturally hydrophilic characteristic of TPS is the main cause of its high tendency in absorbing water. Degradation of the hydrated TPS may result in deterioration of its mechanical performance and durability [4, 5].

* Corresponding author: ismailibrahim@unimap.edu.my

Therefore, the properties of TPS need to be improved to be more comparable to commodity plastics for use in packaging application. In conjunction with this, the TPS has been reinforced with several types of cellulose fibers. Previous research indicated that good dispersion of cellulose fiber leads to the improvement in the mechanical properties of the TPS composite, such as tensile and tear strength [6, 7]. This shows that homogeneous dispersion of cellulose fiber in the TPS matrix can significantly influence the mechanical properties of the TPS/cellulose biocomposites. In previous work, the ultrasonicated cellulose was added into the TPS matrix in dry form. This caused low dispersion efficiency of cellulose in the TPS matrix, due to re-agglomeration of the ultrasonicated cellulose after the drying process [5]. We believed that the use of cellulose in suspension form which is added directly into the TPS suspension upon the ultrasonication process can result in improved dispersion of cellulose in the matrix.

In this project, TPS biocomposites films were produced by reinforcing the matrix with the ultrasonicated cellulose fiber. Several ultrasonication duration were used to isolate the fiber which were; 1, 2, 3 4 and 5 hours. The effect of the ultrasonication duration on crystallinity of the cellulose and the TPS/cellulose biocomposites was analysed. Furthermore, the most optimum ultrasonication duration to produce the highest value of TPS/cellulose biocomposite tear strength was determined.

2 Materials and methods

2.1 Materials

The starch (corn-based) was supplied by Sigma-Aldrich, code number S4126 in 2 kilogram poly bottle packing. This corn starch contains approx. 73% amylopectin and 27% amylose. The microcrystalline cellulose powder was obtained from Sigma-Aldrich, with product code number of C6288. It was produced from cotton liners with particle size of ~20μm. The cellulose acts as filler in the TPS biocomposite. The glycerol or also known as glycerine was used as plasticizer and supplied by the HmbG Chemicals with the code, C0347-91552409. Sodium bicarbonate from HmbG Chemicals (C0735-2134330) was used as expanding agent for the production of the TPS film.

2.2 Methods

2.2.1 Plasticization of starch

The plasticization of starch was done to form the thermoplastic starch (TPS). The process starts with mixing the corn starch with distilled water in the ratio of 1:20. After that, the plasticizer, glycerol was added to the starch solution by the ratio of 1:50. The mixture was continuously stirred at 75-85°C until gelling. The starch plasticization process completed after 45 minutes.

2.2.2 Ultrasonication of cellulose

The cellulose suspension was prepared by mixing the powder form of cellulose with the TPS solution in a ratio of 1:100. The cellulose was isolated through the ultrasonication process by using Branson Digital Ultrasonic Disruptor/Homogenizer, Model 450D with 20% amplitude, 20 seconds pulse on and 10 seconds pulse off for 1.0 hour, 2.0 hours, 3.0 hours, 4.0 hours and 5.0 hours respectively. The process was done to isolate and disperse the cellulose prior being added into the TPS solution. Long period of ultrasonication could result in overheating and degradation of cellulose, therefore the ultrasonication of cellulose was done in the ice bath.

2.2.3 Preparation of thermoplastic starch/cellulose composite

The prepared cellulose suspension was added into the TPS solution. The mixture was stirred continuously at 75-85°C until it is became thick and almost transparent. After that, 0.5 grams of sodium bicarbonate powder was dissolved with 30mL of distilled water and added into the mixture. The mixture was continuous stirred to about 5 minutes. Following, the casting process took place; mixture (biocomposite suspension) was then poured into a non-stick sheet mold and place into the oven at 45°C for 24 hours drying process. The dried film was removed from the mold and cut accordingly for testing. The formulations of the TPS biocomposites are shown in Table 1.

Table 1. The formulation of thermoplastic starch/cellulose biocomposites

Ultrasonication of Filler	Materials	
Isolation Period (hour)	Thermoplastic Starch (Matrix) (wt%)	Cellulose (Filler) (wt%)
-	100	-
0	95	5
1.0	95	5
2.0	95	5
3.0	95	5
4.0	95	5
5.0	95	5

2.2.4 X-ray diffraction (XRD)

The X-ray diffractometer (Bruker D2 Phaser, USA) was used to identify the cellulose and composite phase formation. The crystalline formation of the thermoplastic starch and the cellulose which are isolated using different ultrasonication time was investigated. Each specimen of samples was placed in the sample holder and scanned by a range of 2θ = 10°- 60° with the scan speed of 2°/min. The crystallinity of samples was characterized through the test.

2.2.5 Tear test

The tear strength was measured by using the Instron Machine, according to ISO 34 -1:1994 Method B(a). The test was run using the angle test specimen (type B) without a nick. Crosshead speed of 500 mm/min was applied. The average tear strength values were determined for each of the samples.

2.2.6 Scanning electron microscopy (SEM)

Scanning Electron Microscope (JEOL JSM-6460LA) was used to analyse the topography and morphology of the non-ultrasonicated and ultrasonicated cellulose (4 and 5 hours). The specimens were coated with platinum by using JFC-1600 Auto Fine Coater (JEOL Ltd., Japan) prior to SEM observation. The specimens were observed under magnification of 1,000X to see changes in the cellulose fibres' shape and size.

3 Results and discussion

The X-ray diffraction (XRD) was used to investigate the crystallinity of the corn starch, TPS and its biocomposites. Figure 1 shows the XRD patterns of the cellulose, corn starch, TPS and the TPS/cellulose biocomposites incorporating the non-ultrasonicated cellulose and 5-hour ultrasonicated cellulose. Table 2 summarizes the crystallinity index values (%) of all the materials which were obtained through the XRD analysis.

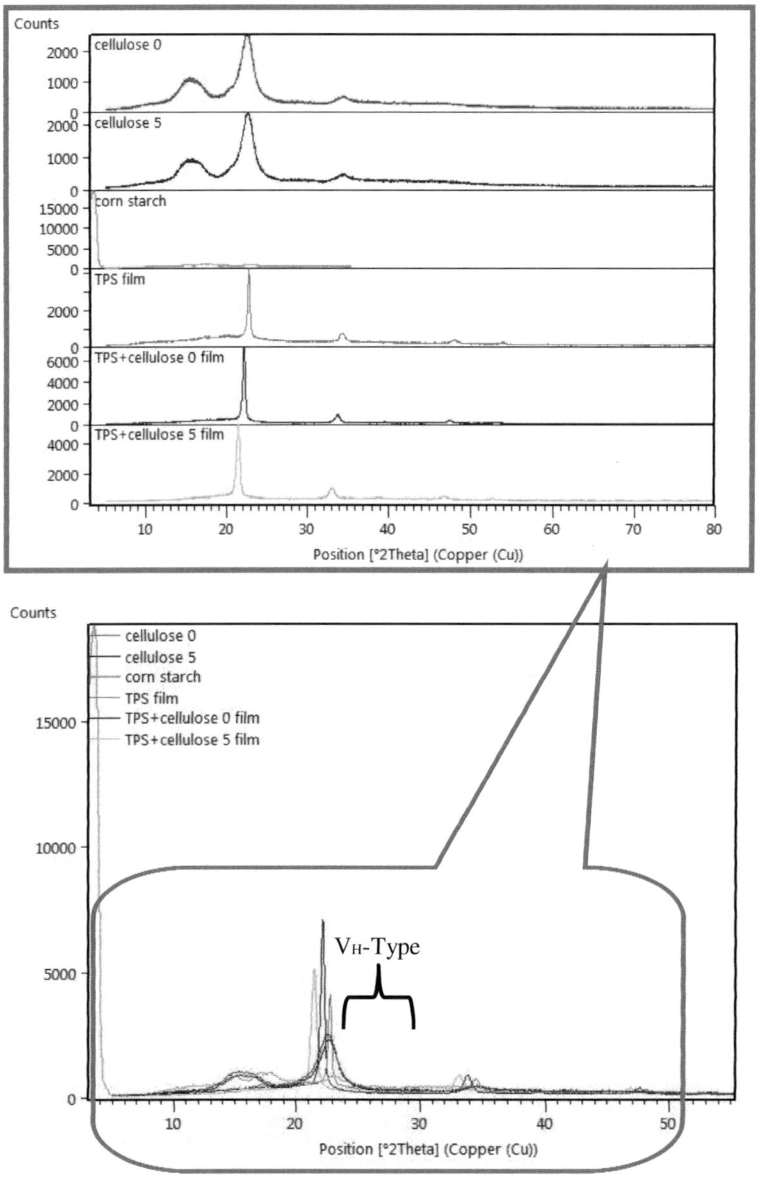

Fig. 1. X-ray diffraction patterns for cellulose, corn starch, TPS and its biocomposites.

Table 2. Crystallinity of the cellulose, corn starch, TPS and TPS/cellulose composites based on XRD analysis.

Sample	2θ (°)	Crystallinity	
		Crystallinity(%)	Amorphous (%)
Cellulose 0 (Without ultrasonication)	15.38, 22.55, 37.46	42.30	57.70
Cellulose 5 (With 5hours ultrasonication)	15.43, 22.70, 34.54	43.10	56.90
Corn Starch	3.24, 5.08, 15.08, 17.52, 23.06	54.70	45.30
TPS Film	22.73, 34.40, 47.96, 53.82	29.80	70.20
TPS/Cellulose 0 Film	22.14, 33.76, 39.30, 47.50, 53.22	49.50	50.50
TPS/Cellulose 5 Film	21.43, 33.07, 38.88, 46.82, 52.62	46.30	53.70

Based on Figure 1, the XRD pattern of cellulose fiber reveals diffraction peaks around 2θ=15.4° and 22.6° which typically represent cellulose type I. Furthermore, there are slight difference between the cellulose without ultrasonication and cellulose undergoes 5 hours of ultrasonication with respect to their peaks (2θ shifted from 15.38° to 15.43°) and its crystallinity index (increased from 42.30% to 43.10%). The obtained results indicated that the ultrasonication process has increased the crystallinity of the cellulose. This was due to the acoustic cavitation produced during ultrasonication treatment. Ultrasound was proved to efficiently formulate fibrillation of cellulose as the coupling of high-power ultrasound into liquids creates extreme shear forces [8]. In other words, the cellulose structure has been re-organized and become more orderly (crystalline) in the portion from longer and thicker fiber into shorter and thinner fiber when the cellulose was ultrasonicated in water medium. The corn starch shows a C-type crystallinity pattern with peaks at 2θ=3.24° (characteristic of B-type polymorphs), 15.08° and 17.52°, (characteristic of both A and B-type polymorphs) and 23.06° (characteristic of B-type polymorphs). After it is converted into a TPS film, there was a typical B-type crystallinity pattern which was clearly observed through 2θ peak = 22.73° (characteristic of B-type polymorphs) [9]. It corresponded to VH-type, and it is mainly characterized by the intense peak at 2θ=22.73°. The VH-type consists of amylose recrystallization induced by lysophospholipids and complex forming agents such as isopropanol and glycerol [10]. Upon the addition of cellulose to the starch matrix, the peaks at 2θ=22.14° and 33.76° (for cellulose without ultrasonication) and 2θ=21.43° and 33.07° (for cellulose with 5 hours ultrasonication), which correspond to the XRD pattern of cellulose I are observed. The results also show that the addition of cellulose significantly increased the crystallinity of the film. Although the starch peak's locations did not change obviously in XRD pattern, the intensities were increased significantly by the addition of cellulose within the TPS matrix. The addition of well-dispersed cellulose fiber can enhance nucleation for TPS matrix crystallization, thus resulting in the increased crystallinity of the biocomposite film [11]. Based on the results in Table 2, 5 hours-ultrasonicated cellulose which reinforced into the TPS matrix has a little drop on its crystallinity (46.30%) as compared to the TPS with the cellulose without ultrasonication (49.50%). The crystalline region of the cellulose fibers has little damaging due to the ultrasonication treatment, meanwhile, it was affected by the high processing temperature (85°C) when the cellulose reinforcing into the TPS, resulting in the inclination of crystallinity. The high crystalline fibers could be more effective in providing better reinforcement for composite materials because of high young's modulus of the crystal region along the longitudinal direction [11].

Based on Figure 2, the tear strength of the biocomposite film sample was significantly increased when the non-ultrasonicated cellulose fiber was added into the TPS. However, the tear strength decreased as the TPS matrix was added with cellulose that being ultrasonicated for 1, 2, 3 and 4 hours. This shows that the application of 1 to 4 hours of ultrasonication did not allow adequate fibrillation of the cellulose for its transformation into reinforcing fiber. As proved through SEM analysis, the 4 hours-ultrasonicated cellulose did not show much difference when compared with the non-ultrasonicated cellulose. However, when the ultrasonication duration was further proceeded for 5 hours, much more fibrillation process could occur for better isolation of the cellulose fiber (see Figure 3a, b and c). The isolated cellulose could provide greater reinforcing effect to the TPS matrix due to its higher aspect ratio fiber (smaller fiber diameter). When the TPS was reinforced with the 5 hours-ultrasonicated cellulose, the tear strength increased from 0.521 MPa to 1.256MPa. It demonstrated 241% increment in tear strength when benchmarked with the pure TPS film. The research indicated that homogenously distributed filler can penetrate within polymeric chains and restrict the slippage of the macromolecules, thereby enhance the tear resistance of the host polymer [12, 13]. The XRD results also indicate that the crystallinity of TPS was increased when it is incorporated with 5-hours-ultrasonicated cellulose. This could also be the reason why the tear strength of the TPS was successfully increased.

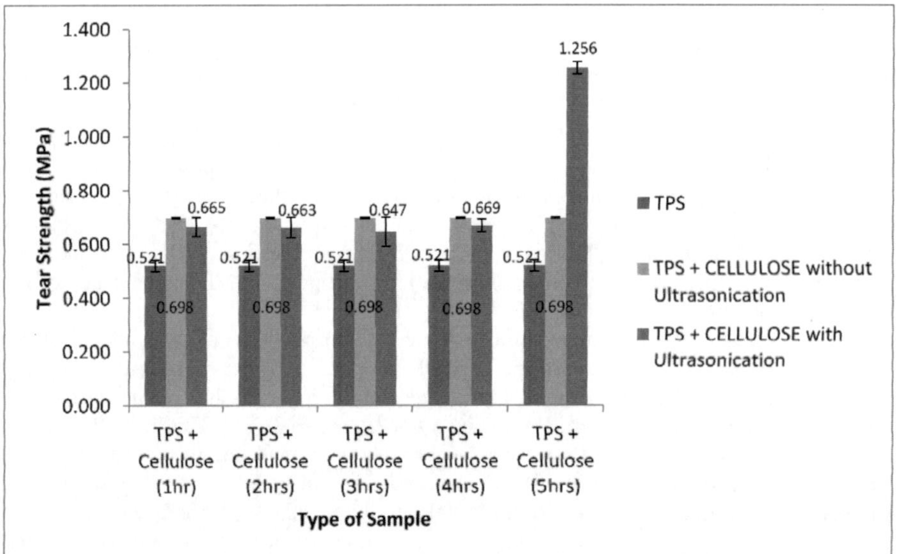

Fig. 2. Comparison on tear strength between the neat TPS and TPS biocomposites

Figure 4.9: Comparison of tear strength between TPS and TPS biocomposite with different ultrasonication period cellulose.

Fig. 3. SEM images (x1000 magnification) of the cellulose fibers (a)non-ultrasonicated (b)4hours-ultrasonicated , and (c)5hours-ultrasonicated

4 Conclusion

The TPS/cellulose biocomposite films were prepared by using 5wt% of cellulose as filler. Prior to the biocomposite production, the cellulose was subjected to ultrasonication process for 0, 1, 2, 3, 4 and 5 hours. The effects of the ultrasonication process on the crystallinity and tear strength of the biocomposite were studied. XRD analysis indicates that there are slight difference between the cellulose without ultrasonication and cellulose undergoes 5 hours of ultrasonication with respect to their peaks (2θ shifted from 15.38° to 15.43°) and its crystallinity index (increased from 42.30% to 43.10%). The obtained results proved that the ultrasonication process has increased the crystallinity of the cellulose. Isolation of the cellulose fibers and enhancement of the the cellulose's crystallinity led to the improvement of the TPS biocomposite's tear strength. The highest tear strength of TPS/cellulose biocomposite film was obtained when the 5hours-ultrasonicated cellulose was employed as filler. It demonstrated 241% increment in tear strength when benchmarked with the pure TPS film.

The authors would like to acknowledge the support from the Fundamental Research Grant Scheme (FRGS) under a grant no: FRGS/1/2019/TK10/UNIMAP/03/2 from the Ministry of Education Malaysia.

References

1. B. Luijsterburg and H. Goossens, H. Conservation and Recycling, E **85**, (2014)
2. R. A. Gross, B. Kalra, Science. E **297,** 5582 (2002)
3. M. Kolybaba, L.G. Tabil, S. Panigrahi, W.J. Crerar, T. Powell, B. Wang, *Biodegradable Polymers: Past, Present, And Future. In ASABE/CSBE North Central Intersectional Meeting* (American Society of Agricultural and Biological Engineers, 2006)
4. S.N.S. Syed Adam, A.F Osman & R. Shamsudin, Sains Malaysiana. E **47**, 6 (2018)

5. A.M.T.L Ashafee, *Effect of Hybrid Cellulose/Bentonite Filler and Dispersing Parameters to Crystallinity, Low Temperature Tear and Tensile Properties of Thermoplastic Starch Composites for Refrigerated Food packaging Application*, (Master Thesis, Universiti Malaysia Perlis, 2017)
6. B. Ruxanda, T. Carmen-Alice, S. Iuliana. BioRes. E **9,** 1 (2014)
7. N.L.I. Zailuddin, A.F. Osman, S. Husseinsyah, Z. Ariffin, F.H. Badrun, Mechanical Properties and X-Ray Diffraction of Oil Palm Empty Fruit Bunch All-Nanocellulose Composite Films, *Proceedings of the Second International Conference on the Future of Asean (ICOFA)*. E **2** (2017)
8. W.S. Ratnayake, R. Hoover, T. Warkentin, Starch-Stärke. E **54**, 6 (2002)
9. E.D.M. Teixeira, C. Lotti, A.C. Corrêa, K.B. Teodoro, J.M. Marconcini, L.H. Mattoso, J. App. Poly. Sci. E **120,** 4 (2011)
10. Y. Chen, C. Liu, P.R. Chang, X. Cao, D.P. Anderson, Carbohydrate Polymers. E **76**, 4 (2009)
11. W. Chen, H. Yu, , Y. Liu, Y. Hai, M. Zhang, P. Chen, Cellulose. E **18**, 2 (2011)
12. Z. Peng, L. Kong, S. Li, Y. Chen, M. Huang, Composites Science and Technology, E **67**, 15-16 (2007)
13. A.F. Osman, G.A. Edwards, T.L. Schiller, Y. Andriani, K.S. Jack, I.C. Morrow, P.J. Halley, D.J. Martin, Macromolecules. E **45,** 1 (2011)

IOP Conf. Series: Materials Science and Engineering **701** (2019) 012046 doi:10.1088/1757-899X/701/1/012046

Enhanced thermal and microstructure properties of Sn-1.0Ag-0.7Cu based lead-free solder with Titanium Oxide addition

Rita Mohd Said[1], *Norainiza* Saud[1,*], *Norhayanti* Mohd Nasir[1], *Mohd Arif Anuar* Mohd Salleh[1], *Mohd Nazree* Derman[1], and *Mohd Izrul Izwan* Ramli[1]

[1]Center of Excellence Geopolymer & Green Technology (CeGeoGTech), School of Materials Engineering, Universiti Malaysia Perlis (UniMAP), Taman Muhibbah, 02600 Jejawi, Arau, Perlis, Malaysia

Abstract. This paper presented the effects on thermal and microstructure properties of Sn–1.0Ag–0.7Cu (SAC107) lead-free solder alloy with the addition of titanium oxide (TiO_2) particles. The SAC107 solders were reinforced with 0.25, 0.5, 0.75 and 1.0 weight percentage (wt.%) of titanium oxide (TiO_2) particles to produce SAC107+TiO_2 solder alloy composite. The solder alloy was prepared by microwave–assisted powder metallurgy route. Based on the results, it showed that the addition of TiO_2 particles influenced the properties of the solder alloy composite by refining the intermetallic compound (IMC) at the bulk microstructure area and slightly decreasing the melting temperature.

1 Introduction

In recent years, many lead-free solder alloys have been developed due to the toxicity and environmental concern of lead-containing solder that widely used in the electronic assemblies [1-5]. This situation has driven many researchers to find suitable lead-free solder alloys such as Sn-Ag, Sn-Ag-Cu (SAC) [6], Sn-Cu and Sn-Zn. The lead-free solder candidates have shown potential to replace the traditional Sn-Pb solder. The use of lead-free solder SAC system is advancing throughout the world with better mechanical properties and wettability [7]. Considerable research efforts into the characteristics of lead-free solder are started to improve lead-free solder technologies as joining materials in the electronics industry as well as their unique physical and mechanical properties have facilitated printed circuit board (PCB) assembly [7-10].

In order to meet the demand for increasingly finer pitch and functional devices, there are some problems need to be resolved for these lead-free solder, such as the formation of large brittle intermetallic compounds (IMC) [11]. When the size of IMC is large, the brittle property of Ag_3Sn and Cu_6Sn_5 in SAC can weaken the properties of lead-free solder joint and crack can easily propagate along their interface [12].

* Corresponding author: norainiza@unimap.edu.my

The brittle Ag_3Sn intermetallic compounds (IMC) in SAC solder will weaken the mechanical properties of the solder joints, yet leading to the failure under stress conditions during the actual service [13]. The low-silver content such as Sn–1.0Ag–0.7Cu (SAC107) formed more β-Sn phase (large β-Sn grains) and decreases the number of Ag_3Sn IMC particles in the bulk solder [14]. It has been proposed that incorporating ceramics reinforcement into these alloys would affect the properties of lead-free solders by refining the microstructure and retarding the IMC growth of the eutectic phase [15-16]. It is well known that ceramic element have excellent properties which are can be improved the properties of SAC107 lead-free solder. The ceramic reinforcement such as titanium oxide (TiO_2) particle was selected to enhance the solder performance due to its various advantages over ferrous and other non-ferrous metallic materials. The high strength-to-weight ratio and excellent corrosion resistance of TiO_2 particles make it selected for reinforcement doped into lead-free solder alloys [17]. Mavoori and Jin have studied TiO_2 and Al_2O_3 particles as reinforcements for a conventional Sn–Pb solder and found significant enhancements of creep and mechanical properties [18].

In this work, SAC107+TiO_2 solder alloy composite properties such as the melting temperature and microstructure properties with varying weight percentages of TiO_2 particles were investigated.

2 Experimental procedure

The SAC107 lead-free solder alloy was mixed with different weight percentages (wt. %) of TiO_2 particles by using planetary mill. The composite solder were compacted into pallet size 12mm x 2mm and the compacted sample was sintered using rapid microwave-assisted sintering technique to grasp the temperature of 185 °C under an ambient condition in 2.45GHz (800W) generated by a 1.27kW SANYO microwave oven. The samples were then reflowed on the copper (Cu) substrate to produce a solder joint.

2.1 Microstructure characterization

Microstructural analysis of the metallographically polished samples of the composite materials were carried out using optical microscope (Eclipse L300N).

2.2 Melting point

The melting point of the solders was investigated using differential scanning calorimeter (DSC) from TA Instruments (model number: 2910). The samples were weighed 5 mg then placed into an aluminum pan, which was then heated with a scanned rate of 10 ° C/min under a nitrogen atmosphere.

3 Results and discussion

3.1 Microstructure analysis

Fig. 1 shows the morphology of pure SAC107 with a varying weight percentage of TiO_2 particles. The addition of a small amount of TiO_2 particles into SAC107 solder has refined the microstructure of SAC107 + TiO_2 composite solder. It showed that with increasing amount of TiO_2 particles in the SAC107 matrix, the β–Sn area become smaller and eutectic phase (Cu_6Sn_5 and Ag_3Sn IMC phase) of the bulk microstructure become finer. The decreasing of the IMCs growth with respect to the addition of reinforcement particle was also reported in the previous study [14, 19-20]. The results suggested that the theory of adsorption phenomenon plays an important role during a solidification process of solder alloy and would greatly affect the microstructure formation [21]. An adsorption prevents the IMC to further growth on the bulk solder area also at the solder/Cu substrate interface. Zhu et al. found that the microstructure of SnBi-0.2Cr solder composite distinctly refined compared to the pure SnBi solder [22].

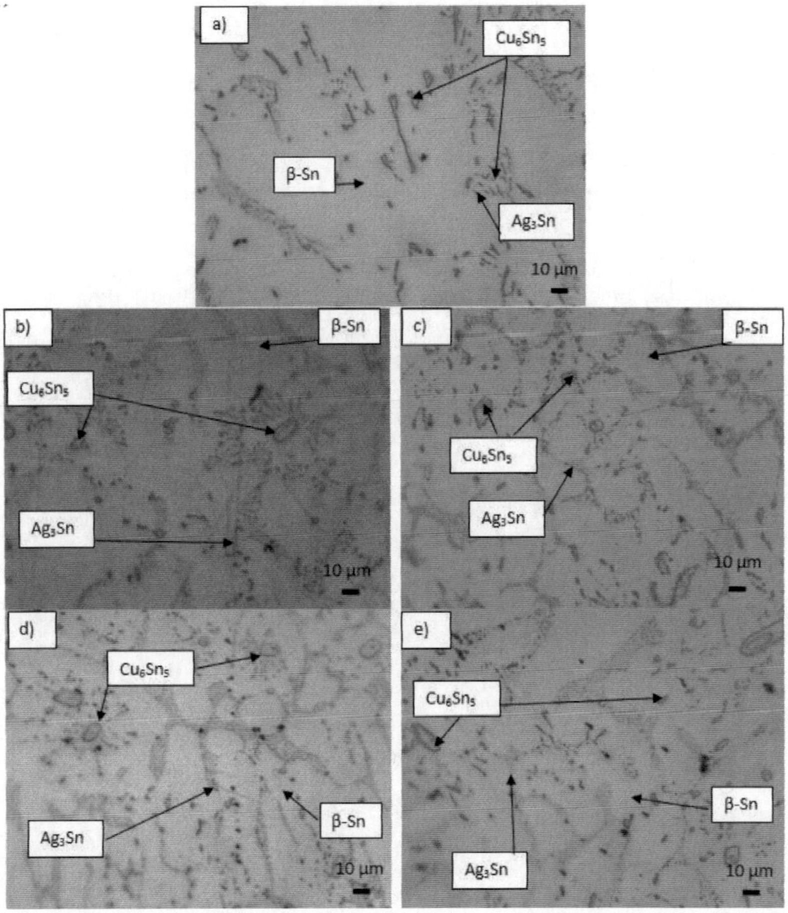

Fig. 1. Optical micrograph of Sn–1.0Ag–0.7Cu with a) 0 wt.%, b) 0.25 wt.%, c) 0.5 w.%, d) 0.75 wt.% and 1.0 wt.% of titanium oxide.

3.2 Melting point behaviour

The effect of TiO_2 addition on SAC107 solder alloy composite melting temperature was showed in Figure 2. The melting temperature of SAC107 + 0.25 wt. % TiO_2, SAC107 + 0.5 wt. % TiO_2, SAC07 + 0.75 wt. % TiO_2 and SAC107 +1.0 wt. % TiO_2 were 216.9 °C, 216.8 °C, 216.7 °C and 216.7 °C, respectively. It is seemed that the melting temperature of SAC107 eutectic was 217.1 °C. Based on the result, reduction in the melting temperature of SAC107 + 0.25 wt. % TiO_2, SAC107 + 0.5 wt. % TiO_2, SAC07 + 0.75 wt. % TiO_2 and SAC107 +1.0 wt. % TiO_2 were 0.2 °C, 0.3 °C, 0.4 °C and 0.4 °C, respectively. These results indicated that the melting temperature of SAC107 composite solder was lower than that of the pure SAC107 solder alloy. These suggested that the reduction of the melting point is possibly attributed to an increase in the surface instability of the composite solder caused by the addition of the TiO_2 particles with a higher surface free energy. It is believed that TiO_2 particles act as additional nucleation sites where it may increase the rate of β – Sn solidification and reduce growth time for Ag_3Sn formation in SAC107 solder alloy composite. The fine microstructure in the bulk solder was also found during the faster of solidification rate [14]. Yang et al. stated that the fast cooling rates produced joints with the best mechanical properties. The reduction in melting temperature of solder composite also reported from the other researcher. Liu et al. [14] have found the melting temperature of Sn–3.8Ag–0.7Cu solder composite with SiC particle were lower than the eutectic solder alloy by about 1 K.

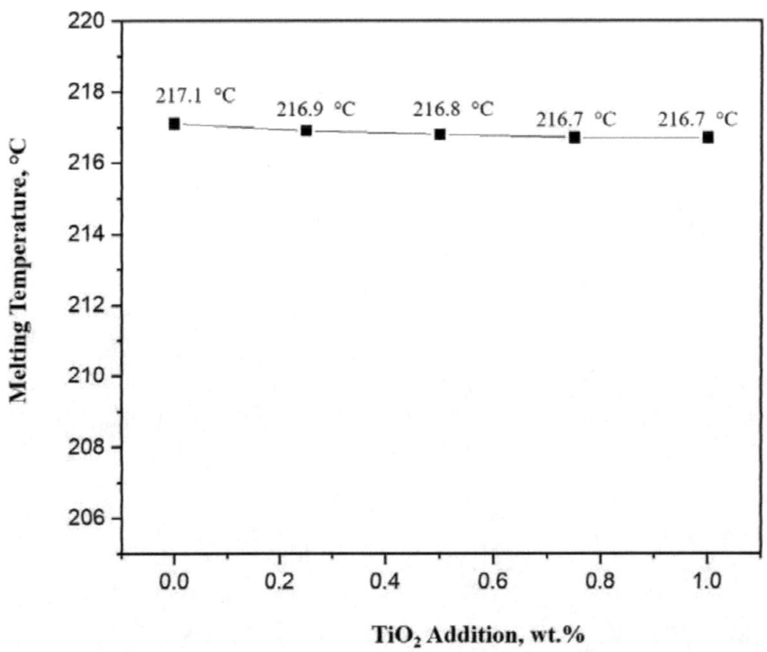

Fig. 2. Melting temperature of the SAC107 solder alloy with TiO_2 addition.

4 Conclusion

The influence of TiO_2 addition on the thermal and microstructure properties of Sn-1.0Ag-0.7Cu (SAC107) lead-free solder alloy was studied in this research. The following conclusions can be drawn:

i. The β–Sn and eutectic area which contain Cu_6Sn_5 and Ag_3Sn intermetallic phase in SAC107+ 1wt.% TiO_2 lead-free solder composite were refined after the addition of TiO_2 particles.
ii. A slight reduction in melting temperature of the SAC107+ 1wt.% TiO_2 lead-free solder composite.

The authors acknowledge with Nihon Superior Co. Ltd, Japan for supporting the research effort through a materials and finance. The author also acknowledges Associate Professor Kazuhiro Nogita from the University of Queensland for his valuable research advises.

References

1. M.I.I. Ramli, N. Saud, M.A.A.M. Salleh, M.N. Derman, R.M. Said, and N. Nasir, Materials Science Forum. **803** (2015)
2. M.A.A.M. Salleh, M.H. Hazizi, M.A.B.M. Abdullah, N. Noriman, R. Mayapan, and Z.A. Ahmad, Advanced Materials Research. **626** (2012)
3. K. Nogita, M.A.A.M. Salleh, S. Smith, Y. Q Wu, S.D. McDonald, A.G.A. Razak, T. Akaiwa, and T. Nishimura, 2017 International Conference on Electronics Packaging (ICEP). (2017)
4. M.A.A.M. Salleh, A.M.M.A. Bakri, F. Somidin, and H. Kamarudin, International Review of Mechanical Engineering.
5. M.A.A.M. Salleh, M.M.A.B. Abdullah, F. Somidin, A.V. Sandu, N. Saud, K. Hussin, S.D. McDonald, and K. Nogita, Revista de Chimie **64**(7) (2013)
L. Yang, S. Quan, C. Liu, G. Shi, Mater. Lett. **253,** 191-194 (2019)
6. H. Nishikawa, Y.H. Tadashi Takemoto, J. Electron. Mater. **38**(12), 2610-2616 (2009)
7. K. N. T. Liang Zhang, Mater. Sci. Eng. R **82,** 1-32 (2014)
8. R.M. Said, M.A.A.M. Salleh, M.I.I. Ramli, and N. Saud, Solid State Phenomena. **280** (2018)
9. M.A.A.M. Salleh, S.D. McDonald, and K. Nogita, Journal of Materials Processing Technology. **242** (2017)
10. K. Nogita, M.A.A.M. Salleh, E. Tanaka, G. Zeng, S.D. McDonald, and S. Matsumura, JOM. **68**(11) (2016)
11. C.-w.H. Liang Zhang, Yong-huan Guo, Ji-guang Han, Yong-wei Zhang, Xu-yan Wang, Microelectron Relia. **52**, 559-578 (2012)
12. F. Somidin, H. Maeno, M.M. Salleh, X.Q. Tran, S.D. McDonald, S. Matsumura, and K. Nogita, Materials Characterization. **138** (2018)
13. A.A. El-Daly, A.F., S.F. Mansour, M.J. Younis, Mater. Sci. Eng. A **578**, 62-71(2013)
14. N. Saud, R. M. Said, M. A. A. M. Salleh, M. N. Derman, M. I. I. Ramli, N. M. Nasir, MATEC Web of Conferences **78** (2016)
15. R. M. Said, M. A. A. Mohd Salleh, M. N. Derman, M. I. I. Ramli, N. M. Nasir, N. Saud, Key Eng. Mater. **700**, 123-131 (2016)

IOP Conf. Series: Materials Science and Engineering **701** (2019) 012046 doi:10.1088/1757-899X/701/1/012046

16. A. O. Takahiro Fujita, Chiaki Ouchi, Hidenori Tajima, Mater. Sci. Eng. A **213**, 148-153 (1996)
17. H. Mavoori, S.J., J. Elec. Mater. **27**(11), 1216-1222 (1998)77
18. P. Y. Ping Liu, Jim Liu, J. Electron.Mater. **37**(6), 874-879 (2008)
19. W. Zhu, W. Zhang, W. Zhou, P. Wu, J. Alloy. Compd. **789**, 805-813 (2019)
20. D.Q. Yu, J.Z., L. Wang, J. Alloys Compd. **376**,170-175 (20014)

IOP Conf. Series: Materials Science and Engineering **701** (2019) 012047 doi:10.1088/1757-899X/701/1/012047

A review of iron nitride based thin films development

Nur Izzati Muhammad Nadzri[1, 1], *Dayang Mas Shufina* Ibrahim[1] , and *Saphuang* Sompon[1]

[1]School of Materials Engineering, Kompleks Pusat Pengajian Jejawi 2, Universiti Malaysia Perlis (UniMAP) Taman Muhibbah, 02600 Arau, Perlis, Malaysia.

Abstract. Thin film technology has taken a lot of area in the industry due to their properties which is better than their parents bulk materials. One of the interesting thin films that has yet to be fully explored is iron nitride based thin films has been attracting attention for the past few years especially in the semiconductor technology. Crystal structure and magnetic properties on the film which depend greatly on the growth method, deposition technique, gas flow, pressure and many more. Suitable parameters allow the presence of required iron nitride based thin films stoichiometry. In semiconductor application, magnetization properties play an important role of developing iron nitride thin films. This will give rise to the formation of strong magnetic moment thus increasing the corrosion resistance of the thin film. However, the difficulties of growing a pure phase of these iron nitrides thin films makes it more interesting to study and recent researches have shown findings that open the possibilities in enhancing properties of this iron nitride based thin films.

1 Introduction

Research interests on the binary nitrides thin films has grown rapidly due to their excellent properties in many areas including semiconductor and micromechanical. However, the commercialization of these type of nitrides is hampered by several challenges, including the instability of the formation and high cost due to the process used to synthesis these nitride is rather a difficult routes and require a system like Electron Beam Physical Vapor Deposition (EB-PVD) or Molecular Beam Epitaxy (MBE) techniques and many more. Recently, transition metal nitrides (TMN) have attracts researches because they are highly electrically conductive, thermally stable and have high melting points and exceptional hardness and corrosion resistance[1]. The use of TMN as coatings has also been successfully explored due to properties like high hardness, biocompatibility and there are recent effort to tailor lattice constant and electrical mechanical and optical properties of this nitrides[2].

TMN is often linked to development of binary thin films. This is because compared to bulk samples, thin films possess better physical and mechanical properties. Thin films is a thin layer material applied to a surface of substrate to be coated or to a previously coating to form layers. The growth of thin films is firmly affected by the substrate which results in the formation of microstructure and determining film properties. One of the interesting traits of thin films is they are usually crystallographically oriented in a specific way (depending on the growth of the thin films) with respect to the underlying substrate which is known as epitaxial thin films. Few binary nitrides have taken an interests in many aspects of technology are boron nitride (BN) and gallium nitrides which is vastly used in the development of UV light applications

Boron nitrides thin films for instance, crystallizes in two forms; a cubic zinc blende and a hexagonal structure. It was found that different crystal structure gives different role in the electronic semiconductor applications. Some of the important applications of boron nitride thin films is as diffusion source. This method can be used to fabricate planar diodes only one photomask by diffusing boron nitrides on the silicon dioxide windows. This will result in the ohmic contact between and nickel can be deposited directly by electroless plating on bare silicon.

Interestingly, the development of iron nitride thin films is fundamentally connected to field of basic research based on phase diagrams. Based on the different properties iron nitride thin films can possess, in this paper we are exploring the new possibilities of enhancing better quality of iron nitride thin film in many areas including semiconductor and micromechanical.

1.1 Iron nitride thin films

Iron nitrides consists of many phases with different crystal structures which results to different properties of each type of iron nitrides. Formation of iron nitrides are normally conducted by the nitriding process which results in phases such as FeN, ζ-Fe2N, ε-Fe3N and γ'-Fe4N. A study conducted by Tesseir et al[3], found that with the formation of different

[1] Corresponding author: n.muhammad-nadzri13@imperial.ac.uk

phases of iron nitrides results in higher wear resistance properties which is usually used in hard coatings application. This is due to the fact that when this binary nitrides are subjected to high temperature, the phase stability improves thus enhancing their properties. One of difficulties of growing nitrides is instability of nitrogen when incorporated with metals (especially transition metals). Therefore, nitrogen percentage needs to be determined in order to obtain preferred stoichiometry of iron nitrides. Concentrations of nitrogen given to iron nitrides will exhibit a ferromagnetic, paramagnetic and non-magnetic characteristics of the material.

During the last decades, iron nitride compounds have been intensively investigated because iron and nitrogen are among the most abundant elements in nature, but also because these compounds are very interesting from a fundamental point of view and industrial applications [4]. Iron nitrides consists of many phases with different crystal structures which results to different properties of each type of iron nitrides. Formation of iron nitrides are normally conducted by the nitriding process which has made this nitrides received attention for many years and has been widely studied due to their ability to improve many things such as surface hardness and wear resistance[5].

The work on the iron nitrides not only limited on bulk samples, extensive research has been conducted on developing iron nitride based thin film in a 3D and 2D. Due to their capability of becoming between ferromagnetic to paramagnetic, iron nitrides thin films is applicable to be used in the semiconductor area especial as magnetic storage data. The crystallographic phase of iron nitrides affected their magnetic properties. Therefore, the desirable to obtain as many pure phases as possible is one of the ways to improve this. Unfortunately, due to their metastable properties, it is significantly challenged to control the phase structure and hence the magnetic properties of iron nitrides.

Figure 1(a-d) shows AFM images of iron nitrides thin films where for AFM images all of the films exhibited uniform granular structure grains with diameter around 30 nm respectively. These grains aggregated together and left many peaks and valleys on the film surfaces which indicates good quality thin films. The only differences from the images are the parameter used to grow the samples[6]. The thin films undergone repeated nucleation process which resulted in a transition from columnar to globular microstructure, a denser microstructure. Furthermore, This results were also supported by a study conducted by Khan et al.,[7] where transmission electron images showed a homogeneous formation pf phases and there were some parts with amorphous iron nitrides matric (Figure 2a and Figure 2b). Figure 2c shows an interface of the film and it does not show any signs of porosity which indicates that the film is densely packed columnar structure with low surface roughness. The last picture of high-resolution transmission electron microscope (HRTEM) shows a lattice spacing between each atomic coulomb around 3.78 nm and there is also a thin amorphous layer with flaccid structure identified by the iron nitride nanoparticles. According Hari et. al.,[8] amorphous inclusions in this films are featureless having no grain boundaries resulting in an expulsion of intrinsic stress and a smoother surface that indicates the structure of the thin films is strongly affected by the film thickness [9–11].

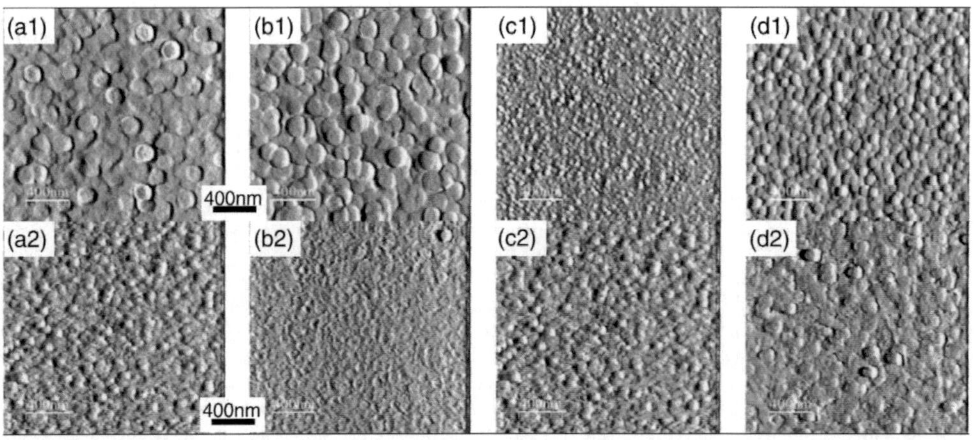

Fig. 1. AFM images of transition between (a1-b2) columnar to (c1-d2) globular microstructure depending on the parameters given to iron nitride thin films[6].

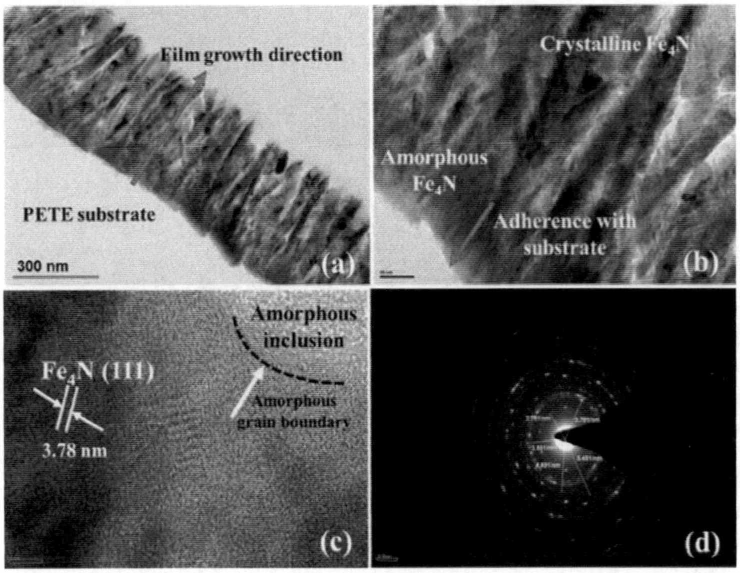

Fig 2. TEM images of 180 nm thick FeN thin films. (a and b) show TEM images, (c) HRTEM image and (d) SAED pattern [6].

Due to properties of iron nitride thin films, there has been vast research happening to synthesis ternary nitrides such as $FeWN_2$. The properties exhibit by iron nitrides sparks the interest among researchers on the possibilities of ternary metal nitrides to be used as epitaxial film layer together with boron nitrides and graphene's via intercalation method. It is also believed that this ternary nitrides iron based thin films materials may exhibit interesting thermal, optoelectronic, semiconducting, magnetic and/or superconducting properties[12–15].

2 Deposition of iron nitrides based thin films

There are a number of techniques in growing thin films, specifically iron nitrides which is by physical vapor deposition (PVD) and/or chemical vapor deposition (CVD) However, in this paper, we will solely focus on the technique by PVD methods as one of the main advantages of PVD is the processes are clean and pollution free compare to CVD process. EBPVD and MBE are widely used because of these techniques produces good quality of films. In PVD, there are variety modes of activation of growth including electron beam, ion beam, laser beam, induction heating and cathodic sputtering[16–18]. Choosing the suitable deposition routes is crucial because it will reflect on thin film quality and it is the most important key for thin film application devices. Thin films have several stages during growth that will then determine the properties of the films such as microstructure, surface, interface, physical and mechanical properties. Thermal deposition or also known as vacuum deposition is the simplest method one could carry for preparing thinner thin films in micron. Two types of source which commonly used are resistive and electron beam source. The target is evaporated by heating the source thus condensing evaporated particles on the substrate. Sputtering is another well known technique which uses radiation for thin film deposition including dc diode (also known as dc sputtering), radio frequency sputtering (RF) magnetron and ion beam sputtering[16–19]. However, many researchers would use to molecular beam epitaxy (MBE) and electron beam physical vapor deposition (EBPVD) techniques.

MBE is a term used to epitaxial growth of compound semiconductor films by process involving the reaction of one or more thermal molecular beams with a crystalline surface under ultra high vacuum conditions. This method is related to vacuum evaporation but offers improved control over the incident atomic or molecular fluxes so that sticking coefficient may take into account and allow rapid changing of beam species. Zuo et. al.[23], succeeded in growing epitaxial boron nitrides (BN) thin films with MBE. They have found appearance of small grains due to heat treatment and also formation of triangular shaped of h-BN domains which is supported both by Raman scattering and x-ray photoelectron spectroscopy (XPS) results This shows the growth route and also with the suitable parameters chose, a high-quality thin film is able to obtain. Due to the better growth deposition with MBE, Sugita et al.[24] and Golden et al.[10], did a study on the magnetic moment on Fe16N2 single crystal thin films and other iron nitride thin films phase by this technique which was rather difficult in growing them with other methods due to their metastable compound state.

Another method of growing iron nitrides thin films is by sputtering, which was studied by Rissanen et al.,[25] and Wang et al.[5]. This technique was selected because nitrogen rich thin film can be produced by controlling their deposition parameters by taking into account the combination between argon and nitrogen gas given to the sputtering system. Furthermore, the film composition showed that there was small amount of oxygen surface contamination. Based on the XPS data (Figure 3 a dan Figure 3b), all samples exhibit almost similar nitrogen content which is an ideal

IOP Conf. Series: Materials Science and Engineering **701** (2019) 012047 doi:10.1088/1757-899X/701/1/012047

value of the phase in iron to nitrogen phase diagram. Figure 4a to Figure 4c show AFM images of iron nitride thin film grown by the sputtering technique. It can be seen that the roughness is low and it formed a solid films under the equilibrium conditions which are predicted to have self-affine surfaces and the roughness can be characterised as dynamic form.

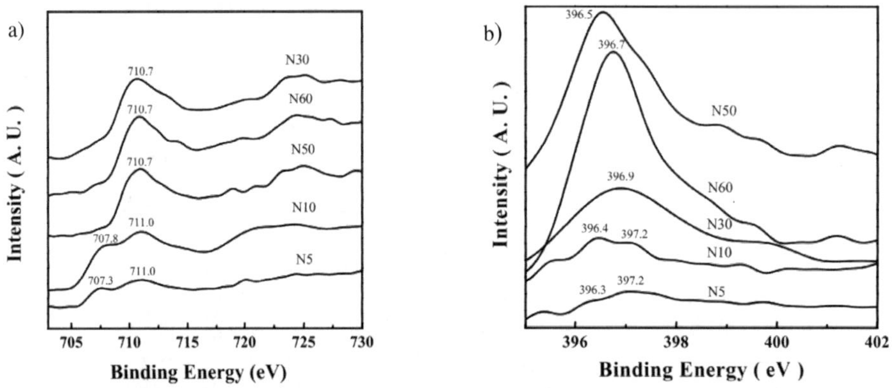

Fig. 3. X-ray photoelectron of (a) Fe 2p spectra and (b) N 1s spectra for samples with different parameter growth deposited by sputtering with different gas flow of nitrogen.

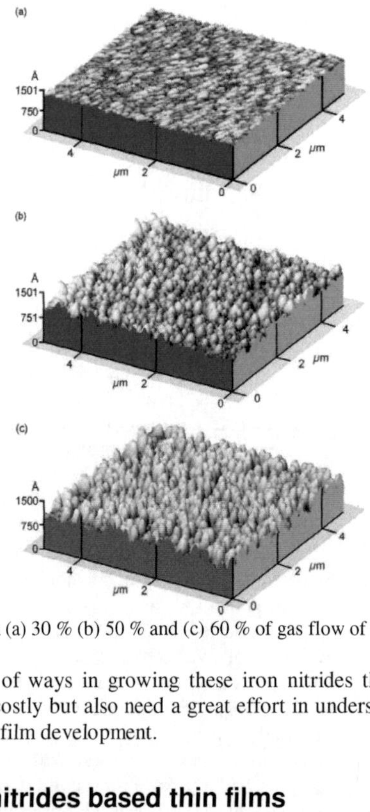

Fig. 4 AFM images of thin film grown with (a) 30 % (b) 50 % and (c) 60 % of gas flow of nitrogen fractions.

There are still many possibilities of ways in growing these iron nitrides thin films. The challenges part is this development of iron nitrides not only costly but also need a great effort in understanding their metastable state in order to enhance their capabilities in the thin film development.

3 Different phases of iron nitrides based thin films

FeN is known to be very sensitive to pressure and the nitrogen flow. Given the suitable temperature and pressure, it is possible to produce this thin film. Higher temperature given to the pre-deposition substrate will introduced ε phase structure. The homogeneity of FeN thin films can be achieved by using higher temperature with half percentage of the overall gas flow. The crystal structure of FeN is cubic which makes silicon (100) as a suitable substrate to use in order to avoid lattice mismatch[26].

A study conducted by Fratczak et al.[4], showed that they were three phases of iron nitrides thin films which are easily produced they are α"-$Fe_{16}N_2$, ε-Fe_3N and γ''-FeN_y. These samples were grown by MBE in ultra high vacuum condition by changing the parameters from getting the purest form (α phase) to epsilon (ε) phase which is mainly in a non magnetic form. It was found that α"-$Fe_{16}N_2$ is present in the ε phase, which is easily formed during the iron and nitrogen deposition. The iron formation as found to be on top of the sample surface which means that nitrogen was not diffusing deep enough. The sample is found experiencing transition from bcc to fcc and hexagonal crystal structure. Therefore, H_2 is added in the gas source in sputtering to lower the chemical potential on the sample surface. However, this resulted with no α" phase formation. The ε-Fe_3N is the easiest phase to obtain while γ''-FeN_y was theoretically predicted. Epsilon phase iron nitride occurred during the mixture of hydrogen and nitrogen in gas flow. The combination process between nitrogen atoms into nitrogen molecules is not possible with the presence of hydrogen due to their chemical factor. The epsilon phase can be in magnetic or non magnetic depending on the nitrogen content incorporated in the compound. The ε-Fe_3N phase has a hexagonal close packed (hcp) crystal structure.

Temperature plays an important role in formation of iron to nitride thin film phases. For example, α'-Fe_8N thin film occurred at temperature at 100°C since it will start to disintegrate around 180°C. It was also reported within this range of temperature, it is possible to grown other iron nitrides thin films such as γ'-Fe_4N and ε-Fe_3N. There was a peak shift in XRD data of α'-Fe_8N with higher power/voltage applied to the deposition. According to Dirba et al.[27] this behaviour indicates the elongation of the c -axis by introducing the nitrogen interstitial as observed in other sputtered Fe-N thin film. The phase formation is greatly influenced by the choice of substrate material and symmetry. Determine the suitable parameters such as nitrogen content and the voltage also play an important role in the development of iron nitride based thin films. Growing sample with higher temperature will increase possibilities of residual presence thus leading to low quality of crystalline thin films.

A much broader region is expected from iron nitride thin films phase diagram and more studies should be conducted in order to understand the iron nitrides thin films properties and characteristics.

4 Properties of iron nitride thin films

4.1 Magnetic properties

Iron nitride is known as magnetic materials. Iron is by far is of interest because it is the cheapest magnetic element with properties like volume-sensitive, with a critical interplay between structure and magnetism. α Fe phase which is in BCC crystal structure is a weak ferromagnetic while γ-Fe (FCC crystal structure) is particularly sensitive to lattice volume, and some conditions may find that this phase are able to range from high moment or low moment or non magnetic at all depending on lattice parameter. Uniform permeability is needed in iron nitride thin films because this will indicate the capability of the thin film to perform in extremely high density magnetic condition as well as in the integrated inductors operating in the gigahertz range. The high permeability of GHz frequency and low energy loss can be obtained by adjusting the component and phase evolution of the thin films. Ferromagnetic nature of Fe4N, for instance, showed presence of magnetic fields which leads to a combination between magnetic and electric hyperfine interactions. The structure of Fe_4N consists of Fe-N and Fe-Fe bonds combine together. This bond is greatly influence by the nitrogen on the electronic structure[25–27].

Previous experiments [28–30] by numerical studies reported the lattice constant is used in the geometry optimization in order to obtain the relaxed structure of thin film. Ferromagnetism in iron nitrides is depending upon nitrogen attractions on more electrons from Fe neighbouring atoms, and these electrons are incompletely screened thus affecting the magnetic moment. The lattice constant also effects the variation of magnetic properties of iron nitride thin films. The variation of magnetic moment also might be owing to the hybridization of the interstitial N atoms with the neighbouring of Fe atoms which will disturb the deformation of the atoms in the structure.

4.2 Corrosion resistance

Corrosion resistance is conducted on nano structured thin film by the EIS technique. EIS technique is a response of a circuit to an alternating current or voltage to a function of frequency. The stability of iron nitride thin films usually increased with decreasing of nitrogen flow rate. The shorter or longer time of formation of passivation layer on thin film surface will determine higher or lower corrosion resistance. Longer formation time for passivation layer indicates to the minimal grain boundary density which usually occur in thinner thin films. Low corrosion resistance usually occurs because higher heterogeneity due to the presence of more nitrogen concentration in the iron nitride film and passive film interface will in low corrosion resistance. Meanwhile, higher corrosion resistance is affected by thin film thickness and the thin film is regarded as more stable phase with higher content of metal nitrides [26, 31- 32].

5 Conclusions

There are still many areas of iron nitrides thin films which is referred to phase diagram that is waiting to be explored. Extensive studies and research shall be conducted in order to enhance the capabilities of iron nitrides thin films not only in semiconductor applications but open up the possibilities to order applications such as hard coating and superconductor materials as with their atomic arrangement and crystal structure iron nitride thin films are able to withstand higher temperature and also possess of higher hardness.

References

1. X. Tian et al., J. Mater. Chem. A, E **3** (2015)
2. L. E. Koutsokeras et al., Surf. Coatings Technol. E **204**, (2010)
3. F. Tessier, M. R. Ranade, A. Navrotsky, R. Niewa, F. J. Disalvo, and C. Mn, Z. Anorg.Allg.Chem. E **627** (2001)
4. E. Z. Fratczak, J. E. Prieto, and M. E. Moneto, J. Allpys Compd. E **586**, (2014)
5. X. Wang et al., Appl. Surf. Sci. E **220** (2003)
6. A. Tayal et al., Surf. Coatings Technol. E **275** (2015)
7. W. Q. Khan, Q. Wang, X. Jin, and G. Yasin, Phys. B Condens. Matter. E **524** (2017)
8. V. Hari Babu, J. Rajeswari, S. Venkatesh, and G. Markandeyulu, J. Magn. Magn. Mater. E **339** (2013)
9. M. Sifkovits, H. Smolinski, S. Hellwig, and W. Weber. E **204** (1999)
10. D. Gölden, E. Hildebrandt, and L. Alff, J. Magn. Magn. Mater. E **422** (2017)
11. P. Prieto, J. F. Marco, and J. M. Sanz, Surf. Interface Anal. E **40**, (2008)
12. R. Niewa and F. J. DiSalvo, Chem. Mater. E **10**, (1998)
13. P. Eklund, M. Beckers, U. Jansson, H. Högberg, and L. Hultman, Thin Solid Films. E **518** (2010)
14. J. M. Mchale, A. Navrotsky, G. R. Kowach, V. E. Balbarin, and F. J. Disalvo. E **2**, 1 (1997)
15. D. S. Bem and H.-C. Zur Loye, J. Solid State Chem. E **104** (1993)
16. G.-K. Li et al., E **589** (2015)
17. L. Thilly, R. Ghisleni, C. Swistak, and J. Michler, Philos. Mag. E **92** (2012)
18. J. A. Venables, G. D. T. Spiller, and M. Hanbucken, Reports Prog. Phys. E **47** (1984)
19. G. R. Etaati, M. T. Hosseinnejad, M. Ghoranneviss, M. Habibi, and M. Shirazi, Nucl. Instruments Methods Phys. Res. Sect. B Beam Interact. with Mater. Atoms. E **269** (2011)
20. P. Delobelle, O. Guillon, E. Fribourg-Blanc, C. Soyer, E. Cattan, and D. Rèmiens, Appl. Phys. Lett. E **85** (2004).
21. S. Y. Kuan and J. C. Huang, Thin Solid Films. E **561**, 43–47 (2014)
22. G. M. Pharr and W. C. Oliver, Mrs Bull. E **17** (1992)
23. Z. Zuo, Z. Xu, R. Zheng, A. Khanaki, J.-G. Zheng, and J. Liu, Sci. Rep. E **5** (2015)
24. Y. Sugita, K. Mitsuoka, M. Komuro, H. Hoshiya, Y. Kozono, and M. Hanazono, J. Appl. Phys. E **70** (1991)
25. L. Rissanen, M. Neubauer, K. . Lieb, and P. Schaaf, J. Alloys Compd. E **274** (1998)
26. L. Rissanen, P. Schaaf, M. Neubauer, K.-P. Lieb, J. Keinonen, and T. Sajavaara, Appl. Surf. Sci., E **138–139** (1999)
27. I. Dirba, P. Komissinskiy, O. Gutfleisch, and L. Alff, J. Appl. Phys. E **17** (2015)
28. A. N. Timoshevskii, V. A. Timoshevskii, B. Z. Yanchitsky, and V. A. Yavna, Comput. Mater. Sci. E **22**, 99–105, (2001).
29. K. Yamaguchi, T. Yui, K. Yamaki, I. Kakeya, K. Kadowaki, and T. Suemasu, J. Cryst. Growth. E **301–302**, (2007)
30. G. Gubert, E. Ribeiro, J. Varalda, A. G. Bezerra, W. H. Schreiner, and D. H. Mosca, J. Alloys Compd. E **725**, (2017)
31. C. Navio, J. Alvarez, M. J. Capitan, F. Yndurain, and R. Miranda, Phys. Rev. B - Condens. Matter Mater. Phys., E **78**, (2008)
32. J.-S. Chen, C. Yu, and H. Lu, J. Alloys Compd. E **625** (2015)
33. C. Sánchez-Hanke et al., J. Appl. Phys. E **99** 1956-1967 (2006)
34. J. M. D. Coey and P. A. I. Smith, J. Magn. Magn. Mater. E **200** (1999)
35. J. M. D. Coey and H. Sun, J. Magn. Magn. Mater. E **87** (1990)

Injection Moulding Recommended Parameters Simulation Analysis of Rice Husk Composite

Haliza Jaya[1], *Nik Noriman Zulkepli*[1], *Mohd Firdaus Omar*[2*], *Shayfull Zamri Abd Rahman*[3], *and Khairul Anwar Abdul Halim*[2]

[1]Faculty of Engineering Technology, Universiti Malaysia Perlis (UniMAP), 01000, P.O. Box, D/A Pejabat Pos Besar, Kangar, Perlis, Malaysia
[2]School of Materials Engineering, Universiti Malaysia Perlis (UniMAP), Kompleks Pengajian Jejawi 2, 02600 Arau, Perlis, Malaysia
[3]School of Manufacturing Engineering, Universiti Malaysia Perlis (UniMAP), Kampus Alam Pauh Putra, 02600 Arau, Perlis, Malaysia

Abstract. Study of recommendation parameters for injection moulding process was carried out with aid of Autodesk Moldflow Insight. Rice husk filled low-density polyethylene was used as the raw material. Standard dog-bone shape mould with ASTM were used throughout the simulation analysis. The investigations were carried out on fill analysis, fill + pack analysis and cool analysis on injection moulded dog-bone shape. From the simulation analysis, the recommended parameters of the injection moulding process for rice husk composites were determined.

1 Introduction

At this date, various field of industries interested in injection moulding has either read the power of flow analysis for injection moulding, or used the technology to aid in the design of injection molded products [1]. It is commonplace for new designs to be numerically simulated or even optimized via computer simulation before the first piece of steel is cut for the mold. Rather than expand time and money building a series of prototypical tools to fine-tune a new product, computer simulation has dramatically affected the way new parts are designed. Designers and process engineers have effectively leveraged the power of simulation to shorten the design process and to eliminate processing problems that once caused reductions in quality and profitability.

Computer simulations offer the tremendous advantage of enabling designers and engineers to consider virtually any molding option without incurring the expense associated with material waste and machine time. The ability to try out new designs or concepts on the computer gives the engineer the opportunity to eliminate problems before beginning production [2]. In addition, the engineer can quickly and easily determine the sensitivity of specific molding parameters on the quality and production of the final product. This computer aided engineering (CAE) virtual design offers tremendous flexibility to the engineer to determine the effects of different gating scenarios, different runner designs, geometric

* Corresponding author: firdausomar@unimap.edu.my

features, and different molding and processing conditions on the moldability and quality of the final product [3]. Furthermore, these CAE tools allows these virtual designs to be completed expediently in a matter of days or even hours rather than the weeks associated with experimental trial and error. Commercially available simulation had been available as early as the mid-1970s in the form of time-shared computer simulation using a lay-flat approach. By the late 1980s a number of different companies such as Moldflow, C-MOLD, and Simcon offered simulation tools for CAE in injection moulding [4]. In addition to the commercially available packages, a number of research codes were developed by universities and independent companies.

2 Methodology

Autodesk Moldflow Insight (AMI) software was used for the simulation process. The purpose of simulation process is to determine the recommended value for the related parameters in injection moulding process. Some of the parameters were selected as an input variable and the others were fixed at the recommended setting. The ranges of input variables were also determined at this stage. The simulation process started by importing a part, a gating system and a cooling line design from modelling software and analyzed based on the process setting that has been set up in the software. In simulation analysis, three analyses were conducted such as fill analysis, fill + pack analysis and cool analysis. All recommended injection moulding parameters were obtained from related result of simulation analysis. The input parameters used in each analysis are listed in Table 1 to obtain the recommended parameters of the thick plat part.

Table 1. List of parameters used in simulation analysis

Analysis	Input parameter	Output parameter
Fill analysis	Material properties Mould temperature (°C) Melt temperature (°C)	Injection profile
Fill + Pack analysis	Material properties Injection profile Mould temperature (°C) Melt temperature (°C)	Packing pressure (MPa)
Cool analysis	Material properties Mould temperature (°C) Melt temperature (°C) Coolant inlet temperature (°C)	Cooling time (s) Packing time (s)

2.1 Import 3D part and gating system

The first step in the simulation process involved importing the part and gating system into AMI software. Two cavities mould were selected according to ISO 3167:2002 (E) for thick flat parts as shown in Fig. 1. Next, the part and gating system were meshed.

IOP Conf. Series: Materials Science and Engineering **701** (2019) 012048 doi:10.1088/1757-899X/701/1/012048

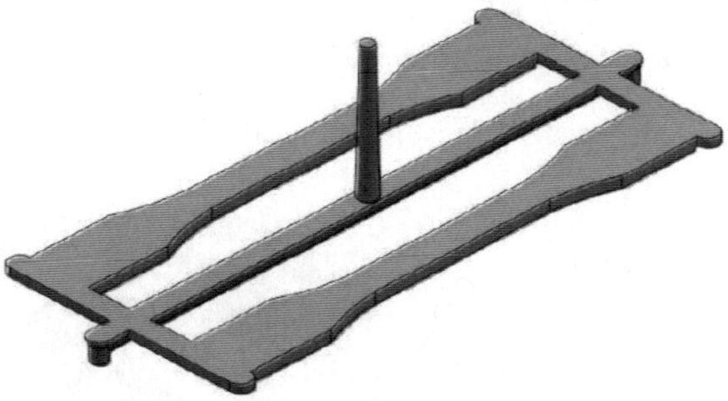

Fig. 1. Thick flat parts with gating system

2.2 Meshing part and gating system

The part and the gating system were meshed in 3D Mesh to obtain more accurate results as shown in Fig. 2. As many as 345914 tetrahedral elements were built in part and gating system, that are connected with 64123 nodes.

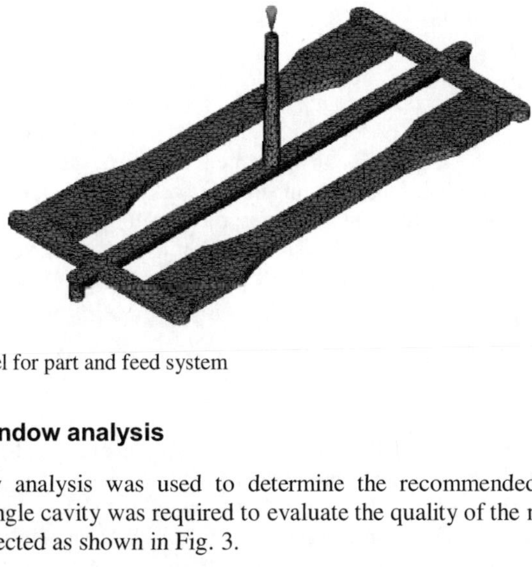

Fig. 2. Meshed model for part and feed system

2.3 Moulding window analysis

Moulding window analysis was used to determine the recommended mould and melt temperatures. A single cavity was required to evaluate the quality of the moulded part based on the material selected as shown in Fig. 3.

Fig. 3. Single part used in moulding window analysis

2.4 Create cooling circuit for core and cavity

Cooling circuit for core and cavity inserts were designed and modelled as shown in Fig. 4. Cooling lines were created using beam elements with 6 mm diameter. Eighty eight beam elements were created for cooling line that is connected using 90 nodes.

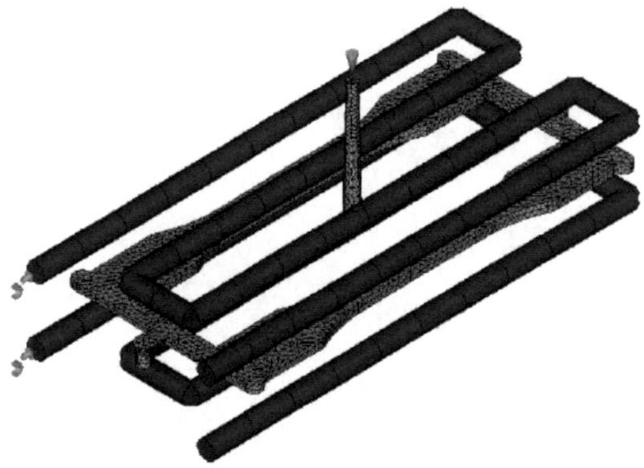

Fig. 4. Cooling circuit design

2.5 Fill and Pack analysis

The input parameters for injection moulding process were set in process setting taskbar. These settings were used as a guideline to calculate the results. The parameters such as melt temperature, packing pressure, mould temperature, velocity/pressure (V/P) switch-over, packing time, and injection profile were set in the process setting. Besides, material properties of mould and injection moulding machine specification were also set in this window.

3 Results and Discussion

3.1 Fill analysis results

The fill analysis was conducted to observe the flow of molten plastic into the mould cavities from the gating system. A fill analysis based on the recommended setting of mould and melt temperature, automatic setting in filling control, V/P switch-over and default packing/holding control, was conducted without consideration on the cooling channel designed. The filling time (s) was observed in this analysis and the results are shown in Fig. 5. The fill time and maximum shear rate obtained in this analysis were used as a reference to determine the filling velocity profile. In addition, the maximum shear rate should not exceed the maximum shear rate allowed in material database. Shear rates are important to the design and moulding of plastic products. As shear rates increase beyond the allowable maximum shear rate, it will increase the stress towards the molecular chains that make up the polymer and eventually damage the molecular chains apart. This will lead to the poor mechanical properties of a product [5].

Fig. 5. Result fill time from fill analysis: 5.498 s

3.2 Fill + Pack analysis results

The packing pressure evaluation was done based on the fill analysis results. From fill analysis results, the maximum injection pressure to fill the cavities is 18.6 MPa. The mass of parts depicted in Fig. 6 were evaluated by increasing the packing pressure until none or only small changing of the part mass occurred. Part mass shows slight increase with the increase of packing pressure until 16.11 MPa. Part mass remained 14.1 g even after 18.9 MPa packing pressure was used. This result shows that 16.11 MPa packing pressure was adequate for the moulded process.

Pack Analysis

Time (s)	Packing (%)	Inj Press (MPa)	Clamp F (tonne)	Part Mass (g)	Frozen Vol (%)	Status
5.619	0.02	1.611e+01	5.41e+00	1.39e+01	10.61	P
5.631	0.04	1.611e+01	5.91e+00	1.39e+01	10.60	P
5.653	0.08	1.611e+01	6.53e+00	1.39e+01	10.65	P
5.693	0.16	1.611e+01	7.29e+00	1.39e+01	10.76	P
5.746	0.26	1.611e+01	7.94e+00	1.39e+01	10.93	P
5.825	0.41	1.611e+01	8.40e+00	1.39e+01	11.22	P
5.992	0.74	1.611e+01	8.65e+00	1.40e+01	11.95	P
6.568	1.85	1.611e+01	8.76e+00	1.40e+01	14.47	P
7.711	4.05	1.611e+01	8.45e+00	1.41e+01	19.18	P
8.923	6.39	1.611e+01	6.24e+00	1.41e+01	23.87	P
9.552	7.60	1.611e+01	3.80e+00	1.41e+01	26.57	P
10.716	9.85	1.611e+01	2.97e-01	1.41e+01	31.28	P
11.951	12.23	1.611e+01	1.06e-02	1.41e+01	36.32	P
13.179	14.60	1.611e+01	1.00e-02	1.41e+01	39.96	P
14.865	17.85	1.611e+01	9.11e-03	1.41e+01	45.41	P
15.576	19.22	1.611e+01	8.92e-03	1.41e+01	47.79	P
15.577	19.23	1.409e+01	7.95e-03	1.41e+01	47.79	P
15.582	19.24	6.040e+00	4.10e-03	1.41e+01	47.81	P
15.586	19.24	0.000e+00	1.22e-03	1.41e+01	47.82	P
15.601	19.27	0.000e+00	1.21e-03	1.41e+01	47.87	P

Fig. 6. Part mass in Pack analysis

3.3 Cool analysis results

Packing time and cooling time were determined in this section. Cool Finite Element Method (FEM) analysis was run with the consideration of cooling channel design and coolant inlet temperature according to the software recommendation. The result of cooling time of the moulded part was obtained from time to time to reach the ejection temperature as shown in Fig. 7. The recommended cooling time is 12.03 s while the recommended time for packing is 10.3 s. From the Cool analysis results, packing time can also be identified. Gate is the smallest area in the mould that the first region will be frozen. When the gate freezes, there are no material allowed to enter the part. Therefore, the packing time is equal to the time of gate freeze as portrayed in Fig. 8.

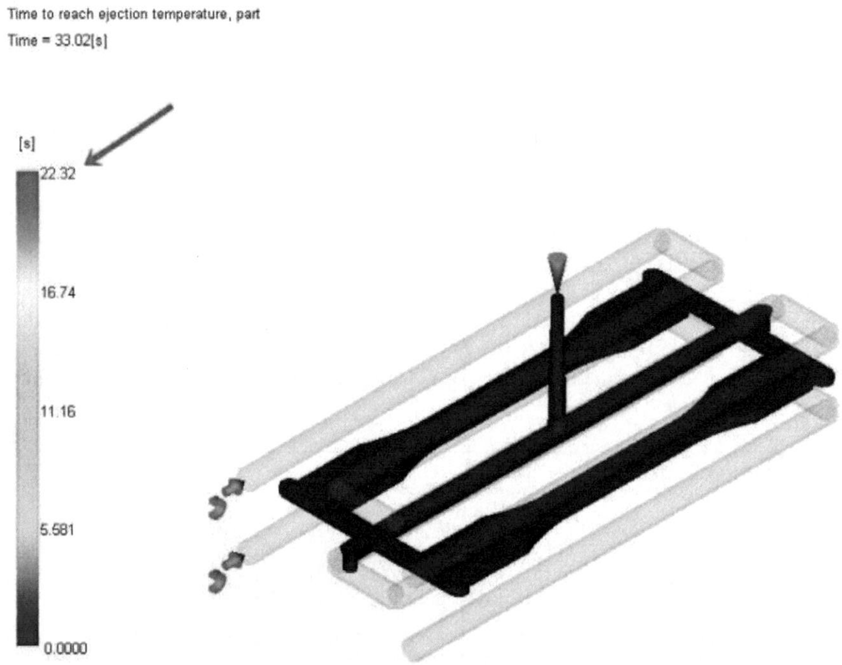

Fig. 7. Recommended cooling time : 22.32 s

IOP Conf. Series: Materials Science and Engineering **701** (2019) 012048 doi:10.1088/1757-899X/701/1/012048

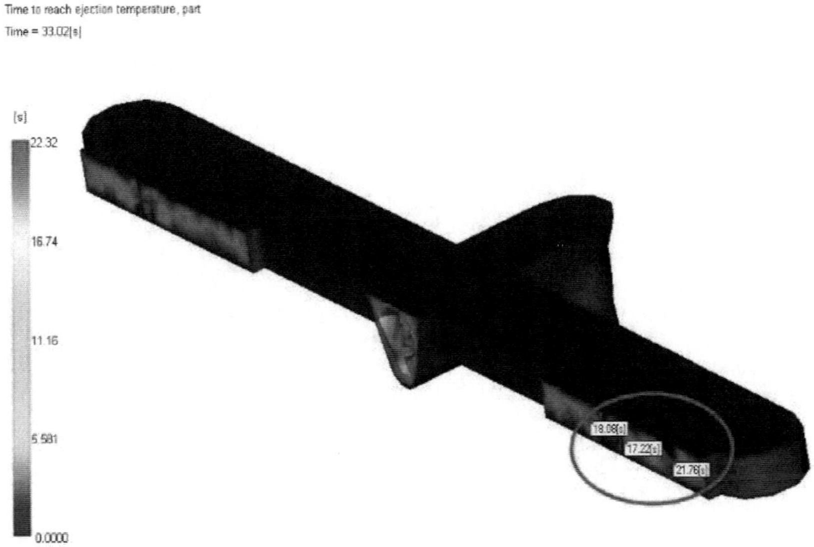

Fig. 8. Recommended packing time: 19.02 s

Based on the simulation analyses, the recommended parameters for the injection moulding process were summarized in Table 2 according to AMI 2012 software. These results were used as a reference in the next step of this study. All the values in Table 2 are satisfy the requirement of injection moulding machine used in this study based on the LDPE properties.

Table 2. Summary of recommended injection moulding parameters via simulation

Parameter	Recommended value
Mould temperature (°C)	50
Melt temperature (°C)	170
Fill time (s)	5.498
Packing time (s)	19.02
Cooling time (s)	22.32
Initial ram speed (mm/s)	8.237
Final ram speed (mm/s)	8.318
Initial ram position (mm)	66
Second ram position (mm)	50.4
P/V switch-over position (mm)	17.3
Cushion (mm)	5
Packing pressure (MPa)	16.11

Conclusion

Three types of analysis were simulated to obtain the recommended setting of machine parameters for moulding the LDPE/rice husk composites parts. Fill analysis was conducted to obtain the recommended filling profile on the molten plastic into the mould cavities. Then, Fill+Pack analysis was performed to determine an appropriate packing pressure of the moulded part. Finally, the packing time and cooling time were observed in Cool Analysis. The results of the recommended parameters were discussed and compiled in Table 2. Therefore, the objective of this study which is to determine the recommended moulding parameters of the LDPE/rice husk composites parts has been achieved.

References

1. B. Benhadou, A. Haddout, M. Benhadouand, H. Bakhtari. Int. J. Mech. Eng. Technol., E **7** (2016)
2. Z. Zhang and B. Jiang. J. Wuhan Univ. Technol. Sci., E **22** (2007)
3. A. Rios, B. Davisand, P. Gramann. *Proc. Compos. Fabr. Assoc. Tampa* (2001)
4. Turng, L. S., M. Peić. *Proc. Ins. Mech. Eng. B J. Eng. Manuf.* E **216** (2002)
5. J. Shoemaker, Hanser Verlag. E **10**, 153 (2006)

IOP Conf. Series: Materials Science and Engineering **701** (2019) 012049 doi:10.1088/1757-899X/701/1/012049

Post-Corrosion Mechanical Analysis of Sn-Zn Alloys: A Short Review

Muhammad Firdaus Mohd Nazeri[1,*], *Mohamad Najmi* Masri[2], *and Ahmad Azmin* Mohamad[3]

[1]School of Materials Engineering, Universiti Malaysia Perlis, 02600 Jejawi, Kangar, Perlis, Malaysia
[2]Advance Materials Research Cluster, Faculty of Bioengineering and Technology, Universiti Malaysia Kelantan, 17600 Jeli, Kelantan, Malaysia
[3]School of Materials and Mineral Resources Engineering, Universiti Sains Malaysia, 14300, Nibong Tebal, Penang, Malaysia

Abstract. Renewed interest in investigating and improving the corrosion performance of the Sn-Zn based alloys were ignited as this alloy produces excellent mechanical properties but suffered from corrosion attack. To support the electrochemical investigations, mechanical analysis is normally performed. This article focuses on the most common mechanical analysis used, including the basic experimental setups, common parameters and selected case studies that are beneficial in understanding the corrosion process.

1 Introduction

Since the banned of lead (Pb) from electronic industry, large numbers of Pb-free Sn-based alloys have been developed to find the best replacement for Sn-Pb alloys [1]. To ensure the Pb-free alloys perform reliably during application, several key performance characteristics have been established for example comparable melting temperature with Sn-Pb alloys existing assembly lines, good wettability and solderability, adequate physical properties and most importantly, environmental friendly [2].

Corrosion was never viewed as critical design factor in Sn-Pb systems. This is contributed by the presence of the stable oxide layer on the surface. At the same time, minor differences in galvanic potential in between Sn and Pb allows small driving force for galvanic corrosion to take place [3]. However, the perspective is changed for the Pb-free alloy systems as the presence of various elements to replace Pb altered the corrosion dynamic [4]. Worse, the advanced recent technologies require very small interconnections in sub-micrometre, presence of corrosion agent possibly aggravates the damage caused by corrosion. Tin-Zinc (Sn-Zn) alloys, especially at the eutectic composition of Sn-9Zn was one of the Pb-free alloys that shows good attributes to succeed Sn-Pb alloys as the mechanical properties of this solder are considered good [5]. However, the large differences in potential between Sn and Zn seriously affects the corrosion performance. Majority of Zn-

* Corresponding author: firdausnazeri@unimap.edu.my

rich phase was reported to be selectively etched producing deep, elongated grooves on the surface [1]. This significantly affected the appearance and functionality of the solder. As part of reliability of Pb-free soldering, renewed interest in the corrosive have arisen around the globe. At the same time, subsequent characterisation methods have mainly been practised to support the electrochemical evaluation on Sn-Zn solders. Such common techniques include mechanical analysis.

In general, subsequent mechanical analysis on the corroded Sn-Zn is important as it gives information on the changes impart by the corrosion attack on the alloy. When equipped with microstructure and phase characterizations, mechanical analysis offer insight on the cause of failure analysis. This allows correct prevention or corrective actions to be carried out to improve reliability. Therefore, based on this short review undertaken in this study, post-corrosion subsequent mechanical analysis of tensile strength measurement is discussed regarding the capabilities of the equipment and the experimental parameters commonly used. Selected case studies associated with these methods are also discussed.

2 Mechanical analysis

Solder connections are constantly subjected to mechanical stresses and strains during operations. Therefore, preventive action importantly requires the understanding and information on the mechanical properties of the Sn-Zn solders especially in harsh conditions such as a corrosive environment and so forth. Currently, data on the mechanical properties of the Sn-Zn solder joint after exposure to corrosion are limited, such as the ultimate tensile strength (UTS) and shear strength. The UTS is one of the most crucial properties for solders. During application, tensile loading through mechanical handling causes bending to the solder joint [2]. In this case, to determine the UTS for bulk Sn-Zn solders, strain rates used are different, for example 0.9 mm/min [6], 3.33×10^{-4} to 3.33×10^{-3} [7], and 10^{-4} to 10^{-2} s^{-1} [8]. The size and shape of the specimen also varied. For instance, Chen et al. [9] used a rod-shaped sample with 16 mm gauge length and 4 mm diameter. Meanwhile, Kim et al. [10] prepared a dumbbell-shaped sample of 25 mm gauge length, 8 mm width and 1 mm thickness. In forming the joint, Cu and Al are among the substrate choices used [11-13]. A butt joint for determining the UTS can be prepared with the Sn–9Zn solder positioned between the Cu plates [14](Fig. 1a). The clamped sample, in this case, is heated at 250 °C for 10 s along with the application of ZnCl-based flux, producing the bond for the end Cu/solder/Cu joint product (Fig. 1b). A crosshead speed of 2 mm/min is then preferably applied to evaluate the tensile strength of the joint [14]. In determining the shear strength, the lap joint is used, adding spacers and a reflow oven for reheating (Fig. 1c), as compared to the butt joint formation [15]. For the measurement of shear strength, the shear speed of 0.3 mm/min or equivalent to a shear rate of 10-2 s^{-1} was reported to be used [12].

IOP Conf. Series: Materials Science and Engineering **701** (2019) 012049 doi:10.1088/1757-899X/701/1/012049

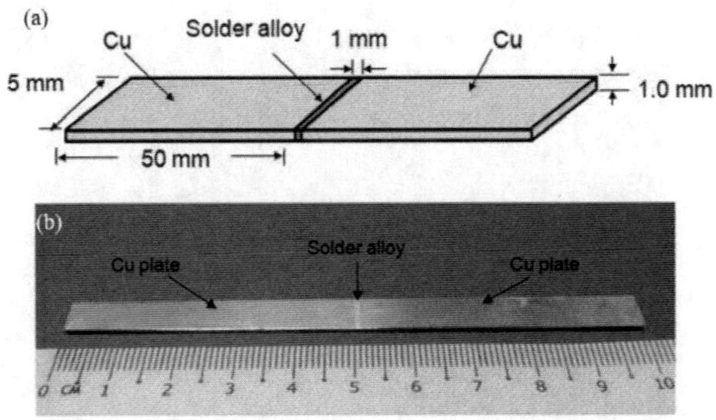

Fig. 1. Schematic diagram of (a) Cu/Sn-9Zn/Cu butt joint arrangement. Reprinted with permission from Ref. [14] and (b) example of Cu/Sn-9Zn/Cu butt joint.

As an example, for the Cu/Sn-9Zn/Cu butt joint, UTS of 45.9 MPa was reported. These values were in the common UTS range for the Sn-9Zn joint (30-80 MPa) [6, 9, 10]. The UTS dropped to 36.4 and 29.5 MPa (Fig. 2) after polarisation in 6 M KOH. The corroded joint also undergoes lesser elongation thereby reflecting a reduction on the ductility. Such reduction in the UTS coop well with the severity of crack initiation points due to the dissolution of Zn and cracks propagations throughout the joint, which explain the failure mechanisms (Fig. 3) [14]. Changes in the ratio of galvanic coupling (between the Sn and Zn) in the solder is a further factor that contributes to the corrosion tendencies of the solder joint [15].

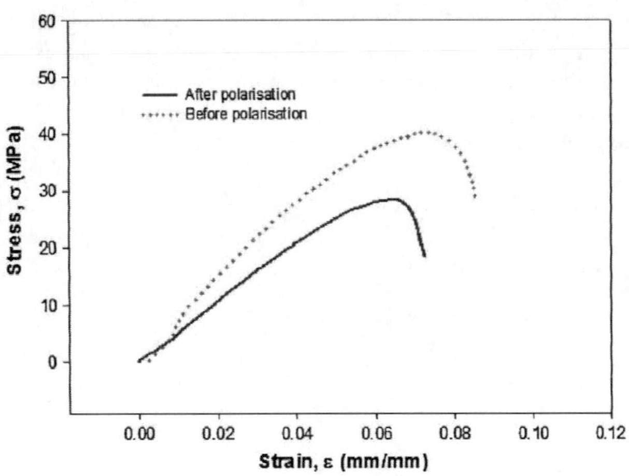

Fig. 2. Stress-strain curve of the Cu/Sn-9Zn/Cu joint before and after polarisation Reprinted with permission from Ref. [14].

IOP Conf. Series: Materials Science and Engineering **701** (2019) 012049 doi:10.1088/1757-899X/701/1/012049

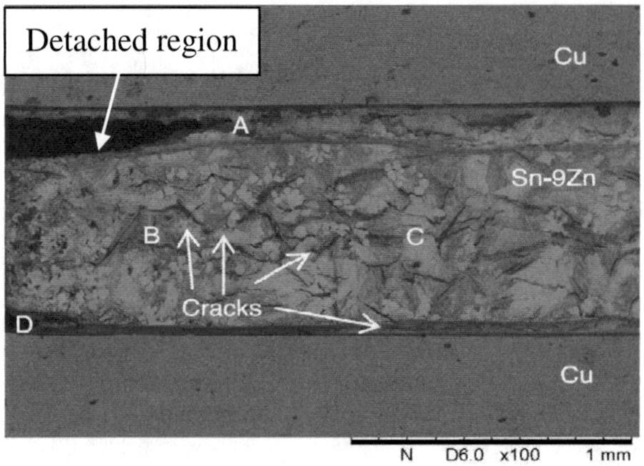

Fig. 3. SEM micrographs of polarised Cu/Sn–9Zn/Cu joints. Reprinted with permission from Ref. [14]

In a separate report, elastic regions from the stress-strain curve for Cu/Sn-9Zn/Cu joints were significantly influenced by the duration of exposure (1, 7, 14, 21 and 18 days) in 3.5 wt. % NaCl (Fig. 4a) [16]. In this case, the length of the flow hardening regions significantly decreased as the immersion time increased and the reduction in the shear strength values correlate well to the severity of the corrosion attack (Fig. 4b).

With the aid of the micrographs and images, changes on the fracture behaviour before (ductile) and after (brittle) corrosion attack are easily explained (Fig. 4c-d). In another case, the higher shear strength of 128.5 MPa was reported for the as-prepared Cu/Sn-9Zn/Cu lap joint [17]. After exposed to 3.5 wt. % NaCl, the shear strength significantly reduced to 70.97 MPa. Again, with the aid of using a micrograph and images, pit formation is noticeable which signals weakening on the structural integrity of the joint. Such selective attack (causing pits) were mainly associated with the dissolution of Zn, which upon shearing caused multiple dimples to occur thereby indicating ductile fracture behaviour [12].

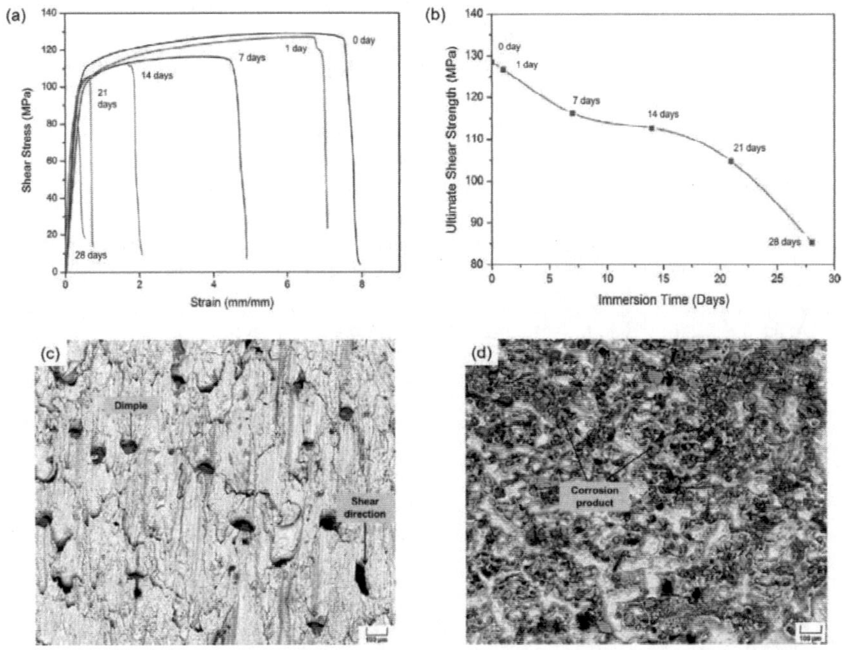

Fig. 4. The (a) stress-strain curve of the Sn-9Zn lap joint, (b) the ultimate shear strength of the Sn-9Zn lap joint, (c) the fracture micrographs of the Sn-9Zn lap joint before corrosion and (d) the fracture micrographs of the Sn-9Zn lap joint after corrosion. Reprinted with permission from Ref. [16].

3 Conclusion

As a subsequent testing to support electrochemical investigation, mechanical analysis required relatively simple set-up. Removal of active material (zinc) during corrosion process was found to significantly affect the mechanical integrity by allowing crack propagations in the solder and reduces the ductility. Furthermore, the duration of exposure also affects the flow hardening region in shear strength. This leads to brittle fracture. For consideration in future research work in post-corrosion study, the effect of solidification time on the grain size and corrosion resistance and modelling/simulations on predicting the corrosion performance should provide insightful discussion.

The authors would like to acknowledge the support from the Fundamental Research Grant Scheme (FRGS) under a grant number of FRGS/1/2019/TK05/UNIMAP/02/5 from the Ministry of Education Malaysia.

References

1. Jumali, N., A.A. Mohamad, M. Nazeri, and M. Firdaus. Materials Science Forum. Trans Tech Publ. 2017.
2. Abtew, M. and G. Selvaduray, Mater. Sci. Eng. R-Rep. **27**, 2000.
3. Song, F. and S.R. Lee. 56th Electronic Components and Technology Conference IEEE. 2006.

IOP Conf. Series: Materials Science and Engineering **701** (2019) 012049 doi:10.1088/1757-899X/701/1/012049

4. Mohd Nazeri, M.F., M.Z. Yahaya, A. Gursel, F. Cheani, M.N. Masri, and A.A. Mohamad, SOLDER SURF MT TECH. **31**, 2019.
5. Jumali, N., M.H. Zainol, A.A. Mohamad, M. Nazeri, and M. Firdaus. Solid State Phenomena. Trans Tech Publ. 2018.
6. Chen, K.-I. and K.-L. Lin, J. Electron. Mater., **31**, 2002.
7. Kim, K.-S., T. Matsuura, and K. Suganuma, J. Electron. Mater. **35,** 2006.
8. Khodabakhshi, F., R. Sayyadi, and N.S. Javid, MAT SCI ENG A-STRUCT., **702**, 2017.
9. Chen, K.-I., S.-C. Cheng, S. Wu, and K.-L. Lin, J. Alloy. Compd. **416**, 2006.
10. Kim, Y.-S., K.-S. Kim, C.-W. Hwang, and K. Suganuma, J. Alloy. Compd. **352,** 2003.
11. Affendy, M.G. and A.A. Mohamad, Int. J. Electrochem. Sci., **7,** 2012.
12. Huang, M., N. Kang, Q. Zhou, and Y. Huang, J MATER SCI TECHNOL. **28,** 2012.
13. Zhao, G., Y. Jing, G. Sheng, and J. Chen, J MATER ENG PERFORM. **25**, 2016. 2037-2042.
14. Nazeri, M.F.M., A.B. Ismail, and A.A. Mohamad, J. Alloy. Compd., **606** : 2014.
15. Huang, M., Y. Huang, H. Ma, and J. Zhao, J. Electron. Mater. **40**, 2011.
16. Affendy, M.G., M.Z. Yahaya, F.C. Ani, and A.A. Measurement **66**, 2015
17. Affendy, M.G. and A.A. Mohamad, Int. J. Electrochem. Sci. **7**, 2012.

IOP Conf. Series: Materials Science and Engineering **701** (2019) 012050 doi:10.1088/1757-899X/701/1/012050

Transient liquid phase bonding for solder-a short review

Norainiza Saud[1,*] and *Rita* Mohd Said[1]

[1]Centre of Excellence Geopolymer and Green Technology (CEGeoGTech), School of Materials Engineering, Universiti Malaysia Perlis (UniMAP), Taman Muhibbah, 02600 Jejawi, Arau, Perlis, Malaysia.

Abstract. The issues on the replacement of high Pb solder for high power module application has been a major concern among researchers since Pb could caused toxicity to the environment. Alternatively, transient liquid phase (TLP) has becomes a potential bonding technology for high power module devices. Typical TLP bonding for solder is a bonding method that using an interlayer of low and high temperature metal. However, a new transient liquid phase bonding which dealing with solder in the form of powder and paste has been developed. Both bonding methods have successfully increased the melting temperature of the bonded alloys higher than the bonding temperature as will be discussed further.

1 Introduction

A challenge for the packaging of the power devices in electronics manufacturing is to find the suitable robust die attach materials. Robust die attach materials is the materials that capable of surviving extreme environment such as high temperature and high pressure. In recent years, the demand for high power electronics module, which requires highly reliable and stable functionality, has been rapidly increasing in particular for applications such as automotive, aerospace, deep-well drilling and energy production industries [1]. Previously, high melting points materials that can withstand high operation temperature (>300 ^oC) are Pb10Sn and Pb5Sn solder. However, the usage of Pb is banned for their toxicity and it has become a challenge for researcher and also manufacturer to find the suitable replacement for Pb-free robust die attach materials. Till now, the successful of finding the high Pb solder replacement is still limited [2]. Nevertheless, the replacement candidates that have gain researcher's attention are Au-Sn, Au-Ge, Zn-Al and Bi based solder alloys [2,3]. However, there are limitations for these solders. Table 1 summarize the strength and weaknesses of the solders.

Melting behaviour and soldering temperature are the critical issues that should be considered in designing the replacement for High Pb solders. A high melting temperature solder alloy would leads to a higher soldering temperature. Higher soldering temperature/parameter could be detrimental to the polymer that being used as die electric materials in the substrate [4]. The melting temperature should be 400 ^oC and above since

*Corresponding author: norainiza@unimap.edu.my

IOP Conf. Series: Materials Science and Engineering **701** (2019) 012050 doi:10.1088/1757-899X/701/1/012050

the operation temperature for high temperature application is more than 300 °C. [5]. Thus, it will leads to the high processing temperature as well as a higher cost to the solder's manufacturer. Not just that, the high temperature during reflow / soldering process can lead to high thermal stresses on the whole electronic circuit, accelerating a range of failure mechanisms to the components and reducing circuit lifetime [2].

Table 1: Strength and weakness of High-Pb solder replacement [2].

Solder	Strength	weakness	Cost
Zn-Al based	does not form IMCs	high corrosion	cheap
Bi based	Good solidification criteria and hardness properties	Poor thermal & electrical conductivity	cheap
Au-Sn based	Narrow solidification range	Formation of brittle phase ζ Au_5Sn	very expensive
Au-Ge based	does not form IMCs, and good thermal conductivity	Too high in hardness and must be alloyed with other elements	very expensive

Therefore a soldering method that could use the existence soldering temperature yet could withstand a high operational temperature has been established [6,7,1]. The method is known as transient liquid phase (TLP) bonding. Transient liquid phase (TLP) bonding or commonly termed as Solid Liquid Inter diffusion (SLID) process is a method that enable a solder joint to be processes at a lower temperature while still resulting in the formation of a joint with a higher re-melting point [3]. This bonding method involved completely consumption of low-melting-point metal and formation of high-melting-point intermetallic [8].

The TLP bonding offers various processing advantages over conventional soldering process including rapid densification, lower soldering or bonding temperatures, reduce microstructural coarsening, void free interface and reduce cost by avoiding the use of expensive pre-alloyed powders [9,10,11].

2 Methodology

2.1 TLP bonding process for solid solder alloys

In the TLP bonding process, typically, a full intermetallic compounds (IMCs) joint can be obtained at low temperature through the diffusion reaction between high-melting metal substrate such as Cu and Ag and low-melting interlayer such as Sn [1,9]. During heating at above the liquidus temperature of Sn, low melting layer (Sn) diffuses into the high-melting metal substrate and form an intermetallic phases [9,10] as illustrated in Fig 1. The heating temperature is just around 280 °C if low melting point metals such as Sn, Bi, In and etc are chosen as the interlayer [2]. The sandwiched layers were then being hold at the bonding temperature until the liquid phase has solidified due to diffusion process and another type of peritectic IMCs (as stated in the phase diagram of the alloy) will form as can be seen in Fig 1 b). Additional heating such as annealing heat treatment will increase the size / number of the peritectic phases as shown in Fig 1 c). Once the TLP bonding process is completed, the melting temperature of the alloys is raised. The resultant alloys will not melt if the same heating temperature is used. It will only remelt if higher heating temperature is used.

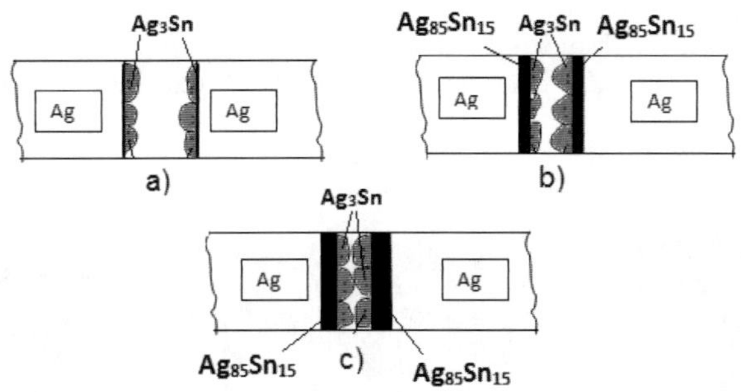

Fig. 1. The schematic illustration interfacial intermetallic formation for TLP bonding for Ag/Sn/Ag layer [9]

2.2 TLP bonding process for solder powder / solder paste.

There is also a research on the TLP bonding process for metal alloys in the form of powder and paste [2] since the application of solder alloys are in the form of solid, powder and paste [4]. Fig 2 illustrate the bonding mechanism of solder powder/paste. Basically, the TLP soldering principles is as follows: the bonding material used a mixture of high and low melting point metal powders. The low melting-point metal powder melts and react with the high-melting-point metal powder to form high-melting-point intermetallic compounds or solid solution and finally subsequent heating will allow the IMCs achieve a high-temperature-resistance bonding [12].

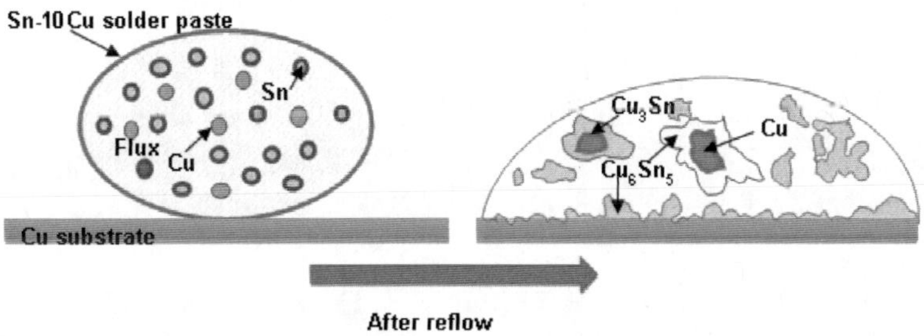

Fig. 2. TLP concept for bonding mechanism for solder powder/ solder paste during heating process.

3 Results

3.1 Intermetallic compound (IMCs)

One of the most important factors that influence solder joint reliability is the formation of IMCs either at the interface of in the bulk solder areas. However, the IMCs formation mechanism by using TLP bonding process is different from the typical IMCs formation during convention soldering process.

Moktari and Nishikawa [3] in their study on TLP of Sn-Bi solder sandwiched with Cu substrate have reported that the formation of Sn-Cu IMC and Bi-rich leads to increase in the melting temperature of the solder joint. It is believed that the addition of Cu particles results in the nucleation and growth of the IMC as can be seen in Fig. 3. From the DSC analysis, once the solder joint bonding temperature and time increase, it allows the Sn and Cu atoms to react and produce a high melting temperature of Sn-Cu IMCs as shown in **Fig.** 4. The valley of 667 and 700 °C represent the melting of the Cu_6Sn_5 and Cu_3Sn IMCs, respectively.

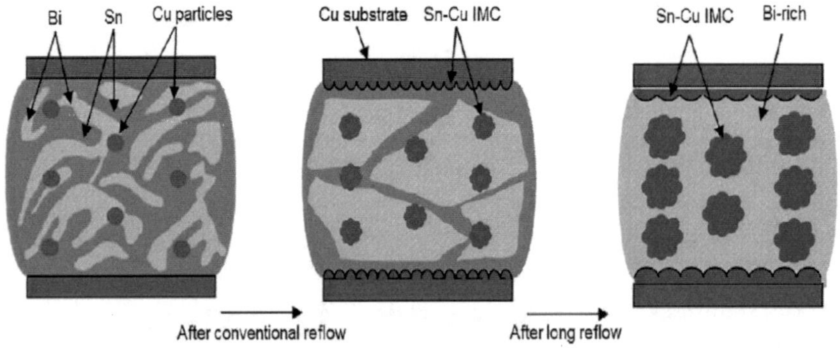

Fig. 3. The illustration of the conversion of the eutectic Sn–Bi to the Sn–Cu IMCs and Bi.[3]

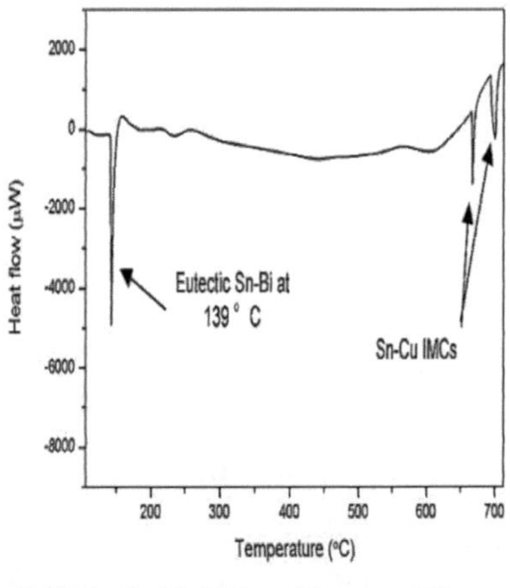

Fig. 4 The DSC peak of Sn-Cu IMCs in Sn-Bi-30Cu solder paste [3]

As in another case, Fig. 5 shows the back-scattered electron (BSE) micrographs of Sn-Cu based solder joints reflowed 235 °C for different bonding times studied by Sun et al, 2019 [13]. As the bonding time increased, Sn continues to be consumed to form the IMCs (cu_6Sn_5 and Cu_3Sn). After 90 min of bonding time, the IMCs at both ends were connected. And the Sn in the liquid phase was became an island-shape/peritectic phase (surrounded by Cu_3Sn and Cu_6Sn_5) after solidification. Afterwards, the island-shape gradually decreased, and a full IMCs solder joint was formed after 150 min. At this point, the alloys has already an elevated melting temperature.

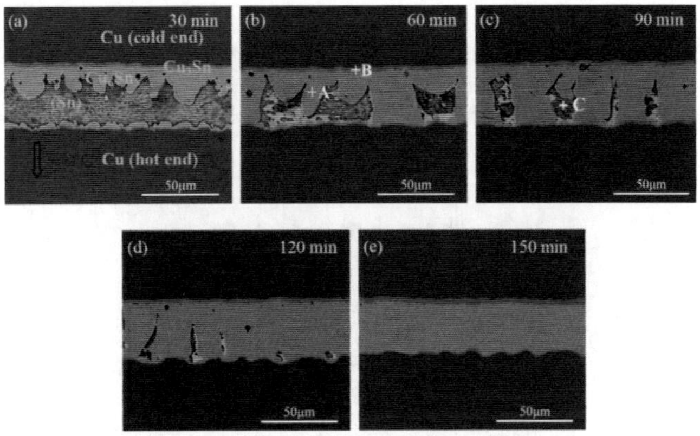

Fig. 5. Cross-section SEM images of Cu-Sn solder joints at 235 °C for; (a) 30 min, (b) 60 min, (c) 90 min, (d) 120 min, and (e) 150 min. [13]

3.2 Peritectic reactions

The Cu-Sn binary system is used here to illustrate the TLP routes for isothermal solidification as shown in Fig 6. Sn-Cu solder with Cu composition beyond 7.6 wt. % will undergo peritectic solidification.

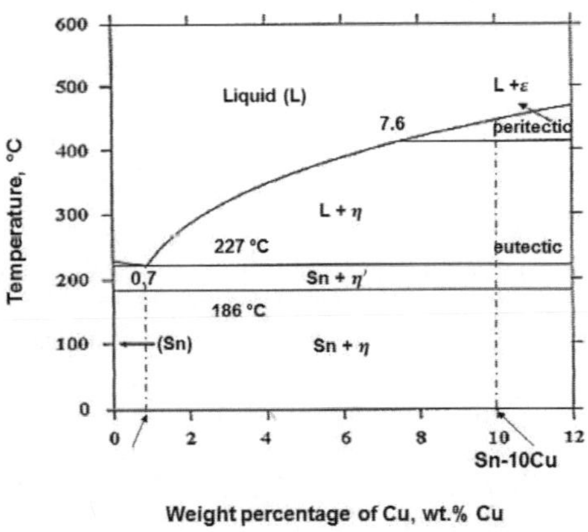

Fig. 6. Sn-Cu alloy phase diagram [17]

Generally, peritectic solidification involved three phases which are primary ε phase, peritectic η phase and liquid phase L. There are three main stages of peritectic solidification which are peritectic reaction, peritectic transformation and subsequent liquid precipitation of peritectic phase from the liquid melt [14]. Peng at el. in his research on peritectic reaction for sn-Cu-Ni solder has chosen Sn-10Cu solder composition as it close to 7.6 wt.% Cu, where it is still classified as a peritectic alloy. During peritectic reaction, the primary ε phase reacts with liquid phase produce peritectic η phase. The nucleation of η phase occurs at the interface between the primary phase and the liquid. Afterwards, a lateral growth of η phase around ε phase takes place as can be seen in Fig. 7. After isolation of primary ε from the liquid, the direct peritectic reaction can no longer take place. Then, the primary ε transforms to the peritectic phase by peritectic transformation. Peritectic transformation is an atomic diffusion control process and normally it takes a longer time to be completed. Due to the sluggishness of long-range solid-state diffusion, the peritectic transformation could not proceed completely. As a consequence, the final microstructure composed of both primary and peritectic phase [14,15,16]. At this time, the melting temperatures for the IMCs shifted to > 415° C.

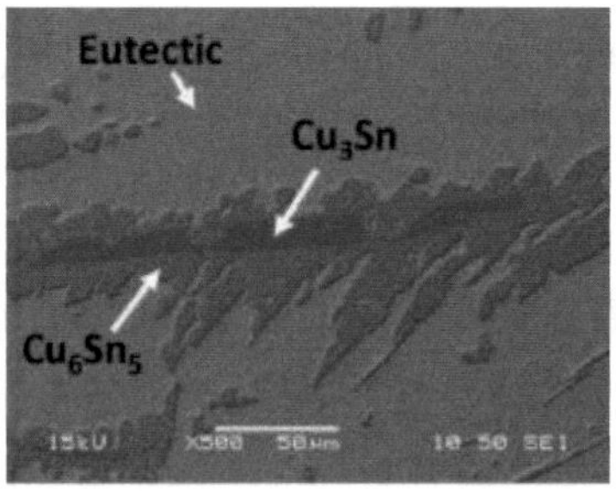

Fig 7. Peritectic microstructure of Sn-Cu based solder. Primary ε (Cu_3Sn) phase surrounded by η (Cu_6Sn_5) phase producing peritectic phase. [18]

From the above discussion, most of the intermetallic phases that results from the TLP bonding process could precipitate between the low and high melting point metals. However, the large difference in the melting point of the involved metals could leads to the formation of kirkendall void. As we know, different diffusion rate of interdiffusion couples could lead to the formation of kirkendall voids [19]. So, having Kirkendall voids right at the joint is also a matter of great concern. Therefore, further investigation on this matter is needed.

4 Conclusion

TLP bonding is a good alternative soldering process for solder that could replace the usage of previous high Pb solder alloys. It results in as-soldered alloys with a higher melting temperature than that used to join the solder. TLP bonding is successfully being applied to either solid, powder or paste solder. However, there is a possibilities on the formation of kirkendall voids at the resultant peritectic IMCs since different diffusion coefficient is involved during the TLP bonding process. Therefore, further study on the physical and mechanical properties of peritectic IMCs is needed for an established experimental data.

References

1. H. Shao, A. Wu, Y. Zhao, G. Zou, L. Liu. Microelectron. Reliab. E **91** (2018)
2. V. Chidambaram, J. Hattel, J. Hald. Microelectron. Eng. E **88** (2011)
3. O. Mokhtari, H. Nishikawa. J Mater Sci: Mater Electron. E **27** (2016)
4. A. Sharif, C.L. Gan, Z. Chen. J Alloy Compd. E **587**, 365-368. (2014)
5. Y. Sohn, C. Moon. Scr. Mater. E **109** (2015)
6. G.O. Cook, C.D. Sorensen. J.Mater.Sci. E **46** (2011)
7. F. Guo, S. Sun, Z. Zhang, H. Chen, M. Li. Microelectron. Reliab. E **80** (2018)
8. A.A. Bajwa, J. Wilde. Microelectron. Reliab. E **60** (2016)
9. S.F. Corbin, D.J. McIsaac. Mat Sci Eng A. E **346** (2003)

IOP Conf. Series: Materials Science and Engineering **701** (2019) 012050 doi:10.1088/1757-899X/701/1/012050

10. X. Qiao, S.F. Corbin. Mat Sci Eng A. E **283** (2000)
11. S. Cheng, C.M. Huang, M.A. Pecht. Microelectron. Reliab. E **75** (2017)
12. H. Feng, J. Huang, X. Peng, Z. Lv, Y. Wang, J. Yang, S. Chen, X. Zhao. Thermochim. Acta. E **663** (2018)
13. L. Sun, M-H. Chen, L. Zhang. J Alloy Compd. E **786** 677-687 (2019)
14. P. Peng, X. Li, J. Li, Y. Su, J. Guo, H. Fu. Sci. Rep. E **6**, 24512 (2016)
15. P. Lu, H. P. Wang. Sci. Rep. E **6**, 22641 (2016)
16. W. Zhai, B. Wei. Mater. Lett. E **108** 145-148 (2013)
17. A. F. Abd El-Rehim, H.Y. Zahran. J Alloy Compd**.** E **695** (2017)
18. S.U. Mehreen, K.Nogita, D. McDonald, H. Yasuda, D. St. John. J Alloy Compd. E **66** (2018)
19. X. Zhao, L. Duan, Y. Wang. J. Eur. Ceram. Soc. E **39** (2019)

Mechanical properties of Sn based lead free solder under different temperature

Norliza Ismail[1], *Maria* Abu Bakar[2,*], *Azman* Jalar[1,2] and *Muhammad Amer* Hariths Jasmin[1]

[1]*Pusat Pengajian Fizik Gunaan, Fakulti Sains & Teknologi, Universiti Kebangsaan Malaysia (UKM), 43600 Bangi, Selangor, Malaysia*
[2]*Institut Kejuruteraan Mikro & Nanoelektronik (IMEN), Universiti Kebangsaan Malaysia (UKM), 43600 Bangi, Selangor, Malaysia*

Abstract. In this study, effect of different temperature condition was investigated towards Sn-3.0Ag-0.5Cu (SAC305) solder wire. Tensile test with 0.1 mms^{-1} strain rate was performed at room temperature (25°C), 60°C, 120°C and 150°C chamber condition. Mechanical properties of SAC305 solder wire were evaluated from stress-strain curves which focused on yield stress, ultimate tensile stress (UTS) and elongation. From the stress-strain curves, there are exhibit the dynamic softening behavior for investigated solder wire. The higher temperature applied during the tensile test, induced the reduction of yield stress and UTS properties. However, elongation properties varied with the temperature condition. Highest percentage of elongation have been indicated at temperature 120°C. Therefore, different temperature conditions applied during tensile test have significant effect to the evaluated mechanical properties of SAC305 solder wire.

1 Introduction

Lead-free solder alloy becomes a replacement of the tin-lead eutectic alloy (Sn-Pb) which has been used for decades in electronic industries. This is due to increasing of environmental and human health concerns over the toxicity of lead. Sn-Ag-Cu solder alloys were considered as one of the most popular lead-free solders because of its good reliability and mechanical properties [1]. Mechanical properties of solder are crucial during the use of electronic products containing this solder material. Electronic components are often exposed to a variety of challenging mechanical loads such as bumps, vibrations and even impacts fall. The soldering connection will also be exposed to heat generated by electronic components during application. This environment brings with it high temperatures and high mechanical loads will alter the structure of the solder connection used in electronic components and the solder connection will have a plastic barrier. Changes in the structure and mechanical properties of this solder material after exposure to possible environment can deteriorate the function of the electronic devices [2]. Therefore, many studies have been done by previous researcher on the mechanical properties of solder [3-5].

* Corresponding author: maria@ukm.edu.my

Tensile test is the most popular and low-cost method to study the mechanical properties of bulk materials. Various important data regarding mechanical properties can be extracted from the tensile test such as young modulus, yield strength, ultimate tensile strength and elongation. From these properties, the strength and ductility of materials can be determined [6]. In the present study, tensile tests of SAC305 solder wire were carried out at various temperature chamber conditions to investigate the effect of test temperature on mechanical properties.

2 Materials and method

Lead-free solder wire of Sn-3.0Ag-0.5Cu (SAC305) with the 2 mm diameter was used in this study. Solder wire SAC305 was provided by RedRing Solder (M) Sdn. Bhd. Solder wire was cutted into 100 mm length and attach to the epoxy resin with 50 mm length both side as shown in Fig. 1. The epoxy resin attached to grip the solder wire during the tensile test. Tensile tests were performed using the Instron universal machine at four different chamber temperatures of 25°, 60 ° C, 120 ° C and 150 ° C. Strain rate of tensile test was setup at 0.1 mms^{-1}. The load cell for this experiment is 1 kN.

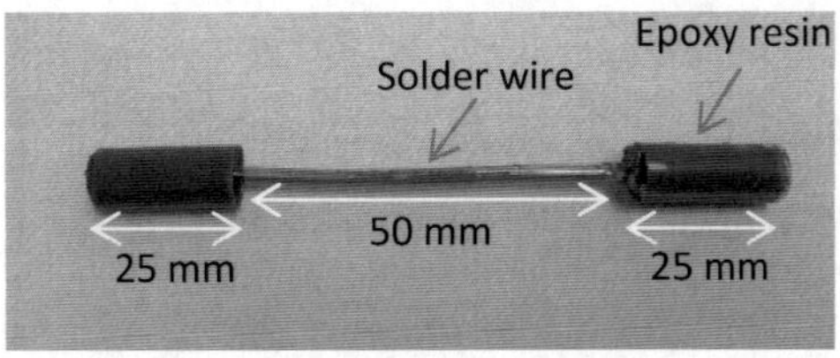

Fig. 1. Solder wire with the epoxy resin attachments.

3 Result and discussion

Fig. 2 shows the stress-strain curves of tensile test on SAC305 solder wire at different temperature. The trend exhibited the decreasing of stress with increasing of temperature. It is revealed that variation of temperature obviously effect the mechanical properties of SAC305 solder wire. In addition, stress-strain curves exhibit the abrupt changed of highest stress value to steady value as mark in the red circle of the curves. This behavior is known as dynamic softening where usually occurred at the initial point of yield stress [7].

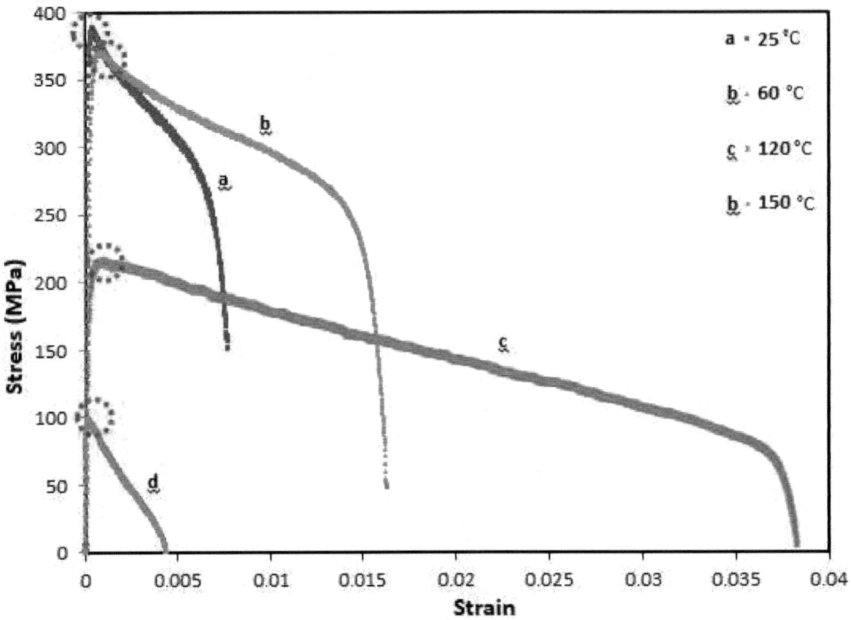

Fig. 2. Stress-strain curves at different temperature condition.

From the tensile test, others mechanical properties evaluated are yield stress, ultimate tensile stress and elongation. Fig. 3 presents the effect of temperature on yield stress. Yield stress decrease with increasing of temperature. Yield stress is known as maximum stress that can be applied before the material begin to change from elastic properties to plastic properties [8]. Parallel with the ultimate tensile strength indicated in Fig. 4, increasing of temperature, decreased the UTS values of tested SAC305 solder wire. UTS values correspond to the maximum stress in stress-strain curve [8]. Fig. 5 displays the trend of elongation with subjected to temperature. The values of elongation is calculated using this following formula:

$$\delta = \frac{L_f - L_o}{L_o} \tag{1}$$

where L_o is the original gauge length, and L_f is the gauge length at fracture. The value of elongation of SAC305 solder wire varied with the temperature condition applied. Percentage of elongation increase with the temperature up to 120°C and then drastically reduce at 150°C. SAC305 solder wire have highest elongation (about ~72%) at 120°C temperature condition. Therefore, it was expressed that at, high ductility SAC305 solder wire was induced at 120°C.

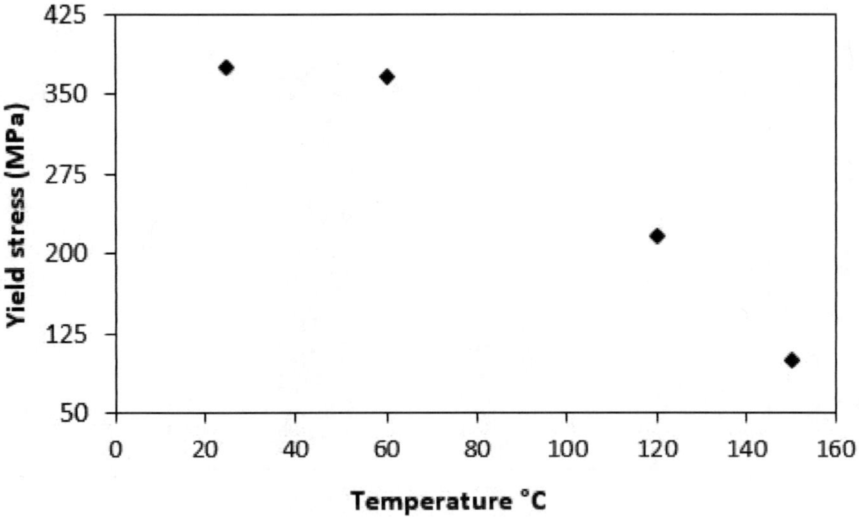

Fig. 3. Effect of temperature on yield stress.

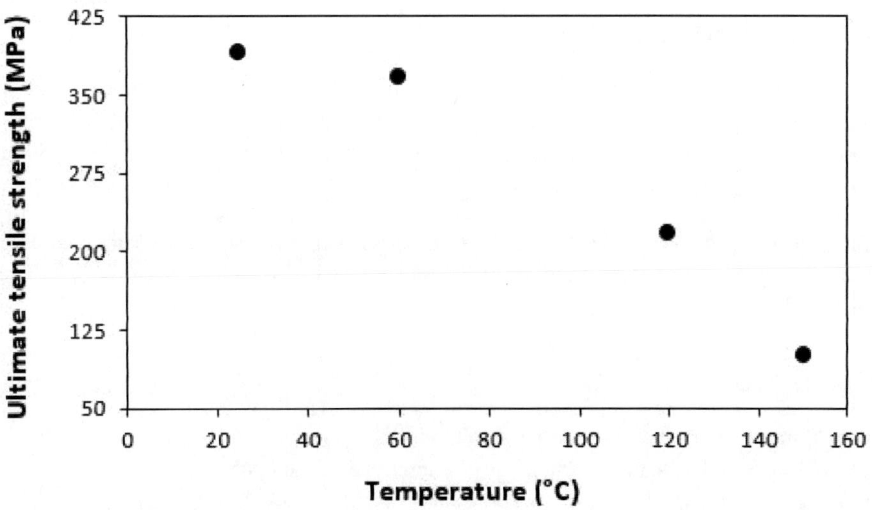

Fig.4. Effect of temperature on UTS.

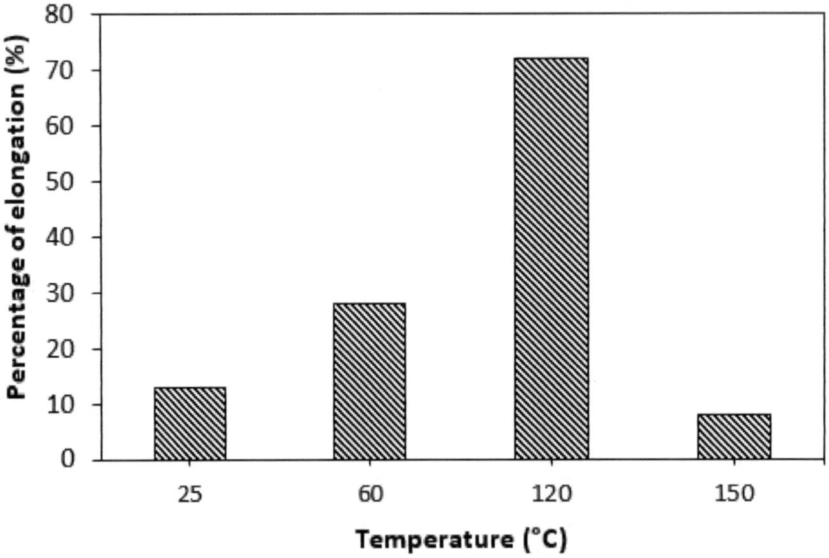

Fig.5. Effect of temperature on elongation at different temperature condition.

4 Conclusion

Variation of temperature condition have a significant effect to the mechanical properties included yield stress, UTS and elongation which evaluated from stress-strain curves of SAC305 solder wire. When temperature increase, yield stress and UTS show the decreasing trend. However, percentage of elongation are varied with temperature.

The authors would like to acknowledge the financial support provided by Ministry of Education, Malaysia (grant number FRGS/1/2019/STG07/UKM/03/1) and Universiti Kebangsaan Malaysia (grant number MI-2018-019).

References

1. L. Sun, L. Zhang. Adv. Mater. Sci. & Eng., (2015)
2. T. Kobayashi, I. Shohji, Y. Nataka. Adv. Mater. Sci. & Eng., (2018)
3. Z. Haidong, I. Shohji, M. Shimoda, H. Watanabe. Mater. Trans. **57**, (2016)
4. I. Abdullah, M.N. Zulkifli, A. Jalar, R. Ismail, M. A. Ambak. J. Elect. Mater. **48** ,(2019)
5. F. Cheng, F. Gao, J. Zhang, W. Jin, X. Xiao. J. Mater. Sci. **46**, (2011)
6. W. D. Callister, D. G. Rethwisch, *Materials ccience and engineering* (John Wiley & Sons, USA, 2014) ,
7. I. Abdullah, M. N. Zulkifli, A. Jalar, R. Ismail. Soldering & Surf. Mount. Tech. **29**, (2017)
8. X. Q. Shi, W. Zhou, H. L. J. Pang, Z. P. Wang. ASME J. Elect. Packg. **121**,(1999)

IOP Conf. Series: Materials Science and Engineering **701** (2019) 012052 doi:10.1088/1757-899X/701/1/012052

Flexural Properties of Al/Floral Foam Sandwich Composite Prepared by Hand Lay Up Process

Zainuddin Firuz[1,*], *Leong* Chit Shing[1], and *Syed Adam* Syed Nuzul Fadzli[2]

[1]School of Materials Engineering, Universiti Malaysia Perlis, Malaysia
[2]Faculty of Engineering Technology, Universiti Malaysia Perlis, Malaysia

Abstract. The flexural properties of aluminum sheet as the facings and floral foam as the core in layered sandwich composite were studied under three-point bending test. Adhesion of core onto aluminum sheets was done by manually spreading mixture of epoxy resin and hardener via hand lay-up procedure. There were four system studied, which are single foam-core sandwich system until four foam-core sandwich system. It was found that single foam-core sandwich has the highest flexural strength over the range of Al/floral foam-core sandwich system. The single foam-core sandwich has a well stress or total load distribution as compared to other three types of sandwiches. The smaller thickness to width ratio of single foam-core system is believed to enable the homogenous stress distribution of loading during testing. In addition, several modes of failure found such as delamination, crack and core failure. As the number of layers of sandwich increase, there were more modes of failure occur, which mostly due to the core failure of floral foam.

1 Introduction

A structural sandwich consists of face panels, a core material and adhesives to join them together. It is the structure composed of two thin, stiff face layers, separated by a thick, mid-layer of low-density materials. The faces protect the relatively vulnerable core material against damage or weathering [1]. The core layer is usually made of cellular polymeric foam materials, such as polyvinylchloride (PVC), polyurethane (PU), polystyrene (PS) while metallic and non-metallic honeycombs, balsa wood or trusses are the most common. In composite sandwiches, the faces are typically composite sandwiches or metals such as aluminum. Face panels are typically 0.3 to 13mm thick. They are chosen on the basis of weight, strength and in term of fabrication. A proper sandwich design based on dense face sheets can optimise compressional, tensional, torsional or flexural properties much more efficiently [2, 3]. Among many others, sandwich panels with metal and fibre reinforced plastics (FRP) skins have been investigated experimentally and analytically by several researchers over the years [4].

* Corresponding author: firuz@unimap.edu.my

1.1 Foam core sandwich panel

Under polymer matrix composites (PMCs), polymer is the medium which holds the reinforcing fibres in place. In order to giving the highest stiffness to weight ratio of common materials design, sandwich construction has been developed and improved to produce even tougher and lighter polymer composites in the future. This requires more reliable predictions about behaviours of polymer especially under extreme conditions.

A sandwich structure must have thin, dense, high strength facings, a low-density core, which is usually thick, relatively weak and a rigid attachment between them. If any one of those key requirements is missing, the structure is not a sandwich [5]. Stuart (1992) states there are no required relationship between the thickness of facings and the thickness of the core. However, in most practical cases, the core is at least three or four times the thickness of the facings and the more efficient sandwich structures are normally observed to have cores an order of magnitude thicker than the facings.

1.2 Adhesive technology

There are different types of adhesives for plastics including cyanoacrylates, urethanes, acrylics and epoxies. Cyanoacrylates are adhesives that cure at room temperature thus, mixing is not required. They produce bonds within a few minutes and some formulations can even achieve 89% of their bond strength after one hour. Cyanoacrylates have excellent shear and tensile strengths. However, they do have poor solvent resistance and poor peel strength. It can attack certain plastics such as polycarbonate (PC) that causes chemical stress cracks

Solvent free products, for instance, acrylic adhesives, form a thermoset polymer bond with plastics. Upon exposure to light, cured acrylics will produce bonds. Its bond strength and curing times can be varied through intensity and exposure to the curing light. Acrylic adhesives produce polymer bonds that have good weather resistance but excessive oxygen exposure during bonding will produce weak bonds.

Urethane adhesive is available in one and two part systems. With the application of a catalyst or heat, curing time is able to be regulated and minimal surface is needed. Adhesive bonds of urethane adhesives are flexible and have good mechanical properties [6]. Epoxies are adhesives that have been used extensively in the automotive industry for body panels and components. Epoxies are usually a two part mixed system, one part epoxy and the other is catalyst or curing agent. By addition of heat, curing times can be accelerated. However, toughened formulations have been developed to improve impact resistance. Adhesive bonds are typically stronger than the substrates and epoxy adhesives offer excellent resistance to various chemicals such as dilute acids, bases, solvents and oils. Besides, these adhesives offer good weatherability.

1.3 Failure modes

The compressed face of a sandwich panel is typically a thin metal sheet which is restrained against its tendency to suffer local buckling by support from the core material. The beams of the sandwich panel are made of face and core materials that yield. Three modes of failure occur: face yielding, face wrinkling and core shear. Debonding is not a significant mode of failure in beams that do not have large flaws in the adhesive layer. Triantafillou and Gibson, 1987 noted that yielding strength of the core may fail because most resin adhesives used in sandwich panels are brittle; they fail by sudden propagation of cracks, causing debonding and delamination (Fig. 1) [7]. Debonding does not happen all the time. Most of the sandwich beams tested failed by face yielding, face wrinkling or core shear, only three failed by debonding and four by local crushing of the core

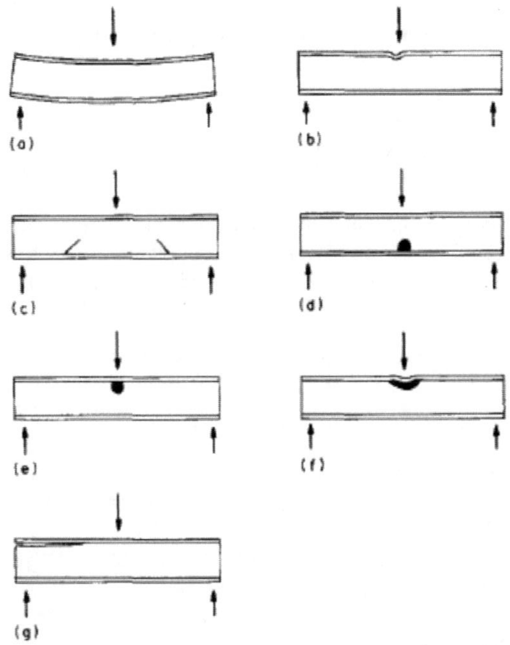

Fig. 1. The failure modes in a sandwich beam with a face that yields and a foam core: (a) face yielding; (b) face wrinkling; (c) core shear; (d) core tensile yield; (e) core compressive yield; (f) core indentation; (g) debonding [7].

2 Methodology

2.1 Materials

Floral foam is the core material for sandwich composite. It is dry and rigid floral foam which is mainly for artificial flower arrangement. For the skin, aluminum sheet is used for sandwich composite. The thickness for each face sheet is approximately 0.35mm. Epoxy resin 331 and hardener 8161 is the two-part adhesive for sandwich composites. The epoxy 331 is clear, with epoxide equivalent weight of 182-192, liquid diglycidyl ether of bisphenol-A (DGEB-A). Hardener 8161 is a clear curing agent (isophorone diamine (IPD)).

Adhesive was the mixture of epoxy hardener and resin with ratio of 2:1. The adhesive acted as the binder for the floral foam to be bonded to aluminum sheets. Epoxy resin and hardener were weighed using analytical balance in the ratio of 1: 2 and mixed well for 2 minutes.

2.2 Hand lay up process

The floral foam core and aluminium sheets were cut into desired dimensions as tabulated in Table 1. Core and aluminium sheet had the same length and width.

Table 1. Core and aluminium sheet dimension.

Element	Dimension (cm)
Length	20
Width	3

The adhesive mixture was spread using a flat blade onto aluminum sheets and slightly pressed the core to the surface of aluminium sheets. The sandwich composite was left for 2 days to ensure the sandwich composite to be cured completely. The sample dimension was prepared as in Table 2.

Table 2. Sample dimension.

Sample	Set Up
Control (Double Aluminium sheets) Length/cm: 20 Width/cm: 3 Thickness/cm: 0.12-0.13	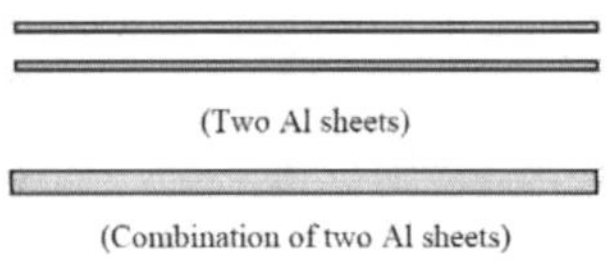
Control (Floral foam) Length/cm: 20 Width/cm: 3 Thickness/cm: 1.3-1.7	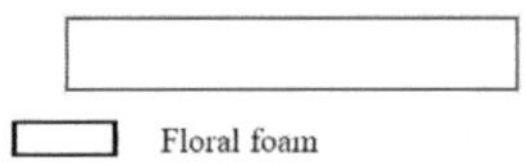
Single layer core sandwich composite Length/cm: 20 Width/cm: 3 Thickness/cm: 1.75-1.80	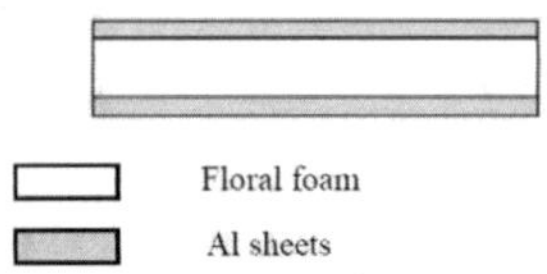
Double layer core sandwich composite Length/cm: 20 Width/cm: 3 Thickness/cm: 3.37-4.00	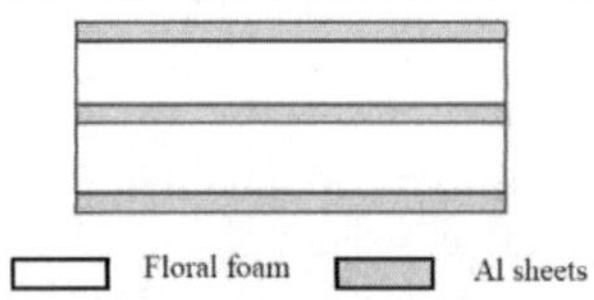
Three layer core sandwich composite Length/cm: 20 Width/cm: 3 Thickness/cm: 5.05-5.30	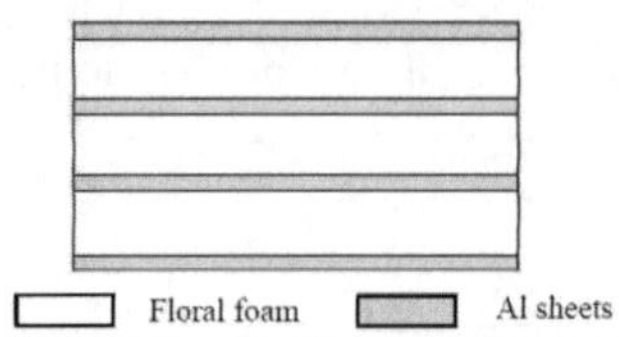

IOP Conf. Series: Materials Science and Engineering **701** (2019) 012052 doi:10.1088/1757-899X/701/1/012052

Four layer core sandwich composite Length/cm: 20 Width/cm: 3 Thickness/cm: 6.8-7.0	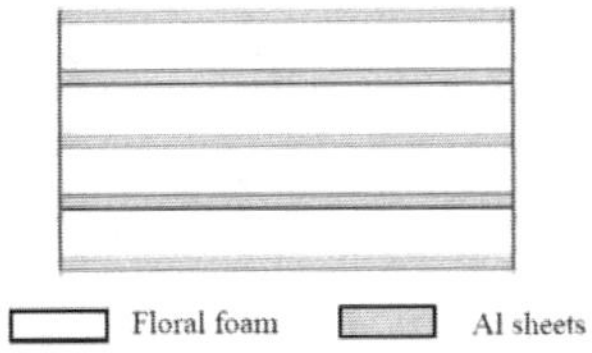

2.3 Flexural analysis

Flexural test was performed to measure flexural strength and flexural modulus according to ASTM D790-10 standard using Universal Testing Machine (Instron 5569) with a 5kN load cell. Crosshead displacement rate of 50 mm/min was constant throughout the test with span of 16 cm. Five samples are prepared for each system.

3 Results and discussions

The floral foam cells were crushed and collapsed during bending and compression, thus its curve assembled many failures where a sharp drop in flexural stress were being observed (Fig. 2). The sudden sharp decrease in the floral foam graph shows the complete failure of the foam core, which represent the complete fracture of floral foam. Single foam core sandwich composite possesses a smoother curve where the flexure stress increased until the 5% strain and started to decline until a broad peak being observed. The broad peak indicated the bending of the foam core reaching an end with the crushed foam near the support. At this point, crushed and collapsed foam cells are no longer support the flexure loading.

Curve for double foam cores sandwich shows a sharp decrease after 10% of flexure strain. It might be due to core failure that tends to occur at the second layer of foam core. The facings of sandwich composite provided the strength for the sandwich until the second failure occurred. Besides, Al sheet which act as the skin was easily debonded from the second foam core. The debond failure mode makes the second foam core vulnerable to crack initiation because the foam core is rigid and brittle.

The curve presented by three foam cores shows significant drop in flexural stress but the flexure strain of the sandwich composite was improved as compared to single foam core and double foam cores sandwiches. Higher flexure loading is needed as the amount of Al sheets increases with every stacked of foam core. Al sheets provide the resistance to bending of the sandwich composite.

The flexural strength of four foam cores sandwich was quite similar to three foam cores. However, its flexural strain increased because of the amount of Al sheets that possesses high ductility. Continuous flexural loading affected the foam core as it failed along the increase in loading but Al sheets exhibit highest modulus of elasticity that helped to support the bending process [8]. The core or skins can fracture under such stress besides the interface between the skins and the core.

IOP Conf. Series: Materials Science and Engineering **701** (2019) 012052 doi:10.1088/1757-899X/701/1/012052

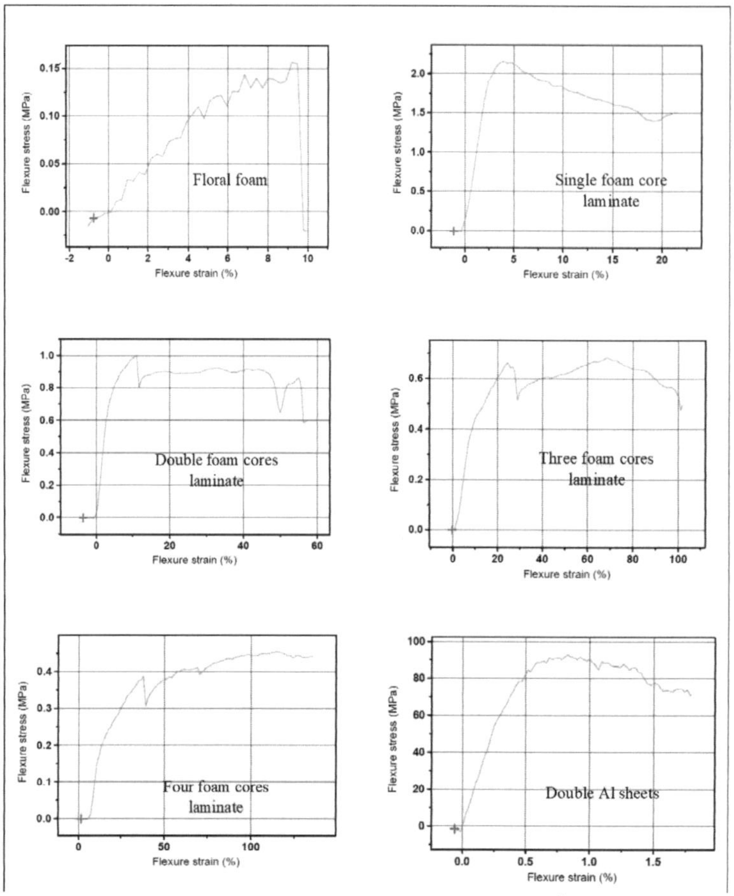

Fig. 2. Load displacement of double Al sheets, floral foam and four different types of sandwich composite.

As shown in Fig 3, single foam core sandwich has the highest flexural strength in comparison to double-foam-core sandwich, three-foam-core sandwich followed by four-foam-core sandwich. This is because of the single-foam-core sandwich has well stress or total load distribution as compared to other three types of sandwiches. Load is not distributed efficiently from the first foam-core layer until the final layer of foam-core as cracks or skin-foam delamination tends to fail the sandwich composite during flexural testing.

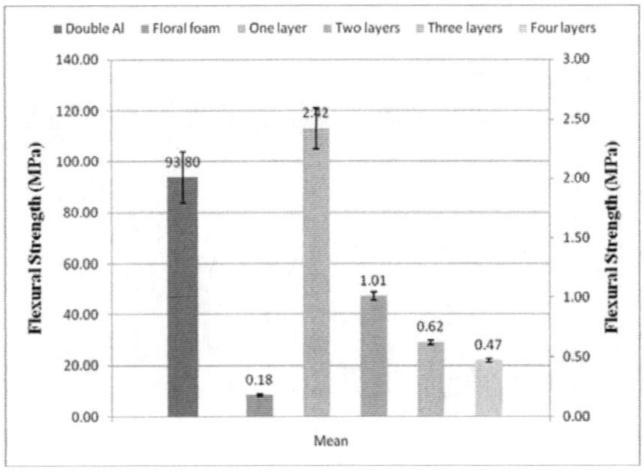

Fig. 3. Flexural strength of Al/PU foam composites.

There is cracking in core of sandwich composite under three-point bending near support as shown in Fig 4. In the region between the loading point and the supporting point, small shear cracks were generated inside the core. The increasing thickness of sandwich composite shows more core failures as compared to double foam-core sandwich. Sandwich composite experienced less flexural modulus as the Al facings limit the bending of sandwich composite. The Al sheets were not stiff enough to prevent any skin indentation and local crushing of the core below the loading point or above the supporting points. The cracks region is where maximum shearing stresses generated under flexural loading [3].

There were small cracks coalesced and formed a single leading crack with further flexural loading on the left side of the sample. Weak adhesion between foam and Al sheets also contributed to this failure mode.

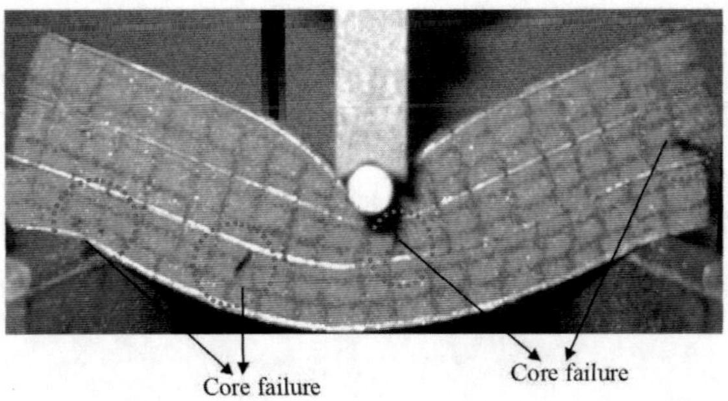

Fig. 4. Deformation of three layers foam core sandwich composite.

In Fig 5, the delamination happens at the first layer of the sandwich composite, where further loading contribute to failure and the second foam core was being forced into continuous delamination. Delamination of the cores and the facing is another important failure mechanism [3][9]. The epoxy adhesive joining the foam cores to polymer matrix composite skins is usually stronger than the foam. Thus, failure generally occurs at the adhesion area.

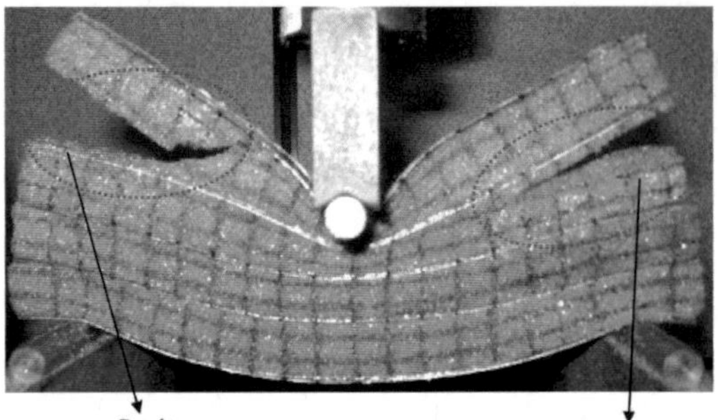

Fig. 5 Deformation of four layers foam core sandwich composite.

This work was financially supported by Minister of Higher Education under Fundamental Research Grant Scheme 9003-00483

References

1. B. Strong, *Fundamental of Composite Manufacturing: Materials, Methods and Applications* (1989)
2. M. F. Ashby, A. Evans, N. A. Fleck, L. J. Gibson, J. W. Hutchinson, H. N. G. Wadley, *Metal Foams – A Design Guide* (2000)
3. L. J. Gibson, M. F. Ashby, *Cellular solids: Structure and Properties 2nd Edition* (1997)
4. F. Fam, T. Sharaf, Compos Struct, **92** (2010)
5. M. L. Stuart, *Handbook of Composite Reinforcements* (1992)
6. F. Zainuddin, H. M. Akil, S. Ahmad, R. Rasid, S. N. F. S. Adam, Key Engineering Materials, (2014)
7. T. C. Triantafillou, L. J. Gibson, Mat Sci Eng, **95** (1987)
8. N. Gupta, E. Woldesenbet, J. Compos Mater, **39** (2005)
9. P. Martin, H. Peter, Compos Struct, **188** (2019)

IOP Conf. Series: Materials Science and Engineering **701** (2019) 012053 doi:10.1088/1757-899X/701/1/012053

Morphology and biodegradability of micro-crystalline cellulose / chitosan films

Song Cheah Chie [1]. *Mohamad Kahar* Ab Wahab[1]*

[1]School of Materials Engineering, University Malaysia Perlis, 02600, Arau, Perlis, Malaysia

Abstract The microcrystalline cellulose (MCC) chitosan film was produced through solution casting with different composition of chitosan and different plasticizers. Glycerol and polyethylene glycol (PEG) was used as a plasticizer in the MCC/chitosan film. Several testing on the biodegradability were conducted such as scanning electron microscope (SEM), moisture absorption and soil burial test. Results show that addition of plasticizer able to improve the miscibility of chitosan on the MCC film. According to moisture absorbtion and soil burial test, the addition of plasticizer showed the MCC/chitosan films were better in water resistance due to the plasticizer was able to facilitate the dispersion of chitosan in MCC films. The morphological analysis also found out that better interaction between MCC and chitosan when polyethylene glycol and glycerol was used as plasticizers in the films.

1 Introduction

Environment and sustainability issues have drawn attention from societies, this has giving outstanding improvements in bio-materials via the development of biocomposites in the field of polymer science [1-5]. Advantages of biocomposites which is not possible for synthetic polymer composite to do is able to depose and compose easily without harming the environment. There is a wide range of application for bio- and green composites, which is from structural to biomedical. Cellulose reinforcement and petroleum-based polymers composites are usually referred to as biocomposites, while composites based fully on naturally derived fibres and biopolymers have been named green composite [6]. However, the intrinsic chemical incompatibility between a hydrophobic polymer matrix and hydrophilic cellulose results in weak interfacial bonding between the cellulosic and biopolymer components, especially in the case of thermoplastic biopolymers [7]. There are many applications in natural fibre-reinforced biocomposites such as packaging, automobiles, building industries and aerospace which uses of product is not required high load carrying capacity [8]. The attractive properties of natural fibre compare to traditional counterparts are free formability, fatigue, substantial resistance to corrosion, low self-weight and relatively high specific strength. However, natural fibres are challenging in selection because of their insufficiency characteristics which are highly anisotropic nature

*Corresponding author: kaharwahab@unimap.edu.my

and high moisture absorption [9]. There has new interest in chitin and chitosan as a matrix in making of plastic film due to their unique properties such as anti-microbial activity, biocompatibility, economic advantages and biodegradability.

Chitosan is from renewable resource and non-toxic polymer which is derivative of chitin, an abundant by product of seafood processing like crab shell, prawn shell waste and lobster shell. In addition, chitosan is the second most abundant biopolymer after cellulose in the nature [10]. Numerous surveys have been done on the studies of film produced from chitosan, for example, chitosan blend with natural polymer or synthetic polymer and modified chitosan. A research showed plasticized maize starch combined with chitosan so as to enhance mechanical properties, water resistance and provide a barrier to oxygen of starch film which is formed by solution casting. Furthermore, chitosan is a natural polymer which acted as biodegradable additive and give anti-microbial properties to the blend [11].

2 Experimental

2.1 Materials

Chitosan act as main component in this chitosan MCC film. Chitosan was manufactured at Sigma Aldrich. MCC was obtained from Sigma Aldrich which follow United States Pharmacopeia (USP) Reference Standard (Material number: 1098388; CAS Number 9004-34-6). Glycerol was used as plasticizer in this film. Glycerol was obtained from HmbG Chemicals (Germany; Code: C0347-91552409). Polyethylene glycol (PEG), also sometimes referred to as polyethylene oxide (PEO) was obtained from Sigma Aldrich. Lithium chloride was obtained from Avantor Perfomance Materials. N,N-Dimethylacetamide is the organic compound which miscible with most other solvents, but it is poorly soluble in aliphatic hydrocarbons. DMAc was obtained from Avantor Perfomance Materials (Germany; CAS number: 127-19-5).

2.2 Preparation of the sample

2.2.1 Preparation of MCC solution

First, the amount of MCC powder, DMAc solution and lithium chloride (dry in oven 24 hours at 110 °C) needed was weight. MCC powder was added into DMAc solution and stirred for 15 minutes. Lithium chloride was then added into DMAc solution and stirred for 45 minutes until MCC powder fully dissolved and form colourless solution.

2.2.2 Preparation of MCC chitosan films with glycerol and polyethylene glycol

Glycerol as plasticizer was added into MCC chitosan solution and then stirred for 15 minutes to get homogeneous solution. Similar steps were used for polyethylene glycol as plasticizer. Then, the solution was casted to the glass plate (50 ml). Finally, took out the film when it fully dried. The formulation of MCC chitosan films were shown in Table 1.

IOP Conf. Series: Materials Science and Engineering **701** (2019) 012053 doi:10.1088/1757-899X/701/1/012053

Table 1. Formulation of MCC chitosan film.

Sample name	MCC: Chitosan ratio	Glycerol (%)	Polyethylene glycol (%)
MCC film	100:0	0	0
MCC/chitosan (95/5) film	95:5	0	0
MCC/chitosan (90/10) film	90:10	0	0
MCC/chitosan (85/15) film	85:15	0	0
MCC/chitosan (80/20) film	80:20	0	0
MCC film	100:0	3	0
MCC/chitosan (95/5) film	95:5	3	0
MCC/chitosan (90/10) film	90:10	3	0
MCC/chitosan (85/15) film	85:15	3	0
MCC/chitosan 80/20) film	80:20	3	0
MCC film	100:0	0	3
MCC/chitosan (95/5) film	95:5	0	3
MCC/chitosan (90/10) film	90:10	0	3
MCC/chitosan (85/15) film	85:15	0	3
MCC/chitosan (80/20) film	80:20	0	3

3 Characterization and Testings

Morphology of the MCC/chitosan films were observed under the SEM model JEOL JSM 6460LA. Crystalline structure and chemical composition of specimen can be determined. The specimens were prepared by mounted it on the aluminum stubs and coat it with the palladium. Specimen need to coat in order to make it conductive so that the electron can react with the specimen to get the morphology needed. Moisture absorption test was used to test the moisture absorption ability of the film. This test was conducted by weight specimens before and after dried in the oven at 45°C for 2 hours. The weight of dry finish product was cut into 4cm x 2cm x 1cm size and test according to ASTM D-570. The specimens were cooled down in the desiccator. Weighed were taken for all the specimens for every 5 days interval in the room temperature. MCC chitosan film was cut into dimension of 3 cm × 3 cm. Containers with few tiny holes at bottom of container were prepared. Moist soil was filled into the container. Specimens were buried from the soil surface with the depth of 5 cm. The specimens were taken out and weighted for first 10 days and followed by 5 days interval after 10 days.

4 DISCUSSION

4.1 SEM morphology

Fig. 1 shows the SEM image of MCC chitosan films. The fracture surface of MCC film was smooth compare with the fracture surface of MCC with addition of plasticizer which was glycerol and polyethylene glycol (PEG). There were more resistant to break was observed due to effect of plasticizer [12,13]. From Fig. 1, the beard formation after tensile test, the surface of MCC with glycerol was prominent as compared to the fracture surface of MCC with PEG. It shows the resistant of the films to break after addition of glycerol as plasticizer. The darker region in the Fig. was chitosan filler which embedded fitly into MCC matrix. This was because chitosan has high compatibility with MCC. Plasticizer only exhibit its effect on MCC matrix and not to the interaction between chitosan and MCC [14]. Hence, there is little different to the MCC chitosan film's fracture surface which due to the plasticizer effect only on MCC matrix. From the SEM image shown in Fig. 1, the fracture surface of MCC chitosan film is rougher compared to MCC film. This due to the addition of chitosan into MCC film which make the void exist in the MCC film filled with chitosan make the film stronger resist to external force. Hence, the tensile strength and break elongation is higher when chitosan added into MCC film.

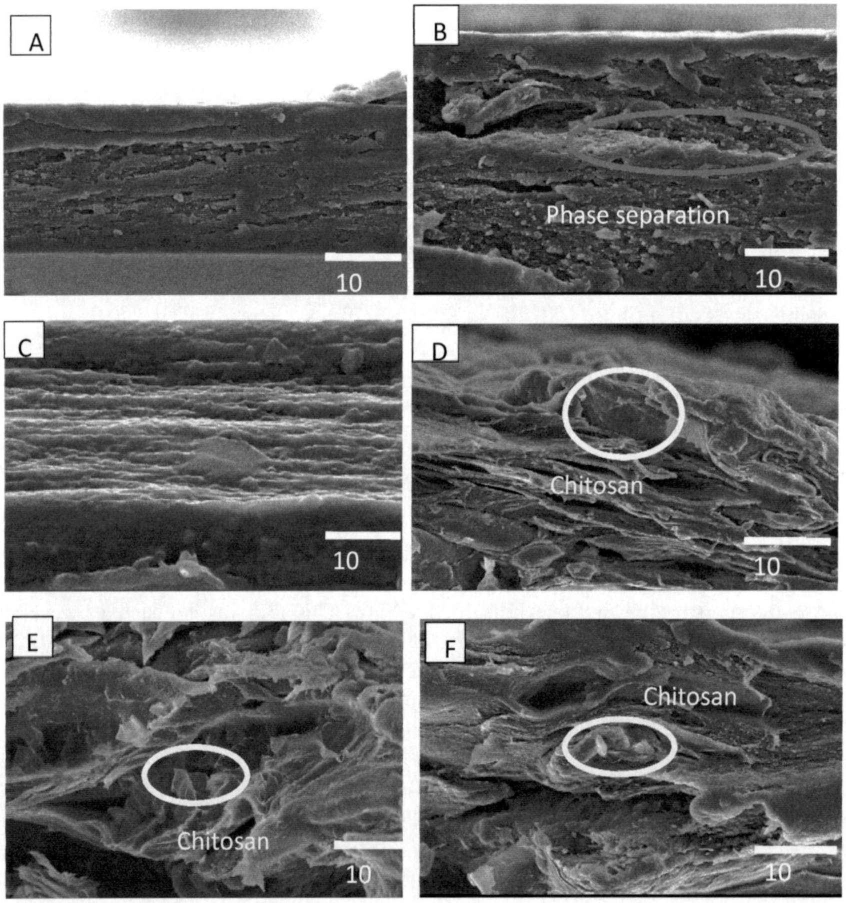

Fig. 1. Image of Scanning Electron Microscope for MCC chitosan films with x2000 magnification. A) MCC film. B) MCC with glycerol. C) MCC with polyethylene glycol (PEG). D) MCC with 10% of chitosan. E) MCC with 10% of chitosan and glycerol. F) MCC with 10% of chitosan and polyethylene glycol (PEG).

4.2 Moisture absorption

Moisture absorption test is carried up to measure the moisture intake by the film after expose to the surrounding in room temperature for days. There were three stages in moisture absorption curve as shown in Fig. 2. The moisture absorption of MCC chitosan

IOP Conf. Series: Materials Science and Engineering **701** (2019) 012053 doi:10.1088/1757-899X/701/1/012053

for the first stage was increase steeply. This was due to void present between the polymer phases. Water molecules hence fill up the void in the film results in increase weight of the film. However, the moisture absorption of MCC chitosan film were increased moderately on second stage. The water molecules was attach to hydroxyl group of the MCC and chitosan. At last, the moisture absorption curve of MCC chitosan film shown in Fig. 2 almost constant at third stage. This was due to void and hydroxyl group of MCC chitosan film almost fully filled and multiple layer of water absorption take place on the surface of the film form at third stage. MCC was very sensitive to moisture absorption. The cellulose crystals with their surface hydroxyls also contribute to moisture uptake. In other words, the density of the hydroxyl groups on the film surface increases with the MCC content in the film [15].

However, this can be improved by adding chitosan as filler. Due to high compatibility of MCC and chitosan, chitosan able to embed well into void of MCC. With the reducing void in the MCC chitosan film, the water molecules which filled up the voids exist in the film will be reduced. The incorporation of chitosan into cellulose film has made cellulose film less water permeable because chitosan could form hydrogen bond with cellulose molecules, thus decreasing water absorbency. Furthermore, the dimension of chitosan formed a rough and compact film structure, and therefore, decreasing water absorbency of cellulose films [16]. Therefore, MCC with 10% chitosan film has 14.56% at day 30 lower moisture absorption compared to that of MCC film which 15.15% at day 30.

Besides, plasticizer PEG and glycerol can replace water in low concentration in order to form interaction PEG or glycerol with MCC. This can be proved by the result where the MCC chitosan film has low moisture absorption when adding plasticizer. During day 30, the moisture absorption for MCC film with PEG and glycerol were 13.86% and 12.36% respectively compared to that of without plasticizer which 15.15%. Plasticizer make MCC chains tightly hold together and make the chain flexible. However, volatility of PEG was higher than glycerol [17]. For example, results shows during day 30, moisture absorption for 10% of chitosan in MCC chitosan film with PEG was 12.86% while 10% of chitosan in MCC film with glycerol only 12.15%. This was because PEG has two hydroxyl group while glycerol only has one hydroxyl group. There was why the moisture absorption of MCC chitosan with PEG ishigher compared to MCC chitosan with glycerol. Since the plasticizer volatility may influence film properties and stability, the less volatility plasticizer is better for use.

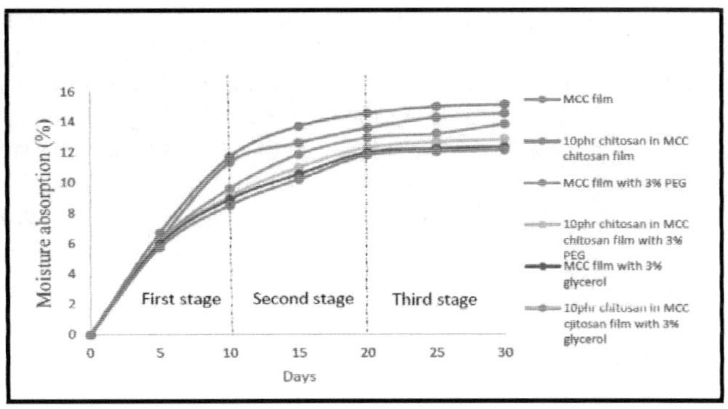

Fig. 2. Moisture absorption of MCC chitosan film in 30 days.

4.3. Soil burial test

Soil burial test is used to measure the degree of film degrade after being expose to the decomposition condition. MCC chitosan film was a biodegradable product that have the degradation properties. Thus, soil burial test was applied to the film in order to know the degree of degradation of the film. Fig. 3 has shown the weight change during the degradation of MCC chitosan film. The weight of MCC chitosan film were initially increase until day 5. This is because the film were absorb water from soil. However, there was different water uptake by the film which due to the function of filler, chitosan and plasticizer. The different water uptake by films from soil were explained in the moisture absorption test. The slower degradation rate in the MCC chitosan film was due to the resistance in water uptake and diffusion through the composite compared to pure MCC, which readily takes up water[18]. After day 5, MCC chitosan films starting experience weight loss over time. 10phr of chitosan in MCC chitosan film having lowest weight loss compared to others. This was because the function of plasticizer and filler, chitosan mostly filled all the voids in the MCC film, thus, the reaction between the film and soil lesser.

On the other hand, pure MCC film was experienced huge weight loss due to the substitution of OH group in MCC with soil and there was many void inside the MCC chains compared to others. From the result, MCC film which adding chitosan as plasticizer having low degree of degradation compared to pure MCC film. This was because chitosan has antimicrobial properties which prevented the microbial attack. Therefore, MCC chitosan film has low weight loss compared to that of pure MCC film. There was degradation mechanism taken place from day 5 on the surface of the film. However, starting from day 10, MCC chitosan film experienced degradation inside the film and begin to break the chains. MCC chitosan film were then having huge weight loss due to chain break and the degradation were faster [19]. After day 15, the MCC film were fully degrade while MCC chitosan film left some residues. Fig. 4 shows the MCC chitosan film condition before and after 5 days. From the Fig. 4 (A), it can be seen clearly that there was some parts of MCC chitosan film started to degrade after 5 days. This results of significant weight loses in MCC chitosan film. There were microbial attack by the microbial colonies on the MCC chitosan film. This can be seen from all MCC chitosan film where yellow colour spot was the microbial colonies. The microbial attack causes the discolouration on the surface of MCC chitosan film. Microbial colonies were microorganisms that will generate the enzymes to causes the breakage of the MCC and chitosan backbone chain [20]. Hence, the film will degrade fully after a certain period. From Fig. 5, MCC film with 3% glycerol was degraded faster as compared to MCC chitosan film after 5 days.

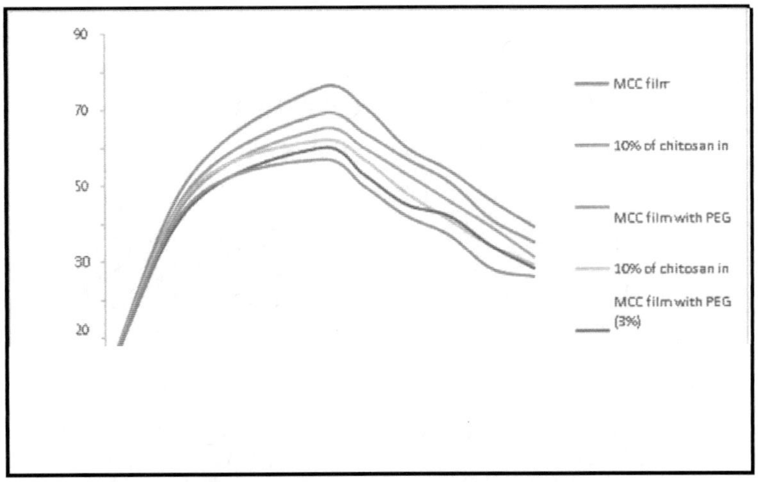

Fig. 3. Weight change of MCC chitosan film (%).

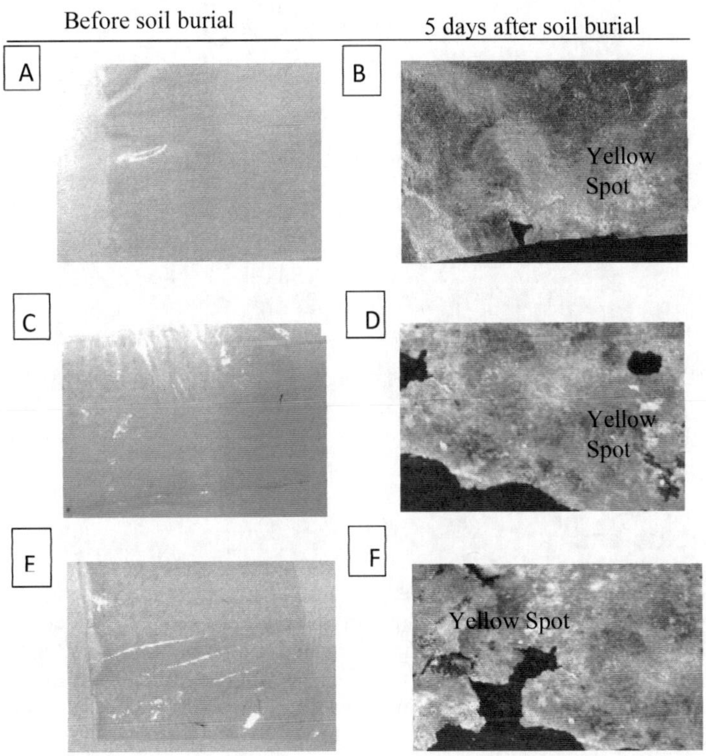

Fig. 4. MCC/chitosan film with 3% polyethylene glycol images from optical microscope before and after 5 days of soil burial.

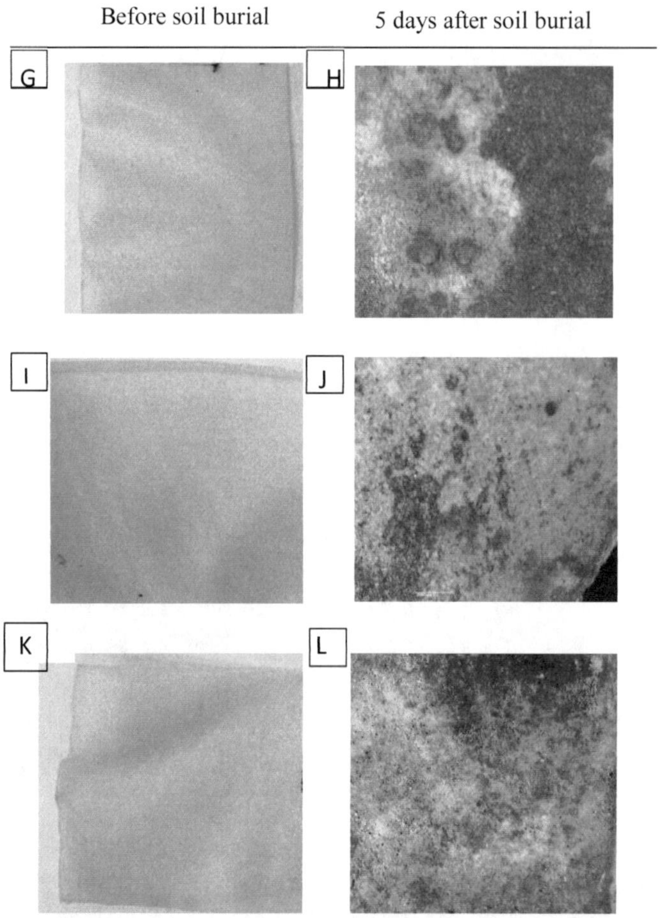

Fig. 5. MCC/chitosan film with 3% glycerol images from optical microscope before and after 5 days of soil burial.

5 Conclusions

In the optimum composition of chitosan in MCC/chitosan film, the weight loss in soil burial test is the lowest due to the chitosan antimicrobial properties. Besides, moisture absorption of MCC chitosan lesser when compared to control film counterparts. This was due to high compatibility of MCC and chitosan which has been proven in Scanning Electron Microscopy images which MCC was embeded well into of chitosan phase. It was then reduced the water molecules filled up the voids exist in the film. From SEM, It was clearly shown that plasticizer enhances the interaction between MCC chains which result in the formation of beards on the surface which resist the external force applied. Besides, the addition of plasticizer has improved the properties of MCC chitosan films.

The authors wish to gratefully acknowledge the School of Materials Engineering, as well as Research and Innovation Department under Universiti Malaysia Perlis and Kementerian Pengajian Tinggi Malaysia for supporting this research project.

References

1. F.L. Mantia, M. Morreale, Compost. Part A: Appl. Sci.Manufac., **42(6)**, (2011)
2. A.H. Mohd Zain, A.W.M. Kahar, I. Hanafi, 2018. J Polym Environ, **26(2)**, (2008)
3. R. J. Walster, R. Rahman, A. W. M. Kahar. IOP Conf. Series: Mater. Sci. Eng., **429**, (2018)
4. D.L. Ngoc, L.T. Saw, N.A. Abdul Rahim, A.W.M. Kahar, X.V.Cao. Polym. Degrad. Stabil., **111**, (2014)
5. A. W. M. Kahar, H. Ismail, A. Abdul Hamid, A J. Therm. Anal. Calorim., **123(1)**, (2016)
6. F. Vilaplana, E. Strömberg, S. Karlsson, Polym. Degrad. Stabil., **95(11)**, (2010)
7. T. Huber, J. Müssig, O. Curnow, S. Pang, S., Bickerton, M.P. Staiger, J. Mater. Sci., **47(3)**, (2012)
8. A.W.M. Kahar, H. Ismail. J. Vinyl Add. Technol. **22(3),** (2016)
9. M. Jawaid, H.P. Abdul Khalil, Carbohyd. Polym., **86(1),** (2011)
10. I. Leceta, P. Guerrero, K. De La Caba, Carbohyd. Polym., **93(1),** (2013)
11. A.O. Ogah, J.N. Afiukwa, A.A. Nduji, Open J. Polym. Chemist., **4(1),** (2014)
12. A.W.M. Kahar, M. Lingeswarran, M.Z. Amirah Hulwani, H. Ismail, Polym. Bull., **76(2),** (2019)
13. H. Salmah, H., Ismail, A. Bakar, J. Appl. Polym. Sci., **107(4),** (2007)
14. N.F. Alias, H. Ismail, A.W.M. Kahar,. Bio-res, **12(4),** (2017)
15. A.W.M. Kahar, N. Sarifuddin, H. Ismail, H., Ira. Polym. J., **26 (2),** (2017)
16. N. Ibrahim, A.W.M. Kahar, D.L Ngoc, H. Ismail, H. Bio-res., **12(2)**, (2017)
17. M.G.A.Vieira, M.A. Da Silva, L.O. Dos Santos, M.M. Beppu, Eur. Polym. J., **47(3)**, (2011)
18. O.S. Dahham, N.Z. Noriman, A.W.M. Kahar, H. Ismail, S.T. Sam, J. Appl. Sci. & Agric., **10(5),** (2015)
19. A.H. Mohd Zain, A.W.M. Kahar, N.Z. Noriman, Procedia Chem., **19,** (2016)
20. A.W.M. Kahar, H. Ismail., N. Othman, J. Appl. Polym. Sci., **128(4),** (2013)

IOP Conf. Series: Materials Science and Engineering **701** (2019) 012054 doi:10.1088/1757-899X/701/1/012054

Preparation and characterization of micro-crystalline cellulose / chitosan films

Song Cheah Chie [1]. *Mohamad Kahar* Ab Wahab[1]*

[1]School of Materials Engineering, University Malaysia Perlis, 02600, Arau, Perlis, Malaysia;

Abstract The microcrystalline cellulose (MCC) chitosan film was produced through solution casting with different composition of chitosan and different plasticizers. This MCC chitosan film was a biodegradable polymer due to all the ingredient was natural resources. MCC and chitosan can be found abundantly in the earth, and thus, can be obtained easily. Several testing were conducted such as tensile test, scanning electron microscope and Fourier transform infrared. The optimum ratio for the MCC chitosan was 90: 10 (MCC: chitosan). At this ratio, MCC chitosan has the highest tensile strength which was 42.16 MPa for the film without adding the plasticizer. MCC chitosan film also having the addition of plasticizer to improve film's properties. Results show that addition of plasticizer able to improve properties of MCC chitosan film such as high tensile strength, high elongation at break and low Young's modulus value. The percentages when compared MCC chitosan film with glycerol as plasticizer than the control film on the optimum ratio for tensile strength was higher 106%; while for film with PEG as plasticizer was higher 102%, elongation at break was higher 121% and 110% for both glycerol and PEG respectively when compared to control film.

1 Introduction

Nowadays, there are many bio-waste from animals and plants that are useful to human being. Scientists and researchers are now more focus on environmental friendly material which is useful and harmless to human health. Therefore, bio-waste such as chitin, cellulose, starch, lignin and so on giving attention from tem to developing and modifying products [1-5]. There are many researchers have savor in environmental friendly and biodegradable materials, which can be obtained from renewable and natural resources. It is due to the advance properties compare to non-environmental friendly films such as low cost, easy availability and biodegradability. biological degradability, the lessening volume of waste and compostability, conservation of fossil-based raw materials, decreasing amount of carbon dioxide release to atmosphere, and also increase utilization of agricultural resources for production of natural polymer [6].

*Corresponding author: kaharwahab@unimap.edu.my

The discovery of renewable sources of polymeric materials has maintained the suitable expansion in economically and ecologically attractive skills. There are many interesting discovery of renewable sources from polymeric materials such as the complete The discovery of biodegradable polymers gave researchers ideas to handle waste disposal problems which created from traditional petroleum derived plastics. Bio-based composites are a class of composite materials that are defined as having at least one of their component phases (i.e. matrix or reinforcement) derived from biomass. Recent heightened research in the development of bio-based composites for solving the escalating problem of end-of-life disposal of plastic materials and reducing number of available landfill sites. There is a development in biocomposites research on the conversion from petroleum-based polymer, for example, polypropylene and polyethylene to naturally derived biopolymer such as starch, chitosan and resin matrices to produce composites [7]. Therefore, present study is focused on microcrystalline cellulose (MCC)/ chitosan film. Chitosan is reinforced into MCC matrix to enhance the film properties. Effect of glycerol and polyethylene glycol (PEG) as plasticizers to MCC chitosan film has also been study. With addition of plasticizer, MCC chitosan film become more flexible and improve the mechanical properties of the films.

2 Experimental

2.1 Materials

Chitosan act as main component in this chitosan MCC film. Chitosan was manufactured at Sigma Aldrich. MCC was obtained from Sigma Aldrich which follow United States Pharmacopeia (USP) Reference Standard (Material number: 1098388; CAS Number 9004-34-6). Glycerol was used as plasticizer in this film. Glycerol was obtained from HmbG Chemicals (Germany; Code: C0347-91552409). Polyethylene glycol (PEG), also sometimes referred to as polyethylene oxide (PEO) was obtained from Sigma Aldrich. Lithium chloride was obtained from Avantor Perfomance Materials. N,N-Dimethylacetamide is the organic compound which miscible with most other solvents, but it is poorly soluble in aliphatic hydrocarbons. DMAc was obtained from Avantor Perfomance Materials (Germany; CAS number: 127-19-5).

2.2 Preparation of the sample

2.2.1 Preparation of MCC solution

First, the amount of MCC powder, DMAc solution and lithium chloride (dry in oven 24 hours at 110 °C) needed was weight. MCC powder was added into DMAc solution and stirred for 15 minutes. Lithium chloride was then added into DMAc solution and stirred for 45 minutes until MCC powder fully dissolved and form colourless solution.

2.2.2 Preparation of MCC chitosan films with glycerol and polyethylene glycol

Glycerol as plasticizer was added into MCC chitosan solution and then stirred for 15 minutes to get homogeneous solution. Similar steps were used for polyethylene glycol as plasticizer. Then, the solution was casted to the glass plate (50 ml). Finally, took out the film when it fully dried. The formulation of MCC chitosan films were shown in Table 1.

Table 1. Formulation of MCC chitosan film.

Sample name	MCC: Chitosan ratio	Glycerol (%)	Polyethylene glycol (%)
MCC film	100:0	0	0
MCC/chitosan (95/5) film	95:5	0	0
MCC/chitosan (90/10) film	90:10	0	0
MCC/chitosan (85/15) film	85:15	0	0
MCC/chitosan (80/20) film	80:20	0	0
MCC film	100:0	3	0
MCC/chitosan (95/5) film	95:5	3	0
MCC/chitosan (90/10) film	90:10	3	0
MCC/chitosan (85/15) film	85:15	3	0
MCC/chitosan 80/20) film	80:20	3	0
MCC film	100:0	0	3
MCC/chitosan (95/5) film	95:5	0	3
MCC/chitosan (90/10) film	90:10	0	3
MCC/chitosan (85/15) film	85:15	0	3
MCC/chitosan (80/20) film	80:20	0	3

3 Characterization and Testings

The mechanical testing used was tensile testing according to ASTM D 882 by using Instron machine. The thickness of the plastic test specimen was prepared according to the ASTM D 6988. The shape of the specimen was in a rectangular shape. Tensile test was used to find the tensile strength, Young's modulus and elongation of the film specimen. There should be at least 3 specimens was use for testing for each batch of the film composition in order to have more accurate result. Structural analysis was performed by Fourier Transform Infrared (FTIR) Perkin Elmer Spectrometer 2000. It was the technique use to get the infrared spectrum. The spectrum was obtained through emission or absorption of the film. The spectra were obtained at resolution of 4 cm^{-1} in the range of 4000-6000 cm^{-1}. The specimens were scan for 32 times for a better resolution on the reading.

4 DISCUSSION

4.1 Structural analysis

FTIR spectroscopy is an analytical technique used to identify mainly organic materials which provides information about the chemical bonds and molecular structure of a material. The existence of a specific chemical bond in any material is indicate from the presence of a peak at a specific wavenumber after scanning the test samples in the infrared light source. Fig. 1 shows the peak of MCC film with and without plasticizer. MCC film with plasticizer exhibit similar peak with that of original MCC film. MCC film with and without plasticizer was compared base on unchained peak, which was at region 2346.16 cm^{-1}. The unchained peak act as reference peak which was used to compare any different peaks occur between the samples due to it will not to be changed whenever what condition for MCC chitosan film. Spectrum band of 3342.9 cm^{-1} was attributed to the O-H stretching vibration of hydroxyl group in the cellulose. The band at 2893 cm^{-1} corresponds to C-H stretching vibration of –CH2– groups. The band at 1644.5 cm^{-1} was assigned for C=O in the aldehyde on the terminal anhydroglucose unit [8]. Peak at 1436.9 cm^{-1} was due to CH2. The 3342.9 cm^{-1} peak from MCC film was shifted a bit to 3328.1 cm^{-1} and 3332.5 cm^{-1} when adding with plasticizer PEG and glycerol. This prove that the interaction bonding between the MCC chains with plasticizer were stronger than that of without adding plasticizer [9]. Band found at 1324.1 cm^{-1}, 1355.4 cm^{-1} and 1323.9 cm^{-1} was associated with C-O stretching vibration of CH2- OH groups. Also a band observed at 1015.4 cm^{-1}, 1014.3 cm^{-1} and 1014.0 cm^{-1} were for -C-O-C- in the cellulose molecules [10].

From Fig. 2, the broad absorption band, which appears in the range of 3600–2800 cm^{-1} is related to collective absorption by both N–H and O–H groups in the polymer, is an indication of the bonding of chitosan with MCC through hydrogen bond [11]. The presence of chitosan is also revealed by its typical peak shown only by treated samples at 1650–1580 cm^{-1} which assigned to N–H bending. A strong absorption peak appears at 1999.1 cm^{-1} for chitosan react with MCC signifies the formation of Schiff base (C=N double bond) between aldehydic carbonyl group of MCC and amino group of chitosan. While C–H were observed at 2892.4 cm^{-1} and 2894.4 cm^{-1}. The absorption peaks at 1643.8 cm^{-1}, 1636.8 cm^{-1} and 1644.3 cm^{-1} were associated with the presence of the C=O stretching of the amide I band, while range from 1450 cm^{-1} to 1470 cm^{-1} was stand for C-H bending, 1356.2 cm^{-1} and 1322.7 cm^{-1} were O-H bending. The peak at 1014.04 cm^{-1}, 1014.5 cm^{-1} and 1019.1 cm^{-1} were assigned for anti- symmetric stretching of C–O–C bridge [12]. The FTIR spectrum of the MCC chitosan film showed that, the broad peak for -OH and -NH groups were shifted from 3342.9 cm^{-1} to 3330.4 cm^{-1}. This shift indicated that interaction between amine groups in chitosan and OH groups in cellulose occurred. The sharp peak of –OH at 3330.4 cm^{-1} indicated the availability of free –OH group of MCC. The carbonyl band of the composite was shifted to a lower frequency, from 1644.5 cm^{-1} to 1643.8 cm^{-1}. This FTIR spectrum showed that the composite may be prepared by the interaction of -NH group of chitosan and –OH group of MCC [13].

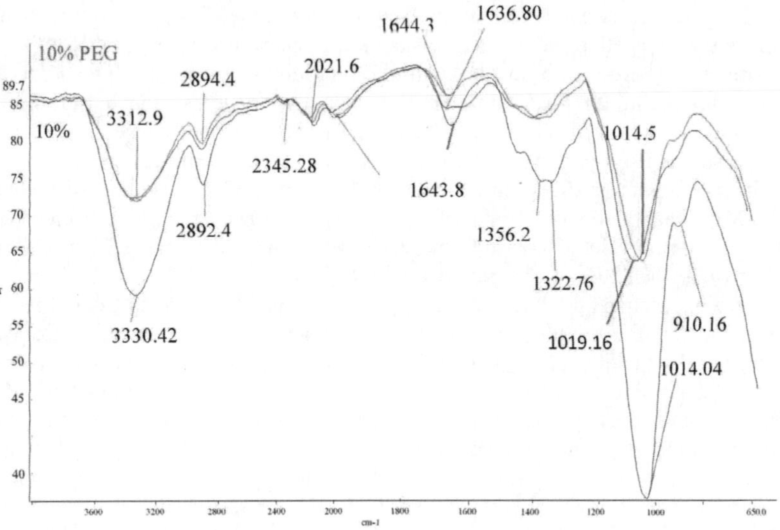

Fig. 1. FTIR spectrum of MCC film with and without plasticizer.

IOP Conf. Series: Materials Science and Engineering **701** (2019) 012054 doi:10.1088/1757-899X/701/1/012054

Fig. 2. FTIR spectrum spectrum of 10% chitosan in MCC chitosan films with and without plasticizer.

4.2 Tensile properties

Tensile test is very important test use to find tensile strength, Young's modulus and elongation at break. Tensile testing is measuring the force required to elongate a specimen to breaking point and determine the material properties from the result obtained. A film needed some tensile strength for its application such as packaging. Besides, it also required certain elongation and modulus to maintain or carry product weight. Fig. 3 shows tensile strength of MCC chitosan film with different ratio of filler, chitosan adding into microcrystalline cellulose (MCC) film. From the graph, it can be seen that tensile strength of MCC film was increased until 10% of chitosan in MCC chitosan film and then decrease the tensile strength. From that, the maximum amount of chitosan that able to enhance MCC film properties was 10%. Tensile value of MCC film is 32.2 MPa increase gradually to 42.16 MPa when load with 10% of chitosan as filler. However, starting from 10% of chitosan, the tensile value decrease from 42.16 MPa to 34 MPa when 15% of chitosan and 27.25 MPa when 20% of chitosan load with MCC film.

The mechanical properties of the composites are determined by several factors, such as nature of the reinforcement fiber, fiber aspect ratio, fiber–matrix interfacial adhesion, and also the fiber orientation in the composites [14]. The improvement in tensile strength at the presence of chitosan was expected as it was very compatible with MCC matrix with high strength properties and as a result it adds stiffness to the composites. In addition, from the graph, the elongation at break for MCC chitosan film also increased with the increased amount of chitosan. This is due to the higher amount of chitosan make MCC chitosan film brittle and easy to break. Furthermore, this improvement in the interfacial bonding also enables a more efficient transfer of stress from the matrix phase to the chitosan phase. This result is consistent with previous findings on cellulose-polymer composite systems [15,16].

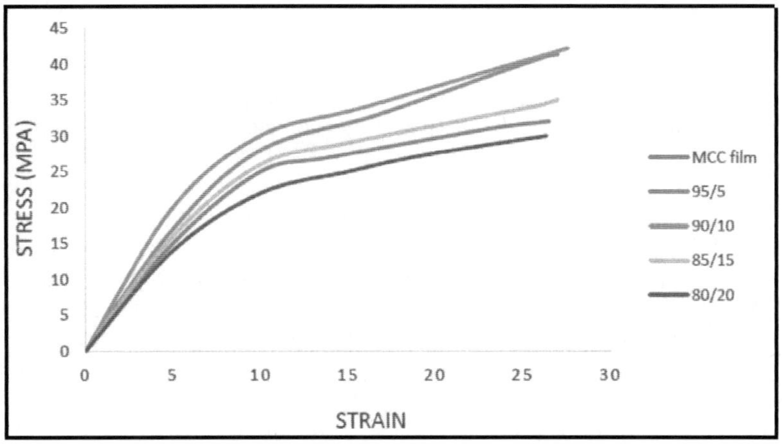

Fig. 3. Tensile strength of MCC chitosan films.

Chitosan was very compatible with MCC rom the formation of hydrogen bond between hydroxyl group of MCC matrix and chitosan filler. Therefore, it able to enhance properties of MCC once it load in MCC film. However, at higher concentration of chitosan, the MCC film will become brittle and easily to break. Hence, the tensile strength will be affected and shows decreasing trend. From Fig. 4, elongation at break of film with 10% of chitosan, MCC chitosan is the highest compare to elongation break of other ratio. Start from 15% of chitosan and beyond the tensile strength and elongation break was decreased and lower than the original MCC film which without adding chitosan. In order to enhance the bonding strength between the cellulosic materials and the matrix polymer, plasticizing agent was used. Fig. 4 shows tensile strength of different ratio of MCC chitosan films with and without plasticizer, polyethylene glycol (PEG). From the graph, tensile strength of MCC chitosan film with polyethylene glycol (PEG) as a plasticizer was higher when compare to MCC chitosan film without plasticizer. There was increased in tensile value from MCC film without chitosan to 10phr of chitosan in MCC chitosan film which is 43.12MPa, the optimum followed by decreasing value to 28MPa at 20% of chitosan in MCC chitosan film.

The same tensile strength result get for glycerol as plasticizer in MCC chitosan film which shown in the graph. There was increased in tensile strength of MCC chitosan film at 10% of chitosan and followed by decreased of tensile value from 44.96MPa, the optimum value at 10% of chitosan to 29.17MPa at 20% of chitosan in MCC chitosan film. This is because PEG and glycerol act as plasticizer which enhanced the tensile properties of MCC matrix. It embedded within the MCC chains to make the film more flexible and tightly hold the MCC chains [17-19]. Hence, MCC chitosan film with plasticizer more resistant to break when the external force applied as compared to MCC chitosan film without plasticizer. Furthermore, the reorganization of the polymeric network due to strong polymer-plasticizer interactions favoured the crystallization process, ultimately leading to the increased chain stiffness and decreased flexibility of the material [20]. However, MCC chitosan film with PEG exhibited lower tensile strength when compare to the MCC chitosan film with glycerol which bigger molecular size of PEG decreased the efficiencies to embed into MCC matrix. Elongation at break is also known as fracture strain which expresses the capability of film to resist changes of shape without crack formation. In general, elongation at break of MCC film will increase as the chitosan filler increased. The optimum elongation at break of MCC chitosan film was at 10% amount of chitosan. As the material's stiffness increased with increasing content of chitosan, and thus the film demonstrated a lower elongation at break in comparison to MCC film.

By adding plasticizer into MCC chitosan film make it becomes more flexible and elastic. Thus, the elongation at break of MCC chitosan film with polyethylene glycol (PEG) as a plasticizer higher compare to that without plasticizer as shown in Fig. 4. From the reference sample, the value of elongation at break was 23.2% while MCC chitosan film with PEG was 24.4%. At 10% of chitosan in MCC chitosan film, the elongation at break for MCC chitosan film without plasticizer was 27.6% while MCC chitosan film with PEG was 30.3%. Although the value of elongation at break decreased after that, the MCC chitosan film with PEG was still higher than that of without PEG. There is same positive result get for MCC chitosan film with glycerol as plasticizer. There was 33.44% elongation at break for MCC chitosan film with glycerol at 10% of chitosan, 121.1% increased as compared to control samples counter parts. Besides, there was improvement of chemical bond between plasticizer with matrix which it make MCC chitosan film resist to break. However, effect of glycerol as plasticizer is more obvious compared to that of PEG. The improvement of break elongation of MCC chitosan film with glycerol was higher than that of PEG. There was because the molecular size of glycerol was small enough and easier to join with MCC chains to poses plasticized effect as compared to PEG.

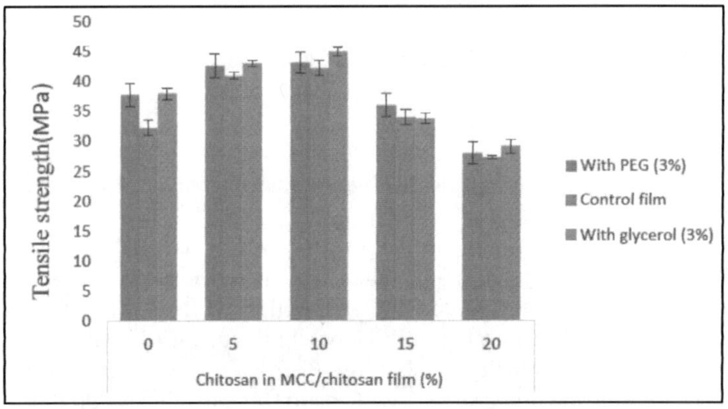

Fig. 4. Tensile strength of MCC/chitosan films with different type of plasticizer.

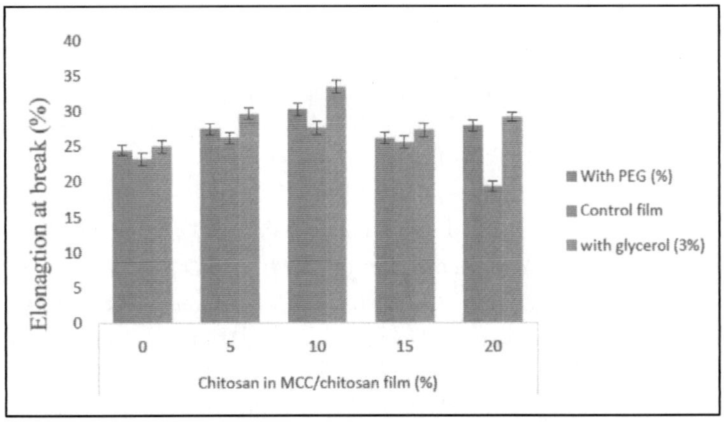

Fig. 5. Tensile strength of MCC/chitosan films with different type of plasticizer.

5 Conclusions

The development and study of microcrystalline cellulose (MCC) chitosan film had been done in simple and economic method. In this study, chitosan had been added into MCC to produce a better properties film with different composition of chitosan added. Incorporation of chitosan in MCC chitosan film will improve mechanical strength of films. The optimum composition of chitosan in MCC chitosan film is 10phr where its tensile strength, elongation at break, moisture absorption and degree of degradation is the highst. However, as the chitosan content increases, the tensile strength and elongation at break decrease significantly due to possible aggregation of chitosan.

The authors wish to gratefully acknowledge the School of Materials Engineering, as well as Research and Innovation Department under Universiti Malaysia Perlis and Kementerian Pengajian Tinggi Malaysia for supporting this research project.

References

1. A.W.M. Kahar, M. Lingeswarran, M.Z. Amirah Hulwani, H. Ismail, Polym. Bull., **76(2)**, (2019)
2. A.W.M. Kahar, H. Ismail., N. Othman, J. Vinyl Add. Technol.**18(3)**, (2012)
3. R. J. Walster, R. Rahman, A. W. M. Kahar. IOP Conf. Series: Mater. Sci. Eng., **429**, (2018)
4. H. Zaman, J. Song, L.S. Park, I.K. Kang, S.Y. Park, G. Kwak, K.B.Yoon. Polym. Bull. **67(1)**, (2011)
5. A. W. M. Kahar, H. Ismail, A. Abdul Hamid, A J. Therm. Anal. Calorim., **123(1),**(2016)
6. N. Ibrahim, A.W.M. Kahar, D.L Ngoc, H. Ismail, H. Bio-res., **12(2),** (2017)
7. F. Vilaplana, E. Strömberg, S. Karlsson, Polym. Degrad. Stabil., **95(11**), (2010)
8. A.W.M. Kahar, H. Ismail. J. Vinyl Add. Technol. **22(3)**, (2017)
9. A.W.M. Kahar, N. Sarifuddin, H. Ismail, H., Ira. Polym. J., **26** (**2**), (2017)

10. D. Trache, A. Donnot, K. Khimeche, R. Benelmir, N. Brosse, Carbohydr.Polym., **104**, (2014)
11. S. Yasmeen, M. Kabiraz, B. Saha, M. Qadir, M., Gafur, S. Masum, Intern. Resear. J. Pure Appl. Chem., **10(4)**, (2016).
12. A.H. Mohd Zain, A.W.M. Kahar, I. Hanafi, 2018. J Polym Environ, **26(2)**,(2008)
13. H. Salmah, H., Ismail, A. Bakar, J. Appl. Polym. Sci., **107(4),** (2007)
14. N.F. Alias, H. Ismail, A.W.M. Kahar,. Bio-res, **12(4)**, (2017)
15. Z.Lin, C. Chen, Z. Guan, S. Tan, X. Zhang, J. Appl. Polym. Sci., **122(4)**, (2011)
16. Z. Xiuju, S. Juncai, Y. Huajun, L. Zhidan, T. Shaozao, J. Thermo. Compos. Mater., **24(6),** (2011).
17. O.S. Dahham, N.Z. Noriman, A.W.M. Kahar, H. Ismail, S.T. Sam, J. Appl. Sci. & Agric., **10(5),** (2015).
18 A.H. Mohd Zain, A.W.M. Kahar, N.Z. Noriman, Procedia Chem., **19,** (2016)
19. A.W.M. Kahar, H. Ismail., N. Othman, J. Appl. Polym. Sci., **128(4),** (2013)
20. H. Liu, R. Adhikari, Q. Guo, B. Adhikari, J. Food Eng., **116(2),** (2013)

IOP Conf. Series: Materials Science and Engineering **701** (2019) 012055 doi:10.1088/1757-899X/701/1/012055

Study on polylactic acid/acrylonitrile butadiene styrene composites filled with micro-crystalline cellulose

Siti Nursyameza Shaharin[1]. *Mohamad Kahar* Ab Wahab[1]*

Contact information: 1: School of Materials Engineering, University Malaysia Perlis, 02600, Arau, Perlis, Malaysia;

Abstract The effect of surface treatment by using silane coupling agent with varying ratio of micro -crystalline cellulose (MCC) on polylactic acid (PLA) / acrylonitrile butadiene styrene (ABS) blend system were investigated. The study had focused on the blend modification onto the mechanical, structural, morphological and biodegradability properties of PLA/ABS blends. The study showed that the tensile properties of the PLA/ABS blends were deterioriated with increasing in MCC content. The interfacial adhesion was approach by using silane treatment caused by the incorporation of MCC and coupling agent. According to the results, the addition of MCC and surface treatment s howed an increasing better in tensile properties as compared to the standardized counterpart. On top of that, the morphological analysis by using scanning electron microscopy (SEM) found out that the interfacial adhesion of MCC was improved in both systems which is with and without silane treatement. Both of the systems was obtained having a weight loss during the soil burial test and the biodegradation process in PLA/ABS/MCC blends as proven in biodegradability analysis.

1 Introduction

Nowadays, plastics are used commonly in our daily life which inflates the necessity of plastics, hence will increasing the demands for plastics. Due to the demands of consumers is high, the production of plastics rapidly over the past few decades. Every year the world production of plastic is estimates to be more than 100 million tonnes. However, most conventional plastic used are polyethylene, polyproylene, polystyrene, polyvinyl chloride and polyethylene terephthalate which are non-biodegradable or their recycling process is either difficult or not economical [1-5]. According to Ibrahim et al. [6] predominant biopolymers such as polylactic acid (PLA), polyglycolic acid (PGA) polycaprolaptone (PCL) and polyhydroxyalkanoate (PHA) has been studied as one of the biodegradable based plastics. Among those biopolymers, PLA was getting most attention as it made from renewable agricultural sources and also can maintain its transparency after processing [8-9].

However, even PLA was selected as good choice in the biopolymer processing, it has to overcome the problems such as low drawability, insufficient toughness and limited gas barrier [10]. To serve as a solution to the aforementioned problems, PLA may be blended with other polymers such as acrylonitrile butadiene styrene (ABS) in order to impart toughness properties on the blend.

*Corresponding author: kaharwahab@unimap.edu.my

IOP Conf. Series: Materials Science and Engineering **701** (2019) 012055 doi:10.1088/1757-899X/701/1/012055

In terms of improving, bio-based content, this work was aims to blend PLA with acrylonitrile butadiene styrene (ABS) and filled with micro-crystalline cellulose (MCC). This may allow for a good balance of performance, from the toughness properties of ABS and bio-based content from PLA. Furthermore, the idea of blending PLA/ABS blends with MCC was introduced as it can reduce the price of highly cost PLA and at the same time can retain equivalent biodegradability [11]. However, the major problem with this blending system is the poor interfacial interactions between the blend's components. Thus by carried the silane treatment onto MMC filler is expected to overcome this blending problems.

2 Experimental

2.1 Materials

PLA was purchased form the Titan Chemicals Corp located in Pasir Gudang, Johor. It is in granules form and has melting temperature at 160 °C. ABS is an amorphous plastic consisting of three monomers; acrylonitrile, butadiene and styrene. The ABS was purchased from Titan Chemicals Corp located in Pasir Gudang, Johor. It is in pellet form and has glass transition temperature at around 105 °C. MCC is a white powdery fibers and high purity cellulose powders. It has a micro size, odourless and combustible. It was purchased from Sigma Aldrich, United States. Silane coupling agent, γ-methacryloxypropyltrimethoxysilane (γ-MPS) with an approximate purity of 97%, was provided by Sigma Aldrich, United States.

2.2 Preparation of the sample

2.2.1 MCC modification with silane coupling agent

MCC was silanized with γ-MPS in a mixture of ethanol/water (80/20) [12]. The γ-MPS was added into the solution and the pH was adjusted to 4 by adding a few drops of acetic acid. The mixture was mechanically stirred for 4 h under ambient temperature. Then, MCC was filtered and washed with ethanol followed by distilled water to drive out any excess silane. Later, the silanized MCC was dried at room temperature for 2 days, and heated in oven at 80°C for another 4h to promote the actual chemical coupling.

2.2.2 Preparation of PLA/jackfruit seed starch blend

PLA/ABS blends were prepared by using heated two roll mill at 170°C with rotation rate of 10 rpm. Firstly, the PLA was added to the nip of the heated two roll mill and processed until it melts. Then, the ABS was added and mixed with PLA until homogeneous. Lastly the MCC filler was added until completion of the mixing. The PLA/ABS/MCC composites was then undergoing the hot press molding to produce a 1 mm sheet. The hot press molding process was take place at 170°C for 6 minutes.

3 Characterization and Testings

The Instron universal testing machine (Model 3366) was to examine the mechanical properties of each PLA/ABS/MCC blend by setting the crosshead speed of 10mm/min under an ambient temperature. An average value was taken for at least 5 samples in order

IOP Conf. Series: Materials Science and Engineering **701** (2019) 012055 doi:10.1088/1757-899X/701/1/012055

to get precise value for tensile strength, elongation at break, and modulus of elasticity for each blend formulation. The morphological studies was carried out on the tensile fracture surface by using the analytical Scanning Electron Microscope (SEM) (Model JEOL JSM-6460 LA) with the supply of accelerating voltage at 10kV. The purpose of this testing was to analyze the topographic imaging of samples such as the fracture surfaces, homogeneity of filler in the PLA/ABS composites and the distribution and dispersion of MCC embedded in PLA/ABS composites. To prevent the charging effect on the surface morphology which would affect the testing results, the fracture samples were required to be coated with platinum layer about 1.5-3.0 nm thickess with aid of utilizing the Auto Fine Coater (Model JEOL JFC 1600). The biodegradability of PLA/ABS composites filled with MCC could be determined by soil burial test under laboratory scale. Rectangle shape samples with the dimension of 3cm x 2cm were prepared from each type of blend ratio. The soil was filled up a container and it was exposed to the environmental condition during the whole testing duration. Gardening soil was used in this experiment. The samples were then buried into the soil with 10cm depth from the soil surface to allow the microorganisms action for 50 days. The buried samples were watered everyday for 1000mL into the container to keep the soil moist. After 50 days, they were then collected and cleaned with distilled water before taking for surface analysis by using optical microscope.

4 DISCUSSION

4.1 Tensile test

Figures 1 (a) – (b) show the effect of MCC loadings on the tensile properties of PLA/ABS blends. PLA is generally known to have high tensile strength, but facing the inferior toughness properties. The addition of ABS was to improve the brittleness of PLA. On the other hands, a polymer matrix with incorporation of filler will increase or decrease the tensile strength depending on the on type of filler. Form the figures it can be seen that the increasing in MCC loadings will increase tensile strength of the PLA/ABS/MCC composites. This due to the incorporation of reinforcing filler on the blends that could form the good mechanical interlocking between filler-matrix prevent the PLA/ABS chains movement [13].

Furthermore, for PLA/ABS composites with silane treatments exhibited higher tensile properties as compared to untreated MCC counterparts as results from the better PLA/ABS – MCC interfacial interaction. The highest tensile strength of PLA/ABS-30% MCC was 22.5 MPa, while for untreated PLA/ABS-30% MCC as merely 20.7 MPa. The better in tensile strength was due to the improvement in interfacial interaction between MCC and PLA/ABS matrix. The Similar trend was observed on the Young's modulus graph for both composites. The Young's modulus for untreated composites at 30% MCC was 1920 MPa and the value was increased for silane treatment composite counterpart to 2175 MPa. The treatment could prevent the chains slippage and provide better PLA/ABS blend's dimensional stability [14]. Furthermore, the formation of interfacial bonding between MCC and PLA/ABS matrix also reduced the flexibility of the blends since the network restricted the movement of blends.

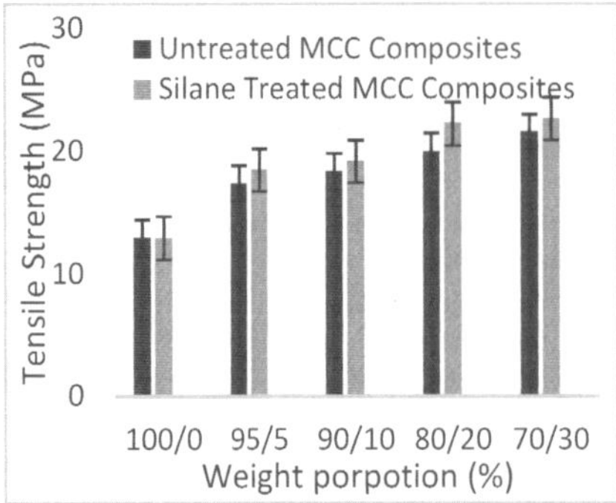

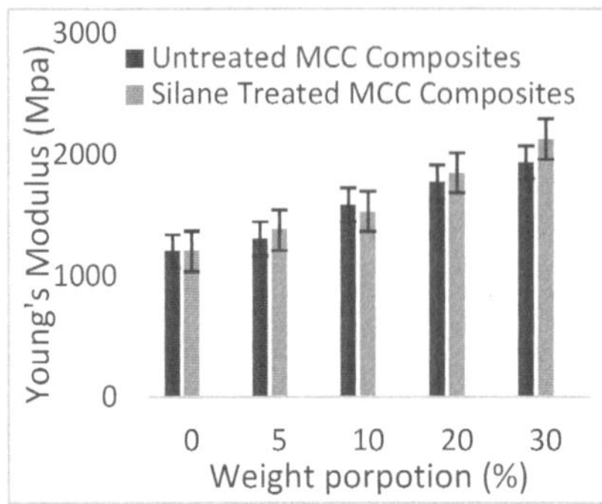

(a) (b)

Figs 1 (a) – (b). Effect of MCC loading on the tensile properties of PLA/ABS composites.

4.2 Morphological analysis

Properties of the polymer composites are strongly dependent on the uniform distribution of the fillers. Hence, Scanning Electron Microscope (SEM) testing of the biocomposites were performed in order to investigate the level of distribution of MCC in PLA/ABS matrix. The SEM micrographs of fractured surfaces of PLA/ABS/MCC composites at 5 % and 30% were shown in Figure 2 and 3. From the morphology of the fracture surface (Figure 2), there is a little compatibility between the PLA/ABS matrix with MCC phase. It could be observed the smooth surface of MCC particles without being adhere with PLA/ABS matrix. This was due to the incompability problem arise from hygroscopic nature of the MCC and PLA/ABS matrix [15]. Meanwhile, there was slight level of agglomeration with 30wt% MCC contents in PLA/ABS matrix. Poor interfacial adhesion have been created between the MCC and the PLA/ABS blend and it contributed to weak interfacial regions [16]. As the MCC content was higher, the poor mixing creating aggromerations of MCC filler and lead to the void formations [17]. In figure 3, MCC particles have better interfacial interaction and imbedded inside thePLA/ABS matrix as a result from the silane treatment. The micrographs were being more compact, and thus the void formations were less for both 5% and 30% MCC composites.

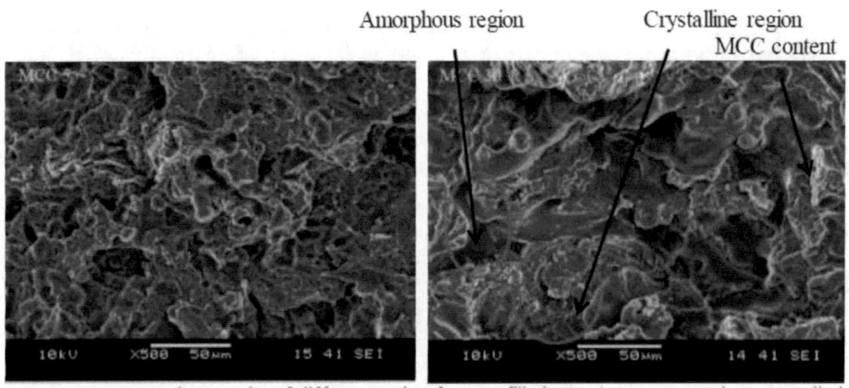

Fig. 2. SEM micrographs of untreated 5% and 30% MCC-filled PLA/ABS composites.

IOP Conf. Series: Materials Science and Engineering **701** (2019) 012055 doi:10.1088/1757-899X/701/1/012055

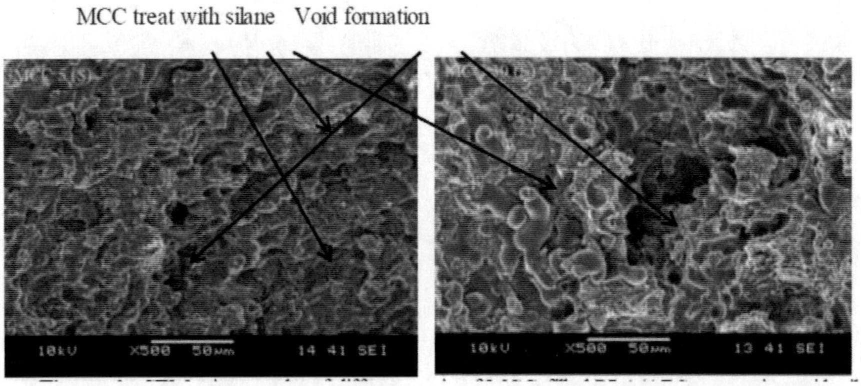

Fig. 3. SEM micrographs of silane treated 5% and 30% MCC-filled PLA/ABS composites.

4.3. Biodegrability analysis

The biodegradability properties is one of the interesting criteria in biopolymers composites. This is due to the service lifetime of the biomaterial must be known to make sure whether it is biodegraded and lost its performance during the application period or not, thus it can perform its optimum usefulness for the certain period for safety. The polymer degradation is an irreversible process that can be induced by several environmental factors. The factors influence the rate of degaradation, including the type of chemical bond in the polymer backbone, hydrophilicity, molecular weight, copolymer composition, crystallinity and presence of low molecular weight compounds [18,19]. PLA is biocompatible and undergo hydrolytic scission to lactic acid when implanted. To study the effect of MCC into the PLA/ABS composites, the MCC loading with 0% and 30% for treated and untreated samples was undergo testing after 50 days burying into the soil. As shown in the figure 4, the increasing of MCC content up to 30wt%, it was apparently more obvious to observe the surface erosion by the microorganisms. The higher the fillers loading, the higher the weight loss in soil burial and could be reduce time to degrade in polymer applications. Incorporation of reinforce fillers and biodegradable polymer, there were more prone to bacterial attack despite presence of ABS in composites. The cellulose were degraded first and leaving vacant space, which increases the surface area for the onwards biodegradation of the polymer matrix. This is due by the formation of voids and cavities, which were more concentrated in the location where the filler attaches to the PLA/ABS matrix. Image of MCC 5% and MCC 30% illustrate that with the addition of silane coupling agent onto MCC, less degradation was observed and that specimen without silane coupling agent had rougher surfaces. This is indicated there was noticeable effect of bacterial action in specimen due to the strong biopolymer-filler interaction by the silane. The poor interaction adhesion was indicated by the large amounts of filler extracted after the soil burial.

IOP Conf. Series: Materials Science and Engineering **701** (2019) 012055 doi:10.1088/1757-899X/701/1/012055

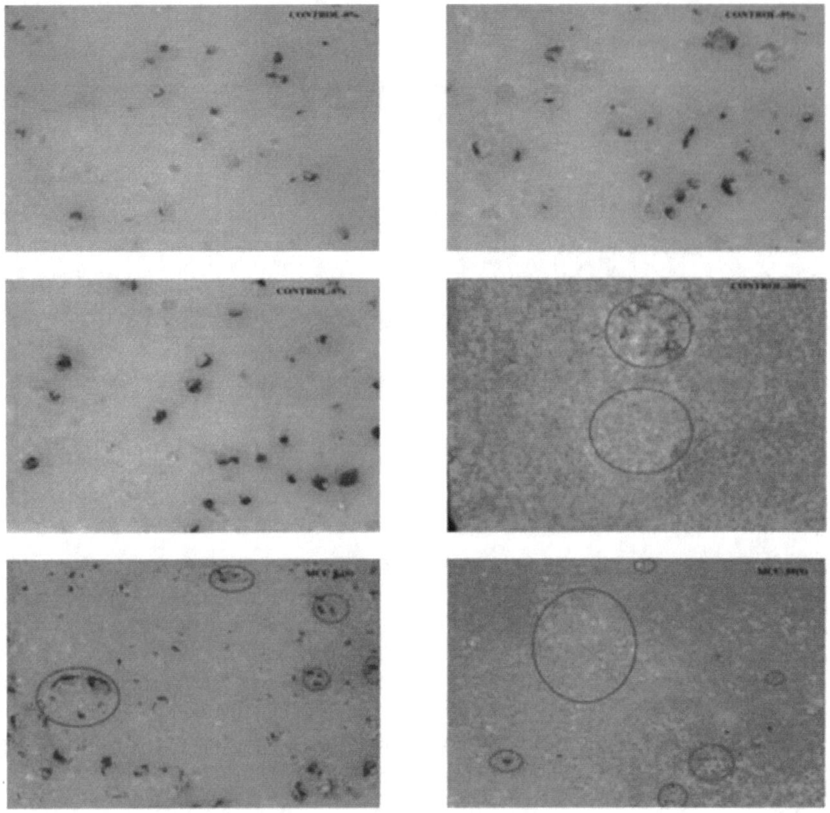

Fig. 4. Optical microscope image for treated and untreated at magnification of x80 for 5% MCC and 30% of compost in soil burial after 50 days.

5 Conclusions

The effect of micro-crystalline cellulose (MCC) loading on polylactic acid (PLA)/acrylonitrile butadiene styrene (ABS) with and without surface treatments by silane coupling agent were investigated. The main purpose of choosing MCC in this study was to produce a partial bio-based plastics with the wish to partially replace the conventional petroleum-based plastics which have some disposing issues. In this study, the incorporation of MCC has been show that by adding reinforce filler capable to improve the tensile properties. The higher the MCC content in the PLA/ABS composite, the tensile strength and Young's modulus show good increament in its value especially for silane treated MCC composites. It is due to the better interacial interaction between PLA/ABS matrix and MMC. In morphology analysis, silane treated MCC was well dispersed and embedded in the PLA/ABS matrix and have less void formation as compared to untreated MCC counterparts. In soil burial test, it was found that increasing in MCC content in the matrix would be enhance the degradation kinetics and thus increased the weight loss. It was observed that more adhering microorganisms colonizing onto the surfaces as more as MCC content was added. As the MCC filler content increased, the formation of voids, cracks and cavities become more prominent.

The authors wish to gratefully acknowledge the School of Materials Engineering, as well as Research and Innovation Department under Universiti Malaysia Perlis and Kementerian Pengajian Tinggi Malaysia for supporting this research project.

References

1. M.T. Ghaleb, Mohd Ishak. J. Appl. Polym. Sci., **123(5)**, (2012)
2. D.L. Ngoc, L.T. Saw, N.A. Abdul Rahim, A.W.M. Kahar, X.V.Cao. Polym. Degrad. Stabil., **111**, (2014)
3. R. J. Walster, R. Rahman, A. W. M. Kahar. IOP Conf. Series: Mater. Sci. Eng., **429**, (2018)
4. H. Zaman, J. Song, L.S. Park, I.K. Kang, S.Y. Park, G. Kwak, K.B.Yoon. Polym. Bull. **67(1)**, (2011)
5. A. W. M. Kahar, H. Ismail, A. Abdul Hamid, A J. Therm. Anal. Calorim., **123(1)**, (2016)
6. N. Ibrahim, A.W.M. Kahar, D.L Ngoc, H. Ismail, H. Bio-res., **12(2),** (2017)
7. L. Elias, F. Fenouillot, J. Majeste, P. Cassagnau, Polym., **48(20)**, (2007)
8. Sumigin, Tarasova, Krumme, Viikna. Proceed. Est. Acad. Sci., **61(3),** (2012).
9. S. Jain, M. Reddy, A. Mohanty, A., Ghosh, M. Misra, J. Macromol. Mater. Eng., **295(8),** (2010)
10. A.W.M. Kahar, H. Ismail. J. Vinyl Add. Technol. **22(3)**, (2016)
11. Mathew, Oksman, & Sain. J. App Polym. Sci., **97,** (2005)
12. A.H. Mohd Zain, A.W.M. Kahar, I. Hanafi, 2018. J Polym Environ, **26(2)**,(2008)
13. H. Salmah, H., Ismail, A. Bakar, J. Appl. Polym. Sci., **107(4),** (2007)
14. A.W.M. Kahar, M. Lingeswarran, M.Z. Amirah Hulwani, H. Ismail, Polym. Bull., **76(2),** (2019)
15. N.F. Alias, H. Ismail, A.W.M. Kahar,. Bio-res, **12(4)**, (2017)
16. J. Clark, J. J. Chem. Technol. Biotechno, **82(7),** (2007)
17. O.S. Dahham, N.Z. Noriman, A.W.M. Kahar, H. Ismail, S.T. Sam, J. Appl. Sci. & Agric., **10(5),** (2015)
18 A.H. Mohd Zain, A.W.M. Kahar, N.Z. Noriman, Procedia Chem., **19**, (2016)
19. A.W.M. Kahar, N. Sarifuddin, H. Ismail, H., Ira. Polym. J., **26 (2),** (2017)

IOP Conf. Series: Materials Science and Engineering **701** (2019) 012056 doi:10.1088/1757-899X/701/1/012056

Thermoplastic Polyurethane (TPU) / Organo-fluoromica Nanocomposites for Biomedical Applications: *In Vitro* Fatigue Properties

Azlin Fazlina Osman[1*] and *Darren* James Martin[2]

[1]Center of Excellence Geopolymer and Green Technology (CEGeoGTech), School of Materials Engineering, Universiti Malaysia Perlis, Arau, Perlis 02600, Malaysia.
[2]Australian Institute for Bioengineering and Nanotechnology, The University of Queensland, QLD 4072, Australia.

Abstract. Poly(dimethylsiloxane) (PDMS) / poly(hexamethylene oxide) (PHMO)-based thermoplastic polyurethane (TPU) nanocomposite was investigated for potential use in biomedical application. Studies on the *in vitro* fatigue behaviour of the TPU and TPU nanocomposite (under physiological saline solution, 37°C conditions) were highlighted in this article. The data were compared with those of commercially available silicone elastomer (Nusil MED 4860). Results indicated that the TPU nanocomposite (2MED-C (2HM)) had greater fatigue properties than the virgin TPU, which provide strong evidence of its greater capacity to withstand cyclic forces than the host TPU when exposed to physiological fluid. This was caused by the presence of well dispersed and impermeable organofluoromica platelets in the TPU matrix resulted in more tortuous path for the physiological fluid diffusion, thereby decreasing the fluid permeability of the polymer. Eventhough the silicone elastomer has lesser hysteresis than the virgin TPU and TPU nanocomposite, its fatigue strength is much lower than those of the TPU nanocomposite. The findings revealed the potential of PDMS/PHMO based TPU nanocomposite to replace silicone elastomer as biomaterial, particularly for implantable biomedical device application.

1 Introduction

Thermoplastic polyurethanes (TPUs) are linear, segmented block copolymers consisting of hard and soft segments. The hard segments are made from diisocyanate and short chain extender molecules such as diols or diamines, are rigid and highly polar. Hard segments have high inter-chain interaction due to hydrogen bonding between the urethane/urea groups [1]. The hydrogen bonding associations within the hard segments of the TPUs act as reinforcing filler for the soft matrix. On the other hand, soft segments consist of long, linear

*Corresponding author: azlin@unimap.edu.my

flexible polyether or polyester chains which interconnect two hard segments. In brief, the hard segments act as multifunctional tie points working as both physical crosslinks and reinforcing fillers, while the soft segments primarily influence the elastic properties of TPU [1, 2]. Thermoplastic polyurethanes (TPUs) are promising candidates for use as biomaterials in the fabrication of a wide range of medical devices due to the relative ease of fabrication, flexibility, good blood contacting properties, biostability and excellent insulation properties [3, 4]. The poly(dimethylsiloxane)(PDMS)/ poly(hexamethylene oxide)(PHMO)-based TPU based on an optimized formulation (Elast-Eon™) from AorTech International Pty Ltd exhibits physical properties comparable to those of medical grade polyether-based TPU materials such as Pellethane™ 80A and is now widely accepted as being amongst the most biostable of all the commercial TPUs [5]. Our previous work demonstrates that the Elast-Eon E5-325 TPU can be tailored to provide advantageous mechanical properties by incorporating organically modified smectic nanosilicates [6-8]. We found that the best surface modification was obtained with the most hydrophobic / dual surfactants system (namely the 75% dimethyldioctadecylammonium chloride (DODMAC) & 25% Choline chloride) [6, 7]. The TPU nanocomposite with organofluoromica (fluoromica surface modified with DODMAC/CC) was then selected for further evaluation on the biostability aspects. To safely utilize these PDMS/PHMO-based TPU nanocomposites in human biological systems, assessing the mechanical performance under specific conditions and environmental influences is also essential, given that the TPU nanocomposite is to be exposed in the human physiological fluid. In vitro tests in liquid media closely mimicking physiological chemical conditions and temperature, are essential for "screening" this TPU nanocomposite for eventual use in implantable devices, and prior to committing to expensive in vivo tests in appropriate animal models. For this work in particular, it is of great importance to assess the time-dependant dynamic mechanical integrity by employing fatigue test. Fatigue is a failure mechanism which occurs when the material is subjected to a cyclic load or repeat stress [9].

In this communication, the *in vitro* mechanical properties of the PDMS/PHMO TPU nanocomposites incorporating organofluoromica were reported and compared to those of the commercial silicone elastomer (Nusil MED 4860). The samples have been subjected to *in vitro* condition by immersing the host TPU and nanocomposite samples in physiological saline solution at 37°C. The *in vitro* fatigue test was carried out to predict the potential improvements in long-term in vivo performance of the TPU nanocomposites for implant applications.

2 Materials and Methods

2.1 Materials

Thermoplastic polyurethane (TPU) was supplied by AorTech Biometerials Pty Ltd (VIC, Australia). This TPU is commercially known as ElastEon™ E5-325 TPU consists of a 1000 g/mol poly(dimethylsiloxane) (PDMS) and 700 g/mol poly(hexamethylene oxide) (PHMO) mixed soft segment in a 98:2 (w/w) ratio, and a hard segment composed of alternating 4,4'-methylene diphenyl diisocyanate (MDI) and 1,4 butanediol (BDO) sequences. The hard segment concentrations is 32.5 wt%. Nusil MED 4860 is a medical grade elastomer with a two-part silicone system in 1:1 mix ratio (Part A:B). Part A consists of 30wt% amorphous silica, while part B is a 5 wt % dimethyl, methylhydrogen siloxane copolymer. When these two components are mixed, Nusil MED 4860 undergoes rapid 5 min curing at 165 °C due to the presence of a platinum catalyst. This material was supplied by CochlearTM Ltd and is commercially available from Nusil (via EIM Medical Consulting Pty Ltd, NSW, Australia).

Somasif ME100 (ME) a synthetic fluoromica (tetrasilicic trioctahedral fluoromica) was supplied by Kobo Products, Inc ((N.J., USA). It is a fine white powder with an average platelet length of approximately 650 nm. This fluoromica was surface modified with dual modification system (75% dimethyldioctadecylammonium chloride (DODMAC) & 25% Choline chloride) to obtain the 'hydrophobic organo-fluoromica) as nanofiller for this study. This dual surface modified fluoromica (organo-fluoromica) was chosen in this study because it has previously resulted in the best mechanical performance when incorporated into the TPU due to its hydrophobic characteristic that can provide more favourable TPU-nanofiller interactions.

2.2 Methods

High Energy Milling Process for Organo-fluoromica (MED-C)

High energy milling on the organo-fluoromica was done using a Netzsch Laboratory Mill type LABSTAR LS1. The 0.4 mm Yttrium-stabilized zirconium oxide (ZrO2) grinding beads were employed as milling media. The milling was done in an ethanol/water mixture at a 1:1 ratio. The organo-silicate was milled for 2 hours to reduce its particle size from ~650nm to ~250nm. The milled MED-C suspensions were then separated by centrifugation using a Beckman Coulter Allegra X-15 benchtop centrifuge with a rotation speed of 4750 rpm for 3 min, and subsequently washed with MilliQ water (18.2 MΩ.cm). The washed clay was dried in an oven at 60 °C overnight and ground by jet milling prior to nanocomposite processing.

Preparation of the TPU Nanocomposite Sheets

E5-325 TPU was used as matrix material in this study. The two hour milled organofluoromica (MED-C (2HM) were used as nanofillers. E5-325 TPU nanocomposite containing MED-C (2HM) was prepared with a 2 wt% nanofiller composition by using melt processing (MP) method. In the subsequent discussion, these nanocomposites are referred to as 2MED-C (2HM). A number denotes the 2 wt% organosilicate loading in the TPU. The first two letters represent the nanosilicate used (ME=Somasif ME100) and the last two letters (D-C) represent the dual DODMAC/CC surface modification. The characters in the brackets; (2HM) denote the nanofiller size reduction process of 2 hour high energy milling. E5-325 TPU nanocomposites were prepared by melt compounding the TPU pellets with 2 wt% organo-fluoromica using a Haake Rheomex OS twin screw extruder (Thermo Scientific, USA). The extrudate was pelletized and dried at ~70 °C for approximately 20 hours prior to being compression moulded. Compression moulding was performed using a hydraulic press. A pair of brass plates was used, with the bottom plate having a rectangular cavity. The moulds were heated to 185 °C, prior to the pellets being added. A "pre-press" was done at 1 kPa for I minute, followed by 5 kPa for 0.5 minutes. Pressure was released to remove any further air bubbles then the samples were pressed for a further 0.5 minutes at 7.5 kPa. The samples were cooled under pressure to about 140 °C using a controlled water flow without releasing pressure. Approximately 1mm thick plaques were produced and they were then annealed under vacuum at 85 °C for approximately 5 hours and left for at least a week to age prior to testing.

In vitro fatigue test

In order to obtain sensible *in vitro* results, the test was performed in an environmental chamber containing phosphate buffered saline (PBS) solution (Ph ~7.4). In addition, 0.02% sodium azide was added to the solution to retard the microorganism growth, and the test was performed at 37°C condition, which approximates human body temperature. The test samples were conditioned for at least 30 minutes in the environmental chamber before proceeding to the mechanical test. Dumbbell samples which were punched from an ASTM D-638-M-3 die were employed for each test. A cyclic tensile load of 500 N was applied axially at 0.2 Hz for 10,000 cycles. Cyclic strain amplitude of 50 % was applied, and the test was recorded using a sinusoidal waveform and a 50 ms data acquisition rate. Hysteresis was calculated on the 5th cycle to 125 % using the equation below [10]:

$$\text{Hysteresis (\%)} = \frac{\text{Area bound by the loading and unloading curve}}{\text{Area under the loading curve}} \text{ X } 100 \qquad (1)$$

Transmission Electron Microscopy (TEM) Analysis

TEM analysis was done to observe the dispersion of organofluoromica platelets throughout the TPU matrix. Thin sections of approximately 80 nm thickness were cut using a Diatome diamond knife on a Leica Ultracut UC6FCS cryogenic ultramicrotome between -80 and -110°C to ensure the polymers were rigid. Sections were picked up using a loop of 2.3M sucrose and mounted on 200 mesh copper grids (ProSciTech, Australia). Grids were placed on droplets of deionized water and then transferred through five washes. The grids were then allowed to air dry in self-closing forceps prior to viewing. Sample was examined at high magnification (93000×) on a Technai F30 FEG TEM (FEI company, Netherlands) operating at 300 kV and images were captured with a Direct Electron LC1100 Lens-Coupled 4k × 4k CCD camera system.

3 Results and Discussion

In vitro (in PBS, 37°C) fatigue properties of silicone elastomer (MED 4860), virgin TPU and TPU nanocomposite

In this work, fatigue test was carried out to determine the behavior of materials in physiological fluid, 37°C under fluctuating loads. Since long term stability *in vivo* is critical for biomedical use, changes in deformation and properties due to fatigue behaviour should also be addressed. It has been discovered that the TPU consists of an elastic portion which stores energy and returns it, and a viscous portion which captures energy and converts it to heat [11, 12]. At 37°C, soft segments are above their glass transition temperature and impart the material its elastic behavior, while hard segments are below their glassy or melt transition temperature and are thought to govern the plastic deformation, high modulus, and tensile strength [11, 12]. This unique TPU morphology evolves with deformation and this evolution is thought to be the primary source of hysteresis and cyclic softening [11-13]. Hysteresis is a measure of absorbed or dissipated energy during any cycle of loading or unloading when material is subjected to repeated loading [11, 12]. Most TPU show hysteresis characteristic, and this can be attributed to a number of factors such as non-affine deformation, irreversible orientation of the hard domains, plastic deformation of the hard domains and energy losses caused by the hard domain orientation [12]. The breakdown and reformation of the TPU mesophase morphology during the cyclic loading are responsible for the stress-softening (Mullin effect) and hysteresis losses. This may somehow contribute to a reduction in mechanical integrity [12, 14]. The area between the two curves is the energy which is not returned but is converted to heat. The hysteresis loop may yield information on fatigue degree [14]. The modification of hysteresis loop with cycle is associated with the stress variation with time, which is depends on the cyclic stability of the tested material. Large modification in hysteresis loop upon high cyclic loading is the indication of cyclic softening or reduction in mechanical strength [14]. For comparison on the cyclic effect, the hysteresis loops from the cycle number of 5, 100, 1000 and 10,000 were obtained from the stress-strain curves during the fatigue test and they are presented in Figure 1.

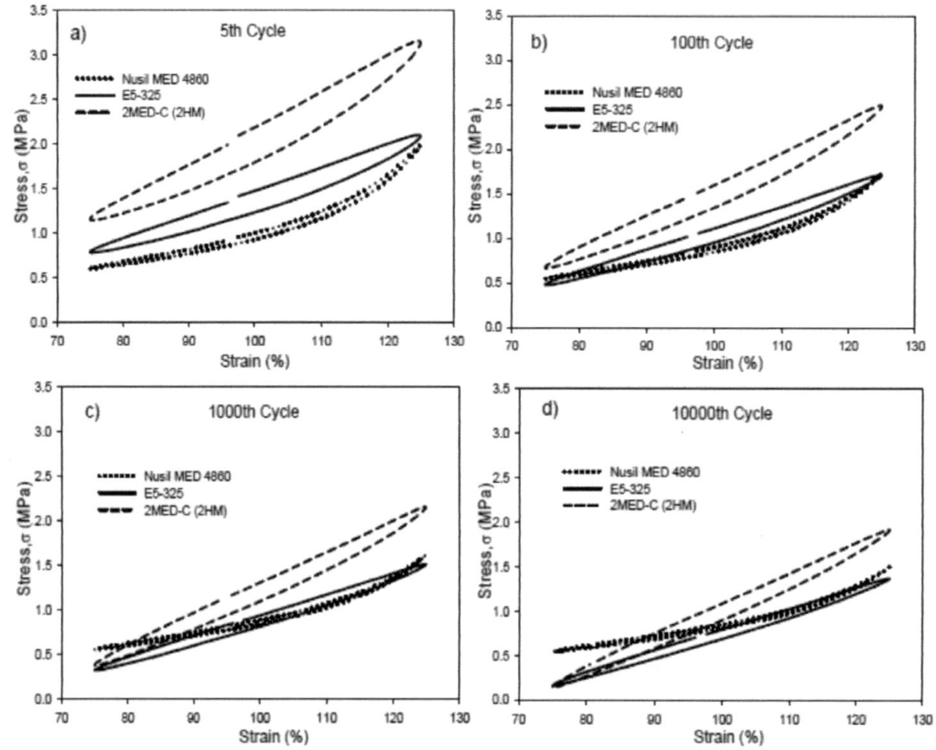

Fig. 1. Hysteresis loops of Nusil MED 4860, E5-325 and 2MED-C (2HM) taken at cycle number a) 5 b) 100 c) 1000 d) 10000

Table 1: *In vitro* fatigue properties of Nusil MED 4860, E5-325 TPU and 2MED-C (2HM)

Number of cycles	Hysteresis loop area (MPa)		
	Nusil MED 4860	E5-325	2MED-C (2HM)
1	4.533606	11.87614	18.95885
5	2.773143	8.884384	14.27935
10	2.530383	8.11088	13.11338
100	2.102186	6.123003	9.957682
1000	1.656158	5.051902	8.130133
10000	0.877921	4.269729	6.851635
Differences in loop area between the 1st cycle and 10,000th cycle	81%	64%	64%

Hysteresis (%) (based on 5th cycle)	7.7 %	19 %	20%

Silicone elastomer generally has lesser hysteresis than TPU due to the absence of structure contributing to plastic deformation and stress relaxation. This can be clearly observed in the Nusil MED 4860, which demonstrate much lower hysteresis than the E5-325 TPU (virgin and nanocomposite). As can be seen from Table 1 and Figure 1, the hysteresis loop of 2MED-C (2HM) shows a larger area than the virgin TPU (E5-325), indicating greater energy dissipation, which is mainly due to the friction of the nanofiller surface and the TPU chains [12]. It is well established that the incorporation of fillers to elastomers causes an increase in hysteresis [11, 14, 15]. Of particular interest, the addition of the organofluoromica into the TPU only resulted in 5 % increase of hysteresis. However, the percentages of reduction in the hysteresis loop area after 10,000th cycle are the same for both E5-325 TPU and 2MED-C (2HM) nanocomposite (see Table 1). This suggests that the incorporation of organofluoromica did not increase the fatigue degree at high cyclic loading (10,000th cycles). Furthermore, 2MED-C (2HM) nanocomposite also performs the highest fatigue strength by demonstrating the highest maximum stress values, both in low and high cycles. For instance, the fatigue strength of 2MED-C (2HM) measured at 10,000th cycle exhibits an increase by 40 % with respect to the virgin TPU. Several findings concluded that dynamic stresses promote chain scission, crack initiation and propagation in other synthetic elastomeric materials when tested under environmental challenged conditions [9, 11, 16]. However, the incorporation of well dispersed nanofiller might retard these processes [13]. As shown in TEM image in Figure 2, 2MED-C (2HM) nanocomposite demonstrates the dispersive and distributive mixing system. Upon deformation, the well dispersed and exfoliated organofluoromica (MED-C (2HM)) platelets might act as nano-barriers for crack propagation, and hence leading to an increase of fatigue resistance. This nano-barrier effect has been explained by Jin et al. [13] who hypothesized that, the incorporation of the organomontmorillonite in the TPU resulted in an increase in fatigue strength due to the formation of 'nanospring-like' system composed of two nanoplatelets and polymer chains, which are intercalated in the two nanoplatelets. Upon cyclic loading, these 'nanospring-like' systems undergo reversible deformation, thereby capable of absorbing deformation energy and preventing the development of fatigue cracks. It is also worth mentioning that when the TPU being immersed in the PBS solution, the permeant (physiological fluid) can result in weakening of hydrogen-bonded hard segment domains and secondary bonding between the TPU chains. However, TEM image in Figure 2 reveals good dispersion of organofluoromica in the structure of 2MED-C (2HM) nanocomposite. This reduce the permeability of the physiological fluid inside the MED-C (2HM) nanocomposite structure due the tortuosity effect, induced by the well dispersed and exfoliated organofluoromica platelets. As a result, the fatigue strength of the 2MED-C (2HM) nanocomposite was superior to the virgin TPU when exposed to PBS solution, 37°C.

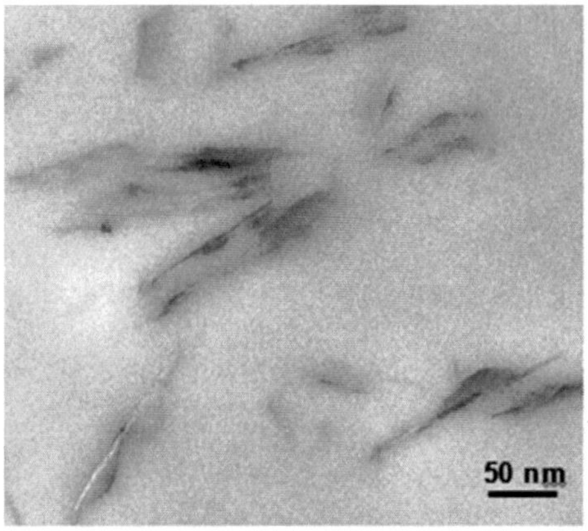

Fig. 2. TEM image of TPU nanocomposite (2MED-C (2HM)) showing well dispersed organofluoromica platelets in the TPU matrix.

4 Conclusion

Our preliminary studies on the *in vitro* fatigue show promising properties of the E5-325 TPU nanocomposite (2MED-C (2HM)), which possess the potential to be further developed for biomaterials. When subjected to the *in vitro* conditions, 2MED-C (2HM) shows the highest fatigue strength by obtaining the highest maximum stress values, both in low and high cyclic loading (up to 10,000 cycles). The well dispersed and exfoliated organofluromica platelets might act as nano-barriers leading to an increase of fatigue resistance. The data reported here are evidence that well-engineered nanofiller systems may assist the TPU performance under the environmental challenge environment. If carefully formulated, this nanofiller can bring positive influence on the viscoelastic properties such as creep and fatigue, in addition to ultimate tensile strength and tear strength. Although, the preliminary *in vitro* fatigue test was performed as more of a proof-of-concept stage, the results gained from this study should provide a useful measure of predicting the E5-325 TPU nanocomposite biostability, essential for the development and design of future biomedical TPU nanocomposites. Thus, these findings may merit further investigation on long-term *in vitro* and *in vivo* mechanical properties and stability under varied environmental ageing for complete investigation on the biostability and biocompatibility of this nanocomposite system.

References

1. M. Szycher. *Szycher's handbook of polyurethanes, 2nd Edition*. CRC Press, Boca Raton, Fluorida, (2012)
2. J.T. Haponiuk, K. Formela, K., *PU Polymers, Their Composites, and Nanocomposites: State of the Art and New Challenges in Polyurethane Polymers: Composites and Nanocomposites*, (2017)
3. M.V. Pergal, M. Balaban, B. Dojčinović, D. Manojlović., *Thermoplastic polyurethane nanocomposites in Thermoplastic Composites: Emerging Technology, Uses and Prospects*, Edited by Ritter, E., Nova Science Publishers, Inc., (2017)
4. D. E. Heath, S.L. Cooper, W.G. Lawrie, *Polyurethane, in Biomedical Science: An Introduction to Materials in Medicine, 3 rd Edition*, Buddy D. Ratner, B.D, Allan S. Hoffman, Frederick J. Schoen, Jack E. Lemons, J. E., Ed. Elsevier , MA USA, (2012)
5. Y. Andriani, I.C. Morrow, E. Taran, G.A. Edwards, T.L. Schiller, A.F. Osman, and D.J. Martin, Acta Biomater., **9**, (2013)

6. A.F. Osman, Y. Andriani, G.A. Edwards, T.L. Schiller, K.S. Jack, I.C. Morrow, P.J. Halley, D.J. Martin. Macromolecules, **45**, (2012)
7. A.F. Osman, P. Halley, D. Martin., Adv. Mater. Res., **795**, (2013)
8. A.F. Osman, K. Jack, G. Edwards, D. Martin, Adv. Mater. Res., **832**, (2014)
9. C. M. Manjunatha, A.A. Chandra & N. Jagannathan, J Indian I Sci., **95**, (2015)
10. Y. Sasaki, M. Yamasaki, & T. Sugimoto, Wood Fiber Sci., **38**, (2007)
11. A. F. Osman, *Biomedical Thermoplastic Polyurethane Nanocomposite: Structure-Properties Relationship*. PhD Thesis, University of Queensland, Australia (2013)
12. C. Prisacariu, *Polyurethane Elastomers: From Morphology to Mechanical Aspects*. Springer Verlag, New York (2011)
13. J. Jin, L. Chen, M. Song, K. Yao. Macromol. Mater. Eng., **291** (11), (2006)
14. G. Frunza, E.N. Diaconescu, *Hysteresis and Mechanical Fatigue. In The annals of University "Dunarea De Jos" of Galati*, (2006)
15. B. Dong, C. Liu, L. Zhang, & Y. Wu, RSC Adv., **5** (22), (2015)
16. G. Lake, Rubber Chem Technol., **68**, (1995)

Coefficient of thermal expansion (CTE) study in metal matrix composite of CuSiC vs AlSiC

Mohd N Arif[1] , *Mohabattul* Z Bukhari[1,*], *Dermot* Brabazon[2] , and *M Saleem* J Hashmi[2]

[1]School of Materials Engineering, Universiti Malaysia Perlis, 02600 Jejawi, Arau, Perlis, Malaysia
[2]School of Mechanical and Manufacturing Engineering, Dublin City University, Glasnevin Campus, Dublin 15, Rep. of Ireland

Abstract. Metal matrix composites are widely used as advanced materials for thermal management applications. In view of the demands in higher temperature products application, materials with low coefficient of thermal expansion (CTE) and high thermal conductivity are required. Copper Silicon Carbide (CuSiC) is highly rated as the best materials selected for this application because it has good thermal characteristics especially in CTE, but its density is almost high for high temperature electronic device application. An aluminium silicon carbide (AlSiC) composite, on the other hand, is a new candidate for electronic device application since it has low density and good CTE characteristics. CuSiC and AlSiC composites were fabricated via powder metallurgy method through variety volume fractions such as 5%, 10%, 15% and 20%. In the present study, the comparison of CTE between CuSiC and AlSiC composites with different volume fraction of SiC particles has been carried out. The CTE of CuSiC and AlSiC composites were investigated and their responses between 25°C to 400°C were studied using horizontal dilatometer. Based on the experiment results, the density of CuSiC composites are decrease while the density of AlSiC composites is increase with increasing in SiC particles content. However, the porosity for both composites is increases with increasing in volume fraction of SiC particles. As of CTE, both composites showed decreased with increasing in SiC particles content that dispersed in their matrix. The experimental CTE values for both composites seem to be much lower than theoretical CTE value, which means that there has been excellent bonding between each particle after sintering process. The particles distribution has been observed to understand more on the aspect of SiC particles homogeneity in copper and aluminium matrix.

1 Introduction

In recent decades, scientists and engineers have been involved in designing a new generation of extraordinary materials from various combinations of properties that cannot be met by pure metal, ceramic and polymer to be used in modern technologies that called composite materials.

*Corresponding author: mzaman19704@gmail.com

Composite has been the most attracted materials to many researchers and become the materials of choice by many researchers due to its low cost, environment friendly and great performance [1]. The morphology, particle distribution, components, degree of crystalline and microstructure texture of the interphase between components can be engineered for various properties of composite materials [2].

Copper silicon carbide (CuSiC) and Aluminium silicon carbide (AlSiC) composites have been chosen due to their great heat properties such as heat dissipation, low CTE and good thermal conductivity. In thermal application, most metal matrix composites facing a mismatch issue in term of differences in CTE of two joining materials that produce stress and strain at the joining interface or in attachment surface. This will negatively affect the performance of the product which eventually lead to failure in operations [3]. Study has been conducted to analyse the effect of volume fraction in Silicon Carbide (SiC) particles reinforcement that will affect the CTE of the copper and aluminium.

CTE study of CuSiC composites were conducted and evaluated by many researchers in which the fabrication of CuSiC composites via powder metallurgy and angular shape SiC were commonly used [4-6]. Traditionally, copper is the good material for heat sink application but it suffers from high CTE and high density, therefore the incorporation of SiC particle into copper matrix will helps in reducing the CTE and density issues [7].

For AlSiC composites, the effect of SiC in aluminium matrix were conducted and evaluated through different fabrication methods [8-13]. Aluminium on the other hand, can resist oxidation under various conditions due to its stable film on its surface. The compatibility between aluminium and SiC has led to a new method development of AlSiC as a great material to be used in thermal application besides CuSiC.

2 Materials and procedures

In this study, powder metallurgy method was used to fabricate CuSiC and AlSiC composites. The powders went through a mixing process, compaction and sintering process. The effect of CTE on CuSiC and AlSiC composites are study with 5%, 10%, 15% and 20% volume percent of SiC reinforced in copper and aluminium.

The weighted powder was placed into a customized mixing container and several stainless steel balls were used to assist the mixing process. Two roll mill machines were used for this process. Since the customized mixing container was used, the mixing speed and mixing time were optimized through try-and-error method. Based on [14], to get best mixing, the optimal rotational speed, N_o of cylindrical mixer is 75% form critical rotational speed where it represent a balance between centrifugal and gravitation forces and can be estimated as Equation (1);

$$N_O = \frac{32}{\sqrt{d}} \tag{1}$$

where d is the cylindrical diameter in meters. So, the calculated mixing speed has been used was 160 rpm as shows in and 4 hours of mixing time for each sample.

Then the powders undergo plastic deformation by compacting them through cold isostatic pressing less than 5 tonnes of load by using hand press machine. These green compact samples were sintered in the tube furnace at 950°C for CuSiC and 580°C for AlSiC for 4 hours sintering time. The temperature of the furnace was slowly increased at 5°C per minute to minimize temperature gradient and prevent thermal shock on the samples. These samples were sintered in argon gas atmosphere. The samples for each volume fraction of SiC were prepared for CTE measurement. The Linseis L75HX1600 horizontal dilatometer was used to measure the CTE of the samples. The temperature range of this analysis was 20°C to 400°C and the heating rate was set at 5°C per minute. The inert gas argon was used to give gas shielding protection for samples from oxidation during the measurement process.

3 Results and discussion

Figure 1(a) shows the micrograph of copper powder at 500x magnification and it demonstrated that the powder particles are in irregular shape and size. However at 2000x magnification, the dendritic shape and non-uniform size of copper powder particle as shown (Figure 1(b)). According to [14], the dendritic shape of copper contributed to high green compact properties due to substantial mechanical interlocking during powder compaction process. It was reported that the density of the sintered product would be closest to theoretical density and low amount of porosity caused by dendritic structure fulfil gap between each particle [15].

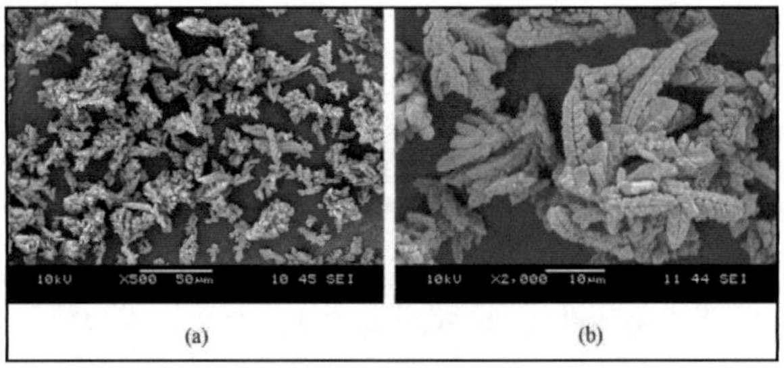

Fig. 1. SEM micrograph of copper powder (a) at 500x magnification and (b) at 2000x magnification.

Figure 2 shows the micrograph of aluminium powder where (a) the image form under 200x magnification while (b) the image form under 500x magnification. Based on observation, the shape for aluminium powder is flake shape. The particle size of aluminium powder was discovered as non-uniform distribution of size between each particle. Unlike dendritic particle, flake shaped particles are more difficult to compact due to minimum contact surface area and low interlocking between each particles. Hence, it can be estimated that there are gaps formed in between aluminium and SiC particle and the total porosity are also high by increasing SiC content into aluminium matrix [14].

Fig. 2. SEM micrograph of aluminium powder (a) at 200x magnification and (b) at 500x magnification.

Based on Figure 3(a), the micrograph demonstrate SiC particles are in angular shape, finely separated particles and inconsistent in sizing. However, at 1000x magnification, SiC particles morphology was found irregular and sharp edge corner as illustrated in Figure

3(b). Theoretically, this type of surface particle of ceramic material is very suitable to fabricate with metal matrix in order to improve the green compact strength [16].

Fig. 3. SEM micrograph of copper powder (a) at 100x magnification and (b) at 1000x magnification.

Figure 4 shows the comparison between true density and theoretical density of the CuSiC composites. Based on Figure 4, it found the true density of CuSiC composites was inversely proportional to the volume fraction of the SiC. The same result was reported by [17] which studies about SiC content effect on the properties of CuSiC composites produced by mechanical alloying. Nevertheless, the average true density of CuSiC composites were much more lower than theoretical density.

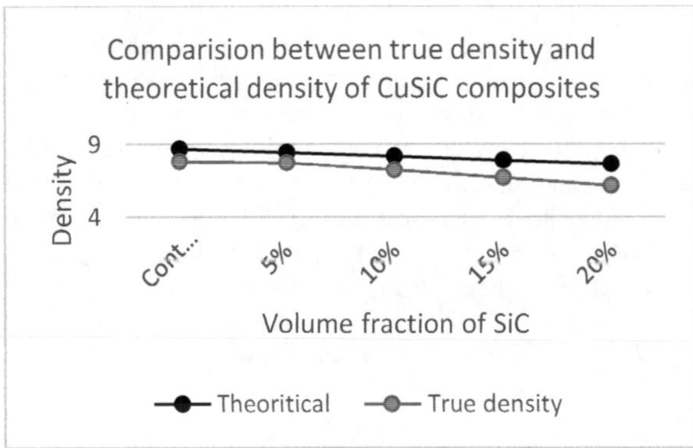

Fig. 4. True density and theoretical density for CuSiC composites.

According to [18], the density of AlSiC composites were increased by increasing the volume fraction of SiC reinforcement. The increasing of AlSiC composite density is due to aluminium reinforced with other material that has higher density for SiC. Based on Figure 5, the true density of AlSiC composites were increased uniformly from 0% to 15% SiC reinforced aluminium matrix and approaching the theoretical density. However, the difference in density value between true density and theoretical density was lower that only 0.118 g/cm^3 in average.

IOP Conf. Series: Materials Science and Engineering **701** (2019) 012057 doi:10.1088/1757-899X/701/1/012057

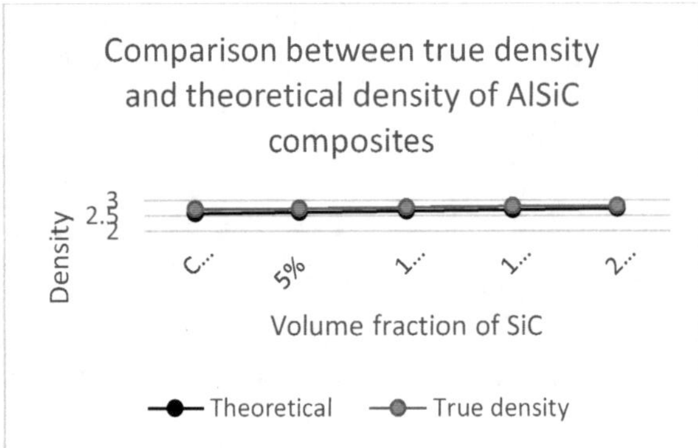

Fig. 5. True density and theoretical density for AlSiC composites.

The porosity of composites with different volume fraction that calculated by using Equation (2);

$$\rho_{bulk} = \left(\frac{D}{D-W}\right) D_w \quad (2)$$

Figure 6 reveals that the porosity of both composites, CuSiC and AlSiC increased with increasing volume fraction of SiC reinforcement. The porosity of CuSiC composites were increasing from 1.7974% to 12.7285% by increment 20% SiC into copper matrix. An increasing of porosity for CuSiC composite also reported by [15] in their studies. The low amount porosity of dendritic CuSiC composite that was expected in characterization of copper powder before was proven after porosity evaluation. Besides CuSiC composites, the porosity of AlSiC composites also increased against increasing amount of SiC. It shows the same result as previous researches. However, the porosity of AlSiC composites is higher compared to CuSiC composites. What was expected in raw material analysis before, the amount of pore in AlSiC is higher due to flake shape of aluminium particles used and this expectation was proven after this evaluation.

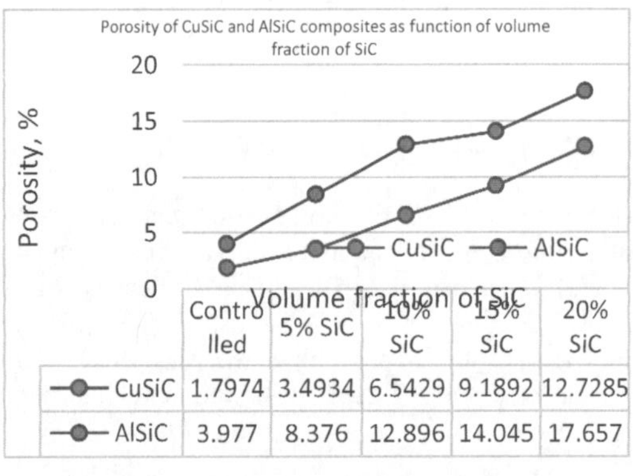

Fig. 6. Porosity of CuSiC and AlSiC composite against volume fraction of SiC.

Figure 7 shows the extracted CTE values of the CuSiC composites based on the raw CTE data. It has been noticed that the CTE of the composites with different compositions increases by increasing in temperature. Based on Figure 10, it shows that Cu-5vol%SiC composite exhibits maximum thermal expansivity compared to other compositions. Generally, the thermal expansivity increases as the temperature increase significantly. With

IOP Conf. Series: Materials Science and Engineering **701** (2019) 012057 doi:10.1088/1757-899X/701/1/012057

additional of SiC particles to copper matrix, there has been reduction in the thermal expansivity at maximum temperature for the different percentage composition of CuSiC composites. Addition of SiC particle with varying volume fraction resulted in the reduction in thermal expansivity of the CuSiC composites. This reduction proved that an additional of SiC particles might affected the expansion behaviour of copper matrix during heating condition. However, the experimental CTE value for each volume fraction of SiC was compared to theoretical CTE value and demonstrated in Figure 8.

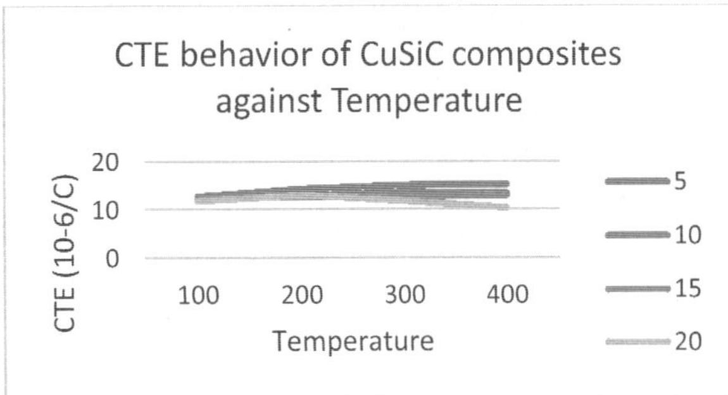

Fig. 7. The CTE values of CuSiC with increasing volume fraction of SiC.

Based on Figure 8, the Cu-5vol%SiC shows highest experimental CTE value which is 14.37 $x10^{-6}/^{o}C$ while the lowest is 11.67 $x10^{-6}/^{o}C$. Since SiC particles impose significant constraint on the copper matrix, the CTE values of the CuSiC composite decreases with the increasing volume fraction of SiC. The decreasing CTE values of CuSiC composites logically as the CTE of SiC is lower than that copper [17]. It has been reported in literature review that the CTE of the CuSiC composites considerably increases by reinforcing SiC particles over [5].

Figure 8 also shows the experimental CTE are compliant with the theoretical dependence. Research [15] dealt if the experimental CTE plot of CuSiC composite shows much higher than theoretical CTE means that, CuSiC composites suffer by weak bonding between copper and SiC particle. But in this study, the experimental CTE was lower than theoretical CTE which mean the excellent interfacial bonding also imposes effective constraint on the expansion of the copper matrix. It can be say that the fabrication process was perfectly done on CuSiC composites due to existing of bonding between copper and SiC particles. This statement also related to shrinkage of CuSiC composite that discuss before. The shrunk of CuSiC composites is almost high because each powder particle undergoing high neck growth while improving it properties [14].

An additional contribution to the decreasing CTE of the CuSiC composites also caused by the growth of porosity. Because of the different CTE between SiC and copper matrix, a CTE mismatch strain is induced when the composites are exposed to higher temperatures [19]. This CTE mismatch strain generates compression and tensile stress field in the matrix and reinforcement respectively. The pores undergo compression stress resulting in the shrinkage of the pore volume. Hence, the overall CTE of the composite decrease. The relationship between the CTE and porosity of the CuSiC composites was illustrated in Figure 9 where the CTE of composites was decreased with increasing porosity in composites.

IOP Conf. Series: Materials Science and Engineering **701** (2019) 012057 doi:10.1088/1757-899X/701/1/012057

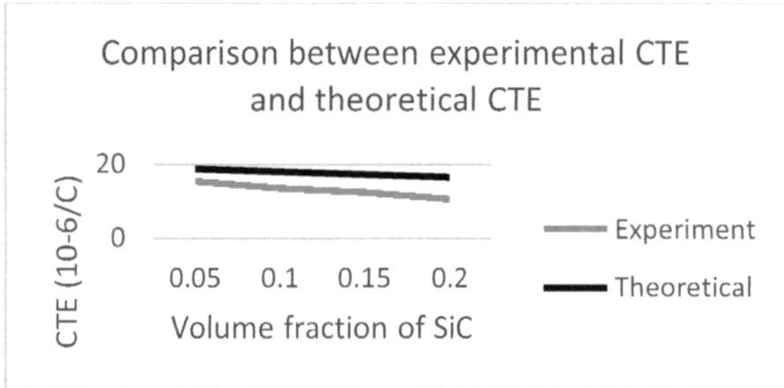

Fig. 8. The average experimental and theoretical CTE of CuSiC composite against volume fraction of SiC.

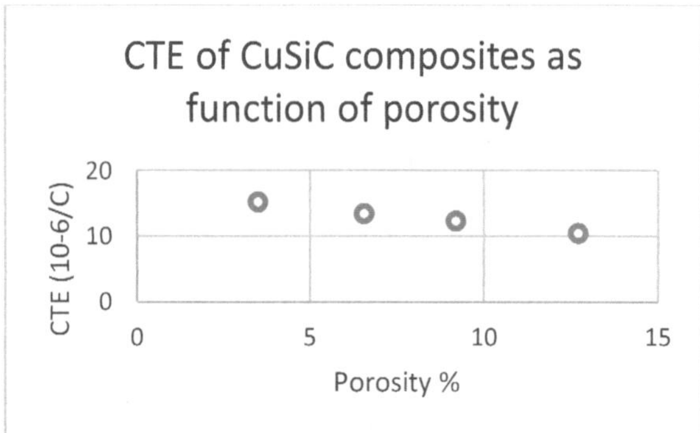

Fig. 9. The relationship between the CTE and porosity of the CuSiC composite.

The extracted CTE values of the AlSiC composites come from raw CTE data is shown in Figure 10. From this figure, it shows the variation of CTE and temperature for different composition of AlSiC composites. It noticed that the CTE of the composites with different SiC particles content increases with the increases in temperature. However, the experimental CTE value for each volume fraction of SiC was compared to theoretical CTE value and illustrated in Figure 11. Back to Figure 10, there is almost consistency in the increase of CTE for different temperature recorded at regular intervals. The Al-5vol%SiC composite exhibited the maximum value of CTE whereas Al-20vol%SiC composite exhibited lowest magnitude of CTE behaviour.

It was stated that the low percentage of SiC content in AlSiC composites appearing CTE behaviour to increase drastically [20]. The reduction in CTE value of AlSiC composites can be attributed to the lower CTE value of SiC particles compared to aluminium matrix. Hence, the ability of SiC particles as reinforcement is effectively constraint the expansion of the aluminium matrix. It also supported by [21], the CTE behaviour will be influenced by the thermal expansion of aluminium and tightened restriction of SiC particles.

The experimental and theoretical CTE of various volume fractions of SiC reinforced aluminium matrix composites were shown in Figure 14. It showed that both CTE of the composites containing SiC particle were declined with the increasing volume fraction from 5% to 20% of SiC particles. The highest experimental CTE value is 18.98 x10^{-6}/ºC while the lowest is 16.93 x10^{-6}/ºC. It agreed well with the accepted theory reported by [13] that the CTE of the composites is determined by the CTE of aluminium and the restriction of SiC particles through interfaces. This statement supported by [22] where increasing percentage content of SiC particles helped to increase the restriction on aluminium matrix, then to reduce the CTE of the composites. However, the experimental CTE of AlSiC composites were much lower than theoretical CTE.

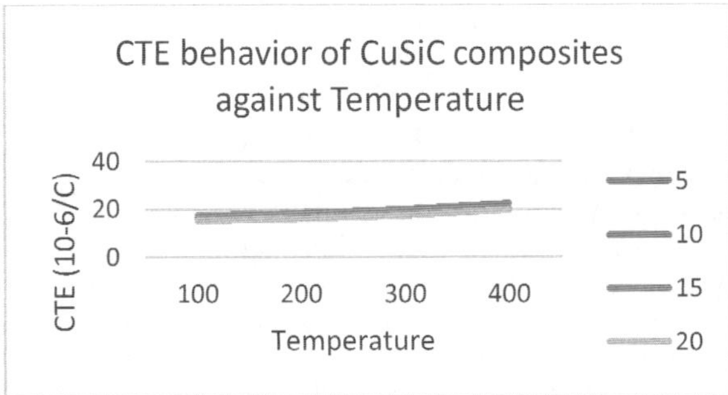

Fig. 10. The CTE values of AlSiC with increasing volume fraction of SiC.

This CTE result indicated that the declines much more dependent on the high particles reinforcement content. Large particles size of ceramic material into metal matrix via powder metallurgy method also resulted in high pore area between reinforcement and matrix [14]. The increasing of porosity in AlSiC composites tends to decreasing of CTE of this composite as demonstrated in Figure 12.

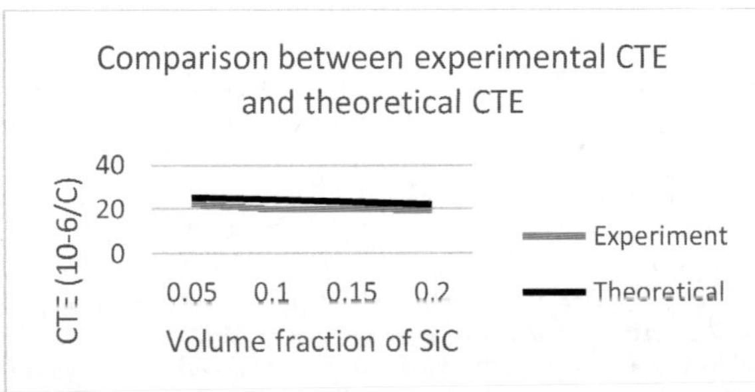

Fig. 11. The average experimental and theoretical CTE of AlSiC composite against volume fraction of SiC.

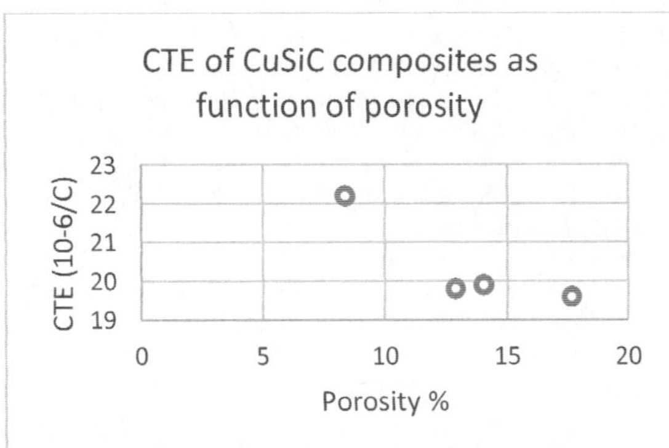

Figure 12: The relationship between the CTE and porosity of the AlSiC composite.

4 Conclusions

The CTE behaviour of CuSiC and AlSiC composites was studied. The composites were fabricated via powder metallurgy method with different volume fraction of SiC particles content. Based on the experimental result and discussions, the following conclusions have been drawn.

i) The density for both composites is following the theoretical density trend. Density of CuSiC composites are decreases and AlSiC composites are increases with increment of SiC particles content. However, the porosity for these two composites is increase by increasing volume fraction of SiC particles.

ii) The use of dendritic particles shape absolutely good fabricated via powder metallurgy method due to best interlocking between powder particles. It also can be reduce the porosity of sintered product. Whilst the flake particles shape shows very high amount of porosity of sintered product and difficult to compact under cold isostatic pressure condition.

iii) The experimental CTE values of CuSiC and AlSiC composites were significantly lower that theoretical CTE values. This result shown CuSiC and AlSiC composites have an excellent bonding inter-particles because SiC particle can restrict the expansion of copper and aluminium matrix. The decreasing of experimental CTE values are also help by porosity of composite itself.

References

1. Rana, R.S., R. Purohit, S. Das, International Journal of Scientific and Engineering Research, **3** (2012)
2. Harris, B., *Engineering composite materials*. IoM (1999)
3. Calata, J.N., et al. IEEE Transactions, **28** (3) (2005)
4. Neubauer, E., P. Angerer, and G. Korb. *Heat sink materials with tailored properties for thermal management*. in *Electronics Technology: Meeting the Challenges of Electronics Technology Progress, 2005. 28th International Spring Seminar,* IEEE (2005).
5. Zhu, D., et al. *Fabrication and properties of SiC/Cu composites for electronic packaging*. in *Electronic Packaging Technology, 2005 6th International Conference* IEEE (2005)
6. Zhu, J., et al. Materials Letters. **61**(13) (2007)
7. Chung, D.D.L., *Materials for Electromics Packaging*. (Butterworth-heinemann, 1995)
8. El-Kady, O. and A. Fathy, Journal of Materials and Design. **54** (2013)
9. El-Kady, O. and A. Fathy, Materials & Design. **54** (2014)
10. Occhionero, M., et al. *Aluminum silicon carbide (AlSiC) for advanced microelectronic packages*. in *IMAPS May, Boston Meeting, Ceramics Process Systems Corp*. 1998.
11. Onat, A., H. Akbulut, and F. Yilmaz, Journal of Alloys and Compounds, **436,** 1 (2007)
12. Singla, M., et al. Journal of Minerals and Materials Characterization and Engineering,. **8**, 06 (2009).
13. Yang, M., et al., Materials chemistry and physics, **99**, 1 (2006)
14. German, R. and R. Iacocca (Springer, 1996)
15. Shu, K.-M. and G. Tu, Materials Science and Engineering: A. **349**, 1 (2003)
16. Harris, B., *Engineering composite materials* (Institute of metals London, 1986)
17. Prosviryakov, A., Journal of Alloys and Compounds. **632** (2015)
18. Huber, T., et al., Composites Science and Technology. **66**, 13 (2006)
19. Hatta, H., T. Takei, and M. Taya, Materials Science and Engineering: A. **285,** 1 (2000)
20. Krishna, S.M., T. Shridhar, and L. Krishnamurthy, *An Investigative Review on Thermal Characterization of Hybrid Metal Matrix Composites* (2014)
21. Arpon, R., et al., Acta Materialia. **51**, 11 (2003)
22. Zhang, Q., et al., Materials Chemistry and Physics. **82**, 3 (2003)

IOP Conf. Series: Materials Science and Engineering **701** (2019) 012058 doi:10.1088/1757-899X/701/1/012058

Corrosion Assessment of Sn-0.7Cu Lead-Free Solder in 1 M Hydrochloric Acid

Tan Xing Cong[1], *Ahmad Azmin* Mohamad[2] and *Muhammad Firdaus Mohd* Nazeri[1,*]

[1]School of Materials Engineering, Universiti Malaysia Perlis, 02600 Jejawi, Kangar, Perlis, Malaysia
[2]School of Materials and Mineral Resources Engineering, Universiti Sains Malaysia, 14300, Nibong Tebal, Penang, Malaysia

Abstract. The corrosion performance of Sn-0.7Cu solder alloy is investigated in 1 M Hydrochloride acid by means of potentiodynamic polarization technique at the scan range of -2V_{SCE} to 1 V_{SCE} was applied to the sample at the scan rate of 1mV/s. Supporting morphological analysis was done by using Scanning Electron Microscope while phase analysis was confirmed by using X-ray Diffraction analysis. Morphological analysis showed two type of oxide layer has formed as corrosion product, which are compact oxide layer and loose rod-like oxide layer after corrosion analysis which confirmed to be SnO and SnO_2. This indicate that complete protection from further corrosion was impossible due to the incomplete passivation layer formation.

1 Introduction

Soldering is one of the metallurgical joining technique that use solder as filler metal for joining two metal . Formation of intermetallic compound is a chemical reaction that bond the solder and metal substrate. It is widely used in electronic industry as by definition, soldering needs relatively low temperature for processing as it contain element with lower melting points [1, 2]. Traditional solder or known as eutectic lead solder have been used for joining alloys at low temperature especially in electronic industry which serve as interconnects that able to provide the conductive path to connect from one circuit to another due to their good combination of material properties, process attributes and low cost.

Restriction of Hazardous Substances (RoHS) originated in the European Union and the usage of dangerous material in electric and electronic product. Lead is one of the 10 hazardous materials that restricted by RoHS and this apply to all European Union has ban from using in electronic products by 1 July 2006[3]. Sn–0.7Cu eutectic alloy is toxic free solder alloy with no lead contained alternative having a melting temperature 227°C. However, alternative solder alloys must need to complete various requirements like chemical or physical properties and cost which mean that the melting temperature, strength and integrity and others should also be similar or better than Sn-Pb alloy, and manufacturing costs must be competitive [4]. The distribution of Cu in Sn-0.7Cu solders is crucial to understand

* Corresponding author: firdausnazeri@unimap.edu.my

the morphology of this binary solder. Two typical regions can be observed in Sn-0.7Cu alloy which are Cu_6Sn_5 and the Zn-rich phase. The bright grey regions shows β-Sn matrix and is primarily solidified, whereas in dark region represents intermetallic compound (IMC). This show the IMC is one of the important factor to enhance the mechanical properties of Sn-0.7Cu solder alloy [5].

A number of literature studies focused on correlations between the microstructure and the properties associated with a number of solder alloys. However, there is a lack of studies of electrochemical corrosion behaviour of Sn–0.7Cu solder alloy when exposed to an acid medium. A solder alloy requires good corrosion behaviour, especially when the solder joints are directly exposed to aggressive medium in an industrial environment [6-8]. Most of the studies developed the corrosion behavior of Sn–Cu alloys in NaCl and KOH solution [7, 9, 10]. In this work, corrosion performance of Sn–0.7Cu solder alloy will be investigated in 1 M HCl by using potentiodynamic polarization. Subsequent characterization techniques of XRD and SEM will be implemented to support the electrochemical findings.

2 Methodology

The Sn-0.7Cu solder was produced into 5 mm diameter and 3 mm thickness pellets. Copper wire was attached on the solder alloy using soldering process. The solder was then mounted with epoxy resin which mixed with epoxy hardener with ratio of 1:1. Mounted samples had undergone grinding and polishing using surface grinder machine to produce a smooth and mirror-like surface.

A potentiostat (Autolab PGSTAT 30) was used for electrochemical corrosion analysis of Sn-0.7Cu in 1 M HCl. For potentiodynamic polarization, scan potential range of -2 V_{SCE} to 1 V_{SCE} was applied on sample at the scan rate of 1mV/s. Saturated Calomel Electrode (SCE) used as reference electrode while platinum work as counter electrode in potentiodynamic polarization analysis analysis.

For characterization, X-ray Diffraction (XRD) model (Shimadzu XRD 6000) was used for phase analysis of Sn-0.7Cu before and after corrosion analysis. The analysis is performed from 20°-90° with the scan rate of 5°/min used. The profile pattern was identified using Xpert Highscore Plus software. Scanning Electron Microscope (SEM), JEOL JSM-6460-LA was used for morphological analysis of sample before and after electrochemical testing.

3 Results and Discussion

3.1 Corrosion Analysis

Potentiodynamic polarization testing of Sn-0.7Cu was carried out in 1M HCl. The cathodic current starts from point A at a potential value of -2.00 V_{SCE} and ended at point B with potential value of -0.53V_{SCE} with the current density decreased sharply from 269 μA to approximately 0.60 μA. The slope intersection of linear region between cathodic region and anodic region represent corresponding corrosion current (I_{corr}) and corrosion potential (E_{corr}) of Sn-0.7Cu solder alloy.

Beyond point C, the current was slightly decrease which representing the initiation of passivation for Sn-0.7Cu at this particular region or in other word is known as passive region. Sn^{2+} ions had deposited on the surface of Sn-0.7Cu solder alloy and forming a passive film. As the applied potential increase, the current density continue reduce until the formation of full oxide film on the surface of the Sn-0.7Cu.

Path CD shows the passive region in polarization study on Sn-0.7Cu. Passive current density (i_P) with approximate value of 0.0063A, is the minimum current required for maintaining the oxide film along passive region [11, 12]. The current is maintained and independent with the increasing of applied potential. This shows that the surface of Sn-0.7Cu solder alloy is covered with passive film. The Sn-0.7Cu solder alloy was remained stable and passive state until the end of the polarization.

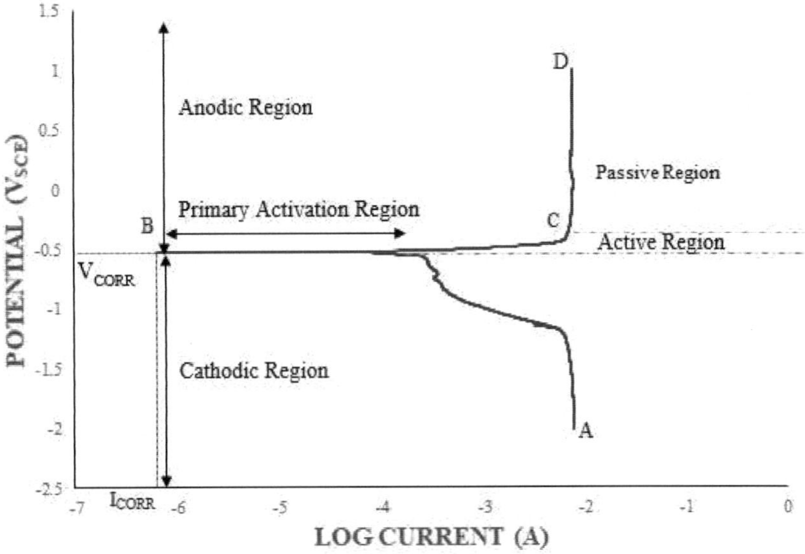

Fig. 1. Potentiodynamic polarization graph of Sn-0.7Cu.

3.2 Phase analysis

X-ray diffraction (XRD) analysis was conducted on Sn-0.7Cu solder alloy before (Figure 2a) and after polarization (Figure 2b). The as-prepared Sn-0.7Cu consist of only two phases, which is body centered tetragonal β-Sn (ICDD: 00-004-0673) and monoclinic Cu_6Sn_5 (ICDD: 03-065-2303). Meanwhile for the XRD analysis after potentiodynamic polarization, two new phases were observed, SnO (ICDD: 00-050-1429) and SnO_2 (ICDD: 00-006-0395). This has proved that the dissolution of Sn is occurred and further react with OH- to form SnO and SnO_2. The obtained result is same with the report of Huang et al. [13] which SnO and SnO_2 were the passivation barrier for the Sn-0.7Cu solder alloy to prevent further corrosion. The equations involved:

$$O_2 + 4e^- + 4H^+ \rightarrow 2H_2O \qquad (1)$$

$$Sn \rightarrow Sn^{2+} + 2e^- \qquad (2)$$

$$2Sn^{2+} + 2OH^- \rightarrow 2SnO + 2H^+ \qquad (3)$$

$$Sn + 2H^+ \rightarrow Sn^{2+} + H_2 \qquad (4)$$

$$2Sn^{2+} + O_2 + 6H_2O \rightarrow 2Sn(OH)_4 + 4H^+ \qquad (5)$$

$$2Sn(OH)_4 \rightarrow 2SnO_2 + 4H_2O \quad (6)$$

3.3 Morphological Analysis

The corrosion product found on Sn-0.7Cu after potentiodynamic polarization (Figure 3). After electrochemical analysis, the surface of the sample looks roughen and expected contributed by to be the formation of SnO and SnO_2.This passivation layer act as an important barrier against the corrosion. In general, the corrosion product is made of arrangement of rod-like corrosion product hence producing two differently distributed passivations layers of compact oxide layer and loose oxide layer. Loose oxide layer was the oxide layer that do not passivate completely. Loose oxide reveals that the corrosion is still incomplete and further oxidation may occur at the certain part [14].

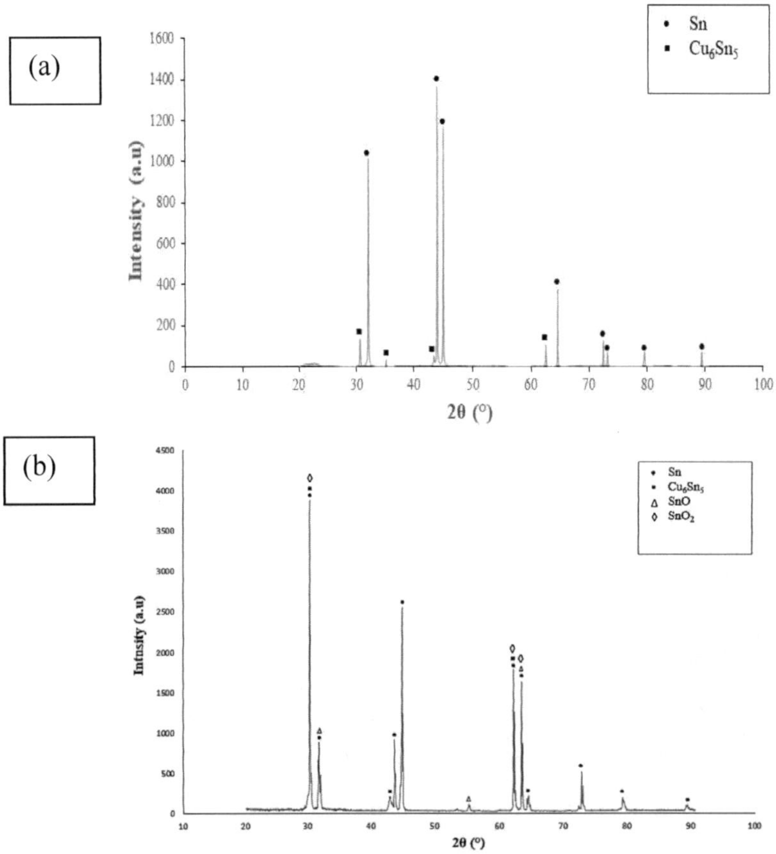

Fig. 2. XRD pattern of Sn-0.7Cu (a) before corrosion and (b) after potentiodynamic polarization.

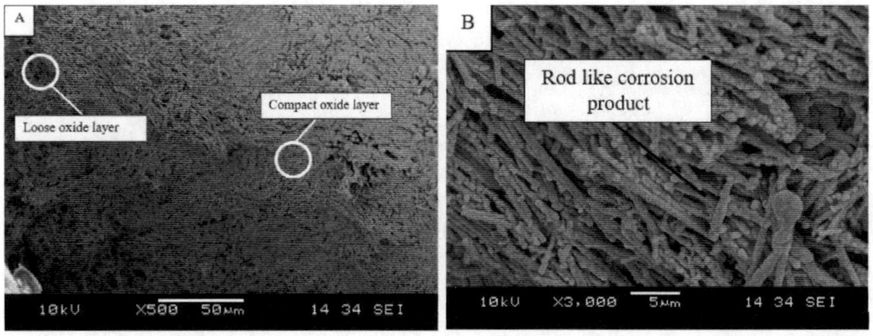

Fig. 3. Scanning Electron Micrograph of Sn-0.7Cu after potentiodynamic polarization at magnification (a) x500 and (b) x 3000.

4 Conclusion

The electrochemical corrosion behavior of Sn-0.7Cu has been investigated through potentiodynamic polarization analysis. In potentiodynamic polarization testing, this solder was shown to produce passivation abilities in 1 M HCl solution. XRD analysis confirmed the existence of passive oxide film which formed by corrosion by-product of SnO and SnO_2 that act as a barrier to increase the corrosion resistance. However, morphological analysis had found some that the passivation layer contained porous region made of loose-oxide formation. This was identified as weak spot for full corrosion protection.

The authors would like to acknowledge the support from the Fundamental Research Grant Scheme (FRGS) under a grant number of FRGS/1/2019/TK05/UNIMAP/02/5 from the Ministry of Education Malaysia.

References

1. Lee, C.C. and J. Kim. Advanced Packaging Materials: Processes, Properties and Interfaces, 2005. Proceedings. International Symposium on. IEEE. (2005)
2. Islam, M., Y. Chan, M. Rizvi, and W. Jillek, J. Alloys Compd, **400**, (2005)
3. Li, Y., K.-s. Moon, and C. Wong, Science, **308**. (2005)
4. Garcia, L.R., W.R. Osório, L.C. Peixoto, and A. Garcia, Mater. Charact., **61**. (2010)
5. Rasid, M., Z. Azwan, M.F. Omar, M. Nazeri, and M. Firdaus. Materials Science Forum. Trans Tech Publ. (2017)
6. Garcia, L.R., L.C. Peixoto, W.R. Osório, and A. Garcia, MATER LETT., **63**. (2009)
7. Osório, W.R., E.S. Freitas, J.E. Spinelli, and A. Garcia, Corros. Sci., **80**. **(**2014)
8. Wang, M., J. Wang, H. Feng, and W. Ke, Corros. Sci., **63**. (2012)
9. Li, D., P.P. Conway, and C. Liu, Corros. Sci., **50**. (2008)
10. Yan, Z. and A.-P. Xian, Metall Mater Trans A, **44**.(2012)
11. Mohd Nazeri, M.F., M.Z. Yahaya, A. Gursel, F. Cheani, M.N. Masri, and A.A. Mohamad, SOLDER SURF MT TECH. **31**.(2019)
12. Nazeri, M.F.M., A.B. Ismail, and A.A. Mohamad, J. Alloy. Compd. **606,** (2014)
13. Huang, H.-z., D. Lu, G.-w. Shuai, and X.-q. Wei, T INDIAN I METALS. **69** (2016)
14. Farina, S. and C. Morando, J MATER SCI-MATER EL. **26**, (2014).

IOP Conf. Series: Materials Science and Engineering **701** (2019) 012059 doi:10.1088/1757-899X/701/1/012059

Investigation on Passivation Behavior of Sn-0.7Cu Solder in Different Polarizing Conditions

Wai Kuan Ho[1], *Ahmad Azmin* Mohamad[2] and *Muhammad Firdaus Mohd* Nazeri[1,*]

[1]School of Materials Engineering, Universiti Malaysia Perlis, 02600 Jejawi, Kangar, Perlis, Malaysia
[2]School of Materials and Mineral Resources Engineering, Universiti Sains Malaysia, 14300, Nibong Tebal, Penang, Malaysia

Abstract. Corrosion behaviour of Sn-0.7Cu solder alloy was evaluated by conducting potentiodynamic polarization and electrochemical impedance spectroscopy (EIS) techniques. The solder was polarized under the scan rate of 1.0 mV/s and with the scan range from -2 to 1 V_{SCE}. EIS test was performed at the frequency of 10000 Hz to 0.1 Hz. The oxide layer was found on the surface of solder by post-corrosion characterization. XRD results indicate that the main corrosion products formed on the solder after both corrosion tests were tin oxide such as SnO and SnO_2. Porous rod-like structure of corrosion product was observed under morphology analysis of Sn-0.7Cu solder. The corrosion product was represented as tin oxide which matched with the result of XRD.

1 Introduction

Lead solder alloy was heretofore extensively used in electronic devices. The development of lead-free solder was invented lately due to environmental and health concern. The solder alloy was switched to harmless substance for soldering. The tin-silver-copper (SAC) solder alloy is highly recommended due to its advantages in mechanical properties and good solderability and can be used as a lead-free alternative [1]. But, the cost of SAC solder was comparatively high as the presence of silver metal. Sn-Cu solder alloy has become a promising alternative on account of low cost, good solderability and great mechanical properties [2]. This in turn allows Sn-Cu alloy to take the spot as the main candidate to replace Sn-Pb in the applications.

Eutectic Sn-0.7Cu has a comparable melting point (227 °C) and good fatigue resistance as compared to Sn-Pb solder. This makes Sn-0.7Cu to be used in existing Sn-Pb production lines without much modifications. Corrosion is one of the critical factors for reliable interconnect system. However, it is often overlooked since Sn-Pb alloys are known to produce stable oxide and the galvanic potential difference between Sn and Pb is small [3]. However, for lead-free solder presence of various alloying elements might contribute differently and should be systematically investigated. In this work, corrosion properties of Sn-0.7Cu will be studied in different polarizing conditions of direct current (potentiodynamic

* Corresponding author: firdausnazeri@unimap.edu.my

polarization) and alternate current (electrochemical impedance spectroscopy, EIS). Further supporting characterizations on microstructure and phase changes will also be reported.

2 Methodology

Pellets of Sn-0.7Cu solder were prepared in 3 mm and 5 mm, thickness and diameter, respectively. Each solder pellets were attached to Cu wire and cold-mounted for ease of handling during grinding and polishing. The samples were ground with SIC paper from grit 400 to 1200 and polished with the aid of 1 micron of alumina solution in order to obtain mirror surface finish. Initial characterization was performed under scanning electron microscopy (SEM) with model of JEOL 6460-LA and Shimadzu X-ray diffraction 6000 diffractometer (XRD). SEM was used for morphological analysis. XRD analysis performed at 2θ values which are ranging from 20° to 90° with a medium scan rate at 5 °C per minute. Cu-Kα radiation was used to investigate the phases exist in the solder pellet and the result was matched with the standard phase identification from International Committee of Diffraction Data (ICDD) X-ray data. For the corrosion analysis, three electrodes compartment cell was prepared. Sn-0.7Cu solder was acted as working electrode while platinum rod and saturated calomel electrode (SCE) were counter and reference electrode respectively. Potentiodynamic polarization test was carried out by immersed the three electrodes into a 1 M hydrochloric acid (HCl) and polarized at the scan rate of 1.0 mV/s and scan ranges in between -2 V_{SCE} to 1 V_{SCE}. Electrochemical impedance spectroscopy (EIS) was performed under the frequency of 10000 to 0.1 Hz and with 10 mV amplitude of AC. The tests were measured by using potentiostat with the model of AUTOLAB PGSTAT 30.

3 Results and Discussion

3.1 Initial Phase Characterization

XRD analysis of Sn-0.7Cu solder before corrosion test revealed the presence of two phases which are the β-Sn (ICDD 01-075-9188) and monoclinic Cu_6Sn_5 (ICDD 01-076-2704) intermetallic compounds (IMC), as shown in Figure 1. The phases obtained was similar in accordance with the previous research where β-Sn matrix was the Sn-rich phase. The phase of Cu_6Sn_5 was produced due to the limited solubility of Cu in Sn. During reflow process, the Cu reacted with the Sn to form Cu_6Sn_5 intermetallic compounds and distributed in the β-Sn matrix [4].

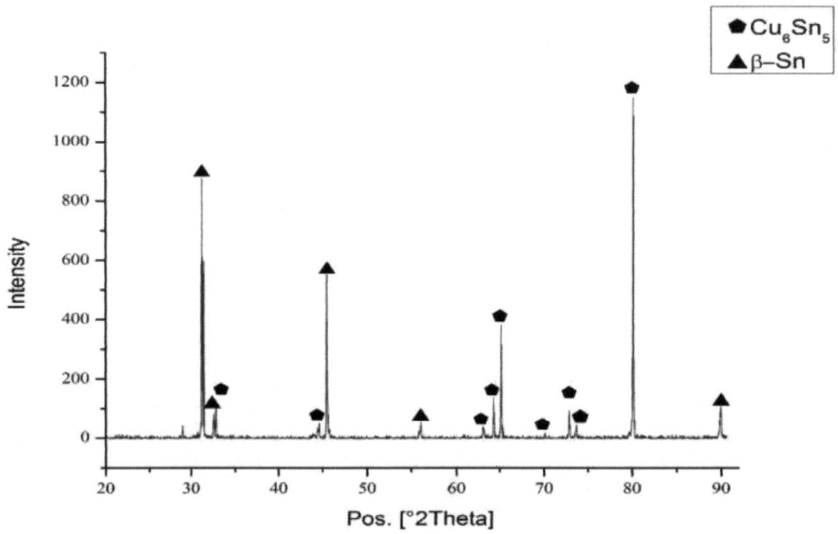

Fig. 1. XRD of as-prepared Sn-0.7Cu solder.

3.2 Corrosion analyses

The potentiodynamic polarization plot for Sn-0.7Cu consisted of two main regions which were anodic and cathodic regions (Fig. 2). In cathodic region, the applied potential increased gradually starting from -2 V_{SCE}. This resulted in the reduction of the current density from point A to B, normally associated with reaction [4]:

$$O_2 + 4e^- + 4H^+ \rightarrow 2H_2O \quad (1)$$

A slightly fluctuation in the curve at cathodic region was basically due to evolve of hydrogen bubble gas. Beyond point B, anodic reaction started with the increasing of current density with potential. This region was considered as the primary active region of the polarization process. Dissolution of active metal occurred at this region. Sn was highly reactive than Cu due to the electrode potential in electrochemical series. Thus, Sn metal will be dissolved by oxygen into metal ions [4]:

$$Sn \rightarrow Sn^{2+} + 2e^- \quad (2)$$

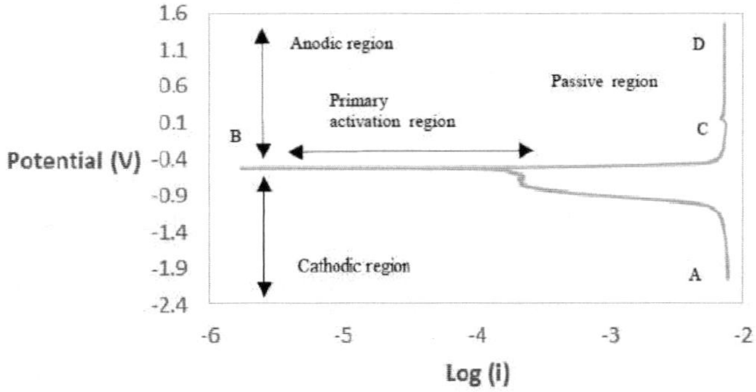

Fig. 2. Potentiodynamic polarization curve of Sn-0.7Cu solder.

The linear region between the point B and C considered as anodic slope while the linear region between point A and B was the cathodic slope. The corrosion potential (E_{corr}) and corrosion current density (I_{corr}) can be obtained from the interception point of both slope as plotted on the polarization curve. The E_{corr} and I_{corr} were -521 mV_{sce} and 0.00017 A/cm^2, respectively. The primary activation region was ended at point C where the current density reached the maximum value. The region beyond point C to point D is known as the passive region. At point C, the passivation potential was the indicator to separate the active and passive region. Passivation current density (I_{pass}) was exploited to describe the passivation behaviour of the solder. The passive region was continued to point D and the current density remained unchanged. The development of the passive region was due to the reaction of Sn^{2+} ions with the hydroxide ions (OH^-) to form oxide on the surface of solder through the reaction [4]:

$$2Sn^{2+} + 2OH^- \rightarrow 2SnO + 2H^+ \quad (3)$$

$$Sn + 2H^+ \rightarrow Sn^{2+} + H_2 \quad (4)$$

$$2Sn^{2+} + O_2 + 6H_2O \rightarrow 2Sn(OH)_4 + 4H^+ \quad (5)$$

$$Sn(OH)_4 \rightarrow SnO_2 + 2H_2O \quad (6)$$

Nyquist plot for EIS measurement consisted of the semi-arc that depicted the corrosion performance of solder (Figure 3). The small semi-arc illustrated the start of dissolution of Sn metal to Sn metal ions. The formation of oxide passive layer was indicated by drop of Z' (imaginary axis). The layer was indicated as the obstacle to prevent the corrosive ions from attacking the Sn-0.7Cu solder. Warburg impedance was existed as diffusion process once the passive oxide layer breakdown. The Warburg impedance was presented as a steep line which connected to the semi-arc and disclosed the corrosion kinetic of the Sn-0.7Cu solder. EIS measurement was determined by the fitting of experimental data with an equivalent circuit. The fitting quality was evaluated by chi-squared ($\chi 2$) values. The interpretation of physical elements of the proposed electrical equivalent circuit as showed in Figure 3a. In the first layer, Rs was the resistance of 1 M HCl solution and R1 and R2 in second and third layer were the resistance of porous and barrier layer respectively. Besides, the CPE1 and CPE2 were related to the capacitance of the porous layer and barrier layer correspondingly. At the second layer of the circuit, R1CPE1 represented the occurrence of metallic corrosion where the charge transfer occurred while R2CPE2 represented the diffusion process.

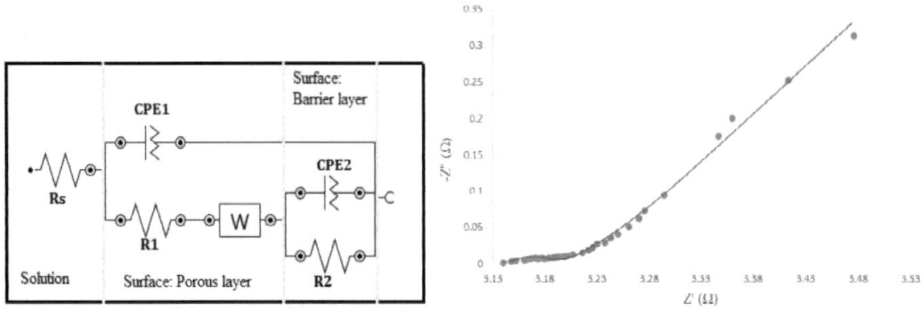

Fig. 3. (a) Equivalent circuit and (b) Nyquist plot for Sn-0.7Cu in HCl.

In this circuit fitting, the chi-squared value was 9.28 x 10^{-4}. Based on the previous research, good fitting of the Nyquist plot was in the range of 10^{-4} to 10^{-3}. The summary of the values of electrical equivalent circuit were summarized in Table 1. R_2 values was indicated that the oxide layer was formed and acted as a barrier to block the corrosion process while CPE_2 value was specified the compactness of the oxide layer formed.

Table 1. Resistance value of electrical equivalent circuit

Element	Value
Rs (Ω cm^2)	3.14
CPE1 (Ω^{-1} cm^{-2} s^n)	0.0764
R1 (Ω cm^2)	0.0589
CPE2 (Ω^{-1} cm^{-2} s^n)	1.5344
R2 (Ω cm^2)	0.8705
χ2	9.28 x 10-4

3.3 Post-corrosion characterizations

The micrographs of Sn-0.7Cu solder polarized in 1 M HCl solution can be clearly observed under SEM as shown in Figure 4. The corrosion products with rod-like structure were loosely distributed on the surface of the solder with different orientations and was not fully covering the surface of the solder in both potentiodynamic polarization (Fig. 4a-b) and EIS (Fig. 4c-d). Tin oxide was found in the corrosion products which transformed from the reaction between Sn and corrosive solution. The existence of the halide ions (Cl-) in corrosive solution, which was destructive to the intact surface of the solder [5]. As a result the passivation film produced were porous and offer relatively weak protection towards the solder in the long term exposure to HCl.

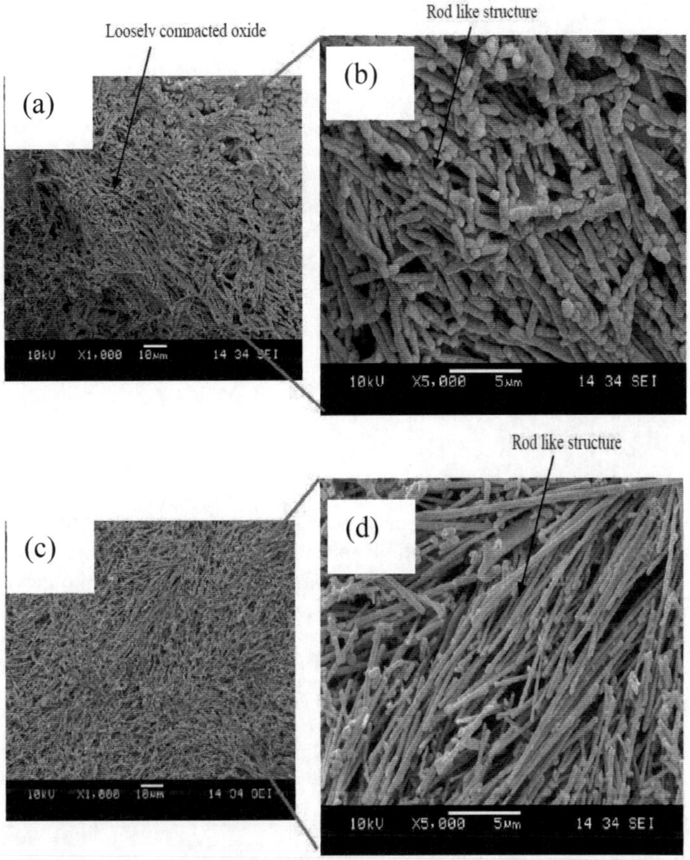

Fig. 4. SEM micrographs of Sn-0.7Cu solder after potentiodynamic polarization at magnification of (a) x1000, (b) x5000, after EIS (c) x1000, (d) x5000.

Phase analysis of Sn-0.7Cu solder after potentiodynamic polarization (Figure 5a) and EIS (Figure 5b) showed four different phases that existed. The corrosion products of SnO (ICDD 01-077-2296) and SnO_2 (ICDD: 01-071-5330) were identified along with the original phase of β-Sn (ICDD 01-075-9188) and monoclinic Cu_6Sn_5 (ICDD 01-076-2704). Presence of comparable phases in both polarizing conditions was due to the similarity of the experiment conditions without altering any parameters.

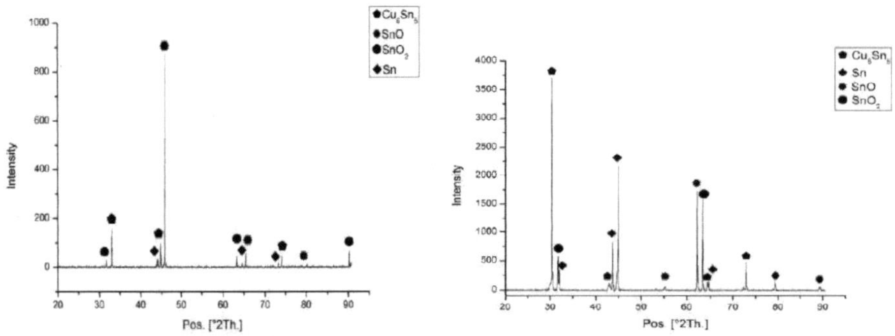

Fig. 5. XRD pattern of Sn-0.7Cu solder after (a) Potentiodynamic polarization and (b) EIS.

4 Conclusion

The electrochemical corrosion behavior of Sn-0.7Cu has been investigated through potentiodynamic polarization and EIS analysis. In potentiodynamic polarization testing, this solder was shown to produce passivation abilities in 1 M HCl solution. This is further supported by the finding in EIS. XRD analysis confirmed the existence of passive oxide film which formed by corrosion by-product of SnO and SnO_2 that act as a protection film to increase the corrosion resistance. However, the nature of the film was found to be porous in both polarizing conditions revealing insufficient protection for prolonged period in this acidic condition.

The authors would like to acknowledge the support from the Fundamental Research Grant Scheme (FRGS) under a grant number of FRGS/1/2019/TK05/UNIMAP/02/5 from the Ministry of Education Malaysia.

References

1. Wang, M., J. Wang, and W. Ke, Mic. Relib. **73**, (2017)
2. Li, G., Y. Shi, H. Hao, Z. Xia, Y. Lei, and F. Guo, J. Alloy. Compd. **491**, (2010)
3. Song, F. and S.R. Lee. 56th Electronic Components and Technology Conference. IEEE, (2006)
4. Rasid, M., Z. Azwan, M.H. Zainol, M.F. Omar, M.N. Derman, M. Nazeri, and M. Firdaus. Solid State Phenomena, (2018)
5. Li, D., P.P. Conway, and C. Liu, Corros. Sci., **50**, (2008)

IOP Conf. Series: Materials Science and Engineering **701** (2019) 012060 doi:10.1088/1757-899X/701/1/012060

Characterization of Electrical Current Stress and Indentation Creep of Carbon Nanotubes-reinforced Low Melting Temperature Sn-58Bi Composites

Kian Lu Sia, Syed Alif Ilyasak Barakbah and Ho Cheng How*

Department of Mechanical, Materials and Manufacturing Engineering, Faculty of Science and Engineering, University of Nottingham Malaysia, Jalan Broga 43500 Semenyih, Selangor, Malaysia

Abstract. Sn-58Bi eutectic alloy emerges as one of the potential lead-free solder material candidates due to its low melting temperature that enables low temperature soldering reflow process during the manufacturing of fine-pitch and low z-height advanced microelectronic devices. Its low melting temperature provides feasible solution for a two-step reflow process in the assembly of 3D-stacked advanced packaging technology, and at the same time reduces dynamic warpage-induced quality and reliability issues. However, miniaturization of these microelectronic devices has triggered significant reliability risk to its solder interconnects due to the requirement of high current density that resulted in severe physical failures, such as electromigration (EM), and thermomigration (TM). To mitigate the risk, carbon nanotubes (CNT) are incorporated into elemental Sn and Bi powders using planetary ball milling technique, and subject to liquid phase melting to form low melting temperature CNT-reinforced eutectic Sn-58Bi composites. Electrical resistance and current density of the eutectic alloy and its composites were investigated, and results show that Sn-58Bi-0.03CNT composite has the lowest electrical resistance and was able to survive 26 times longer duration of current stressing, even with the supply of higher stressed current. Creep performance of eutectic Sn-58Bi system was also examined in this study due to its high homologous temperature of 0.72 at room temperature. Marginal improvement in indentation creep performance was seen in the composites.

1 Introduction

Under the Restriction of Hazardous Substrates (RoHS), the use of lead (Pb) has been prohibited in the electronic packaging since July 2006 [1]. Alternative solders based on Sn-Ag-Cu lead-free materials were an immediate replacement of conventional Sn-Pb solders since then. While adverse reliability impacts were reported in these lead-free solder joints [2-3], its higher melting temperature (T_m~217°C) requires higher soldering reflow temperature, as compared with that of eutectic Pb-Sn solder (T_m~183°C). This translates to a higher risk of thermally induced damages in the microelectronic devices, such as moisture escaped related "pop-corning" phenomenon, as well as dynamic warpage of the polymer substrate [4].

In fact, recent development of coreless technology in polymer package substrate for meeting the performance demand in advanced microelectronic device has further increased dynamic warpage risk [5]. Kim et. al. [5] has demonstrated the correlation between dynamic warpage behaviour and the soldering reflow temperature of a typical flip-chip packaging technology. In this case, the warpage of the component was the highest at the temperature range of around 240°C that used for conventional lead-free solders metallurgy. Based on this learning, it highlights the opportunity to innovatively reduce the dynamic warpage of the electronic device through an introduction of low melting temperature solder metallurgy that requires a lower soldering reflow temperature. While Indium is extremely expensive, the use of eutectic Sn-58Bi alloy appears to be a promising candidate based on its low eutectic temperature of 139°C. However, it has been reported that the addition of Bi in Sn-3.7Ag solder resulted in reduction of its elongation and creep properties [6-7]. This signifies the concern of using eutectic Sn-58Bi alloy in high performance microelectronic devices, as high homologous temperature ($T/T_m > 0.5$) at service environment may cause significant degradation to its creep performance.

Furthermore, the trend with continued pursuit of miniaturization and improved performance of microelectronic devices also triggered the requirement of higher input/output (I/O) density, thus smaller solder interconnects with increased reliability risk is anticipated. The increment of current density through solder interconnections could easily result in EM [8,9]. This could lead to the bottleneck in carrying high current density due to several physical failures including EM and TM. According to International Technology Roadmap for Semiconductors 2007 (ITRS 2007) [10], EM is considered an upper constrain of high current density packages in wafer-level packaging (WLP) for Microelectromechanical systems (MEMS). In eutectic Sn-58Bi solder system, mass accumulation of Bi in eutectic Sn-58Bi was observed in an accelerated test of EM when it is under current density of 5×10^3 A/cm^2 at 75°C [8]. This results in Bi segregation-induced brittle behaviour at intermetallic compound of solder interconnections [11].

To address the above concerns, multi-walled CNT are incorporated into Sn and Bi elemental powders using planetary ball milling technique, and subject to liquid phase melting to form low melting temperature CNT-reinforced eutectic Sn-58Bi composites. While CNT are known to possess superior Young's modulus and electrical conductivity [12], it is the objective of the present study to characterize the effect of CNT reinforcement on electrical current density and indentation creep performance of low melting temperature CNT-reinforced Sn-58Bi composites.

2 Experimental details

2.1 Sample preparation

Based on the established method that used by Dele-Afolabi et al. for the mixing of CNT-reinforced Sn-5Sb solder composites [13], planetary ball milling technique was carried out at 400rpm with ball-to-powder ratio of 20:1 for up to 2 hours to prepare the mixture of Sn-58Bi, Sn-58Bi-0.01CNT, and Sn-58Bi-0.03CNT from high purity (99.9% purity) Sn and Bi elemental powders, and multi-walled CNT with an average diameter of ~26nm. Figure 1 shows the micrographs and energy dispersive x-ray spectrum of as-received Sn, Bi and multi-walled CNT. The mixtures were then melted in a crucible at 300°C with the addition of soldering flux to form eutectic Sn-58Bi monolithic solder alloy, 0.01wt.%CNT-reinforced Sn-58Bi and 0.03wt.%CNT-reinforced Sn-58Bi composites, Figure 2.

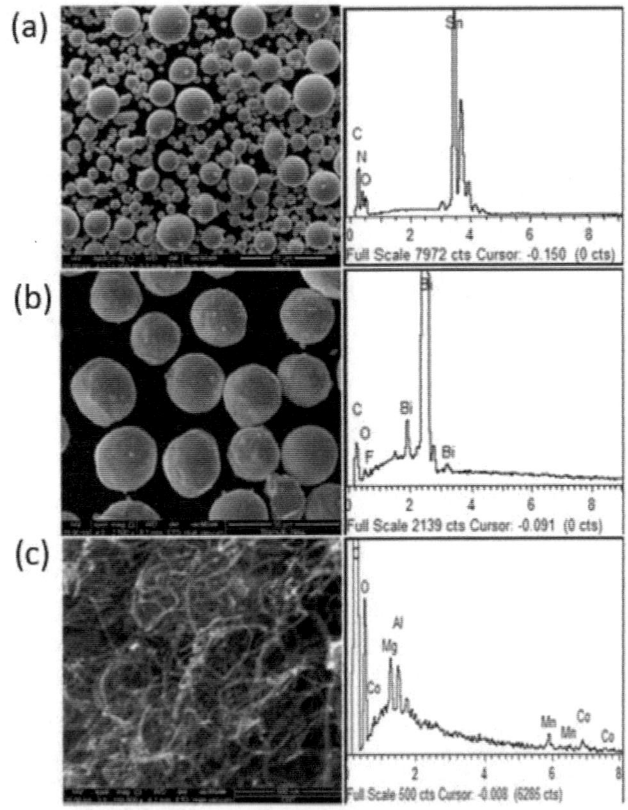

Fig. 1. SEM micrographs and EDX spectrum of as-received (a) Sn elemental powder; (b) Bi elemental powder; (c) multi-walled CNT

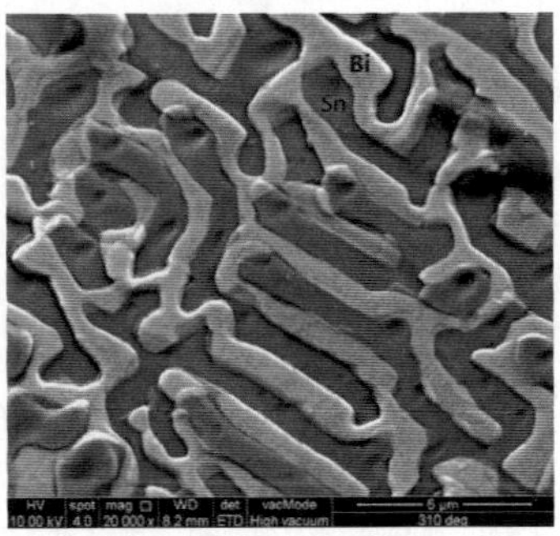

Fig. 2. SEM micrographs of eutectic Sn-58Bi solder alloy

2.2 Electrical current stressing

The V-shape solder joint technique that developed by Yue et al. [14] for detecting electromigration under direct current stressing of up to 1.5×10^4 A/cm^2 was referred, but the fabrication technique and the area determination for the definition of current density were not clearly defined. Hence, an improved alternative experimental setup, together with appropriate design of test coupon was carried out in this investigation. Figure 3 shows the design of the test coupon, whereby a copper trace of 0.03mm thick x 0.3mm width x 20mm length was printed on a single layer of flame retardant graded 4 (FR4) printed circuit board (PCB) with two copper pads at both ends. A hole of 0.4mm diameter was drilled through the center of the trace to electrically isolate the trace so that solder sample could be melted to form a conductive solder bridge between the traces.

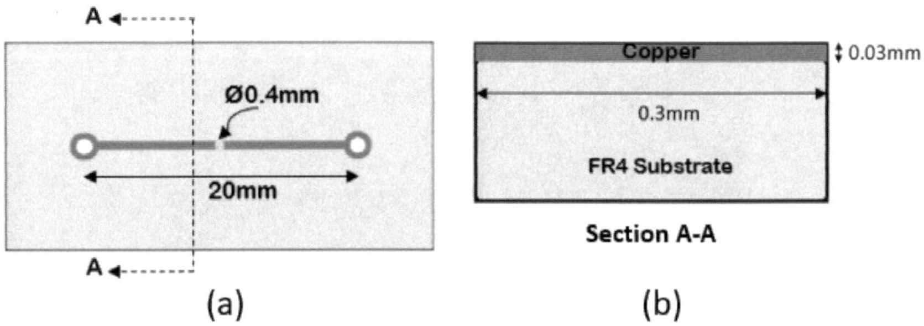

Fig. 3. Test coupon for electrical current stressing experiment; (a) the design of copper trace with two pads at both ends and a hole of 0.4mm diameter at the center of the trace; (b) section A-A of the copper trace

Figure 4(a) shows the actual experimental setup of the test coupon where the pads of the copper trace were soldered with American Wire Gauge (AWG) 26 copper wire using Sn-0.7Cu solder to connect to the DC power supply (GW INSTEK GPS-3303) via crocodile clips. Sn-0.7Cu solder of higher melting temperature (T_m=227°C) was selected in this case to ensure the failure location for the test sample would occur at low melting temperature eutectic Sn-58Bi solder bridge, instead of at solder pad connection. 1mg of solder sample was then transferred to the hole and reflowed under a soldering rework station with the presence of soldering flux to form the solder bridge in order to complete a close-loop circuit. The nature of surface tension of molten solder bridge went through self-alignment to form a controlled shape, Figure 4(b).

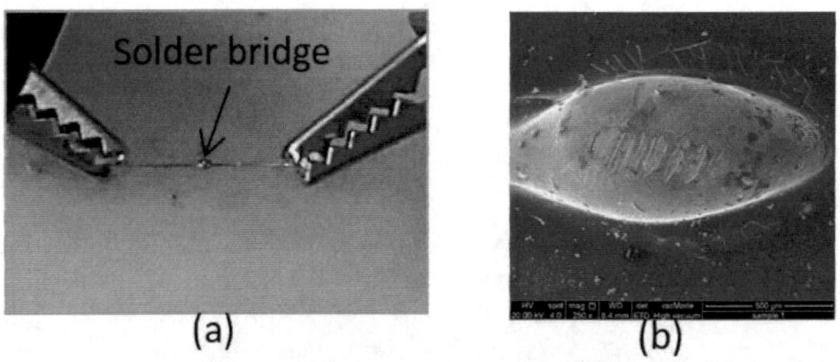

Fig. 4. (a) Actual experimental setup of the test coupon where the pads were soldered with copper wire and connected to a DC power supply via crocodile clips; (b) SEM micrograph of the solder bridge (i.e. solder sample) of self-aligned shape due to its surface tension at molten stage

2.3 Indentation creep test

Using a Vickers hardness tester, the hardness indentation was made on the polished samples at room temperature. The indentation was made with a pyramid indenter under a constant load of 10N. Both the hardness value and the diagonal length of the impression were recorded for each dwell time. In this study, the dwell time of the indentation was taken as a parameter, and its variation was made within a range of 10s to 60s. For each sample, a minimum of four indentations were taken, and the distance between adjacent indentations were at least 2.5 times the diagonal length of the impression, in accordance to BS 6286: 1982 [15].

3 Results and discussion

The electrical current response for test coupons of various solder samples were plotted against voltages, as demonstrated in Figure 5(a). Based on the gradient, electrical resistance of 0.138Ω, 0.135Ω, and 0.132Ω were found on eutectic Sn-58Bi monolithic alloy, Sn-58Bi-0.01CNT, and Sn-58Bi-0.03CNT composites, respectively. From the trend, the incorporation of CNT into Sn-58Bi solder alloy resulted in lower electrical resistance, hence improving the conductivity of composites solder bridge by allowing more current to pass through it at constant voltage. This result matches the finding that obtained in other study [8].

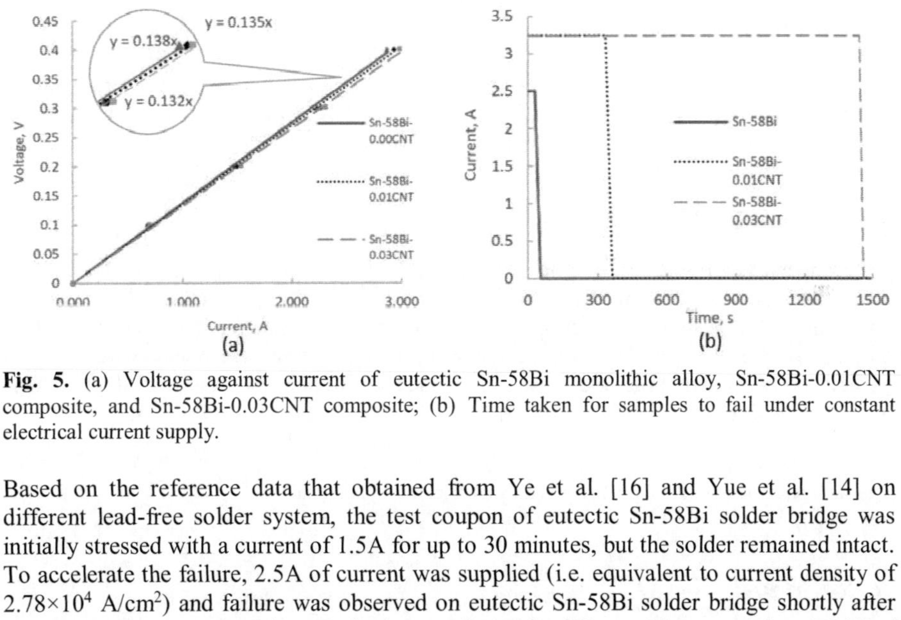

Fig. 5. (a) Voltage against current of eutectic Sn-58Bi monolithic alloy, Sn-58Bi-0.01CNT composite, and Sn-58Bi-0.03CNT composite; (b) Time taken for samples to fail under constant electrical current supply.

Based on the reference data that obtained from Ye et al. [16] and Yue et al. [14] on different lead-free solder system, the test coupon of eutectic Sn-58Bi solder bridge was initially stressed with a current of 1.5A for up to 30 minutes, but the solder remained intact. To accelerate the failure, 2.5A of current was supplied (i.e. equivalent to current density of 2.78×10^4 A/cm^2) and failure was observed on eutectic Sn-58Bi solder bridge shortly after 55s, but not for the composites. To further accelerate the failure of the composites, 3.24A of current, which equivalent to current density of 3.6×10^4 A/cm^2, was supplied to Sn-58Bi-0.01CNT and Sn-58Bi-0.03CNT composites solder bridges. Results show that Sn-58Bi-0.01CNT composite fails after 330s of stressed duration, whereas Sn-58Bi-0.03CNT fails after 1440s, as illustrated in Figure 5(b). Even with higher magnitude of the supplied current, Sn-58Bi-0.03CNT composite clearly showed significant improvement in its capacity in carrying high current density of 26 times longer duration than its monolithic alloy.

As for the indentation creep performance investigation, indentation length was found to increase with dwell time for all solder materials tested, Figure 6(a). By comparing with the monolithic alloy, the composites exhibit lower indentation length. Based on power law relationship that applied for an indentation creep of a solid [17], for each dwell time, the same set of data in Figure 6(a) was plotted into indentation creep rate, as illustrated in Figure 6(b). In this figure, it is noted that for all the materials, the indentation creep rate decreases rapidly initially, and the rate of reduction trending down substantially with further increased in dwell time. This indicates the occurrence of primary creep in the initial stage where higher indentation creep rate is expected, and the indentation creep rate reduces subsequently to suggest a steady-state creep. This observation is in good agreement with that observed by Zaki et.al. [18]. By focusing on steady-state creep, marginal improvement in indentation creep rate was observed on the composites with 0.01wt%CNT and 0.03wt% CNT reinforcement, as compared with its eutectic Sn-58Bi monolithic alloy.

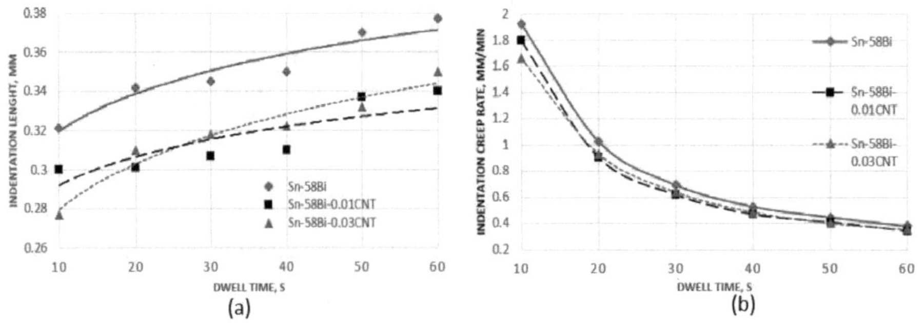

Fig. 6. (a) Indentation length increases with dwell time; (b) Indentation creep rate was higher in the monolithic alloy than that of the composites.

4 Conclusions

Planetary ball milling and liquid phase melting techniques were used to synthesize low melting temperature eutectic Sn-58Bi solder alloy and its CNT-reinforced composites. Using in-house developed test coupon and experimental setup, the presence of CNT in eutectic Sn-58Bi solder alloy was found to improve its electrical conductivity, and its capacity to withstand higher electrical current density. In addition, marginal improvement in indentation creep rate was also seen in CNT-reinforced eutectic Sn-58Bi composites.

The authors would like to acknowledge Noor Hashimi Mohamad Nor, Muhammad Nasuha Bin Mazha, and Khairani Hasuna binti Jaapar for their laboratory support.

References

1. John H. Lau, Journal of Electronic Packaging, **128**, 297-301 (2006)
2. Ming-Hung Shu, et. al., Microelectronics Reliability, **52**, 2690-2700 (2012)
3. David B. Witkin, Materials Science and Engineering A, **532**, 212-220 (2012)
4. Zequn Mei, et. al., Hewlwtt-Packard Journal, Article 10 (1996)
5. Jinho Kim, et. al., *IPC APEX EXPO Proceedings* (2012)
6. Liang Zhang, et.al., Microelectronics Reliability, **52**, 559–578 (2012)
7. Lu Shen, et. al., Materials Science & Engineering A, **561**, 232-238 (2013)
8. Y. C. Chan and D. Yang, Prog. Mater. Sci., **55**, 428–475 (2010)
9. F. Wang, L. Liu, D. Li, and M. Wu, J. Mater. Sci. Mater. Electron., **29**, 21157–21169 (2018)
10. W. Arden et al., *International Technology Roadmap for Semiconductors* (2007)

11. Q. Li, N. Ma, Y. Lei, J. Lin, H. Fu, and J. Gu, Journal of Electronic Materials, **45**, 5800–5810 (2016)
12. Z.A. Ghaleb, M. Mariatti, Z.M. Ariff, Composites: Part A, **58**, 77-83 (2014)
13. T. T. Dele-Afolabi, M. A. Azmah Hanim, M. Norkhairunnisa, H. M. Yusoff, and M. T. Suraya, J. Alloys Compd., **649**, 368–374 (2015)
14. W. Yue, H. Qin, M. Zhou, X. Ma, and X. Zhang, Acta Metall. Sin., **48**, 678–686 (2012)
15. H.C. How and T.N. Baker, Wear **232**, 106-115 (1999)
16. H. Ye, C. Basaran, and D. C. Hopkins, Int. J. Solids Struct., **41**, 4939–4958 (2004)
17. R. Goodall, T.W. Clyne, Acta Materialia **54**, 5489-5499 (2006)
18. Hesham Mohaned Zaki, et. al., Materials Sciences and Application, **2**, 1076-1082 (2011)

IOP Conf. Series: Materials Science and Engineering **701** (2019) 012061 doi:10.1088/1757-899X/701/1/012061

Experimental measurement for non-spherical bulk materials flow behaviour in rectangular enclosure

Anis Farhanah Mohd Fadzir[1], *Siti Ilyani* Rani[1,*], *Jolius* Gimbun[2]

[1]Faculty of Chemical Engineering Technology, TATI University College, 24000, Malaysia
[2]Center of Excellence for Advanced Research in Fluid Flow (CARIFF), Universiti Malaysia Pahang, 26300, Malaysia.

Abstract. Dust control is a major concern in food industries since dust emission is generally unavoidable during bulk material handling. Obtaining a better understanding of the way particles are pulled away from a stream of particles and the generation of a dust cloud will assist in the design of more efficient dust control systems. In this paper, the particles-air interactions during free fall of bulk powder is reviewed. An experimental investigation was performed to elucidate the effect of conveying velocity and particle properties on velocity profiles. The Laser Doppler Anemometry (LDA) technique was employed to measure the particle velocity during free fall. Results revealed that the particle properties (particles with flaky shapes and lower bulk density) had a great influence on dust cloud generation as compared to the conveying velocity. The results presented here may facilitate the development of a quantitative tools for dust control system design that deal with a falling stream of bulk powder.

1 Introduction

The transportation of bulk powder is a vital process in food industries since numerous raw materials and products are in powdered or particulate form. Operation involving particle handling and processing, airborne dust may arise from the dispersal of powder and could cause air pollution, contaminate industrial buildings which may affect the employee health and can trigger an explosion. In operation involving free falling of bulk materials such as in silo, a dust cloud can be formed since some of the particles are broken away from the particle stream to the surrounding air, and hence form the first airborne dust. Continuous airborne dust may be generated due to the re-suspension and re-dispersion of smaller particles that rebounded away from the particle stream during the impact of a free falling stream onto a surface. Figure 1 illustrates the dust generation for a free falling stream of bulk materials.

* Corresponding author: ilyani@tatiuc.edu.my

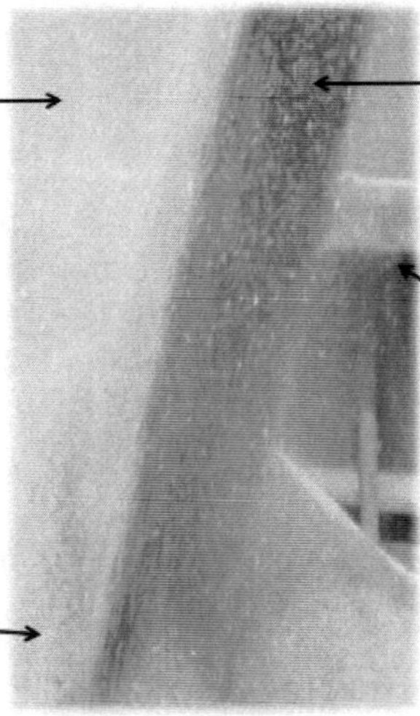

Fig. 1. Dust generation for a falling stream of bulk materials.

Understanding about the mechanism of particle settling is important to elucidate the dust generation during free falling, and hence assist in the design of effective dust control systems and risk assessment. Numerous studies have attempted to explain the motion of the particle stream based on spherical particle [1-3]. In 1995, Cooper and Arnold [4] found that dust cloud generation was due to the dust released from the impact onto stockpile. Ogata [5] reported that the free falling velocity of a single particle in an unbounded fluid was lower than a particle in a powder jet and the velocity increased with an increasing mass flow rate. The study by Wypych et al. [6] examined the air entrainment of a free falling bulk materials. In 2009, Ansart et al. [7] studied the dispersity of free falling particles with a particle image velocimetry (PIV) camera and their analysis had attempted to explain important theories and experimental significant on gas–solid flows, air entrainment models and dust control. Li et al. [8] carried out a numerical study of the effects of the conveyor chute on air entrainment. Numerical investigation of dust cloud generation during silo filling by using Computational Fluid Dynamics (CFD) provides deep insights of particle/air dynamics [9-11]. A considerable amount of literature was published on air entrainment in free falling bulk materials. The development studies of air entrained by falling bulk material were traced by [12].

Although there were many research about the motion of free falling single spherical particles in an ambient fluid, air entrainment in free falling bulk materials and settling of a spherical particle stream, only a few of them focused on the motion of non-spherical particles in the particle stream during free falling. Therefore, this study aims to experimentally investigate the effect of particle properties by means of shape, bulk density and size on velocity profile during free falling. The study also explores the influence of conveying velocity on the particle motion. The study findings are beneficial to the general understanding about dust cloud formation process; hence, assist in dust control and risk assessment in industrial processes.

2 Methodology

The experimental set-up is comprised of a square Perspex test enclosure of 60 cm by 60 cm cross section and a height of 100 cm as shown in Figure 2(a). For the experiments, materials were loaded into a hopper with a square opening outlet of 65°, then discharged into an enclosure by a screw conveyor positioned under it. Different conveying velocities were used (0.9 m/s, 1.5 m/s and 2 m/s) to produce different mass flow rates of powder. The experimental method used in this test was based on the assumption that the mass of bulk material was maintained approximately constant for each powder. For each experiment, the free falling stream of particles falls into the enclosure that was prefilled with white, dense, non-irritant, non-flammable and non-hazardous fog for particle tracing. Figure 2 (b) shows the enclosure with five various vertical positions and the axial direction for measurement.

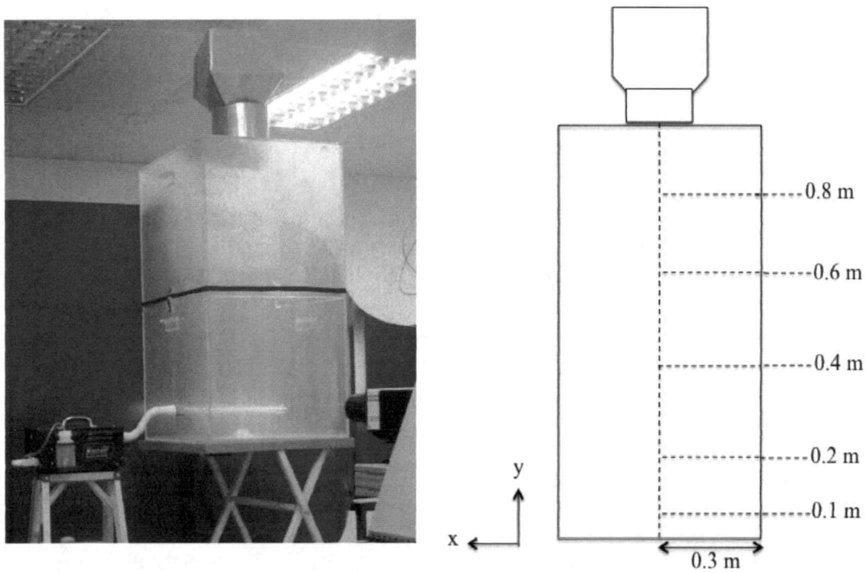

(a) (b)

Fig. 2. (a) Experimental arrangement (b) Various positions at the enclosure for measurement.

The experimental testing for this study utilised the Laser Doppler Anemometry (LDA) to measure the velocity profiles of particles in the particle stream. The calibration was undertaken prior to commencement of measurement to ensure the continuous laser was specifically focused at the required point. In this setup, two 100 mW of DopplerLite diode-pumped solid-state (DPSS) laser were used with 491 nm and 514 nm wavelengths. The laser beam was divided by a Brag Cell to obtain two subdivision beams that cross outside the sensor. The crossing area was known as the probe volume. The two parallel subdivision beams were focused by a lens that intersected at a point in the falling particle stream flow, forming interference fringes. Since the particle stream flow traversed the intersection volume, the FiberLite receiving optics received light scattered from both beams that produced light intensity fluctuations. The photomultiplier converted the light intensity fluctuations to the electrical signals, which were then converted to the velocity information in the Burst Spectrum Analyser (BSA) processor. The processing results were conducted by using BSA Flow Software. This LDA system is comprised of a fully automated traverse system for positioning the optical system at different levels, as depicted in Figure 2 (b).

IOP Conf. Series: Materials Science and Engineering **701** (2019) 012061 doi:10.1088/1757-899X/701/1/012061

The materials used in the experiment are summarised in Table 1, and were selected as they have different particle shapes in nature. These materials were selected to classify the effect of particle shape on the velocity profile of a particle stream. The morphology of dust is given in Figure 3, which were taken by scanning electron microscope (SEM).

Table 1. Summary of materials.

Material	**Shape**	**Average particle size (μm)**	**Bulk density (kg/m^3)**
Castor Sugar	Crystal	336.3	891.2
Oat	Flaky	423.4	273.0

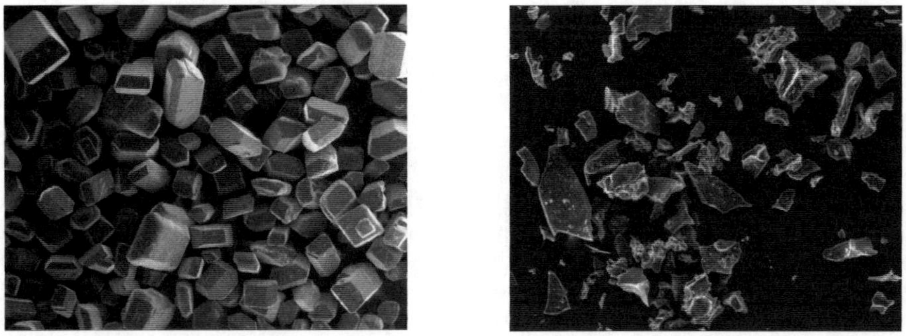

(a) (b)

Fig. 3. Photos of dust taken with a scanning electron microscope. (a) Castor sugar (b) Oat

3 Results and Discussion

Mean velocity of particles during free fall provides an important information on structures of particle streams and dust migration. In this study, the main influence factors on the dust cloud generation, such as the particle characteristics (shape, average particle size as well as bulk density) and conveying velocity, were experimentally investigated.

3.1 Influence of Particle Properties

The schematic diagram of particle streams during free fall is illustrated in Figure 4. The drag force acting on a single particle in the particle stream is not that significant on a single particle in stagnant air. However, free falling of particles may induce downward flow of air and hence may affect the particle drag and dispersion. The free fall drag force acting on the particle is affected by the air velocity and shape of the particle. Generally, drag acting on a spherical particle is mostly uniform, but there is a high tendency of nonuniform drag force on flaky particles. The drag forces acting on falling particles are influenced by the inter-particle distance within the particle cloud. During particle falling, the inter-particle distance is very small. Therefore, very little drag force acts on the majority of particles as compared to the drag force on the exterior sides of particle flow.

Fig. 4. Illustration of clustered particles of castor sugar (crystal) and oat (flaky) in the core stream.

The air flow pattern around a single particle is contorted by the moving clustered particles. The air flow pattern around Particle A on the left side can be gradually developed in the opposite direction of the particle movement, generating turbulent wakes, which will modify the turbulence of air. Hence, the pressure on the left side of Particle A will gradually decrease. However, the air flow pattern on the right side of Particle A, may not be well developed due to the influence of Particle B. Therefore, the drag force on the left side of Particle A is larger than that on the right side, which may possess a tendency of clockwise rotation. The drag forces on Particle A are different than those of Particle B due to its shape and size. With further downstream, the particle flow begins to break up into smaller clusters, which increase the exposure of the powder into the air. Once the drag force achieved a certain value, Particle A will be pulled away from the stream core of the stream and escape into the surrounding to become an airborne dust.

Assuming the same phenomenon, the uneven drag forces will cause Particle C to be pulled away from the core of the stream. The drag force will induce the air into the falling particle stream and the particle flow will appear to be largely broken up during free fall. Due to the motion of particle free falling, the voidage of the particle stream will increase with increasing drop height and more surrounding air will be induced into the stream core. Therefore, the particle flow increases rapidly in this regime and will cause a gradual reduction in the core stream diameter. Since the core of the particle stream disappears, forces acting upon a particle in the flow become equivalent to each other, which makes the particle velocity close to terminal velocity. Therefore, the particle movement decelerates until it stops.

3.2 Influence of Conveying Velocity

The mean velocity plotted in Figure 5 clearly shows the influence of particle properties in dust cloud generation. It was observed from Figure 5 (a) that the flow of castor sugar in the core stream increased with further downstream as a result of moving clustered castor sugar particles. Higher bulk density (891.2 kg/m^3) of castor sugar particles (in crystal shape) with smaller voidage prevented the drag force to disrupt the particle flow. Hence, the particle flow at the core stream underwent accelerations with increasing conveying velocity. In Figure 5 (b), there is a clear trend of decreasing flow of oat particles with increasing drop height. Oat particles conveyed with 2.0 m/s decelerates at 0.1 m from the bottom of the enclosure due to the terminal velocity. Low bulk density (273.0 kg/m^3) and far inter-particle distance of oat particles (in flaky shape) are the important factors to encourage drag force action in generating a higher dust emission during free falling, regardless of particle size.

It was observed in Figure 5 that the measured mean velocities of the LDA results had significant impact on particle characteristics and showed a slight impact on conveying velocity for different enclosure positions. The mean velocity plotted in Figure 5 clearly shows that particles in the core region (centre) had a higher velocity than the dilating region (from the wall), which has smaller particles. Also, the width of the curve from 30 cm (the centre of enclosure) increases towards the bottom. The finding is consistent with findings of past studies by Ansart et al. [7], where the moving layers of the fluid experience friction with the nearby stagnant fluid. This friction initiates the movement of nearby layers, which causes the spread of particles. The spread of the particle causes a reduction in the mean bulk density. The reduction in bulk density causes a reduction in the intensity of motion force of plume particles, due to the difference between bulk density and air density, and finally the reduction in particle velocity. The development of particle velocity is due to the reduction of the particle volume fraction with increasing drop height. It was found that when the particle volume fraction decreases, the air friction on each particle increases, which leads to reduced particle velocity.

It is appears from Figure 5(a) that the increment of conveying velocity did not show any significant difference on the particle velocity in all positions, except a slight velocity difference for 0.9 m/s of conveying velocity. After comparing the trends displayed in Figure 5 (a), it can be seen that the highest conveying velocity (2.0 m/s) produced the highest downward mean velocity towards the core of particle stream. Figure 5(b) is quite revealing in several ways. The results suggested that the mean velocities at the core stream were proportional to the speed of conveying velocities at positions 0.8 m, 0.6 m and 0.4 m. However, at 0.2 m from the bottom, the mean velocity of free falling particles at 0.9 m/s and 1.5 m/s conveying velocities were decreased towards the centre due to terminal velocity. What is interesting in this data is that the mean velocity of particle conveyed by the highest conveying velocity (2.0 m/s) at the bottom of the enclosure (0.1 m), reduced about 20% from the top due to the particle-fluid interaction, such as the drag force resulting in the deceleration of particle flow.

IOP Conf. Series: Materials Science and Engineering **701** (2019) 012061 doi:10.1088/1757-899X/701/1/012061

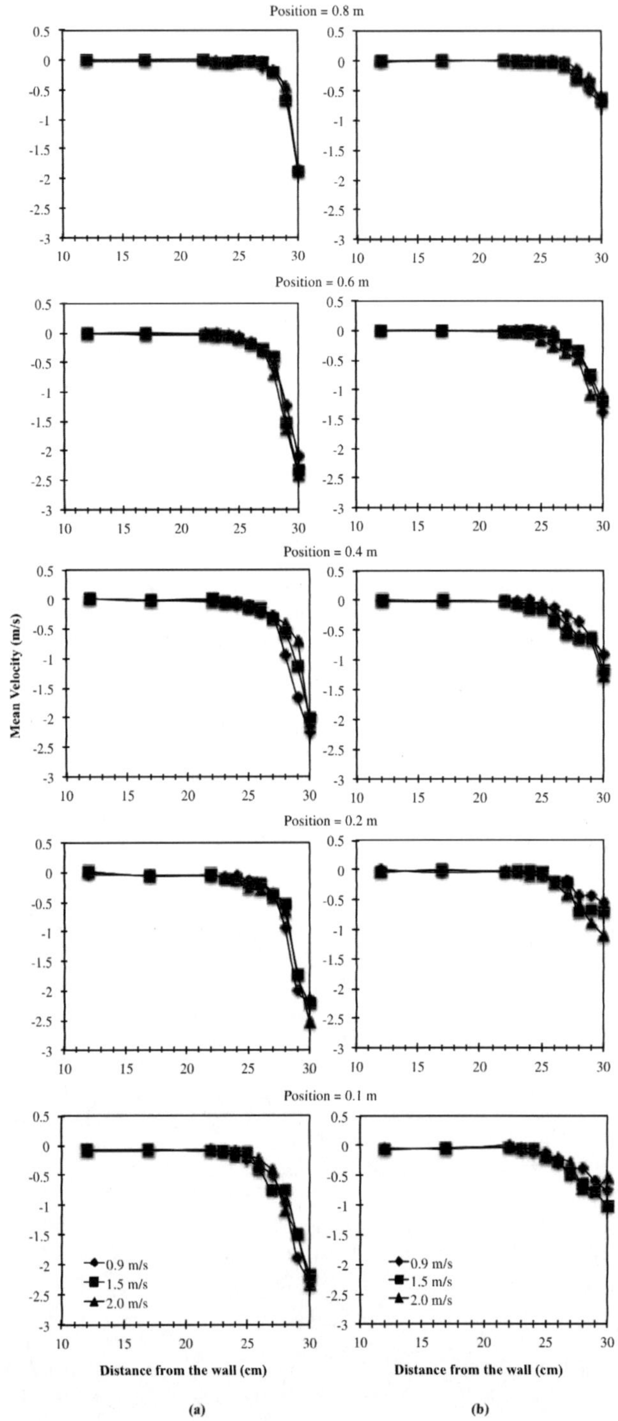

Fig. 5. Measurement of mean velocity at various positions for (a) castor sugar and (b) oat.

4 Conclusion

The effect of particle density and particle shapes on the dust cloud formation and particle velocity was elucidated using a noninvasive LDA technique. The result shows that particle shapes greatly influence the particle falling velocity. Oat particle which has a flaky shape has about 60% less free falling velocity compared to castor sugar which is mostly in smooth cube shape. The flaky particles also tend to generate a larger dust cloud due to non-uniform draf forces acting on the particle surface.

The authors are greatly grateful to the Ministry of Higher Education (MOHE), Malaysia for the generous financial contribution rendering this work possible by FRGS grant, FRGS/1/2015/TK02/TATI/03/1. Sincere thanks to TATI University College for the support, which makes this important research viable and effective. The authors are also thankful to Associate Professor Dr. Jolius Gimbun and his team from Centre of Excellent for Advanced Research in Fluid Flow (CARIFF), Universiti Malaysia Pahang for their help and constructive discussion on experimental setup and LDA measurement.

References

1. M. Crawford, *Air pollution control theory* (McGraw-Hill Book Company, New York, 1976)
2. G. Gouesbet, A. Berlemont, Prog. in Energy and Combustion Sci. **25**,133–159 (1999)
3. A.A Esmaili, T.J. Donohue, C.A. Wheeler, W.M. McBride, A.W Roberts, I. J. of Mineral Processing **142**, 82-90 (2015)
4. P. Cooper, P.C. Arnold, Kona Powder Part. J. **13**, 125 (1995)
5. K. Ogata, K. Funatsu, Y. Tomita, Powder Tech. **115**, 90–95 (2001)
6. P. Wypych, C. Dave, C. Paul, Chemical Eng. and Processing: Process Intensification **44**, 323–326 (2005)
7. R. Ansart, A. Ryck, J.A. Dodds, M. Roudet, D. Fabre, C. Francois, Powder Tech. **190(1-2)**, 274–281 (2009)
8. X. Li, Q. Li, D. Zhang, B. Jia, H. Luo, Y. Hu, Adv. Powder Tech. **26**, 236-243 (2015)
9. A. Klippel, M. Schmidt, O. Muecke, U. Krause, J. of Loss Prevention in the Process Industries **29**, 22-137 (2014)
10. S.I. Rani, B.A Aziz, J. Gimbun, Process Safety and Env. Protection **96**, 14-21 (2015)
11. L.L.L. Waduge, S. Zigan, L.E. Stone, A. Belaidi, P. García-Trinanes, Process Safety and Env. Protection **105**, 262-273 (2017)
12. X. Li, Q. Wang, Q. Liu, Y. Hu, Powder Tech. **291**,159-169 (2016)

IOP Conf. Series: Materials Science and Engineering **701** (2019) 012062 doi:10.1088/1757-899X/701/1/012062

Effect of Surface Preparation on the Isothermal Oxidation of Fe-33Ni-19Cr Alloy

N Parimin[1,*] and Jemahri[1]

[1]School of Materials Engineering, Universiti Malaysia Perlis, 02600 Arau, Perlis, Malaysia

Abstract. The effect of surface preparation on the isothermal oxidation behavior of Fe-33Ni-19Cr alloy was study in this paper. The surface preparation of the test samples was obtained by grinding process using different grit of silicon carbide abrasive paper, namely P600, P800 and P1000. The surface area and weight of the sample were measured before the oxidation test, by using Vernier caliper and analytical balance. The ground sample of Fe-33Ni-19Cr alloys were isothermally oxidized at 1000°C for 150 hours in laboratory air. The weight of oxidized samples was recorded after the isothermal oxidation test at every 25 hours exposure duration in order to determine the oxidation kinetic of all samples. The oxide surface morphology of Fe-33Ni-19Cr alloy was characterized by scanning electron microscope. The oxidation kinetic of all samples displayed a weight gain pattern as the exposure duration increase, followed a parabolic rate law. While the result of surface morphology shows that the lowest grit of surface finish, P600, display less oxide exfoliation occurred.

1 Introduction

The Ni-based alloys used for high strength and high temperature applications which also called superalloy was developed to use in turbine engine and high performance at high or raise temperature application for mechanical properties [1]. Even they are expose in long term at temperature above 650°C, they are specially appropriate choice for these demanding applications because of their ability to retain most of their strength. Combination of great low temperature ductility, high strength and high surface stability, make these alloys becomes more versatile [1]. Ni-based superalloy was generally used for a number of applications which include turbine discs of jet engines, as well as pipe lines in oil and gas fields due to the excellent high temperature oxidation resistance, corrosion resistance and mechanical accomplishment [2]. Oxidation resistance of this superalloy and bond covering materials in high temperature situations is a great extent directed by their capacity to specially shape an adherent oxide scale on the metal surface. These alloys are also suitable to work under unstable environmental conditions. Furthermore, the Ni, Cr and Co, are element which represent the base of these systems. Carbide and other compounds are formed because of the existing of the minor addition of the elements like B, C, Ti, Nb, Al and Fe [3]. Unfortunately,

* Corresponding author: noraziana@unimap.edu.my

at the temperature above 900°C, the damaging of the microstructure components may occur from formation of nitrides, carbides and other phases.

Fe-Ni-Cr alloy is a Ni-based alloy, has excellent mechanical properties with high temperature oxidation resistance, used in the power plants application [4]. Generally, for this alloy, the development of protective chromia scales that form on the surfaces by the oxidation of Cr during elevated temperature exposure can act as protective oxides layer [5]. The existence of a critical Cr content to form a continuous chromia scale has been investigated. At high temperature exposure, the formation of a protective Cr_2O_3 scale is required to avoid degradation by severe oxidation processes for alloys. A number of investigations have been carried out to improve the understanding of the growth mechanism of Cr_2O_3 scales.

However, the oxide will exfoliate at elevated temperature exposure if the protective oxide layer did not well adhere to the alloy surface. The initial oxide layer formation was determined by good surface preparation. The optimum groove mark induced by surface preparation will enhance the formation of protective oxide layer. The grinding effect during surface preparation results the cold working effect on the subsurface regions of the alloy, which assist diffusion. Additionally, the subsurface region will form fine grain structure which increase the grain boundary area. This would affect the higher short circuit diffusion of ion species through grain boundary area. As a result, rapid initial oxide layer was form on the alloy surface which act as a barrier and protect the alloy [6-7]. On the other hand, the ground surface sample was developing a micro scratch that provides a base structure for oxide nucleation. Primarily, another researcher would also grind the sample surface to enhance the suitable surface roughness [6-11]. Therefore, this study will be focused on the investigation of the oxidation behaviour of Fe-33Ni-19Cr alloy with different surface preparation.

2 Experimental

The material used in this study was as-received Fe-33Ni-19Cr alloy with measured chemical composition using optical emission spectrometer in weight percent (wt%) as follow: 32.51 Ni, 18.90 Cr, 0.078 C, 0.053 Al, 0.489 Ti, 0.001 S, 0.315 Si, 0.556 Mn, 0.007 P, 0.082 Cu and balance Fe. The test samples were cut to a rectangular shape with nominal dimensions of 10mm x 10mm x 3mm. The test sample was undergoing a series of grinding process at different grit size of silicon carbide abrasive paper, namely P600, P800 and P1000. The surface area and weight of all samples were measured before the oxidation test. The test samples were arranged in a quartz boat for the oxidation test. The isothermal oxidation test was run at 1000°C for 150 hours in laboratory air using LT furnace. The oxidation test was run discontinuously at 25, 50, 75, 100, 125 and 150 hours. The oxidized samples weight was measured after each interval duration to determine the oxidation kinetic. The surface morphology of the oxidized samples were investigate using scanning electron microscope (SEM).

3 Results and Discussion

3.1 Oxidation Kinetics

The graph for total weight change per surface area as a function of time is shown in Fig. 1. The results show that increasing oxidation time was increase the weight gain value. P600 sample shows lowest weight gain pattern, followed by P800 and P1000 samples. P600 samples recorded a lowest weight gain value at 25 hours exposure which is 0.3242 $mgcm^{-2}$, compared with P800 and P1000 samples which are 0.3620 $mgcm^{-2}$ and 0.4045 $mgcm^{-2}$, respectively. The highest weight gain value noted at 150 hours exposure showing an

IOP Conf. Series: Materials Science and Engineering **701** (2019) 012062 doi:10.1088/1757-899X/701/1/012062

increasing weight gain recorded for all samples which are 0.7996 $mgcm^{-2}$, 0.8986 $mgcm^{-2}$ and 0.9873 $mgcm^{-2}$ for P600, P800 and P1000 samples, respectively.

The analysis of the oxidation rate law of each sample was done by plotting a double log plot of weight change per surface area (w/A) data as a function of time, based on equation (1), where x is weight change per surface area, t is time and m is constant. The m values of 1, 2 and 3 represent linear, parabolic and cubic oxidation, respectively. The m value for all samples recorded at 2.07, 2.02 and 2.09 for P600, P800 and P1000, respectively, denoted that all samples approximately follows parabolic rate law (m = 2). This identification indicates that the oxide growth rate for all samples was followed diffusion-controlled mechanism.

The parabolic rate constant was identified from equation (2) where x is weight change per surface area, t is time in seconds, K_p represent parabolic rate constant and C is a constant. The parabolic rate constant has been interpreted from square of weight change over time. From the calculated value, the highest K_p value was determined from P1000 sample, which is 2×10^{-6} $mg^2cm^{-4}s^{-1}$, indicating the highest oxidation rate take place on that sample. While the lowest oxidation rate occurred on P600 sample with the parabolic rate constant of 1×10^{-6} $mg^2cm^{-4}s^{-1}$. P600 sample denoted the best oxidation resistance compared to P800 and P1000 samples.

$$log\, x = 1/m\, log\, t + c \quad (1)$$

$$x^2 = K_p\, t + c \quad (2)$$

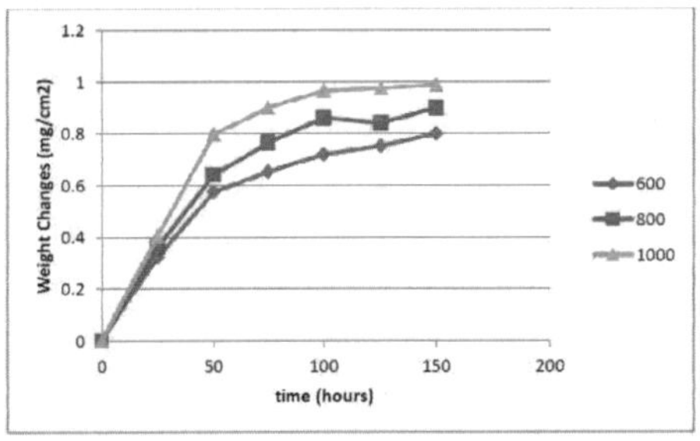

Fig. 1. Oxidation kinetics of Fe-33Ni-19Cr alloy at 1000°C.

3.2 Surface Morphology

Fig. 2 shows the SEM image of oxidized Fe-33Ni-19Cr alloy after exposure for 100 hours. The images display the continuous and uniform oxide scale formed on the alloy surface with visible of oxide exfoliation and overgrown porous oxide structure. Numerous oxide exfoliation area was recorded on the P1000 sample compared to P600 sample.

Fig. 3 show the SEM images of oxidized Fe-33Ni-19Cr alloy after 150 hours exposure. SEM image of P600 sample in Fig. 3 (a) exhibited the small oxide exfoliation area surrounded by porous oxide structure. This porous oxide has a wobbly structure which is has a tendency to exfoliate during further exposure duration. Whereas, SEM image of P1000 sample in Fig. 3 (b) exhibited a large oxide exfoliation area compared to P600 sample. The formation of porous oxide structure also visible on the alloy surface. On the other hand, it was seen in the SEM image, the flaky oxide structure near the oxide exfoliation area with evidence of crack formation on the structure. This flaky oxide also has a tendency to exfoliate causing the exposure of the protection oxide layer of the alloy. When the oxide exfoliates, it will expose the based metal and reduce the protective barrier. This will cause a fresh oxide will continue to grow to seal the unprotected area, hence further increase the weight of the sample. This will lead to increase in the weight gain pattern as recorded on the P1000 sample as displayed in Fig 1.

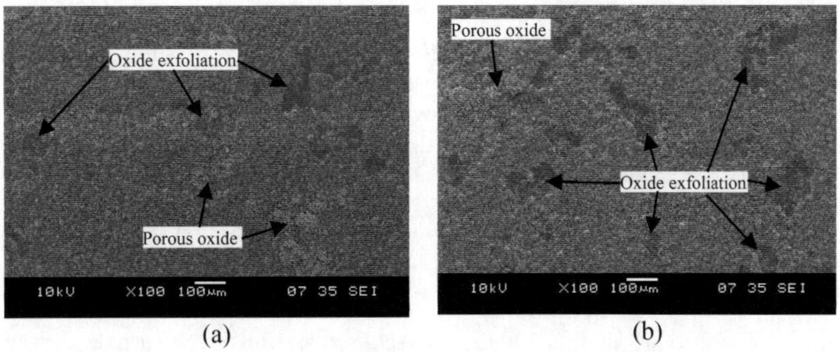

Fig. 2. SEM images of oxidized Fe-33Ni-19Cr alloy after 100 hours: (a) P600, (b) P1000.

On the other hand, the effect of surface preparation has affected the formation of oxide scale on the surface. Oxidized samples that has been grind using P600 grit encounter less oxide exfoliation compared to P1000 sample. This phenomenon was proved that the surface preparation prior the isothermal oxidation test has been success due to the good adhesion of oxide scale on the alloy surface. This finding was supported by [6], which stated that the rougher surface develops a more resistant to scale spallation. Their research finding shows that thermal cycling would result in repetitive spallation of the oxide scales which initiated the reduction of chromium content in the metal surface. In addition, the temperature and surface treatment play an important factor in the high temperature condition.

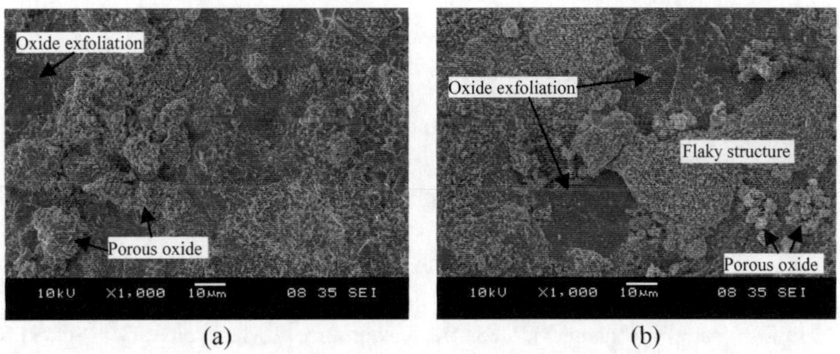

Fig. 3. SEM images of oxidized Fe-33Ni-19Cr alloy after 150 hours: (a) P600, (b) P1000.

During surface preparation, the grinding process will introduce the cold working effect on the subsurface regions of the alloy, contain a large number of dislocations which promote diffusion of metal ion species. It also causes the alloy regions to recrystallize, hence refine the grain at the subsurface regions. As the grain refined, it will increase the volume of grain boundary, which provides rapid diffusion routes for chromium, which allows the metal ion to diffuse to the alloy surface and form a continuous and protective oxide layer. This phenomenon may be eliminating the internal precipitation due to scale formation that prevents the access of gas to the metal. Even though the scale exfoliate, it forms rapidly by the relatively rapid diffusion of chromium across the grain boundaries to the surface [6]. This mechanism cannot be achieved in the alloys which had been chemically etched, because of the removal of the surface regions which had been mechanically damaged. This fact was generating the idea to grind the alloy surface prior the oxidation test, without etching [7]. On the other hand, the ground surface sample was developing a micro scratch that provides a base structure for oxide nucleation. Principally, the sample surface of the oxidation test was ground to an optimum surface finish to enhance the suitable surface roughness to promotes the oxide scale growth.

4 Conclusion

The effect of surface preparation at three different grit size of silicon carbide abrasive paper on the oxidation behaviour of Fe-33Ni-19Cr alloy has been study. All samples were obeyed a parabolic rate law, indicating a diffusion-controlled oxide growth rate. P600 sample exhibited the lowest weight gain pattern and parabolic rate constant value compared with P800 and P1000 samples, consequently presented good oxidation resistance. The surface morphology of P600 sample recorded a continuous oxide scale formed on the alloy surface with visible of oxide exfoliation and porous oxide structure. Similar observation was recorded for P1000 oxidized sample surface, except that the area and number of oxide exfoliation is larger compared with P600 sample.

The authors would like to thank the Ministry of Education Malaysia for the research fund under the Fundamental Research Grant Scheme (FRGS) (Project Code. FRGS/1/2016/TK05/UNIMAP/02/4).

References

1. H. K. D. H. Bhadeshia, *Nickel Based Superalloys* (University of Cambridge, 2003).
2. D. Chellaganesh, M. A. Khan, J. T. Jappes, S. Sathiyanarayanan, High Temp. Mat. & Pro. (2017)
3. F. A. Pérez-González, J. H. Ramírez-Ramírez, M. Terock, N. F. Garza-Montes-De-Oca, U. Glatzel, R. Colás, Cor. Eng., Sci. & Tech. **51**(7), (2016)
4. Z. Yang, J. T. Lu, Y. Yuan, X. B. Zhao, Y. F. Gu, H. F. Yin, Y. Y. Dang, Mat. High Temp. **33**(2), (2016)
5. L. Liu, Z. Yang, C. Zhang, M. Ueda, K. Kawamura, T. Maruyama, Cor. Sci. **91**, (2015).
6. M. Hänsel, C. A. Boddington, D. J. Young, Cor. Sci. **45**(5), (2003)
7. D. Oquab, N. Xu, D. Monceau, D. J. Young, Cor. Sci. **52**(1), (2010)
8. T. J. Pan, Y. S. Li, Q. Yang, R. F. Feng, A. Hirose, Cor. Sci. **53** (6) (2011)
9. A. Palani, B. T. Lu, L. P. Tian, J. L. Luo, Y. C. Lu, J. Nuc. Mat. **396** (2-3), (2010)
10. K. T. Voisey, Z. Liu, F. H. Stott, App. Surf. Sci. **252**(10), (2006)
11. G. R. Holcomb, D. E. Alman, Scrip. Mat. **54**(10), (2006)

IOP Conf. Series: Materials Science and Engineering **701** (2019) 012063 doi:10.1088/1757-899X/701/1/012063

Investigation of Isothermal Oxidation on Heat-treated Fe-40Ni-24Cr Alloy at 950°C

Noraziana Parimin[1,*], and Singh[1]

[1]School of Materials Engineering, Universiti Malaysia Perlis, 02600 Arau, Perlis, Malaysia

Abstract. In this research, Ni-based superalloy which is Fe-40Ni-24Cr alloy was experienced a heat treatment process to alter the grain size of the sample to observe its influence on the oxidation properties. The alloys were heat-treated at temperatures of 950ºC, 1050ºC, and 1150ºC for 2 hours followed by water quench. The heat-treated alloys were undergoing an isothermal oxidation test at 950°C for 150 hours in laboratory air. The weight gain of oxidized heat-treated alloy was recorded in order to determine the oxidation kinetic of the test samples. Scanning electron microscopy (SEM) technique was employed in this study to analyze the oxidation behaviour of heat-treated samples. Grain size measurement shows that the grain size increases as the heat treatment temperature increases. The oxidation rate of heat-treated Fe-40Ni-24Cr alloy obeys the parabolic rate law indicating the diffusion-controlled oxide growth mechanism. The fine grain size of alloy heat-treated at 950ºC possess a better oxidation resistance and has a lower oxidation rate compared to coarse grain samples. The surface morphology indicate that the fine grain heat-treated alloy at 950°C shows uniform oxide layer formed on the surface, while alloy heat-treated at 1050°C and 1150°C shows the uneven oxide growth, oxide flaking and oxide spallation.

1 Introduction

In this modern world, high performance aircraft jet engine could not function without the discoveries and advances in superalloys. Some superalloys are used at cryogenic or room temperatures despite it is developed for high temperature use. Superior heat and oxidation resistance are necessary during its applications and the material must alongside retain its mechanical reliability and microstructure [1]. Nickel (Ni)-based superalloys are used for elevated temperature components of gas turbines in industrial and aerospace applications due to their excellent thermal and surface stability at elevated temperatures [2-3]. Over time, several changes have been made in the manufacturing of Ni-based superalloys to improve the processing methods and to optimize the chemical composition and microstructure [4]. Nickel-chromium (Ni-Cr) and iron-nickel-chromium (Fe-Ni-Cr) alloys are the distinctive heat resistance Ni-based alloys. Fe-Ni-Cr alloys are resistant to oxidation and heat that was developed for high temperature oxidizing environments. The Fe-Ni-Cr alloys have excellent oxidation resistant in many severe environments as it forms a protective film of chromium

* Corresponding author: noraziana@unimap.edu.my

oxide, Cr_2O_3 [5]. They also have good resistance to oxidation and sulfidation and has good strength at high temperatures. They have a maximum operating temperature of 1205°C. Microstructures and the mechanical properties of Ni-based superalloys can be modified by heat treatment. The mechanical properties such as strength, fatigue and creep resistance could be improved with suitable heat treatment process. Heat treatment is normally used to obtain the optimal combination of size, morphology and distribution of γ precipitates that leads to an enhanced mechanical performance. This is because the strengthening of Ni-based superalloys is strongly influenced by coherent precipitation of γ' phase in the γ matrix [6]. Therefore, the aim of this study is to investigate on the oxidation behaviour of heat-treated Fe-40Ni-24Cr alloy at 950°C. The grain size of the heat-treated alloys at different heat treatment temperatures have a significant impact on the oxidation kinetic and oxide scale formation of the alloy.

2 Methodology

Fe-40Ni-24Cr alloy was used as the material in this study. The chemical composition in weight percent was measured using optical emission spectrometer (OES), which are: 40.45 Ni, 24.11 Cr, 0.05 C, 0.08 Al, 0.03 Ti, 0.44 Si, 0.7 Mn, 0.01 P, 0.11 Cu, 0.25 Mo, 0.17 Co, 0.44 Nb, 0.05 W and balance Fe. The Fe-40Ni-24Cr alloy was experienced a series of heat treatment process at different temperature, namely 950ºC, 1050ºC, and 1150ºC for 2 hours soaking time, followed by water quench. These samples are denoted as HT950, HT1050 and HT1150 for alloy heat-treated at 950ºC, 1050ºC, and 1150ºC, respectively. The grain size of the heat-treated alloy was measured using linear intercept methods ASTM E112. The heat-treated alloys were then undergoing a series of grinding process. The grinding process was carried out so that the samples have a good surface finish and accurate dimensions required so that the oxidation layer can be formed evenly on the sample. The heat-treated Fe-40Ni-24Cr alloy were grind until P600 grit surface finished. The dimension and weight of the heat-treated samples was measure before the oxidation test. The weight of oxidized heat-treated sample was measure after the oxidation test to determine the kinetic of oxidation. The ground samples are then arranged in a quartz boat for the oxidation test. The heat-treated alloy was undergo the isothermal oxidation test at 950°C for 150 hours. The oxidation test is carried out in a LT furnace in laboratory air. The oxidation test was run discontinuously for every 25 hours interval durations until 150 hours to record the kinetics of oxidation of all samples. The oxidized heat-treated samples were characterized in terms of oxidation kinetics and surface morphology using scanning electron microscope (SEM) technique.

3 Results and Discussion

3.1 Heat Treatment of Fe-40Ni-24Cr Alloys

The heat treatment process on Fe-40Ni-24Cr alloy indicate that the higher the heat treatment temperature, the larger the grain size of the sample. HT950 sample which has been heat-treated at 950°C indicates the average grain size of 71.4 μm, recorded a fine grain size. Meanwhile, the average grain size of the HT1050 sample which has been heat-treated at 1050°C is 91.3 μm. HT1150 sample which has been heat-treated at 1150°C obtained an average grain size of 105.1 μm, indicating the coarse grain structure. The results show an increasing trend of the average grain size of the alloy's microstructures as the heat treatment temperature increases.

3.2 Oxidation Kinetic of Oxidized Fe-40Ni-24Cr Alloys

Fig. 1 shows the graph of weight change per surface area versus time that was oxidized at 950°C for 150 hours. The graph shows an increasing trend of the sample weight as the oxidation time increases. The oxidation kinetic curve for HT950 sample shows the lowest weight gain, followed by HT1050 sample. Whereas, the oxidation kinetics curve of HT1150 sample shows the highest weight gain pattern.

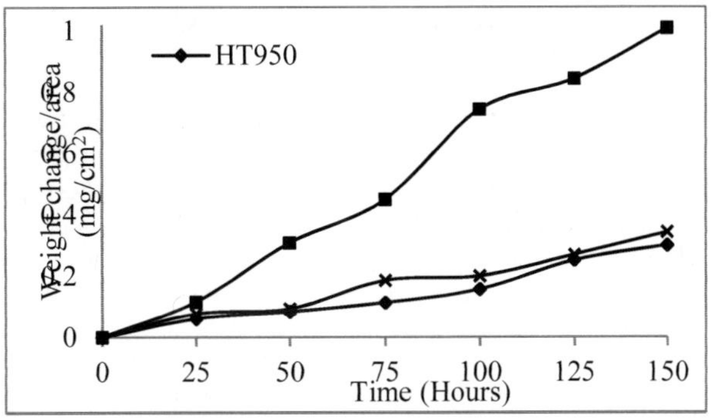

Fig. 1. Oxidation kinetic of heat-treated Fe-40Ni-24Cr alloy at 950°C for 150 hours.

HT950 sample which has the fine grain structure exhibit lowest weight gain and thus has a high oxidation resistance. Hence, this fine grain sample shows a lower oxidation rate. This is because, fine grain size has higher grain boundary area which act as a diffusion path of outward metal ion migration from based metal to metal-gas interface. The higher ion diffusion rate on fine grain sample was developed a rapid initial oxide layer that formed on the alloy surface that protects the alloys' surface from further hostile oxide formation. When the whole alloy surface was covered with the thin oxide layer, the formation of oxides was slowly decrease as the oxide layer itself act as a barrier for further reaction with the oxidizing environment. This mechanism was explained about the weight gain pattern of fine grain HT950 sample as shown in Fig. 1, which exhibited the steadily and lowest weight gain.

The oxidation kinetic curve of HT1050 sample displays a moderate weight gain. Similar observation was noticed for this sample, as it developed a low weight gain value comparable with HT950 sample. On the other hand, the HT1150 sample which has coarse grain structure, indicates the highest weight gain per surface area value. The weight gain for coarse grain structure increases due to the continuous oxide growth during the exposure to the working temperature. The higher weight gain of this coarse grain sample due to the possibility of continuous oxide growth that forms a protective oxide layer at the alloys surface. From the weight gain data, it can be note that thick oxide layer was formed on the alloy surface of HT1150 sample as it recorded a high weight gain value compare with HT950 and HT1050 samples.

The double log plot of graph in Fig. 1 has been done to determine the oxidation rate law of all samples. Equation (1) was use to analyze the data where x is weight change per surface area, t is time and m is constant. The m value with 1, 2 and 3 represent linear, parabolic and cubic oxidation rate laws, which can be determined from the double log plots. From the analysis that has been done, it was confirmed that all samples follow the parabolic rate law. The parabolic rate constant was determined from parabolic rate law of equation (2), where K_p denote parabolic rate constant. The parabolic rate constant value was calculated from the square of weight change per surface area as a function of time. The analysis recorded a parabolic rate constant for HT950, HT1050 and HT1150 are 5.336×10^{-7} mg^{2}cm^{-4}s^{-1},

5.933 x10^{-7} $mg^2cm^{-4}s^{-1}$ and 1.943 x 10^{-6} $mg^2cm^{-4}s^{-1}$, respectively. The higher value of parabolic rate constant indicates a higher oxidation rate of the samples. Therefore, from the values obtained, HT1150 sample has the highest oxidation rate while HT950 sample has the lowest oxidation rate that served its high oxidation resistance.

$$log\ x = 1/m\ log\ t + C \quad (1)$$

$$x^2 = K_p\ t + C \quad (2)$$

3.3 Surface Morphology of Oxidized Fe-40Ni-24Cr Alloys

Scanning electron microscope technique was used to analyzed the surface morphology of oxidized surface. Two samples from 25 hours of oxidation and three samples from 150 hours of oxidation were analyzed. Fig. 2 (a) shows the surface morphology of HT950 sample oxidized for 25 hours. The SEM image shows the continuous oxide layer was formed on the sample surface with evidence of small area of oxide spallation. Fig. 2 (b) shows the surface morphology of HT1150 sample oxidized for 25 hours. The SEM image displays the formation of oxide spallation at numerous parts with formation of void on the spallation area. The void tends to formed underneath the oxide scale which is exposed when the oxide exfoliates from the sample surface. On the order hand, the image also display the formation of overgrown oxide scale which has the permeable structure. This overgrown oxide scale has a tendency to spall during prolong exposure to high temperature oxidation due to the wobbly structure. Oxide spallation occurs in this coarse grain sample due to the weak adhesion of oxide scale on the alloy.

(a)

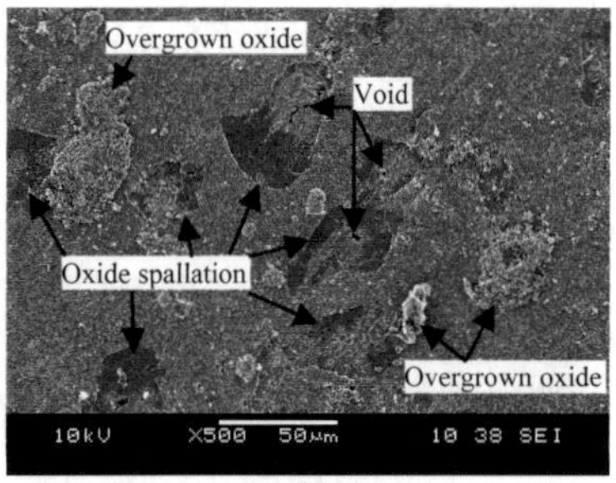

(b)

Fig. 2. SEM image of oxidized Fe-40Ni-24Cr alloy after 25 hours; (a) HT950; (b) HT1150.

Fig. 3 shows the surface morphology of heat-treated Fe-40Ni-24Cr alloy oxidized for 150 hours. Fig. 3 (a), (b) and (c) shows the SEM images for HT950, HT1050 and HT1150 samples, respectively. All samples indicated the formation of continuous oxide layer on the alloy surface. On the other hand, evidence of oxide spallation, overgrown oxide scale and voids also formed on the alloy surface. The main different is that the area of oxide spallation for HT950 sample was smaller compared to HT1050 and HT1150 samples. The formation of oxide spallation on HT1150 samples was occurred on a large area indicating that this sample was formed a loose oxide scale that tend to spall after prolong exposure at high temperature condition. The number of voids also increase as the area of oxide spallation increase. The void was formed underneath the spall oxide which was exposed due to spallation mechanism. This indicates that the oxide scale forms on the oxide structure contains pores below it. The pores was occurred due to rapid metal ion diffusion that develops an open passage towards the oxide surface. The outward metal ion diffusion was enhanced by the grain boundary area

IOP Conf. Series: Materials Science and Engineering **701** (2019) 012063 doi:10.1088/1757-899X/701/1/012063

on the alloy. When the diffusion rate is high to developed an oxide layer, it will generate the voids along the grain boundary area. This void was unstable, which is, as the oxidation test duration increase, it will act as a break point of oxide spallation. When the oxide spall, it will expose the surface of the alloy and disturb the protection behaviour. This phenomenon will generate a fresh oxide scale to reformed on the exposing area which tend to thicken the oxide scale. This will lead to increase in the weight gain of the sample as indicate in the Fig. 1. The formation of numerous and large area of oxide spallation represented the low protection behaviour of oxide scale developed on the alloy surface.

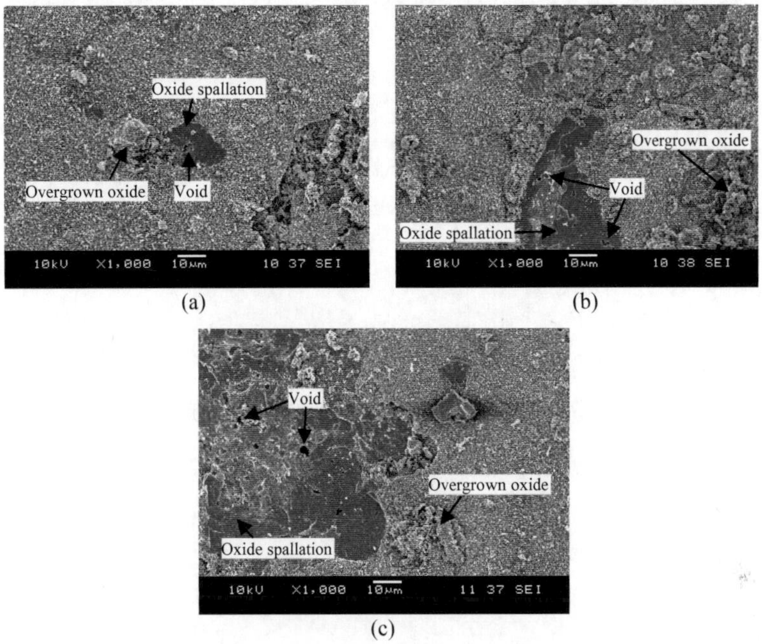

Fig. 3. SEM image of oxidized Fe-40Ni-24Cr alloy after 150 hours; (a) HT950; (b) HT1050 and (c) HT1150.

4 Conclusion

Based on this study, it can be concluded that the increase in the heat treatment temperature of the alloys leads to increase the grain size. Fe-40Ni-24Cr alloy which is heat-treated at 1150°C (HT1150) indicate the coarse grain size of 105.1 μm, whereas the alloy heat-treated at 950°C (HT950) indicate the fine grain size of 71.4μm. The oxidation kinetics of all heat-treated oxidized samples recorded an increasing weight gain pattern as the exposure time increase, with followed a parabolic rate law. HT950 sample exhibited the lowest weight gain pattern and lowest oxidation rate compared to HT1050 and HT1150 samples, hence presented a superior oxidation resistance. The surface morphology analysis from scanning electron microscope shows a continuous oxide scale formed on the samples surface after 25 hours and 150 hours exposure with evidence of oxide spallation, overgrown oxide scale and voids. HT1150 samples recorded a numerous voids formation with larger oxide spallation area.

The authors would like to thank the Ministry of Education Malaysia for the research fund under the Fundamental Research Grant Scheme (FRGS) (Project Code. FRGS/1/2016/TK05/UNIMAP/02/4).

References

1. M. H. Thammaiah, *Isothermal Oxidation Comparison of Three Ni-Based Superalloy* (2016)
2. J. Jiang, G. Xiao, Y. Wang, Y. Liu, J. Alloys & Comp. **784**, (2019)
3. C. N. Athreya, K. Deepak, D. Kim, B. de Boer, S. Mandal, J. Alloys & Comp.,**778** (2019)
4. M. Konter, M. Thumann, J. Mat. Pro. Tech. **117** (3), (2001)
5. A. Col, V. Parry, C. Pascal, Cor. Sci. **114**, (2017)
6. C. Li, Z. Yuan, Y. Fan, S. He, W. Xuan, X. Li, Y. Zhong, Z. Ren, J. Mat. Pro. Tech. **246**, (2017)

AUTHOR INDEX